Handbook of Microscopic Anatomy

Continuation of Handbuch der mikroskopischen Anatomie des Menschen
Founded by Wilhelm von Möllendorff
Continued by Wolfgang Bargmann

Edited by A. Oksche and L. Vollrath

Henning Schmalbruch

Skeletal Muscle

With 129 Figures

Springer-Verlag Berlin Heidelberg NewYork Tokyo

Handbook of Microscopic Anatomy

Volume II/6: Skeletal Muscle

Privatdozent Dr. H. Schmalbruch

Københavns Universitet, Panum Instituttet, Neurofysiologisk Institut,
Blegdamsvej 3 C, DK-2200 København N

Professor Dr. Dr. h.c. A. Oksche

Institut für Anatomie und Zytobiologie der Justus Liebig-Universität, Aulweg 123, D-6300 Giessen

Professor Dr. L. Vollrath

Anatomisches Institut der Johannes Gutenberg-Universität, Saarstraße 19–21, D-6500 Mainz

ISBN-13:978-3-642-82553-8   e-ISBN-13:978-3-642-82551-4
DOI: 10.1007/978-3-642-82551-4

Library of Congress Cataloging in Publication Data. Schmalbruch Henning, 1938- Skeletal muscle. (Handbook of microscopic anatomy; vol. II/6) Bibliography: p. Includes indexes. 1. Striated muscle – Anatomy. 2. Histology. I. Title. II. Series. [DNLM: 1. Muscles – anatomy & histology. 2. Muscles – pathology. QS 504 H236 Bd. 2 T. 6] QM571.S36 1985 611'.73 85-12642
ISBN-13:978-3-642-82553-8 (U.S.)

2122/3130-543210

Preface

This volume is intended to cover research in the field of muscle morphology since publication of the previous edition by Häggquist in 1956. The development of new techniques, coupled with an intensified interest in muscle, has resulted in a vast literature which no single person could review, especially within the limitations of one volume. When I accepted the flattering offer to write a new edition, I quickly abandoned any hope of a comprehensive review. Instead, I tried to consider, within my limits, those lines of research which I believe to be important for the understanding of mammalian and ultimately human muscles under normal, experimental, and pathological conditions. It would be naive to suggest that muscle can be adequately described in purely morphological aspects; I would characterize the results of my effort as "muscle as seen with the eyes of a morphologist".

It gives me pleasure to acknowledge the help of several colleagues who read and commented on drafts of individual chapters: Dr. Brenda Eisenberg, Chicago; Dr. Else Nygaard, Copenhagen; Dr. Stefano Schiaffino, Padova; Dr. Michael Sjöström, Umeå; Dr. Lars-Erik Thornell, Umeå. None of these individuals can be held responsible for any error or obscurity that persists. Indeed, without their assistance there would have been more. I also thank those colleagues who allowed me to include their published and unpublished material; their names, and also those of the publishers who kindly granted copyright permission, are given in the individual figure captions.

I am indebted to Mrs. E. Fischer and Mrs. M. Løvgren for their patience in typing the successive versions of the manuscript, Mr. F. Riis for preparing the original diagrams, and Mr. A. Djørup, engineer, for his assistance with the word processor.

I owe a deep debt of gratitude to Mrs. M. Bjærg for many years of cooperation in the laboratory. Mrs. Bjærg double-checked all references, and we sincerly hope that the list of references will still be useful when future achievements have antiquated this review. I am grateful, too, to the publishers for expert processing of this monograph.

I gratefully acknowledge the financial support of my work by the Danish Medical Research Council.

I want to dedicate this book to the memory of my friend, Gustav G. Knappeis, Offenbach-Germany 1899 – Hørsholm-Denmark 1981.

København HENNING SCHMALBRUCH

Contents

A. General Overview

Skeletal muscles develop force and cause movement. The parenchymal cells of the muscle tissue are multinucleated syncytia, the muscle fibres, which may be more than 10 cm long. Most muscles consist of one set of fibres, i.e. the fibres do not act in series. Each fibre contains serially arranged sarcomeres the length of which changes from 1.3 to 3.5 μm during shortening and stretching, demonstrating the variable length of the muscle fibre.

The contractile properties of the muscle fibres and its function in the body determine the internal architecture of a muscle. The load and the shortening velocity are inversely related. The maximum force during isometric contractions depends on the cross-sectional area, and short muscles develop force with little energy. Prestretching increases the force of a muscle fibre; maximum force is produced at the "optimum sarcomere length", which in situ is attained when the joint is in a midposition such that the sarcomere lengths of the agonist and antagonist are about the same. If the maximum shortening velocity per sarcomere at zero load is given, the shortening velocity of the muscle depends only on the number of sarcomeres in series; hence, long muscles are best suited for rapid (or long-range) movements.

Parallel-acting muscle fibres or chains of fibres must have the same number of sarcomeres – otherwise they would shorten at different speeds. The fibres insert in a staggered fashion; they form a parallelogram together with the tendon sheets at which they insert. The endplate is close to the middle of a fibre, and the endplates of a muscle are usually concentrated in narrow endplate zones. A muscle may have the form of only one parallelogram and be unipennate, or of several parallelograms and be multipennate. Correspondingly, it may have one or several endplate zones. The volume of a shortening muscle fibre remains constant; the angle of insertion increases during contraction, and the parallelogram becomes wider to provide space for the thickening of the muscle fibres. This protects the blood vessels and the intramuscular nerve branches. Fusiform muscles with converging fibres do not exist; the fibres would be sheared off from their insertion by the increase in circumference during shortening.

The coarse collagen bundles of the perimysium are arrayed in a way that they do not interfere with the displacement of the fascicles of muscle fibres in relation to each other; the collagen bundles only hold the muscle fibres together. Each individual muscle fibre is surrounded by a fine network of helically wound collagen fibrils which are part of the sarcolemma. They are slack at rest length, but become taut and longitudinally oriented when the fibre is stretched; there is then an increased resistance to stretch. The myofibrils, along the entire length of the fibre, are mechanically linked across the plasma membrane to the collagen fibrils of the sarcolemma which, even when the continuity of the myofibrils is interrupted, is able to transmit force.

Muscle fibres are the effectors of motor units. Several fibres receive synaptic inputs from the same motoneuron, and each fibre, at least in adult mammals, is innervated by one and only one motoneuron. A terminal motor axon branch forms one cholinergic synapse at the middle of the fibre, the neuromuscular junction, or motor endplate. This position ascertains rapid and synchronized activation of all fibres of the motor unit. Polyneural innervation of the same fibre by several motoneurons in mammalian skeletal muscles occurs only during development; multiple innervation, i.e. several branches of the same motoneuron terminating at several synapses on the same myofibre, is found in intrafusal fibres and in some fibres of extraocular muscles. Intrafusal fibres are both multiply and polyneurally innervated. Polyneural and multiple innervation are not uncommon in avian and amphibian muscles.

Maximal tetanic activation of a muscle blocks its blood supply. The slow soleus muscle of cat develops tetanus at rather low innervation frequency, when the oxygen supply still exceeds the oxygen consumption. The main function of this muscle is postural; it usually is submaximally innervated and is most resistant to fatigue. Fast muscles have to shift to anaerobic metabolism and subsist on the glycogen stores of the fibres; these muscles fatigue and are best suited for intermittant activity. Muscle fibres are specialized with respect to the preferred pathway of metabolism. Fibres with high oxidative capacity are rich in mitochondria and triglycerides and are supplied by many capillaries ("red" fibres); fibres poor in mitochondria are rich in glycogen and glycolytic enzymes ("white" fibres). Slowly contracting fibres work preferentially oxidatively, whereas fast contracting fibres may be glycolytic or mixed glycolytic and oxidative. In small mammals and in the small extraocular or laryngeal muscles, fast fibres may be rich in oxidative enzymes as well. Most vertebrate muscles consist of a mixture of fibres of different types.

Slow and fast muscle fibres of mammals are almost always twitch fibres; they conduct action potentials, give an all-or-nothing response, and are focally innervated at only one endplate. Slowly contracting fibres in avian and amphibian muscles may be "slow" (tonic) fibres which do not conduct action potentials, respond to repetitive stimulation with a graded contraction, and are innervated at multiple sites along their length. These fibres occur in extraocular muscles and in muscle spindles of mammals as well; whether they are present in other mammalian muscles is debatable.

The contractile machinery of skeletal muscle fibres consists of the proteins, actin and myosin, contained in thin and thick filaments, respectively. Actin monomers are globular and are within the thin filaments arranged in a double helix. Two sets of thin filaments, each about 1 μm long, are linked by a Z disc in an antiparallel fashion. The myosin molecule has a double-headed tadpole structure. Several hundred myosin molecules constitute a thick filament about 1.5 μm long. The tails of the molecules form the shaft of the filament, and are arranged such that within each half-filament the head regions are directed towards the filament ends. At regular intervals of 14.3 nm, the head regions stick out from the filament shaft. Neighbouring thick filaments are in their middle linked by sets of 3–5 short bridges constituting the M line. Thick and thin filaments are laterally arranged. A sarcomere consists of two I bands composed of thin filaments bound to a Z disc on one side, and freely extending into the A band on the other side. The A band is composed of thick filaments. Contraction involves the stepwise interaction of the myosin heads with the actin monomers of the thin filaments which thereby are pulled into the thick filament array. The interacting filaments develop force, and if the load is smaller than the maximum force of the fibre, they slide along each other, the distance between the Z discs decreases, and the sarcomere shortens. The interaction of myosin and actin in resting muscle fibres is depressed by tropomyosin and troponin, which are regulatory proteins localized with actin in the thin filaments. Free calcium ions inhibit this depressor action and thus initiate contraction. Calcium is liberated from the cisterns and tubules of the sarcoplasmic reticulum. Relaxation involves re-uptake of calcium by the sarcoplasmic reticulum and disengagement of the myosin heads from their binding sites at the actin monomers. Both the myosin-head action during contraction and the storage of calcium against a concentration gradient at rest are energy-consuming processes, and both the myosin heads and the reticulum membranes are sites of ATPase activity.

The signal for the release of calcium and, hence, for contraction arises at the endplate under the influence of acetylcholine liberated from the motor axon terminal; it spreads as action potential along the plasma membrane and into the interior of the fibre along the T tubules, narrow invaginations of the plasma membrane. The T tubules make contact with the sarcoplasmic reticulum at the triadic junctions. How the depolarization of the T tubule membrane induces calcium release from the sarcoplasmic reticulum is as yet obscure.

The specialization of individual muscle fibres into fast and slow and fatiguable and fatigue-resistant fibre types is expressed by the abundance or scarcity of membranes of sarcoplasmic reticulum and T system, and of mitochondria. There are also differences in the structure of the myofibrils, the M and Z lines, the motor endplates, and the plasma membranes. Due to the complex structure of the myosin molecule, several myosin isoenzymes are known, some of which characterize fast and slow fibres of different origins.

Multinucleated muscle fibres arise by fusion of mononucleated myoblasts. It is now widely agreed that, before fusing, the differentiation of myoblasts involves several steps. Nuclei of muscle fibres are irreversibly postmitotic, and it has been shown that the myoblasts stop synthesizing DNA and commence transcribing RNA for muscle-specific proteins before they actually fuse. How the terminal step of myoblast differentiation is controlled is a matter of controversy. Myofibrils first occur after fusion, but if fusion is experimentally prevented, the most mature myoblasts start to produce actin and myosin filaments. It is probably the large size of the muscle fibre that makes it necessary for the cell to become multinucleated, which keeps the ratio of DNA to protein above a certain limit. Cardiac muscle cells are smaller and usually mononucleated, but in some species the nuclei of hypertrophic hearts become polyploid. The nuclei of skeletal muscle cells invariably are diploid. Because they do not synthesize DNA, growth or regeneration of muscle fibres always necessitates proliferation of myoblasts; some of their daughter cells mature and then fuse with the muscle fibre. Satellite cells are the reserve myoblasts of adult muscles.

Whereas most muscles are heterogeneous with respect to the metabolic and contractile properties of their muscle fibres, all fibres of a motor unit are of the same type. Thus, there are fast and slow, and fatiguable and fatigue-resistant motor units within the same muscle. The motor units are sequentially activated during voluntary contractions. Slow and fatigue-resistant units tend to be activated first, whereas fast and fatiguable units have a high threshold for activation. This is a result of the synaptic organization of the anterior horn cells. Peripheral and central influences may alter the apparently stereotyped sequence of activation, depending on the motor task.

This short overview of the essential facts and problems of muscle illustrates that morphological, biochemical, and physiological studies have contributed to a unified picture, and that it is impossible to see muscle tissue under one aspect only. This is the greatest advance since publication of the previous edition of this handbook volume. Thirty years ago, seemingly insurmountable gaps existed between the autolytic muscle fibres studied by anatomists, *Muskelbrei* and actomyosin threads produced by biochemists, and the living muscles investigated by physiologists. Progress was initiated by electron microscopy which showed the complexity of the fibre structure, and provided a "framework" for the results obtained using different techniques. The extent to which research on muscle contraction had gone astray before the array of myofilaments was clarified has become history.

B. Microanatomy of Muscle

I. The Array and Length of Skeletal Muscle Fibres

Skeletal muscle fibres are thread-like syncytia of fused mononuclear cells; they are connected at both ends to tendons, which in turn are connected to the skeleton, and in some instances, also to the connective tissue of the skin. The tensile strength per unit area of the tendon tissue is about 100 times greater than that of muscle tissue; hence, the cross-sectional area of the muscle is usually larger than that of the tendon. Because the weakest part of a chain determines its strength, muscle fibres must have the same cross-sectional area throughout their length. If fibres are tapered or wedge-shaped, the attachment of the tendon is staggered such that the decreasing number of myofibrils is compensated for by an increasing number of collagen fibrils running parallel to the muscle fibre. Fibres or strands of fibres working in parallel must be of the same functional length, i.e. the number of sarcomeres must be the same. These two demands – equal cross-sectional area along the fibre length and equal number of sarcomeres in fibres working together – determine the array of muscle fibres within a muscle. The basic unit is a parallelogram formed by the fibres and the tendon sheets at which they more or less obliquely insert. Muscles can be described as either simple or complex pennate, depending on whether the entire muscle consists of only one or of several parallelograms. This principle of muscle architecture was first noticed by NIELS STENSEN (NICOLAUS STENO) in 1667. Truly fusiform muscles – tapering muscle fibres inserting in a small area – do not exist. The array of fibres in a parallelogram does not imply that all fibres run from tendon to tendon; it is not the fibre length but the number of sarcomeres in series that must be equal. The fibres of most human muscles run through the muscle and are not arranged in series; only in the sartorius muscle do the fibres terminate within the muscle belly, and the gracilis muscle consists of two interdigitating sets of fibres each terminating in the midregion of the muscle (SCHWARZACHER 1959; COËRS and WOOLF 1959). In bovine and porcine muscles, however, muscle fibres terminating within the muscle belly are common (SWATLAND and CASSENS 1972). The diaphragm of rat consists of radially arranged fibres running all the way from the tendinous centre to their insertion at the rib cage; whether this is the case in man as well is unknown.

A muscle in which a few sarcomeres act in series is best suited to develop force at little energy expense; the force depends on the total cross-sectional area only. When the maximum shortening velocity of a sarcomere (intrinsic speed of shortening) (CLOSE 1965, 1972) is given, the shortening velocity of

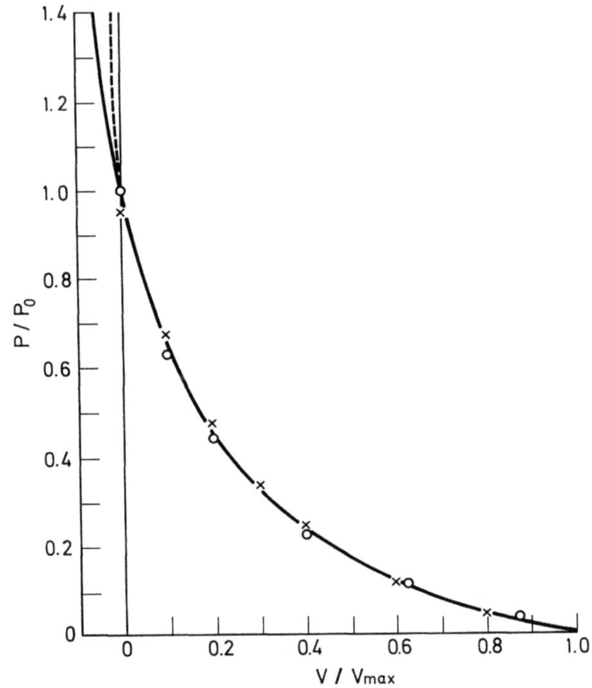

Fig. 1. Force-velocity relation of frog skeletal muscle at 0° C, describing how the speed of shortening increases with decreasing load. The continuous line according to HILL's (1938) equation: $P/P_0 = (1 - V/V_{max})/(1 + P_0/a)V/V_{max})$, where P_0 is the maximum tetanic force and V_{max} the maximum shortening velocity without load; P and V are load and shortening velocity, respectively. P_0/a is a constant which is different for different muscles. It is about 4 for frog muscle fibres at 0° C, 2–5 for the human brachial biceps muscle at 37° C, and around 10 for slow invertebrate muscles (for references, see REICHEL 1960). The shape of the curve distinguishes different muscles because also P_0 and V_{max} differ. The measured data for lengthening contractions with loads larger than P_0 deviate from the equation (*dotted line*). The *crosses* and *circles* represent the results of calculations based on two different theories (for details see HUXLEY 1974). (From HUXLEY 1974, with permission of the author and the Physiological Society)

the muscle depends on the number of sarcomeres working in series. Hence, long muscles are best suited for fast movements. The shortening velocity of a muscle decreases with increasing external load, according to the hyperbolic equation formulated by HILL (1938), and approaches zero at isometric contractions. Loading a contracted muscle with a weight exceeding its maximum isometric force causes a lengthening or eccentric contraction (Fig. 1).

The length of muscle fibres may be determined by measuring isolated individual fibres or small fibre bundles, or by tracing the action potential of an individual fibre along the length of the muscle. The human brachial biceps muscle consists of fibres 10–13 cm long (BUCHTHAL 1961). The fascicles of muscle fibres in the human sartorius and gracilis muscles are 41–49 cm and 20–29 cm long, respectively; the fibres of both muscles measure 7–18 cm in length (SCHWARZACHER 1959). The greatest fibre length has been described by LOCK-

HART and BRANDT (1938) who isolated a 34-cm-long fibre fragment from a sartorius muscle. The fibres of the rabbit anterior tibial muscles are 40–60 mm long, which amounts to 60%–80% of the length of the muscle belly (CRAWFORD 1973). The average fibre length in the teniussimus muscle of cat is only one-fifth of the muscle length (ADRIAN 1925).

A living muscle fibre may contract or be stretched, which is reflected by the varying height of the individual sarcomeres. These may reversibly shorten to less than 1.5 µm and be stretched to 3.5 µm. Thus, the functionally relevant information is the number of sarcomeres in series. These are either counted in the microscope, or the number of sarcomeres is computed from the fibre length and the average sarcomere length. The soleus muscle of cat consists of fibres 30–50 mm long, each containing 12,000–15,000 sarcomeres (TABARY et al. 1972). The fibre length in the soleus muscle of adult rats is 12 mm and in the extensor digitorum longus muscle it is 10 mm, with 5,000 and 4,000 sarcomeres, respectively (CLOSE 1964). The masseter muscle of rhesus monkey consists of fibres with 4,000–6,000 sarcomeres; the fibres of the temporal muscle have an average of 7,000 sarcomeres (MAXWELL et al. 1981).

To standardize the length of a muscle fibre and its sarcomere length, both are related either to the equilibrium length or to the in situ length. Equilibrium length is the length at which the living muscle at rest does not yet develop force when passively stretched, but is also no longer slack (LINDHARD 1931). Alternatively, it may be defined as the length which a resting stretched muscle attains when it is released. The two procedures do not give identical results because of the hysteresis of the length-tension diagram. In situ length, often also called "rest length", is defined by the joint being in a midposition such that agonist and antagonist are stretched to about the same degree. In frog (BUCHTHAL 1942) and guinea pig muscles (HØNCKE 1947), a sarcomere length of 2.2–2.3 µm corresponds to the equilibrium length; the in situ length is 10%–30% above equilibrium length (HILL 1950a; BUCHTHAL and KAISER 1951). The degree of stretch at which a tetanically stimulated frog semitendinous fibre develops maximum tension (optimum sarcomere length) usually corresponds to the in situ length (BUCHTHAL 1961), i.e. to 120% equilibrium length (Figs. 2, 3). By contrast, RAMSEY and STREET (1940) measured, also in frog semitendinous fibres, maximum force at 100% "rest length" when the fibre was "just taut". This "slack length" corresponds to a sarcomere length of 2.05 µm (HUXLEY 1974). The optimum sarcomere lengths are larger in mammalian muscles: rat soleus and extensor digitorum longus muscles 2.50–2.55 µm, mouse brachial biceps muscle 2.8 µm, cat flexor hallucis longus muscle 2.7 µm, cat soleus muscle 2.8–3.1 µm (CLOSE 1972). The difference between cat and rat soleus muscle is striking; it may relate to the fact that the maximum sarcomere lengths to which the muscles can be stretched in vivo, by dorsiflexion of the foot, are 2.8 µm in rat and 3.3 µm in cat (SCHMALBRUCH 1971).

The neuromuscular junctions of a muscle may be stained using various methods; most often the histochemical reaction for acetylcholinesterase is used. Each fibre has one endplate which is close to the midportion of the fibre. The endplates are mostly confined to a narrow endplate zone (COËRS and WOOLF 1959; SCHWARZACHER 1959). The longitudinal dispersion of the endplates within a

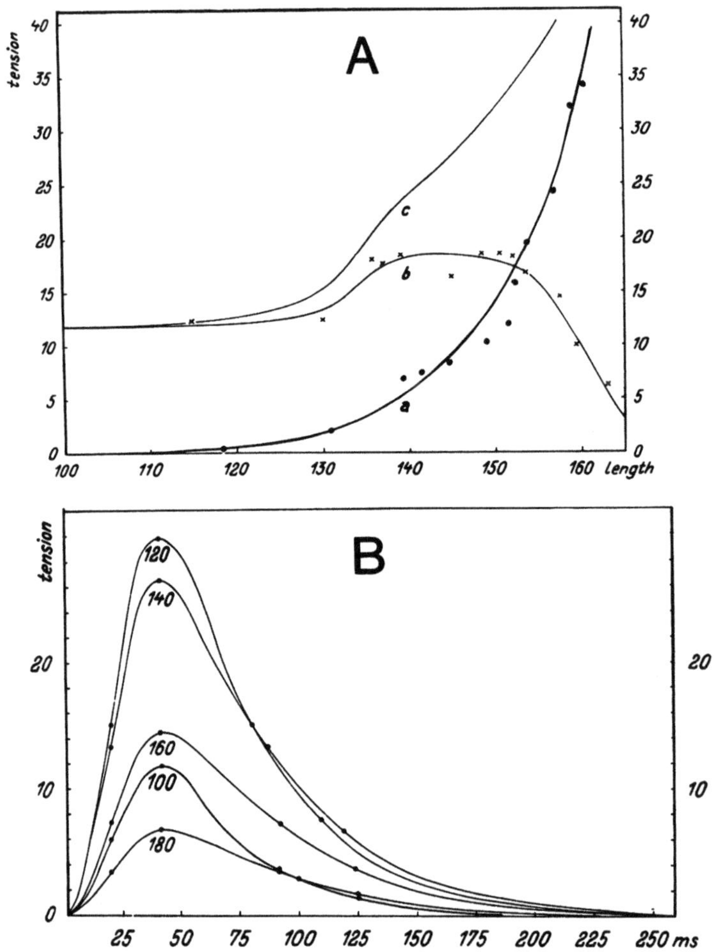

Fig. 2A, B. The relation between resting tension, fibre length, and twitch force. (Frog semitendinous muscle, single fibre, 12°–14° C.) **A** Length-tension diagram at rest (curve *a*). Note that the resistance to stretch increases with increasing length (equilibrium length = 100). Curve *c* describes the isometric twitch tension the fibre reaches when it is stimulated at different lengths, the extra tension produced is given by curve *b*. Maximum force is developed when the fibre is pre-stretched to 135%–140% equilibrium length, the force declines when the fibre is overstretched. *Ordinate*, tension in mg. **B** Isometric twitches of a single fibre at different degrees of pre-stretch (equilibrium length = 100). This fibre reached maximum force at 120% equilibrium length. *Ordinate*, tension in arbitrary units. The figure depicts the type of experiments from which curves *b* and *c* in **A** have been obtained. (From BUCHTHAL 1942, with permission of the author and the Danish Royal Society)

fascicle is at most 0.5 mm (COËRS and TELERMAN-TOPPET 1977). The width of the endplate zone of the human brachial biceps muscle is <15% of the fibre length (BUCHTHAL 1961). Individual living fibres have been isolated from the flexor digitorum brevis muscles of rats; the fibres are 0.7–1.2 mm long and all endplates are placed near the middle of the fibres (BEKOFF and BETZ 1977). The fact that the endplate is situated at the midportion of a muscle

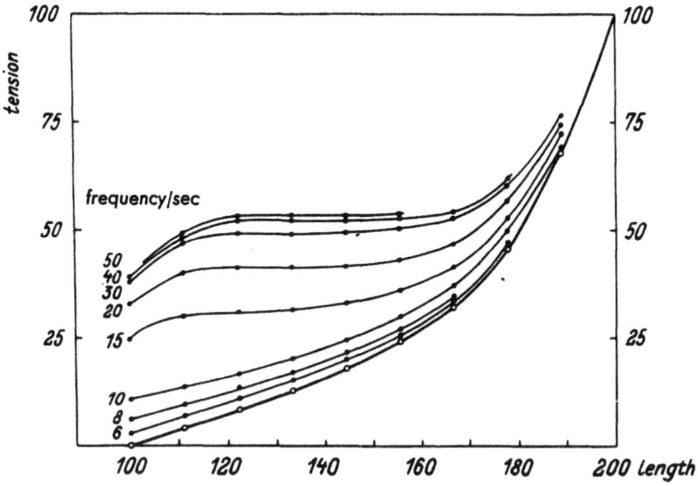

Fig. 3. The relation between resting tension, fibre length, tetanic force, and stimulus rate. (Frog semitendinous muscle, single fibre, 12°–14° C.) Length-tension diagram (*open circles*) and the isometric tensions that are produced by tetanic stimulation at different degrees of pre-stretch (*closed circles*). Maximum force is reached at 50 Hz and 120% equilibrium length. The diagram also illustrates the possible role of the innervation rate for force gradation in situ (see Chap. H). *Ordinate*, tension in arbitrary units. (From BUCHTHAL 1942, with permission of the author and the Danish Royal Society)

fibre ensures the fastest activation of all its contractile material. Nevertheless, in the highly complex flexor carpi radialis muscle of cat, endplates have also been found over the outer fibre segments (GALVAS and GONYEA 1980). The physiological implications of this finding are obscure.

The course of the endplate zone across the muscle reflects the way the muscle is attached to its tendons. The endplate zone of simple pennate muscles is almost straight, in complex pennate muscles it may take a complicated course and different compartments of a muscle may have unconnected endplate zones (Fig. 4). The sartorius muscle of man consists of muscle fibres in series and the endplates are scattered over the entire length of the muscle. Two endplate zones are found in the gracilis muscle of man, rat, and cat, each belonging to different sets of muscle fibres interdigitating across the midportion of the muscle (JARCHO et al. 1954; CHRISTENSEN 1959; COËRS and WOOLF 1959). The myotendinous junctions of some vertebrate muscle fibres react for acetylcholinesterase as well (see Sect. C.IV.3d); erroneously, this has been taken as evidence that motor nerve endings and short fibres occur at the ends of the human brachial biceps muscle (MCCOMAS et al. 1982; KERESHI et al. 1983). AQUILONIUS et al. (1984) prepared longitudinal cryosections 20 μm thick from entire muscles of adult subjects and stained the endplate region for acetylcholinesterase. The course of the endplate zone was reconstructed from serial sections. In the brachial biceps muscle, the endplates were distributed as a 5–10-mm wide band which was slightly V-shaped and passed through the middle of both heads. In the anterior tibial muscle, the endplates formed a parabola with its apex

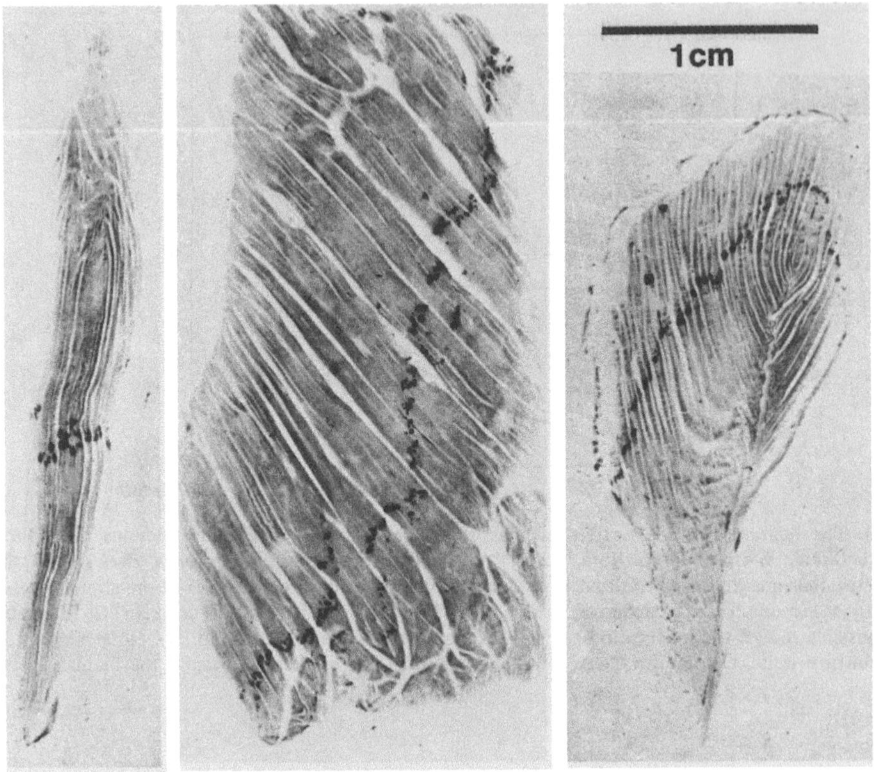

Fig. 4. Infant muscles. The endplates are stained for acetylcholinesterase to demonstrate the course of the endplate zone. *Left:* Extensor hallucis longus muscle of infant aged 1 month. Long pennate muscle. *Middle:* Frontal section of vastus internus muscle, infant aged 8 months. Muscle with parallel fibres. *Right:* Flexor carpi radialis muscle of infant aged 3 months. Complex pennate muscle. Note that in all muscles the endplates are roughly in the middle of the fibre bundles. (From Coërs and Woolf 1959, with permission of the authors and Blackwell Publishers).

at the proximal end of the muscle. The endplate zone was not always as sharply demarcated as in the brachial biceps muscle. In the sartorius muscle, the endplates were scattered throughout the muscle and there was no real endplate zone.

In birds and amphibia, endplates may be found along the entire length of a muscle, either because short fibres are arranged in series, or because one axon forms many endplates (multiple innervation), or because several axons terminate on the same fibre (polyneural innervation).

II. The Diameter of Skeletal Muscle Fibres

The fibre diameters of skeletal muscle fibres range from 10 to 100 μm; under pathological conditions, fibres of up to 200 μm diameter may be found.

Measuring the diameter of a muscle fibre is not without problems. The best way of determining the true diameter would be to measure it in isolated living fibres. The only data for mammals are from rat flexor digitorum brevis muscle, which consists of small fibres about 1 mm in length, with a diameter of 30–40 µm (BEKOFF and BETZ 1977). Fixation or freezing of a living muscle fibre for histology induces contracture and shortening; the diameter increases correspondingly because the fibre volume remains constant. Muscle fibres have to be fixed at a defined length either by fixing them in vivo by perfusion, or by holding the muscle or a bundle of fibres at constant length. This limits the usefulness of "needle biopsies" of human muscle. The classical advice to let the muscle "rest", i.e. die before it is fixed, is obsolete because this causes autolytic changes of the muscle fibres. Many fixation procedures, all embedding procedures, and even slow freezing, result in shrinkage. The diffusion gradient during fixation may produce shrinkage and swelling at the same time. One frequently observes that, in muscles fixed by immersion, the fibres close to the surface of the sample are thicker than those in the centre; this difference disappears when the same muscle is fixed by iso-osmotic vascular perfusion or by quick freezing. Vascular perfusion with high osmolarity solutions causes shrinkage around large blood vessels; shrinkage or swelling of the tissue also depends on the perfusion pressure (KAUFMANN 1981). Rapid freezing at constant length without fixation gives the best preservation, but staining and dehydration of sections might reduce the cross-sectional area of the fibres as well. As a rule of thumb, in the best paraffin sections, the muscle fibres have 30% smaller diameters than in sections from freshly frozen material. This means that, under "optimal" conditions, the cell volume shrinks by 50%, under "less-than-optimal" conditions the shrinkage might be even larger. To give an example, the fibres of the brachial biceps muscle of man had a mean diameter of 23 µm after paraffin embedding (ETEMADI and HOSSEINI 1968), in celloidine-embedded material they were 48 µm thick (BARIN-BAUM 1963), and in freshly frozen material they measured 50–70 µm (BROOKE and ENGEL 1969).

The cross-sectional profile of muscle fibres is rarely circular. Therefore, the best measure of the fibre size would be the cross-sectional area. To obtain a reliable mean value, planimetry of about 200 fibres is necessary. This was most tedious before digitizer tablets became available. Nevertheless, the difference between the true cross-sectional area and the area of a circle calculated from the mean of the largest and smallest diameters of a fibre is negligible compared with the effects of specimen preparation. Some workers determine only the smallest diameter to exclude obliquely sectioned fibres which cause an overestimate of the fibre area. This approach, however, results in an underestimate, because fibres have elliptical cross-sectional shapes. Fibre diameter measurements in the microscope give a rough estimate only. Expensive automated devices working with a scanning beam often are not able to distinguish adjacent fibres. Reliable fibre measurements have to be done on micrographs or projections; ruler and pen, or an electronic system, are equally suitable as long as the quality and the size of the sample are sufficient.

The variation of fibre diameters within a given muscle is of practical interest because the scatter of diameters increases in a diseased muscle or under experimental conditions. The normal scatter of fibre diameters in unfixed frozen-sectioned muscles of man does not exceed 20% (SD expressed as % of mean; own unpublished data). Most workers have measured the diameters of different fibre types separately; their results are reviewed in the context of fibre types (Chap. D, Tables 5–9).

III. The Number of Fibres of a Muscle

In muscles of small experimental mammals, it is feasible to count all fibres within a cross-section passing through the widest girth of the muscle, provided all fibres really pass through the widest girth. This may be so in simply built muscles, but it is certainly not the case in complex muscles in which the tendons of origin and insertion overlap at midbelly. The "deficit" will be larger in muscles in which the angle of insertion is acute, than in muscles with a blunt angle of insertion. More fibres pass through the middle of the belly of the same muscle when it is shortened than when the muscle is stretched. BURKE et al. (1974) estimate that in the soleus muscle of cat, only 80% of the fibres are contained within a cross-section through midbelly; the deficit in complex muscles is much larger.

Entire human muscles are difficult to prepare for histology. Different approaches have been used to overcome this problem: (a) The muscle is longitudinally divided and each bundle is separately assessed. (b) The muscles of stillborn infants are used. (c) The fibre number is calculated from the mean fibre size and the total cross-sectional area of the muscle.

The use of stillborn infant muscles (CHRISTENSEN 1959) implies that there is no postnatal increase in fibre number. This is correct for human muscles: the 100,000–120,000 fibres found in the adult sartorius muscle are present in fetuses of about 25-cm crown-rump length (gestational age, 180 days) (MONTGOMERY 1962; STICKLAND 1981). Rodent muscles, however, are immature at birth, and the fibre number may double during postnatal development (CHIAKULAS and PAULY 1965) (see Sect. G.V.1). It is possible to determine fibre size and muscle area in live human muscles and to calculate the fibre number; volunteers undergoing a training program, or patients, can be studied with the help of the needle-biopsy technique and ultrasound (IKAI and FUKUNAGA 1968) or CAT scanning (HÄGGMARK et al. 1978). Several sources of error affect the results. The mean fibre size in the sample may not be representative for the entire muscle, or the degree of stretch may be different in the histological sample and in the muscle when the cross-sectional area is measured. The angle of fibre insertion in atrophic muscles might be more acute than in hypertrophic muscles, and at the same muscle length the fibre number at midbelly may appear larger in hypertrophic than in atrophic muscles (see above).

LEXELL et al. (1983a, b) sectioned entire lateral vastus muscles obtained at autopsy and determined the fibre size in multiple sites. There was no relation between mean fibre size and total cross-sectional area of the muscle in different

Table 1. Number of fibres in muscles of baboon and man. *In vivo*, the diameter of the muscle fibres was assessed in needle biopsies and the total fibre number was calculated from the size of the fibres and the total cross-sectional area of the muscle as determined by ultrasound- or CAT-scanning; *count*, the number of fibres was determined in cross-sections of muscles obtained at autopsy

Species	Muscle	Number of fibres	Method	References
Baboon	Med. gastrocnemius	229,852	Count	WRAY (1962)
	Abductor pollicis	9,363		
Man	Quadriceps femoris	1,730,000	In vivo	YOUNG et al. (1982)
	Rectus femoris, stillborn	186,292	Count	CHRISTENSEN (1959)
	Lat. vastus	464,000	Count	LEXELL et al. (1983a)
	6 males 19–37 years	393,000–550,000	Count	LEXELL et al. (1983b)
	6 males 70–73 years	278,000–423,000		
	Sartorius		Count	MONTGOMERY (1962)
	Newborn	101,000		
	Adult	116,000		
	Fetal 25 cm	120,000	Count	STICKLAND (1981)
	Newborn	222,424	Count	CHRISTENSEN (1959)
	Gracilis, newborn	144,933		
	Semitendinosus, newborn	508,219		
	Gastrocnem., newborn	1,505,000		
	Med. gastrocn., adult	1,000,000	Count	FEINSTEIN et al. (1955)
	Anter. tibial, adult	270,000		
	Brachiorad., adult	130,000		
	Brachial biceps			
	Newborn	580,000	Count	CHRISTENSEN (1959)
	Adult	434,500 (SD 110,500; n=42)	Count	BARIN-BAUM (1963)
	Adult	199,240-316,243	Count	ETEMADI and HOSSEINI (1968)
	Adult	382,700-641,000	In vivo	NYGAARD (1980)
	Brachial triceps			
	11 females 26 years	272,000–455,000	In vivo	SCHANTZ et al. (1983)
	10 males 27 years	364,000–600,000		
	Dorsal inteross. I	40,500	Count	FEINSTEIN et al. (1955)
	Lumbricalis I	10,300		
	Opponens pollicis, newborn	79,080	Count	CHRISTENSEN (1959)

subjects, indicating that there is considerable interindividual variation. The fibre number did not differ between the right and left legs of patients with unilateral quadriceps atrophy after knee injury. Hence, muscle atrophy is not due to fibre loss, but to atrophy of the individual fibres (YOUNG et al. 1982). No difference in fibre number was found in the quadriceps femoris muscle of untrained and trained subjects, and of bodybuilders. This indicates that muscle hypertrophy is completely accounted for by transverse growth of the individual fibres, rather than by hyperplasia (SCHANTZ et al. 1981). By contrast, TESCH and LARSSON (1982) found more muscle fibres in bodybuilders than in normal subjects, and concluded that either the fibre number had increased or the bodybuilders were genetically selected. The fibre number of the triceps brachii muscle was the same in trained males and females (SCHANTZ et al. 1983; NYGAARD et al. 1983). The number of fibres in the lateral vastus muscle of elderly males was smaller than in young males, suggesting that fibres are lost in old age (LEXELL et al. 1983b).

The force of a muscle is correlated to its cross-sectional area (MAUGHAM et al. 1983; SCHANTZ et al. 1983). These studies which use computer assisted tomography (CAT) confirm a previous ultrasound-scanning study by IKAI and FUKUNAGA (1968), and show that increased strength is always due to muscle hypertrophy rather than to an increase of the specific force of the muscle fibres.

In muscles of experimental animals subjected to exercise, several authors found fibres bisected by clefts, as well as branching fibres; according to these authors, this indicates that the number of muscle fibres increases by longitudinal splitting (ROWE and GOLDSPINK 1968; EDGERTON 1970; GONYEA 1980; HO et al. 1980; SILBERMAN et al. 1983). The hypothesis that mature muscle fibres can divide, and that this is the growth mechanism of muscle (WEISMANN 1861) is now obsolete; branching fibres, as in tissue culture, result from incomplete fusion of myoblasts and myotubes during regeneration, and indicate fibre damage (SCHMALBRUCH 1976a, b; DRENCKHAHN and LÜLLMANN-RAUCH 1979; LANDON 1982) (see Sect. G.I).

The fibre numbers in human and baboon muscles are listed in Table 1. Data for experimental animals are contained in Table 5 (see also Chap. H, Table 12).

IV. The Connective Tissue of the Muscle

From a descriptive point of view, one can distinguish the epimysium (which surrounds the entire muscle), the perimysium (which encircles somewhat angular fascicles), and the endomysium (which encircles individual muscle fibres). The endomysium is so delicate that even after the use of special light microscopic stains it is often not visible. It contains the capillaries and the terminal nerve twigs.

Fascicles are subdivided into primary fascicles of 10–100 fibres; the delineation is not always distinct. WALLS (1960) wants to reserve the term perimysium for the tissue around the smallest fascicles. (This is merely a matter of nomenclature.) The perimysium can also be described as septa between different parts of the muscle, rather than as sheaths around fibre bundles. Large blood vessels and nerves run only in septa where muscle spindles are also found (Figs. 5, 6).

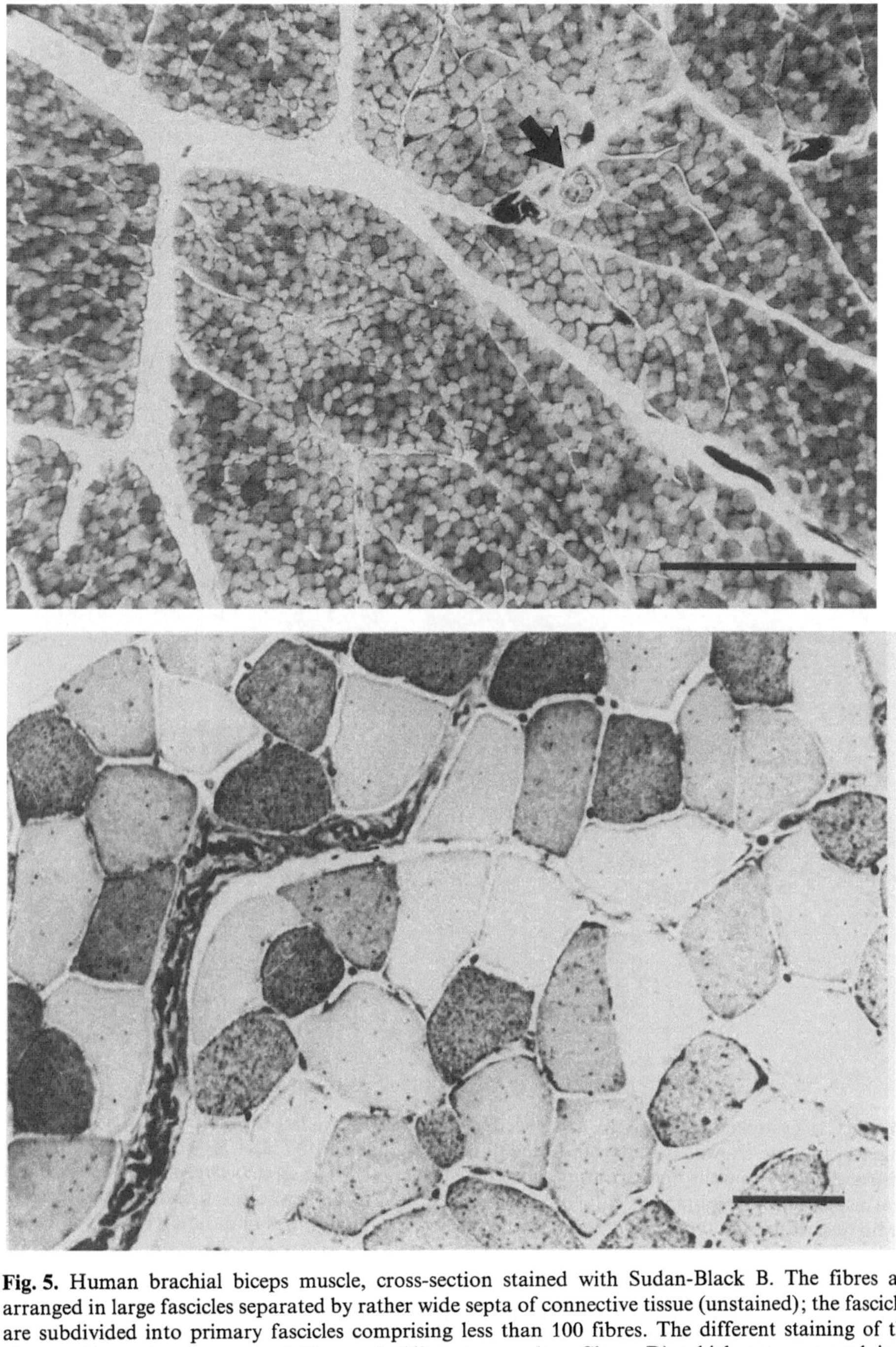

Fig. 5. Human brachial biceps muscle, cross-section stained with Sudan-Black B. The fibres are arranged in large fascicles separated by rather wide septa of connective tissue (unstained); the fascicles are subdivided into primary fascicles comprising less than 100 fibres. The different staining of the fibres reflects the presence of fibres of different type (see Chap. D) which are arranged in a checkerboard pattern. A bundle of myelinated nerve fibres (below) runs between primary fascicles to supply the individual muscle fibres (see also Fig. 14). One muscle spindle is seen (*arrow*). Bars, *top* 500 μm; *bottom* 50 μm

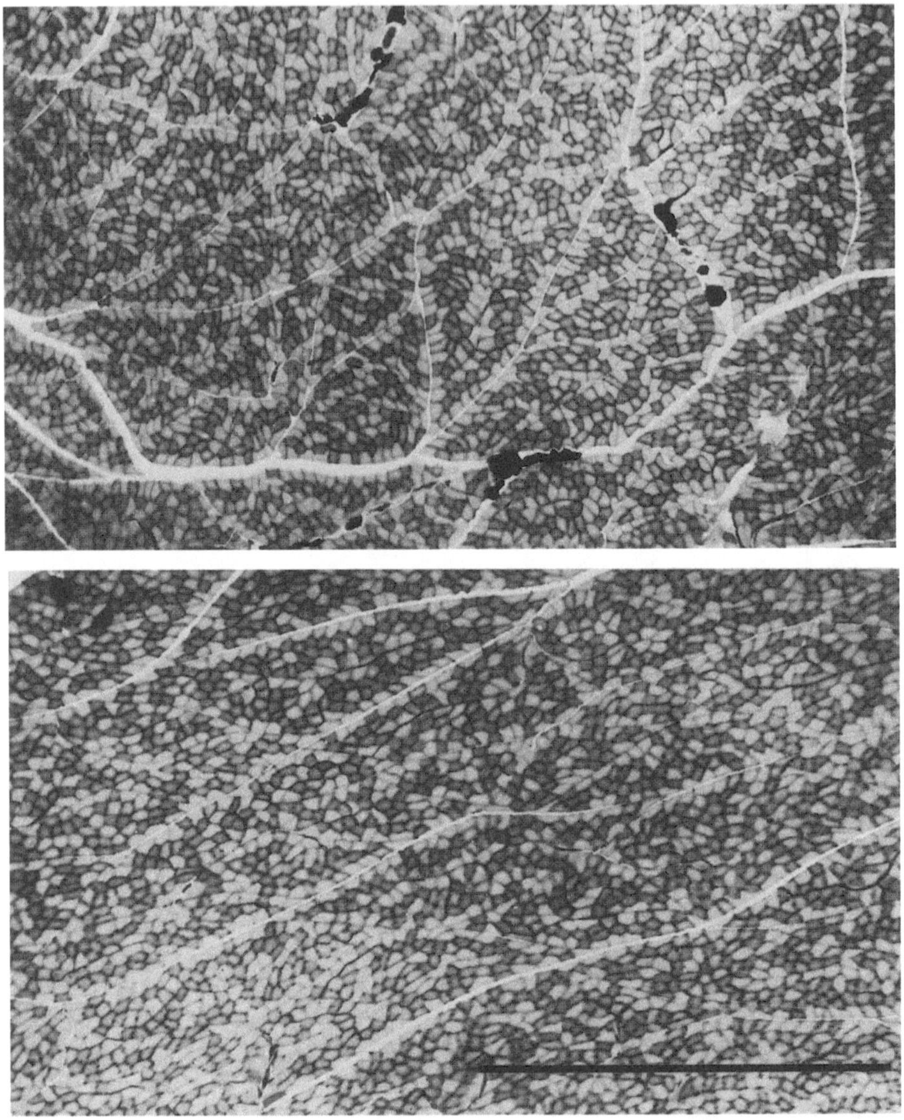

Fig. 6. Cat gastrocnemius muscle, cross-section stained with Sudan-Black B. Two regions of the same section are shown to demonstrate the different outlines of the fascicles, reflecting the internal displacements within the muscle during shortening and stretch. The septa between the fascicles ("neutral displacement membranes", Fig. 9) contain few fat cells (*black*). Also this muscle is heterogeneous with respect to fibre types and shows a checkerboard pattern of fibres of different type. Bar, 1 mm

1. Endomysium

Adjacent fibres are linked by the endomysium in such a way that in stretched and shortened muscles the sarcomeres tend to remain in register (BUCHTHAL and KNAPPEIS 1940). The endomysial collagen fibrils remain attached to a fibre when it is isolated and contribute to its mechanical properties. These collagen fibrils are part of the sarcolemma (see Sect. C.IV).

The breaking force per unit area of living frog muscle fibre is only one-third above its maximum tetanic force (3.3–3.7 kg/cm²); the breaking force of individual fibres and of the entire muscle is about the same (4–5 kg/cm², 20° C) (WALTER 1947; CASELLA 1951). Thus, the breaking force of the muscle lies in the fine network of collagen fibrils of the sarcolemma, and not in the epi- or perimysium, which are removed when the fibres are isolated. The breaking force of a human muscle is 9 kg/cm² (WÖHLISCH et al. 1926), 50%–100% above the maximum force. Data for individual fibres are not available.

The resistance to stretch is unaltered in rat muscles at rest when the epimysium is circumferentially sectioned (STOLOV and WEILEPP 1966). This supports the notion that the coarse connective tissue contributes little to the stiffness of the resting muscle. The myofibrils of a locally injured muscle fibre retract and an empty segment of the sarcolemma is formed (SPEIDEL 1939). The muscle fibre is still able to develop 70%–100% of its original force (STREET and RAMSEY 1965). It appears as if the sarcolemma is mechanically linked to the myofibrils (see Sect. C.II), and that the force developed by the still-intact fibre segments is transmitted by the collagen layer of the sarcolemma across the damaged region. The breaking force of the wall of the empty fibre segment is 500 kg/cm² (STREET and RAMSEY 1965), which is close to the breaking force of a tendon (600 kg/cm²; CRONKITE 1936; ELFTMAN 1944). The empty segment lengthens immediately after injury, and the fibre becomes slack. The length of the empty segment increases by about 50% when the fibre is adjusted to its new equilibrium length. This suggests that, in intact fibres at equilibrium length, the collagen envelope is slack, and that the myofibrils carry the resting tension. The initial phase of the length-tension diagram of a muscle fibre (Figs. 2, 3) probably represents the tension of the resting myofibrils; at about 150% equilibrium length the slope of the diagram becomes steep and the collagen envelope increasingly contributes to it (BUCHTHAL 1942; CASELLA 1951). By contrast, RAMSEY and STREET (1940) attributed the tension of a resting muscle fibre to its sarcolemma, rather than to its myofibrils. PODOLSKY (1964) succeeded in peeling off the sarcolemma of isolated frog muscle fibres ("skinned" fibre preparation) and found that below a sarcomere length of 3.2 μm the elastic modulus of a skinned fibre is identical with that of an intact fibre.

Two models have been proposed for the texture of the collagen fibrils surrounding the individual muscle fibres. PETERSEN (1924) assumed that collagen fibrils, which are inextensible by forces produced by the muscle fibre, form a stocking-like structure, and that they are arranged in two opposite spirals. The pitch of these spirals should increase during stretch. The difficulty with that model is that the surface of the muscle fibre increases during stretch because its volume remains constant, whereas the surface of a taut envelope does not increase (RUSKA 1954). The elastic modulus of the sarcolemma is seven times larger in longitudinal than in transverse direction (RAPOPORT 1973). From longitudinal and transverse length-tension diagrams, FIELDS (1970) and FIELDS and FABER (1976) calculated that the collagen fibrils run at a 55° tilt to the axis of the muscle fibre, and form an inextensible envelope which is slack at equilibri-

Fig. 7

Figs. 7, 8. Replicas of the surface of frog sartorius muscle fibres fixed at 2.2 μm and 2.8 μm sarcomere length, respectively. The sarcomere spacing is visible through the sarcolemma, in Fig. 8 the Z discs are also visible within the shrunken I bands. At short sarcomere length, individual collagen fibrils run obliquely in relation to the fibre axis; they attain a more parallel course in the stretched fibre. On top of the collagen bundles are coarse bundles of collagen fibrils (*F*) which obviously are not stretched together with the fibre. These bundles probably belong to the "neutral connecting threads" of the perimysium (see Fig. 9). Bars, 5 μm (From SCHMALBRUCH 1974)

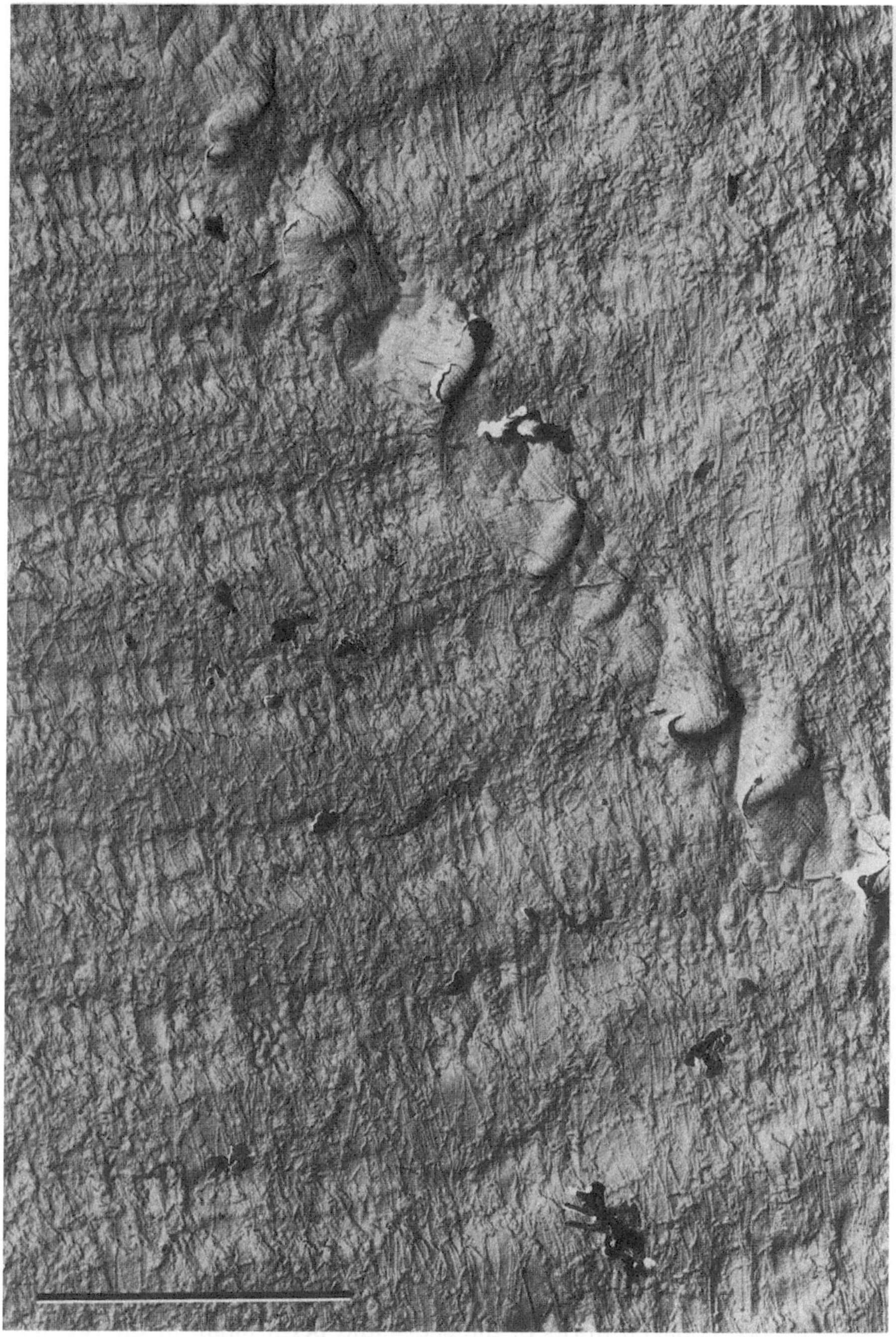

Fig. 8

um length. The tension rises steeply when the increasing surface of the stretched muscle fibre reaches that of the envelope (at 140% equilibrium length). This is in agreement with the results of the single fibre studies of BUCHTHAL (1942). A network of helically arranged collagen fibrils has been demonstrated in replicas of the surface of frog muscle fibres (SCHMALBRUCH 1974); the pitch is about 45° at 2.2-µm sarcomere length and many fibrils are slack. Between 2.8- and 3.1-µm sarcomere length, the collagen fibrils run more or less parallel to the axis of the muscle fibre and slack has disappeared (Figs. 7, 8). This demonstrates that the collagen fibrils are slack at equilibrium length *and* that their orientation changes during stretch.

2. Perimysium

Replicas of the surface of the muscle fibres, furthermore, show large obliquely running bundles of collagen fibrils, the orientation of which does not change with stretch (SCHMALBRUCH 1974) (Figs. 7, 8). They probably belong to the perimysium. Together with the findings described above, this supports a model for the relative displacement of muscle fibres during shortening as proposed by FENEIS (1935) (Fig. 9). The fascicles of a muscle, in a staggered fashion, obliquely insert at its tendons, and there are gaps between the areas of insertion. When the muscle shortens and the angle of insertion becomes less acute, the gaps between the fascicles widen and provide space for the increase in fibre diameter. The spaces between the fibre bundles also prevent the interfascicular blood vessels and nerves from being squeezed. FENEIS (1935) calls this mechanism the "unfolding of the muscle". The large collagen bundles of the perimysium are arranged parallel to the tendons and obliquely to the muscle fibres, and hence do not resist movement of the fascicles. They are "neutral connecting threads" and bind the fascicles to septa of perimysium, which thus act as "neutral displacement membranes". The advantage of such an arrangement is that it minimizes internal resistance of the muscle, and creates regions with minimal displacement around the large blood vessels and nerves. The geometry of the tendon insertion, and the necessary internal displacements probably determine the array of "neutral displacement membranes", i.e. the fascicle architecture (Fig. 6).

ROWE (1981) studied the connective tissue of muscles of rats, rabbits, sheep, and cattle using scanning electron microscopy and found a dense network of collagen fibrils and coarse fibres arranged in a criss-cross pattern. ROWE surmises that the coarse perimysial collagen fibres play a role in force resistance. It is obvious that the resolution of scanning electron microscopy (BOYDE and WILLIAMS 1968; ROWE 1981) does not enable elucidation of the array of the finest collagen fibrils forming the sarcolemmal part of the endomysium.

The crimped collagen fibrils of the endo- and perimysium of the rat soleus muscle have a unimodal diameter distribution with a mean of 30 nm. The ordered collagen fibrils close to the myotendinous junctions (microtendons) have the same diameter distribution, whereas fibrils of the tendon range in diameter from 20–250 nm (MOORE 1983). MOORE (1983) distinguishes the areolar intra-

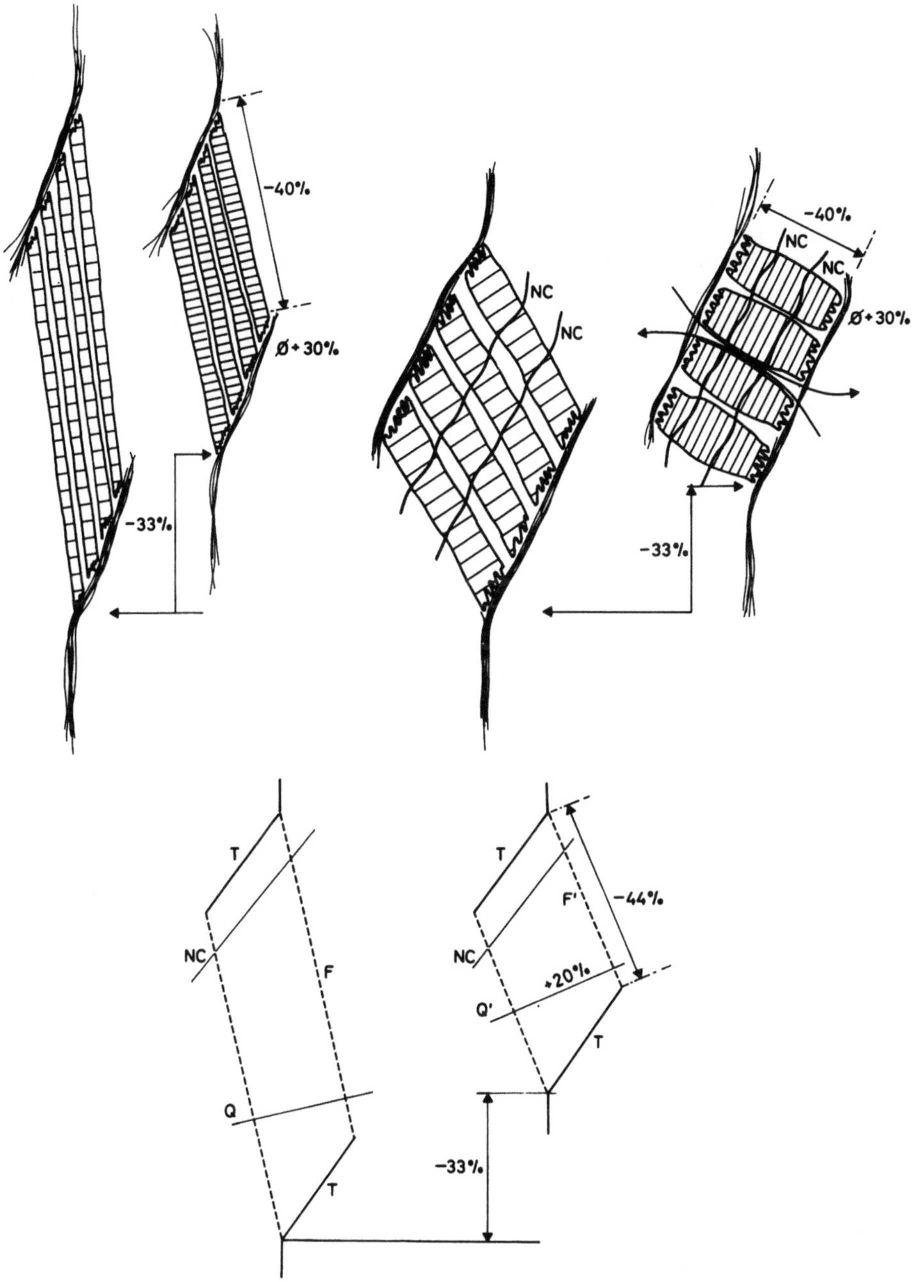

Fig. 9. Diagram showing the geometrical changes in a shortening muscle, and the displacements of the fibre bundles in relation to each other (after FENEIS 1935). The muscle fibre bundles (*F*) insert obliquely at tendon sheets (*T*), and muscle fibres and tendon sheets form a parallelogram. The length of the fibres in the three examples given is 60%–70% of the length of the muscle belly. To shorten the muscle by about 33%, the fibres have to shorten by 40%–44%. The fibre volume remains constant and the diameter of the muscle fibres increases by 30%. At the same time, the angle of insertion becomes less acute and the muscle "unfolds". This increases the width of the parallelogram (*Q*) by 20%, which provides some of the space for the increase in muscle fibre diameter. The length of the tendon sheets remains constant, and collagen bundles running parallel to the tendon sheets remain slack; these collagen bundles act as "neutral connecting threads" (*NC*), which hold the fibre bundles together without interfering with shortening or stretch of the muscle. The fascicles slide along each other along "neutral displacement membranes" (*bent arrows*), the septa of connective tissue between the fascicles (see also Fig. 67)

muscular component of the connective tissue from the closely packed and highly oriented collagen of the microtendons and tendons, and disputes the justification of the classical terms epi-, peri-, and endomysium.

The tendons are part of the series-elastic element of muscle. Brisk and rapid contractions require the tendon to be stiff, otherwise rise in tension is delayed. Therefore, to perform a rapid (ballistic) contraction, the limb is positioned in such a way that the tendons are prestretched. A remarkable example of the opposite principle has been found in the superficial metapatagialis muscle of pigeon. This muscle inserts in the skin, and works in series with smooth muscle cells and elastic tendons (Hikida and Peterson 1983).

V. The Vascular Supply

Skeletal muscles are richly supplied by blood vessels which form an anastomosing network inside the muscle. The first detailed description was given by Spalteholz (1888).

Myrhage describes a "basic unit" for the vascular supply of cat hindlimb muscles (Myrhage 1977; Myrhage and Eriksson 1980) (Fig. 10). Each muscle is supplied by one or several main arteries entering the muscle independently of the nerve. The blood supply of long muscles is in a segmental fashion, and the proximal, medial, and distal parts each have their own main arteries. The main artery branches into "primary" arteries after it has entered the muscle; these run perpendicularly or obliquely to the fibre direction and cover a certain territory (Fig. 11). The width of the primary arteries ranges from 80 to 360 µm; it is largest in the soleus muscle, which has a high perfusion rate and during activity relies on oxidative metabolism (see below). The primary arteries branch several times in a dichotomous manner and finally give rise to "secondary" arteries running parallel to the direction of the muscle fibres and representing the centres of the basic units. The width of the secondary arteries is 60–100 µm. One secondary artery may connect to two primary arteries forming an arterio-arterial anastomosis. The distance between the parallel running secondary arteries is about 1 mm in the soleus muscle and 1.5–2 mm in other muscles. Primary and secondary arteries are accompanied by veins, usually one artery by one vein, but in the soleus muscle the primary veins may be in duplicate. The secondary arteries and veins give off terminal arterioles and venules, running transversely. The terminal arterioles and venules do not accompany each other. Each terminal arteriole is connected to the two adjacent terminal venules by several capillaries running parallel to the longitudinal axis of the muscle fibres. Adjacent capillaries are linked by short transverse shunts. Each secondary artery and its lateral branches, formed by the terminal arterioles, constitute the "basic unit" of the arterial tree. Each unit supplies a cylinder of muscle tissue with a cross-sectional area of about 1 mm^2. A small cat muscle, the tenuissimus muscle, contains only one "basic unit" supplied by one artery running the entire length of the muscle.

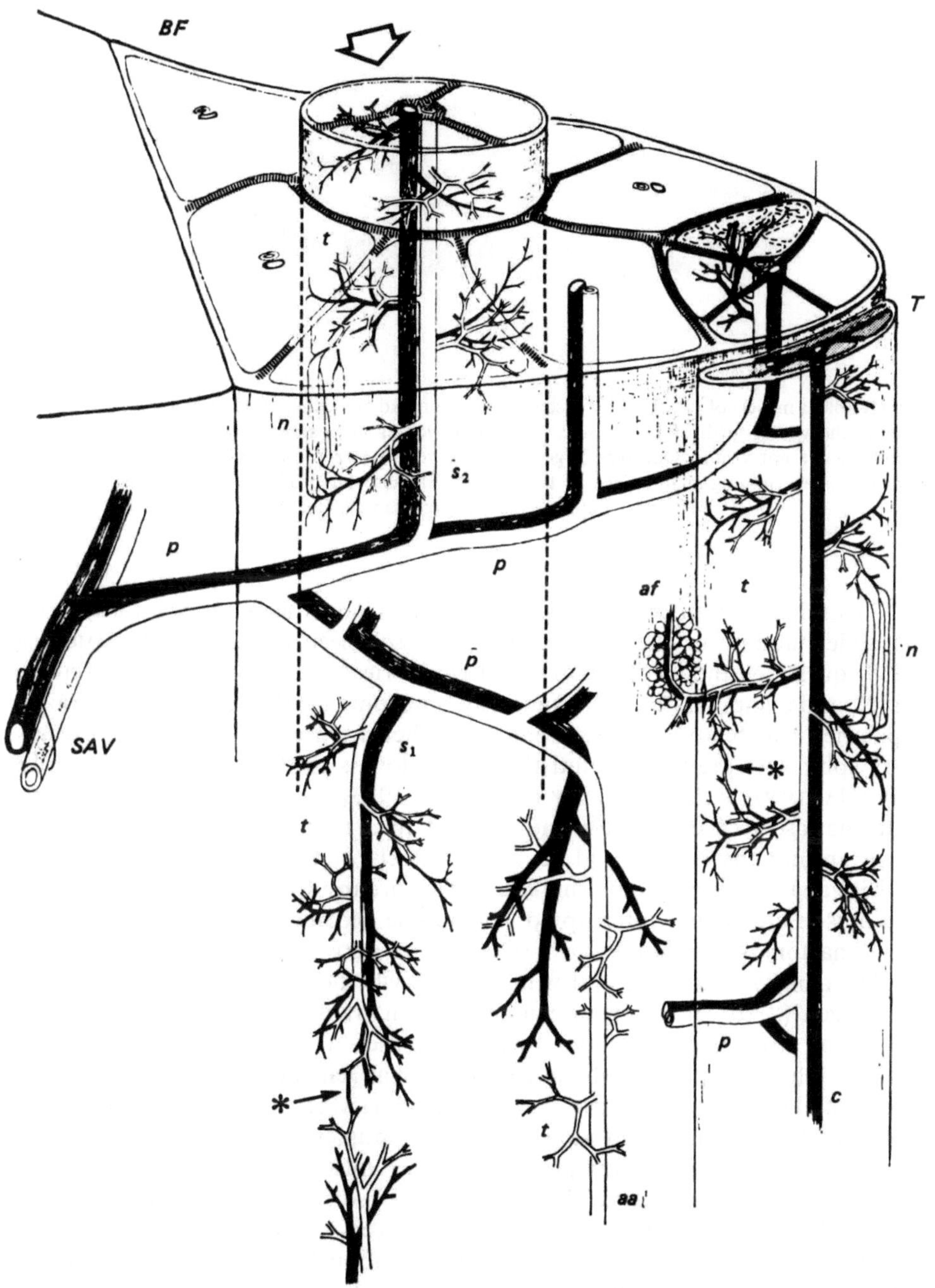

Fig. 10. The architecture of the vascular supply of cat muscles. To the right, the tenuissimus muscle (*T*) is shown which contains only one "basic unit" supplied by a central vessel (*c*); thicker muscles like the biceps femoris (*BF*) consist of many "units". The supplying main artery and vein (*SAV*) give off primary arteries and veins (*p*), running transversely and obliquely to the fibre axis. The primary vessels of different muscle segments are longitudinally connected by secondary arteries and veins (s_1, s_2); the arteries may form arterial anastomoses (*aa*). The secondary vessels run in the centre of cylindrical "basic units" (*open arrow*), comprising fibre bundles <1 mm in diameter. At different levels, terminal arterioles and venules (*t*) arise, longitudinally connected by capillaries (*n*). The arterioles supply the adipose tissue between the fascicles as well (*af*); the arterioles may form arteriolar anastomoses (*stars*). (From Myrhage and Eriksson 1980, with permission of the authors and Cambridge University Press)

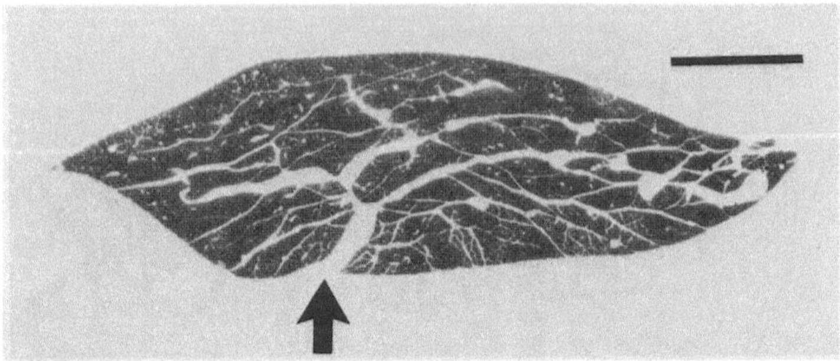

Fig. 11. Soleus muscle of rat, fixed by vascular perfusion and embedded in plastic, 3-μm cross-section through the widest girth of the muscle. The blood vessels are ballooned and appear empty; the endothelia are not visible. One sees large supplying vessels (*arrow*) which branch and give rise to primary vessels running in the plane of the section. The circular profiles of longitudinally running secondary blood vessels are seen as well. Bar, 1 mm

This scheme is an oversimplification, but it nonetheless appears useful for the understanding of the array of blood vessels seen in sections. HAMMERSEN (1968) questions the idealized array of capillaries as described by SPALTEHOLZ (1888), and demonstrates dichotomously branching capillaries and long straight capillaries, which cross the next venule and then connect to an arteriole. Precapillary sphincters are lacking and the blood flow is controlled by the arterioles only (HAMMERSEN 1968; HOLLEY and FAHIM 1983). Primary arteries may penetrate the fascia of the muscle and connect to arteries of the loose connective tissue around the muscle and to blood vessels of the tendons. The capacity of these anastomoses is quite small, and they are insufficient if the main arteries to a muscle are acutely interrupted; the muscle becomes revascularized through these anastomoses within 2 weeks (CLARK and BLOMFIELD 1945).

The density of arteries and veins reflects whether a muscle depends mainly on oxygen ("red" muscle), or whether it can transiently live on its glycogen stores and work anaerobically ("white" muscle). This difference is even more obvious when the pattern of the capillary arrangement is considered. Already RANVIER (1874) demonstrated that the capillaries of the rabbit soleus muscle take a tortuous course, and are connected by numerous transverse shunts with sinusoid dilatations; the capillaries of "white" muscles run straight, and transverse shunts are less frequent than in the "red" soleus muscle (Fig. 12).

It is difficult to count capillaries in routine cross-sections for light microscopy and to distinguish closed capillaries from mononuclear cells (HAMMERSEN and APPELL 1976). In the older literature, the number of capillaries per unit cross-sectional area varies enormously, even for the same muscle of the same species, and the data exceed those in recent studies (Table 2) by up to a factor 10 (for a review; see *Handbook of Circulation,* 1959; HAMMERSEN 1968). Some of the micrographs published suggest that interstitial nuclei may have been counted as capillaries.

As endothelial cells of rat, cat, and rabbit react intensely for alkaline phosphatase (BOURNE 1943), they can be identified using histochemistry. An alterna-

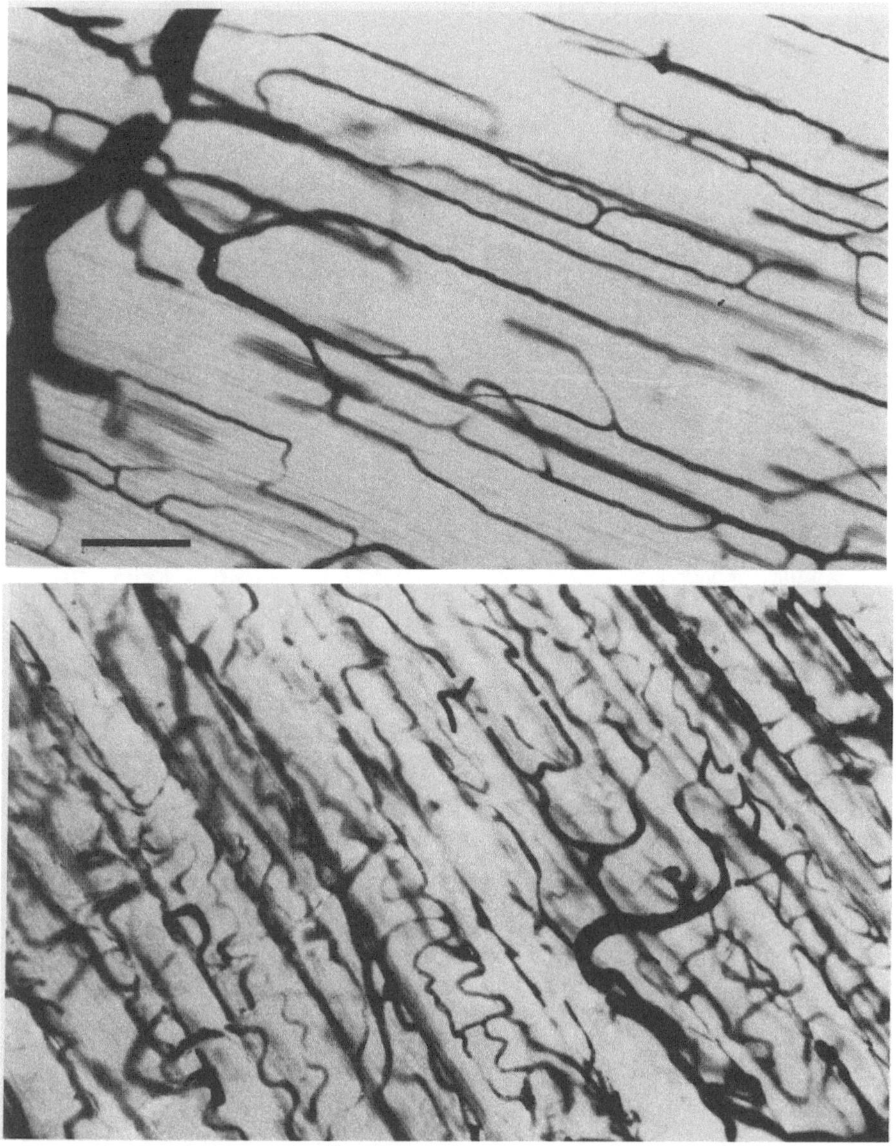

Fig. 12. Gastrocnemius (*top*) and soleus muscle (*bottom*) of rat. Longitudinal frozen sections, 100 µm thick. The blood vessels have been injected with Indian ink. The capillaries run parallel to the muscle fibres: their course is straight in the gastrocnemius and meandering in the soleus muscle. Bar, 100 µm

tive method is to inject the capillaries with Indian ink (e.g. HAMMERSEN 1968), or to balloon them by perfusion fixation (SCHMALBRUCH 1971) (Fig. 13). Also fluorescent dyes have been used (VETTERLEIN and SCHMIDT 1983). Provided the perfusion or injection technique is adequate and done in the living animal, all capillaries are identified. An adequate perfusion of human muscles is not

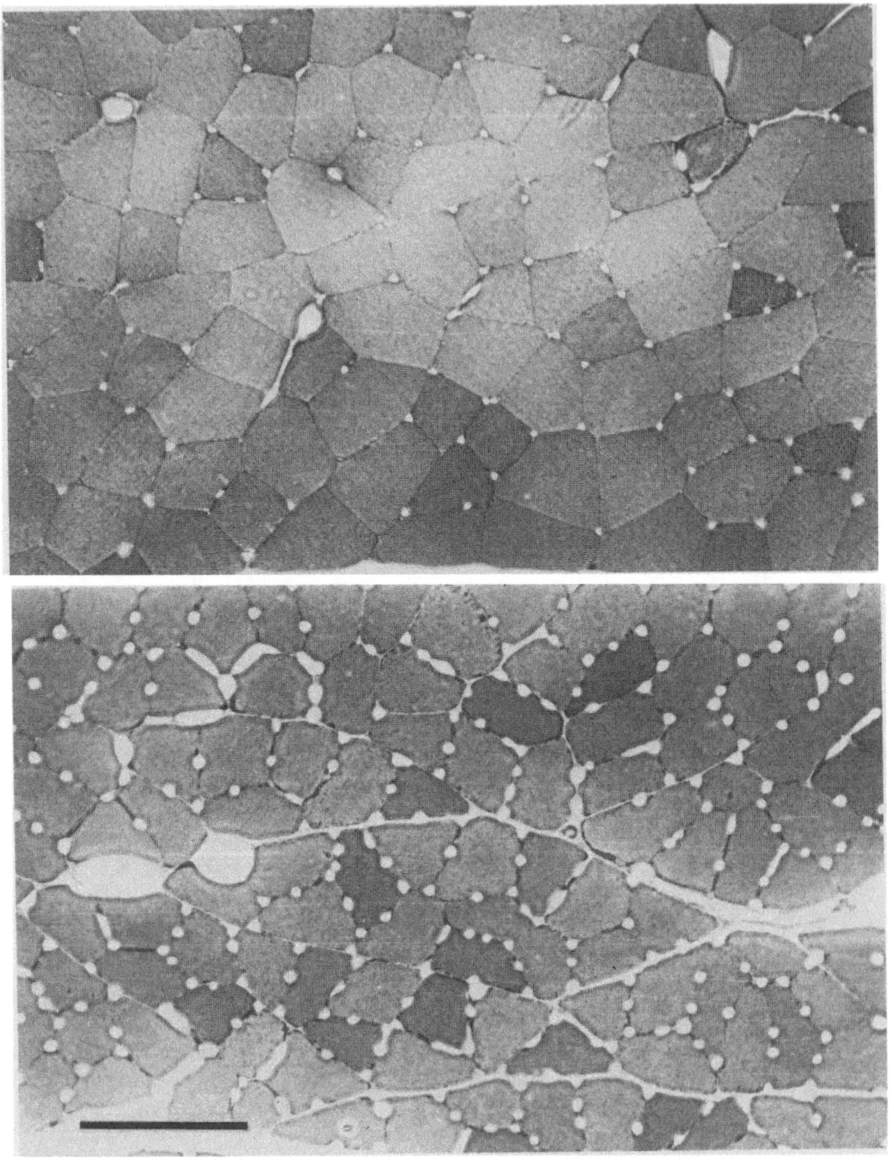

Fig. 13. Gastrocnemius (*top*) and soleus (*bottom*) muscle of rat. The muscles were fixed by vascular perfusion, embedded in plastic, and 3-μm cross-sections were stained with p-phenylenediamine and photographed with phase-contrast optics. All capillaries are clearly discernible. Note the high density in the soleus muscle as opposed to the gastrocnemius muscle. Bar, 100 μm

possible, and the endothelial cells do not react as distinctly for alkaline phosphatase as in animal muscles. ANDERSEN (1975) introduced a PAS method for capillary counts in human muscles. The capillary basal lamina reacts with the periodate-acid-Schiff (PAS) stain, and is readily identified when the glycogen contained in muscle fibres is digested before the stain is applied. The reliability of this method has been disputed (BRODAL et al. 1977), and inconsistent results have been attributed to it. Nevertheless, counts made using electron microscopy and the PAS method gave identical results (NYGAARD and SCHMALBRUCH 1980).

The basic array of blood vessels is the same in all muscles. Hence, the incidence of capillaries on a cross-section through a muscle (Table 2) is a measure of the flow resistance of the vascular bed, and also of the perfusion rate in vivo. The density of capillaries in the soleus muscle of cat is $1{,}600/mm^2$ muscle fibre cross-sectional area. The corresponding value for the medial head of the gastrocnemius muscle is 600 (SCHMALBRUCH 1971). Thus, the number of capillaries per unit cross-sectional area muscle fibre is 2–3 times greater in the soleus than in the gastrocnemius muscle. The blood flows at rest are 20 and 9 ml/min × 100 g muscle, respectively (2.2:1); the maximum blood flows are 118 and 46 ml/min × 100 g (2.5:1) (REIS et al. 1969).

During isometric contractions, the intramuscular pressure rises and impairs the blood supply; in human calf muscles it is completely interrupted at 15% maximum force (BARCROFT and MILLEN 1939). The same limit for the blood circulation during sustained isometric contractions has been found in dog gastrocnemius muscle (HIRCHE et al. 1970). The blood flow through the cat gastrocnemius muscle increrases during repetitive electrical stimulation, reaches its maximum at 4 Hz, and decreases at higher rates of stimulation. The oxygen consumption exceeds the oxygen delivery at 8 Hz, but tetanus does not develop below 16 Hz. Therefore, the gastrocnemius muscle cannot at the same time maintain a contraction and a sufficient blood supply. The blood flow through the soleus muscle at 12 Hz is about 50% of its maximum. Tetanus develops at 8 Hz because the rise time of the twitch is longer than in the gastrocnemius; even at 20 Hz, oxygen delivery still exceeds oxygen consumption. The natural innervation rate of the soleus muscle rarely exceeds 20 Hz; hence this muscle can maintain a contraction and still work aerobically. The gastrocnemius muscle is occasionally innervated at frequencies of 50–60 Hz, at which the blood supply is completely shut off. This muscle has to shift to an aerobic metabolism during contraction, which requires that contractions are regularly interrupted in such a way that the glycogen stores can be refilled (FOLKOW and HALICKA 1968). Also during isotonic contractions the blood flow is more impaired in "white" than in "red" cat muscles (BONDE-PETERSEN and ROBERTSON 1981).

Blood flows at the same rate through human muscles and cat muscles. Rest values from 14 (brachial biceps) to 26 (quadriceps) ml/min × 100 g have been found; the maximum is 2–2.5 times the flow at rest (MISHRA and HAINING 1980). In the anterior tibial muscle of young boys, the maximal blood flow is 46 ml/100 g/min (LEINONEN et al. 1979).

Muscles of small mammals may predominantly work either aerobically or anaerobically; in man the differences are less pronounced and all muscles contain fibres with different oxidative capacity. Fibres rich in oxidative enzymes are surrounded by more capillaries than fibres rich in glycolytic enzymes (ROMANUL 1965). This explains why in muscles composed of fibres rich and poor in oxidative enzymes, the capillaries are unevenly distributed (HAMMERSEN 1968) (Fig. 13). The blood flow at submaximal exercise correlates with the percentage of oxidative slow-twitch fibres in the muscle (FRISK-HOLMBERG et al. 1981).

Table 2. Number of capillaries in cross-sections of skeletal muscles. *LM*, light microscopy; *EM*, electron microscopy

Species	Muscle	Cap./mm² tissue cross-section	Cap./mm² fibre cross-section	Cap./fibre	Cap. around each fibre	Technique	References
Rat	Ant. tibial		575-2,180	1.2–2.0		Perfusion, Epon, LM	SCHMALBRUCH (1971)
	Soleus		2,880	2.8			
		920		2.6		Fluorescein-perfusion, frozen sections	VETTERLEIN and SCHMIDT (1983)
Rabbit	Med. gastrocn.		920	1.2		Perfusion, Epon, LM	SCHMALBRUCH (1971)
	Soleus		1,620	2.6			
	Sartorius	300		1.7		Indian ink-perfusion	HAMMERSEN (1968)
	Gracilis	219		1.5			
	Adduct. magn.	244		1.6			
	Semitendinosus	247		0.9			
	Ext. digit. long.		346[a]	1.4		Alk. phosphatase, frozen sections	BROWN et al. (1976)
	Flexor. hall. long.		221[a]	1.4			
	Soleus		370[a]	2.1			
	Vocalis			1.0		Alk. phosphatase	HALL-CRAGGS (1968)
			3,400	0.7		Perfusion, Epon, LM	SCHMALBRUCH (1971)
	Thyrearytenoid		860	0.7			
Cat	Med. gastrocn.		590	1.1			
	Soleus		1,590	1.9			
	Sartorius	–		0.8		Indian ink-perfusion	HAMMERSEN (1968)
	Rectus femoris	811		0.9			
	Tenuissimus	657		0.9	2.7–3.4	Alk. phosphatase and EM	MYRHAGE (1978)
	Biceps fem.	617		0.8	1.5–2.0		
	Lat. gastrocn.	570		0.8	1.9–2.8		
	Med. gastrocn.	695		1.1	3.3–3.8		
	Soleus	948		1.9	4.8–5.0		
	Inferior oblique						
	Orbital layer	2,354		1.2		Alk. phosphatase	LENNERSTRAND (1980)
	Global layer	895		1.2			

Man							
Children 0–11 years	Brach. biceps, Lat. vastus, Peroneal		1,051			EM	JERUSALEM et al. (1974)
Adults							
Untrained males	Lat. vastus	510		3.8		Araldite, LM	SCHÖN (1978)
Marathon runners		660		5.3			
Untrained males	Lat. vastus	305		1.8	4.4	EM	BRODAL et al. (1977)
Trained males		425		2.5	5.9		
Untrained and trained males		317 (259–390)	760–1,000	1.0–2.4	2.0–7.0	PAS, LM	
Untrained females		301		1.1	3.0	EM	INGJER and BRODAL (1978)
Trained females		404		1.7	4.4		
Untrained females	Brach. biceps	305		1.4	3.4–4.3	PAS, LM	NYGAARD (1981)
Trained females		528		2.5	5.8–6.4		
Untrained males		398		1.8	4.4–5.2		
Trained males		435		2.4	5.9–6.1		
Untrained males	Quadriceps femoris	329		1.4	3.0–4.2	PAS, LM	ANDERSEN and HENRIKSSON (1977b)
Trained males		395		2.0	4.2–5.5		
Untrained males		369	1,030	1.8	4.5	EM	INGJER (1979)
Trained males		589	1,510	3.1	7.3		
Male body builders		320		2.6	5.9–7.6	PAS, LM	SCHANTZ (1982)
Male rowers		559		3.1			LARSEN and FORSSBERG (1980)
Male ice-hockey players	Lat. vastus		760–1,020		4.3–5.7	PAS, LM	GREEN et al. (1981)
	Lat. gastrocn.		920–1,160		5.1–5.2		

[a] Calculated from mean fibre size and number of capillaries per fibre

The total capillary density increases during physical exercise training programs, together with the maximal oxygen uptake (ANDERSEN and HENRIKSSON 1977b; BRODAL et al. 1977; INGJER and BRODAL 1978; INGJER 1979). More capillaries and a higher oxidative capacity of the fibres have been found in marathon runners than in untrained subjects (SCHÖN 1978). The average number of capillaries around a fibre in a given muscle increases with the mean fibre size, probably to maintain a constant maximum diffusion distance (MYRHAGE 1978; BRODAL et al. 1977; SCHANTZ 1982).

The density of the capillaries increases correspondingly when a fast predominantly glycolytic muscle is experimentally transformed into a slow predominantly oxidative working muscle (BROWN et al. 1976; DODD et al. 1980). This is achieved either by nerve transplantation or by chronic electrical stimulation imitating the innervation pattern of a slow "red" muscle (see Sect. D.V).

The fibres which are rich and poor in oxidative enzymes in most heterogeneous skeletal muscles of mammals form a checkerboard pattern. This reflects the intermingling of the motor units (see Sect. H.III). Only in porcine skeletal muscle (DAVIES 1972), and to a lesser extent in rabbit anterior tibial muscle (JAMES 1972a, b), is the distribution of fibre types non-random and fibres rich in oxidative enzymes form bundles or subfasciculi. This is believed to ensure a more efficient blood supply of the oxidative muscle fibres.

Skeletal muscles are devoid of lymph vessels.

VI. Nerve Supply

1. Composition of Nerve Branches to Muscles

One or several nerve branches enter the muscle close to its endplate zone. A nerve innervating a skeletal muscle contains afferent and efferent nerve fibres. All efferent fibres of the somatic system and all afferent fibres from muscle spindles and Golgi tendon organs are myelinated. The non-myelinated fibres belong to the autonomic nervous system and supply blood vessels (BOYD and DAVEY 1966), or they are afferent and arise from physiologically but not morphologically identified pain receptors, and from unknown receptors that ultimately connect to the cardio-respiratory system (MCCLOSKEY and MITCHELL 1972; MITCHELL et al. 1977).

The diameters of the myelinated nerve fibres supplying muscles of adult man, cat, and baboon form a bimodal histogram with a trough of 7–9 μm. The small afferent fibres and the efferent γ-fibres to intrafusal muscle fibres form one peak, whereas the large afferent and efferent nerve fibres form the second peak.

2. The Number of Motor Units and Its Determination

The number of large efferent fibres reflects the number of motoneurons innervating a muscle, and hence also the number of motor units in this muscle

(Chap. H). The number of efferent or afferent fibres in a nerve can be obtained by eliminating the sensory or motor fibres by de-afferentiation or de-efferentiation, by labelling the motor axons or motoneurons with specific stains or retrograde tracers, or by using electrophysiological methods.

ECCLES and SHERRINGTON (1930) were the first to determine the number of efferent motor fibres. They excised the appropriate dorsal root ganglia of cats to eliminate the afferent fibres from nerves to limb muscles, and assumed that the surviving fibres innervated the muscle fibres. The figures obtained, and also the later data from the sartorius muscle of rabbit (VAN HARREVELD 1947) still include the thin fusimotor fibres of the γ-system, the function of which was unknown at that time. The experiments have been repeated in cat (REXED and THERMAN 1948; BOYD and DAVEY 1966, 1968), rat (GUTMAN and HANZLIKOW 1966), and baboon (WRAY 1969) (some data in Table 5).

For man, the only data were drawn from patients with poliomyelitis when complete de-efferentiation had occurred, and the number of afferent fibres was subtracted from average counts of all fibres in normal nerves. The fibre deficit after de-efferentiation was assumed to give the number of motor fibres (FEX and WOHLFAHRT, cited in FEINSTEIN et al. 1955). Experimental de-efferentiation in animals is more reliable because the normal control nerve can be obtained from the same animal. The error, compared with optimal de-afferentiation, is $<5\%$ (cat, BOYD and DAVEY 1966; baboon, WRAY 1969). In many de-afferentiation experiments, the number of motor fibres is underestimated because it is often impossible to excise a dorsal root ganglion without damaging the ventral root. BOYD and DAVEY (1968) and WRAY (1969) compared the results after de-afferentiation and de-efferentiation. A motor deficit of up to 15% is consistently found in cats after de-afferentiation, in baboon (WRAY 1969) and rat (SCHMALBRUCH 1984) it may be even greater because ganglion and ventral root are supplied by the same blood vessels. Another source of error is the fact that nerve fibres begin to branch several cm proximally to the entrance into the muscle (ECCLES and SHERRINGTON 1930; EDDS 1950; FERNAND and YOUNG 1951; WRAY 1969); therefore, the number of large efferent nerve fibres may exceed the number of motoneurons by 50%.

For the majority of cases, 50%–60% of all large myelinated fibres in a muscular branch are efferent. In large nerves, still containing afferent fibres from the skin or from joints, only 20%–30% of the myelinated fibres are motor axons (BOYD and DAVEY 1968; SCHMALBRUCH 1984).

Ventral but not dorsal root fibres are stained using the histochemical reaction for acetylcholinesterase after prolonged incubation. About half of the myelinated axons of peripheral nerves show a positive reaction. These axons are assumed to be efferent (ZENKER and HOHBERG 1973; GRUBER and ZENKER 1978; MÜNTENER et al. 1980; GOTTSCHALL et al. 1980a, b).

Another method is based on the retrograde axonal transport of horseradish peroxidase (KRISTENSSON and OLSSON 1971); about 2 days after intramuscular injection, application to a cut nerve stump, or intranervous injection, the perikarya of the motoneurons (and of the dorsal root ganglion cells) have accumulated enough peroxidase to be shown histochemically. The cells are then counted in serial sections. Not only α-motoneurons but also γ-motoneurons are labelled.

The difficulty with this procedure is either that not all cells are labelled, or that the tracer has spread to adjacent muscles and thus labels other neurons as well. Also fluorescent compounds have been tried as retrograde tracers. These have sometimes the disadvantage of being toxic; the number of labelled cells tends to be smaller than after labelling with peroxidase (BRUNNER et al. 1980; ILLERT et al. 1982; SCHMIDT et al. 1983).

McCOMAS et al. (1971) introduced an electrophysiological method of determining the number of motor units. It is based on the fact that the compound action potential of a muscle increases stepwise in response to graded changes in the electrical stimulus to the nerve. If one assumes that each increment in action potential is due to the excitation of one additional axon, the total number of motor units is obtained by dividing the average increment into the compound action potential evoked by a maximal stimulus. The electrophysiological counts in rat and monkey compare reasonably well with histological fibre counts after de-afferentiation (EISEN et al. 1974; PEYRONNARD and LAMMARRE 1977). The general applicability of the method has been criticized (for review, see BUCHTHAL and SCHMALBRUCH 1980).

3. The Terminal Innervation Ratio

When the nerve enters the muscle, the bundle of nerve fibres and also the individual axons branch; each bundle of nerve fibres runs more or less transversely in the perimysium until it reaches the fascicle of muscle fibres it innervates (Figs. 5, 14). Then it breaks up, and individual nerve fibres enter the endomysium of the fascicle. The axon loses its myelin sheath immediately before its terminal arborization on a muscle fibre. Axons that have left the bundle, and have entered the fascicles of muscle fibres, do not branch any more. As a rule, each subterminal axon innervates one muscle fibre. When a muscle fibre loses its innervation, adjacent nerve fibres may sprout, and a sprout may reinnervate this muscle fibre. The average terminal innervation ratio (TIR) in normal human muscles is not 1:0 as one would expect; a mean of of 1.2 muscles fibres are innervated by one subterminal axon. Subsequent to nerve injury, this value may increase to 2.3 (COËRS et al. 1973a, b). The intravital methylene blue staining has proved to be the most reliable method for demonstrating the arborization of the intramuscular nerves (COËRS and WOOLF 1959).

During the first half of the century, metal impregnations were popular for showing the course of the smallest nerve fibres. A wealth of non-existent structures was demonstrated, among others the vegetative innervation of skeletal muscle fibres. These techniques came into disgrace with the advent of electron microscopy, but during the last decade they have experienced a renaissance, in particular for showing the development of muscle innervation (see Sect. G.III.2d). The techniques are basically identical to those used during the pre-electron microscope era, and the findings should be interpreted with the utmost care.

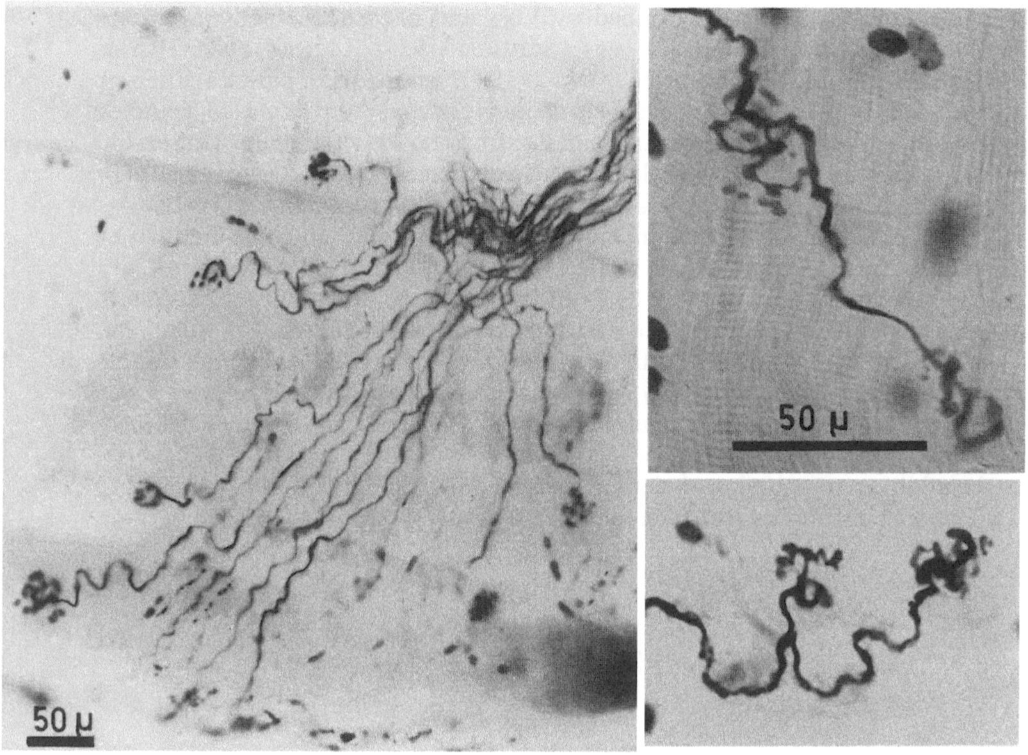

Fig. 14. Human muscle, supravital methylene blue staining to show the terminal nerve branches. A bundle of axons, after entering a primary fascicle, splits up and each terminal axon branch innervates one muscle fibre (*left*). The endplates are distinct. Only rarely the axon branches again and supplies two endplates on two muscle fibres (*right*). Terminal branching is the rule in muscles that had been partially denervated because surviving axons sprout and reinnervate the denervated muscle fibres. (Micrographs courtesy of Prof. Coërs, Brussels)

VII. Muscle Spindles

Muscle spindles are sensory organs of the muscle. They consist of a capsule containing thin intrafusal muscle fibres which are innervated by efferent (fusimotor) γ-axons; in their midregion the intrafusal muscle fibres are in contact with thick sensory axons. In cat and monkey, up to 30% of the motor axons innervate both extra- and intrafusal fibres (β- or skeleto-fusimotor axons) (Bessou et al. 1965; Jami et al. 1980; Murthy et al. 1982) (see Sect. F.II.4b).

Muscle spindles lie in the perimysium of the muscle, and tend to be concentrated in the muscle belly where the nerve enters (Barker 1959). They are particularly frequent in small muscles of the hand and foot, and in dorsal neck muscles. The number of spindles per unit weight muscle in numerous human muscles has been determined by VOSS and his school; large variations have been described (FREIMANN 1954; SCHULZE 1955; KÖRNER 1960; HOYER 1963; SMITH and MARCARIAN 1966; VOSS 1971). Even for small muscles different authors arrive at different results. HNIK and ZELENA (1961) find 13 spindles in rat soleus muscle, whereas ARENDT and ASMUSSEN (1974a) find 31–37. Spindle counts in muscles of experimental animals have been reviewed by COOPER (1966), ARENDT and ASMUSSEN (1974a), and BUCHTHAL and SCHMALBRUCH (1980).

The number of myelinated nerve fibres to a muscle, and the number of spindles within this muscle are related. One finds about 10 nerve fibres per muscle spindle. Because roughly half of the myelinated nerve fibres of a muscle are motor axons ($\alpha+\gamma$), there is about one spindle for every three motor units. This relation varies much less than the number of spindles per unit muscle weight. Small muscles with many spindles per weight muscle tend to have fewer muscle fibres per motor unit than do large muscles with few spindles (COOPER 1966).

In addition to spindles, skeletal muscles contain Golgi tendon organs, and few Pacinian corpuscles (for references, see RICHMOND and STUART 1985).

C. Skeletal Muscle Fibres

I. The Contractile Apparatus

1. Cross-Striation

More than 80% of the volume of an adult muscle fibre is occupied by myofilaments. There are two types: (a) thick myosin filaments, 1.6 μm long and 12–14 nm in diameter, and (b) thin actin filaments, 0.98 μm long and 5–7 nm in diameter (frog, PAGE and HUXLEY 1963). The myofilaments are oriented in the fibre axis and are overlappingly arranged in transverse register such that bands occur consisting of thick or thin filaments, or of both. Light microscopy therefore shows a distinct cross-banding pattern; a faint longitudinal striation due to the sarcoplasmic elements between bundles of filaments may be seen as well.

The cross-banding pattern of skeletal muscle has often been studied in muscle of invertebrates in which the band spacings are 3–5 times longer than in vertebrate muscles. The results in the older literature have been reviewed by HÄGGQUIST (1931, 1956) and SCHMIDT (1937). It is interesting to emphasize that already SCHMIDT (1937) critically discussed the optical artefacts unavoidable in a system of bands and lines in specimens several times thicker than the band spacings.

The bands composed mainly of thick filaments show high birefringence in polarized light and are called A bands (for anisotropic). The former name of this band is Q (for German *Querscheibe*). The bands composed of thin filaments show weak (SCHMIDT 1937) birefringence and are named I bands (for isotropic). Each I band is bisected by a narrow Z line (for German *Zwischenscheibe*). In the middle of each A band is a lighter H zone (formerly Q_H) (for Hensen, a German histologist) bisected by the faint and narrow M line (for German *Mitte*) (Fig. 15).

In shortened muscle fibres C lines (for contraction) occur within the A band (see below). N bands (for German *Nebenscheibe*) are sometimes observed within the I band of stretched muscle fibres on both sides of the Z lines. The N bands do not reflect structures of the contractile system but are due to the mitochondria localized at the I band level which project into the cross-banding pattern (HÄGGQUIST 1931). The N bands seen by light microscopy should not be confused with the N_1 and N_2 bands in electron micrographs (see below).

The repeating unit of the banding pattern is the sarcomere which extends from Z line to Z line. This definition relates to the fact that at the ends of the muscle fibre the cross-banding pattern terminates where one would expect a Z line.

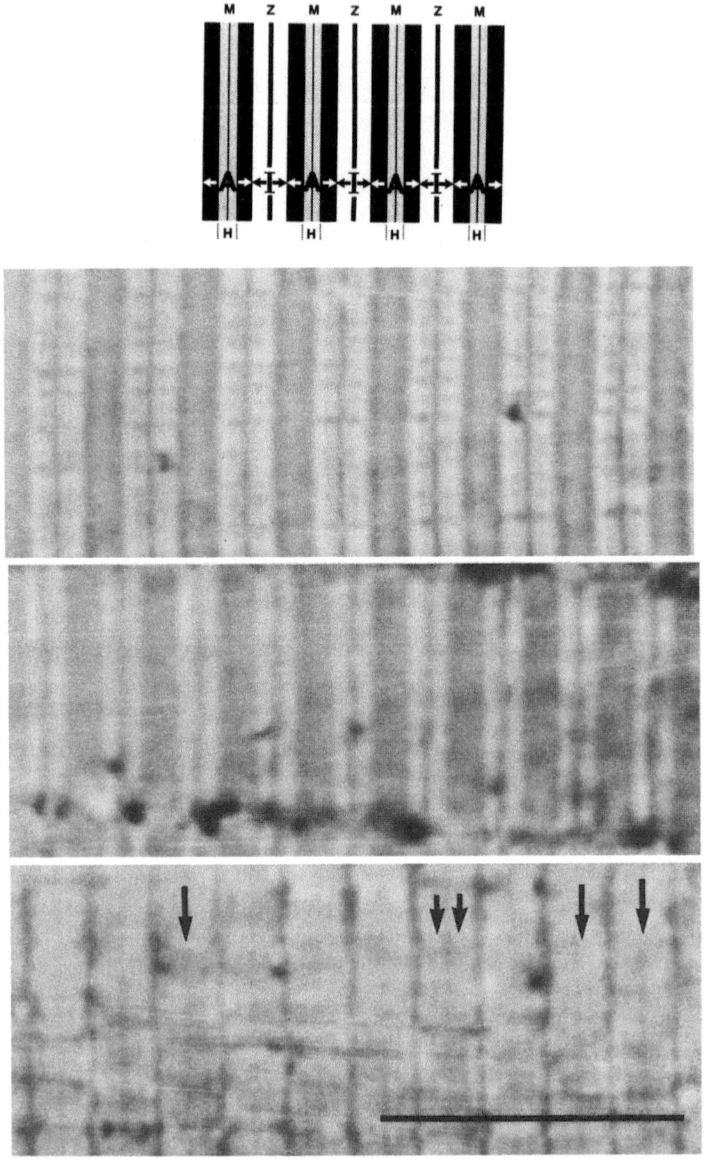

Fig. 15. Scheme of the cross-banding pattern of vertebrate skeletal muscle, and phase-contrast micrographs of 1-μm plastic sections of fibres of the human medial vastus muscle fixed at different sarcomere lengths. The presumed sarcomere length of the scheme is 2.7 μm, and it has been printed to fit approximately the magnification of the micrographs. Bar, 10 μm. At 3.1-μm sarcomere length (*top*), the I bands appear 1.6-μm wide, and the H zones are distinct. The M line is not visible. At 2.5-μm sarcomere length (*middle*) the I band width has decreased and H has disappeared. At 2.1-μm sarcomere length (*bottom*) the I band has almost disappeared and C_M lines have formed midway between the Z lines (*long arrows*). Additional faint lines are present midway between Z and C_M (*short arrows*). Electron micrographs (Fig. 16b) reveal that these lines represent the ends of the overlapping I filaments. The accuracy of the measurements of the band widths (not of the sarcomere lengths) is limited by the resolution of the optical system (objective Zeiss Plan-Apochromat 100, aperture 1.3), which is not better than 0.25 μm

2. Myofibrils

Contraction involves the reversible interaction of the myofilaments, which is controlled by Ca^{2+} ions stored in, and released from, the sarcoplasmic reticulum; the electrical signal for Ca^{2+} release to initiate contraction is conducted by a transverse tubular system (T system) from the sarcolemma into the interior of the fibre. The energy for contraction and also the energy necessary to store Ca^{2+} ions within the sarcoplasmic reticulum against a concentration gradient is obtained from splitting ATP. ATP is produced by oxidative phosphorylation within mitochondria, and also by glycolysis within the cytosol. The time allowed for diffusion of Ca^{2+} and ATP probably limits the maximum distance between the interacting filaments and the respective sarcoplasmic organelles. The mass of contractile filaments is therefore subdivided by strands of sarcoplasm, which delineate bundles of myofilaments corresponding to the myofibrils, the smallest structures within the contractile material that can be visualized using light microscopy. The diameter of a myofibril is usually less than 1 µm.

It is easy to isolate long myofibrils from insect flight muscles which have been the favourite objects of early light microscopists. These muscles are extremely rich in mitochondria which swell in isolated fibres and separate myofibrils. It is difficult to isolate long myofibrils from other insect muscles and in particular from mammalian muscle fibres; the separation of the mass of contractile filaments by sarcoplasm is less pronounced, and it is inconstant and variable along the length of the muscle fibre (Figs. 16, 21). MacKay and Harrop (1969) describe the myofibrils in rat muscle fibres as a three-dimensional mesh rather than as a series of separate strands. Electron microscopy of cross-sections of properly fixed muscles does not show a subdivision into myofibrils at the A band level because sarcoplasmic elements occur mainly at the I band level (Figs. 17, 18). Frozen cross-sections of unfixed muscles do not show myofibrils, unless the sarcoplasmic elements are specially stained; this suggests that myofibrils are not cell "organelles". The distinct myofibril pattern in routine cross-sections for light microscopy is the result of swelling and disruption of the sarcoplasmic elements. Hence, myofibrils when seen by light microscopy are artefacts reflecting the array of sarcoplasm.

The size and shape of the myofibrils in cross-sections, i.e. the array of sarcoplasm, may vary in different muscle fibres. This is most obvious in amphibia; cross-sections of twitch fibres show well delineated thin myofibrils, whereas in slow tonic fibres the myofilaments are subdivided in poorly defined rather large fields (Fibrillenstruktur vs Felderstruktur, Krüger 1929, 1952) (Chap. E). This difference is best seen in specimens which according to electron microscopic standards are poorly fixed. Mammalian skeletal muscles contain only twitch fibres, but the cross-sectional image of the fibres may vary as well (Krüger 1952; Bubenzer 1966; Schmalbruch 1971). This is due to the presence of fast- and slow-twitch fibres (see Chap. D) with different amounts of sarcoplasm. Often the swelling of the sarcoplasm is inhomogeneous and the myofibrils aggregate into groups (fields of Cohnheim). The difference in myofibril pattern between fast- and slow-twitch fibres is far less distinct than between twitch and slow fibres, and it is definitely unsuited for identifying fibre types (Fig. 18).

Fig. 16. A Human medial vastus muscle, fixed at 2.9-μm sarcomere length. Note branching myofibril (*open arrow*). **B** Rat extensor digitorum longus muscle (glycerinated), fixed at 1.8-μm sarcomere length. In both micrographs, the approximate length of the I filaments is indicated (*arrows*). The H zone in the stretched muscle (**A**) is poorly defined because the section is too thick (see Fig. 21). The M lines and the pseudo-H zones are distinct. In **B**, the thin filaments overlap and form a C_M line. It appears threefold because M and pseudo-H are seen through the overlapping thin filaments. The three-line pattern probably corresponds to the three-line pattern seen in Fig. 15, *bottom*. The greater distance between the lines in the light micrograph may reflect a diffraction phenomenon. In both samples, the sarcoplasm is artefactually swollen and separate myofibrils have formed (for comparison, see Figs. 17, 18, 21). Bar, 1 μm

Glycerination destroys all membrane elements of a muscle fibre, but leaves the contractile apparatus intact. Hence, glycerinated myofibres or myofibrils are useful tools for studying the mechanism of muscle contraction in vitro. They can experimentally be induced to relax or to contract.

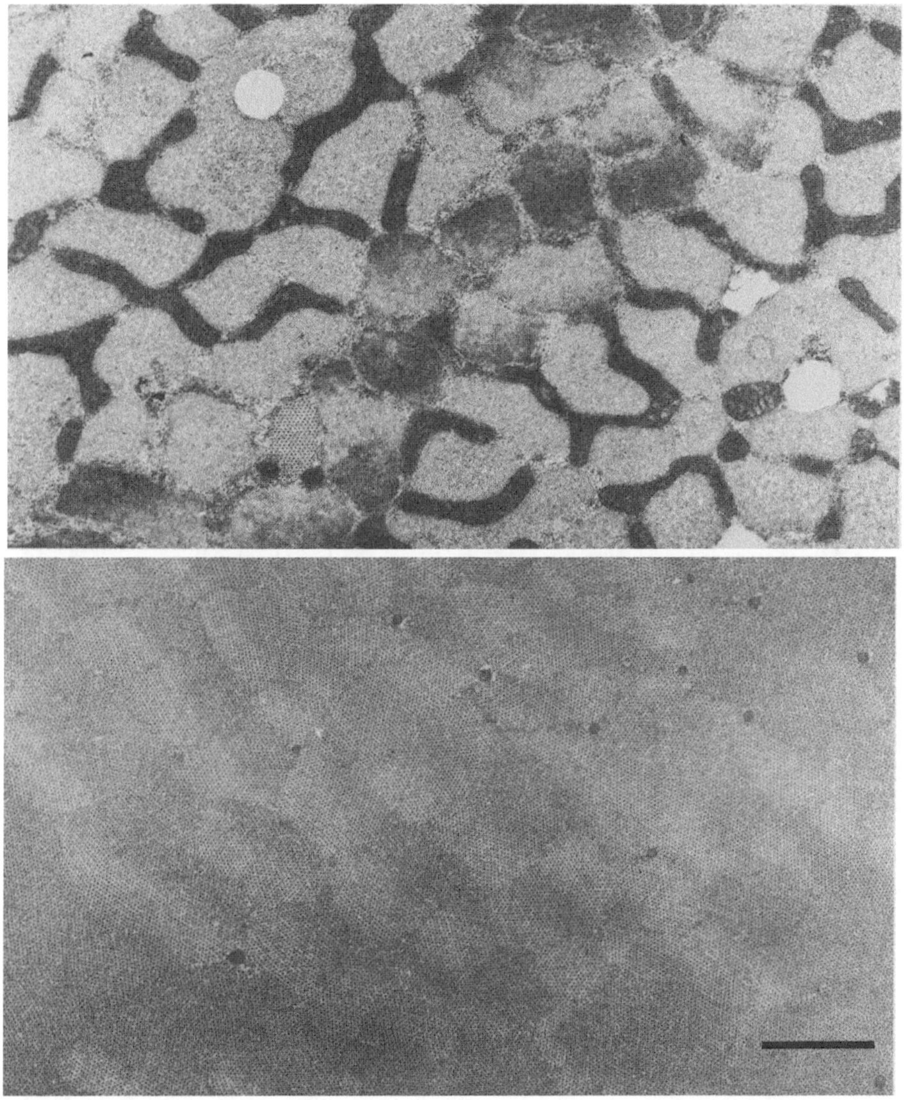

Fig. 17. Human deltoid muscle. Cross-sections through I (*top*) and A band (*bottom*). Note absence of myofibril formation at the A-band level in properly fixed specimen. The electron-translucent vacuoles represent triglyceride droplets. Bar, 1 μm

3. The Arrangement of Myofilaments in Sarcomeres

Electron micrographs of cross-sections show different images at different levels of the sarcomere because the band pattern of skeletal muscle fibres arises from two sorts of filaments being arranged in register (HALL et al. 1946). Actin filaments 5–7 nm in diameter spaced out in a rather irregular way are seen

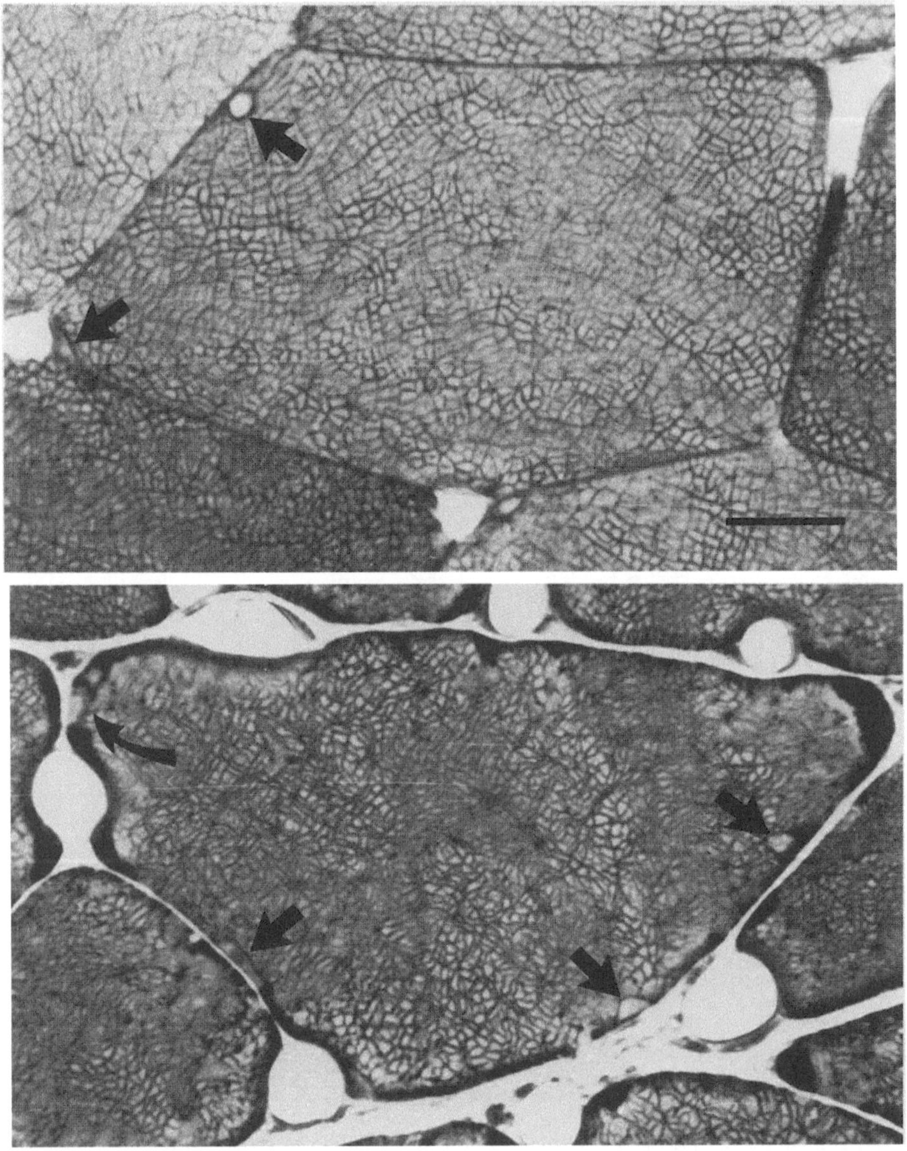

Fig. 18. Gastrocnemius (*top*) and soleus muscle (*bottom*) of rat fixed by vascular perfusion and embedded in plastic; 3-μm cross-sections were stained with p-phenylenediamine and photographed with phase-contrast optics. Separate myofibrils are absent, but the array of sarcoplasm is seen because the membranes within mitochondria are stained. This is more distinct when the plane of focus passes through the I bands than when it passes through the A bands (see also Fig. 17). The array of sarcoplasm is different in the fast-twitch (*top*) and slow-twitch (*bottom*) fibres; this may give rise to "myofibrils" of different shape, when the fibres disintegrate in routine specimens. Note also the difference in capillary density. The dark material beneath the sarcolemma of the soleus muscle represents masses of subsarcolemmal mitochondria. *Straight arrows*, myonuclei; *bent arrow*, possible satellite cells. Bar, 10 μm

in the I band; close to the Z line the array is orthogonal. Cross-sections through the central part of the A band show only thick filaments arranged in a hexagonal pattern. The centre-to-centre spacing is about 40 nm; it varies due to shrinkage of the specimen and compression during sectioning (HUXLEY 1957), and it also depends on the fixatives used (DAVEY 1973). The volume of the filament array is constant, independent of the degree of shortening or stretch; therefore the interfilament distance increases in shortened and decreases in stretched fibres (CARLSEN et al. 1961; BRANDT et al. 1967) (see Sect. C.I.7a). Cross-sections through the ends of the A band show a double aray of thick and thin filaments. The thick filaments form the same hexagonal lattice as in the middle of the A band, and thin filaments are positioned between the thick filaments such that one thin filament lies symmetrically between three thick filaments. In consequence, there are twice as many thin filaments than thick ones (Figs. 19, 20). The lattice of thick filaments is usually complete, but single thin filaments may be missing (insect muscles, SQUIRE 1981; human muscle, Fig. 19). This is probably due to the fact that not all actin filaments are of equal length; in rat many may be less than 0.25 µm long (TRAEGER and GOLDSTEIN 1983).

There are relatively more thin filaments in muscles of insects and other invertebrates than in vertebrate muscles. The ratio of thick-to-thin filaments tends to be larger in muscles with fast contractions (1:3 in insect flight muscles) than in muscles with slow contractions (1:7 to 1:10). The lengths of the filaments and thereby the sarcomere lengths are often greater than in vertebrate muscles; the sarcomere length may be up to 10 µm (for references see SMITH et al. 1966; HAGOPIAN 1966; REGER 1967a, b; REGER and COOPER 1967; HOYLE 1967, 1969; ZOBEL et al. 1967; FAHRENBACH 1967; HAGOPIAN and SPIRO 1968; FRANZINI-ARMSTRONG 1970a; HAYES et al. 1971; DEWEY et al. 1973; PRINGLE 1974).

Longitudinal sections show only thin filaments within the I band; within the A band both thick and thin filaments are seen. The thin filaments run from the Z line through the lateral part of the A band and terminate at the border of the H zone. The H zone is identical with the central part of the A band, which consists of thick filaments only (Fig. 21). Very thin longitudinal sections reveal different patterns within the lateral part of the A band. Either thick and thin filaments alternate, or between two thick filaments are two thin filaments (HUXLEY 1957). One may even find four thin filaments between two thick ones, provided the section is very thin. The explanation is found in Fig. 20. Sections comprising only one thick-filament layer will show the 1:1 relation if they follow the 1,1 projection plane, but the 2:1 relation in the 1,0 plane. The 4:1 relation may occur in sections about 20 nm thick running midway between the 1,0 and 1,1 plane. If a thin section is not exactly longitudinally oriented a Moire pattern may appear (ACHATZ 1968). Moderately thin longitudinal sections within the overlap zone only show alternating thick and thin filaments or no filament pattern at all (Fig. 23).

In the middle of the A band, the thick filaments are connected by cross bridges. These occur in 3–5 regular sets and form the M line (KNAPPEIS and CARLSEN 1968). The morphology of the M line is different in physiologically different muscle fibres (see Sect. C.I.14 and Chaps. D, E, and F).

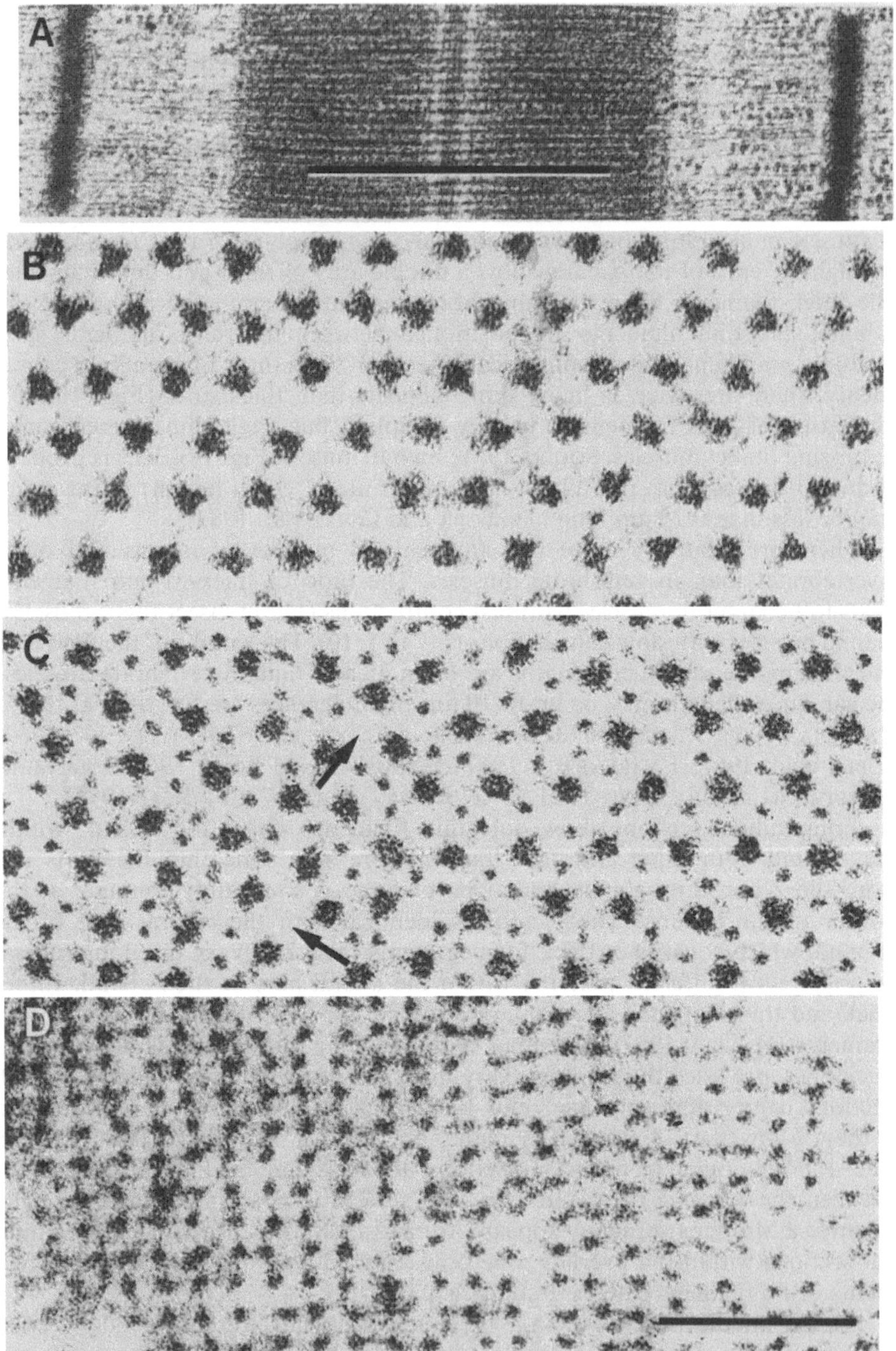

Fig. 19 A–D. Human brachial biceps muscle. **A** Longitudinal section showing one sarcomere. The muscle fibre is stretched and A and I bands, Z and M lines, and the pseudo-H zone are distinct. The section is too thick to show the overlap of thick and thin filaments and the H zone. Bar,

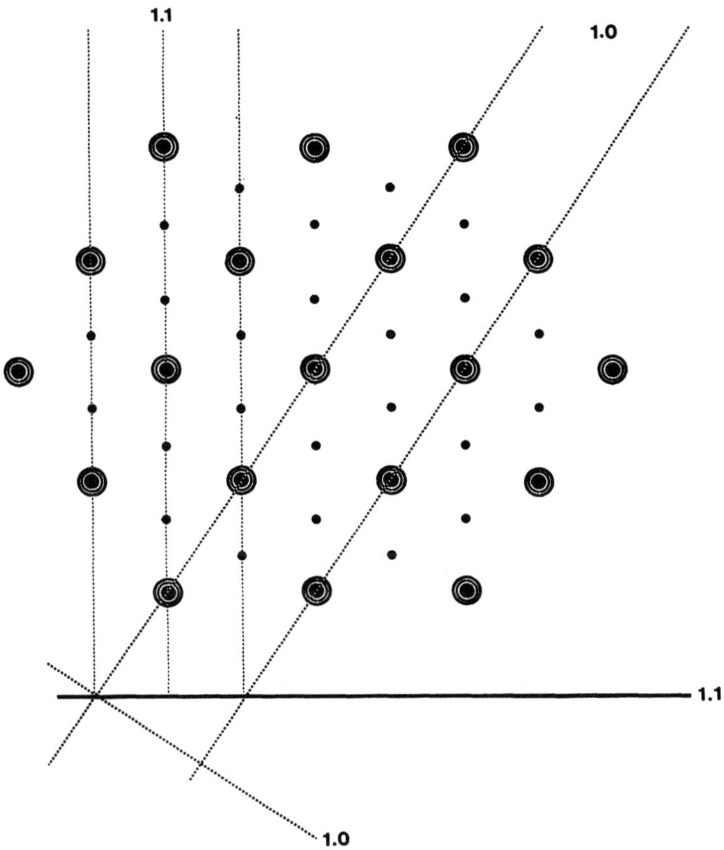

Fig. 20. Diagram of an end-on view of the thick and thin filaments in the overlap zone of the A band (corresponding to Fig. 19C). Two lattice planes of the crystal array of the thick filaments are indicated. The 1,1 projection has a narrower spacing than the 1,0 projection. In longitudinal sections comprising several filament layers, the thick filaments will be most distinct in the 1,1 projection because they project together with the thin filaments, in 1,0 projections the interposed thin filaments will blur the image. Thin longitudinal sections in the 1,1 plane comprising only one layer of thick filaments will show alternating thick and thin filaments; alternatingly, one thick and two thin filaments will be seen if the section is in the 1,0 plane. Very thin longitudinal sections oriented between the two planes may show four thin filaments between two thick ones, and thin sections in the 1,1 plane passing between the thick filament layers show only thin filaments (see Figs. 21, 22, 23)

---

1 µm. **B** Cross-section through the middle of the A band. Thick filaments form a hexagonal array; in places M bridges are visible. The micrograph is oriented such that one of the three 1,1 crystal lattice projections is roughly vertical, and that one of the three 1,0 projections is horizontal (see Fig. 20). **C** Cross-section through the overlap zone, same orientation as in **B**. The thick filaments are in the same hexagonal array, and each thick filament is surrounded by six thin filaments. Several thin filaments are "missing" (*arrows*), which is in harmony with the observation that many are less than 1 µm in length (see text). The material between the filaments probably represents myosin cross-bridges. **D** Cross-section through the I band close to the Z line. The thin filaments are in an orthogonal array with 20- to 25-nm spacing, in places Z disc matrix is deposited. Bar (for **B–D**), 0.1 µm

Fig. 21. Rat anterior tibial muscle, 2.5-μm sarcomere length. Two sarcomere planes are shown. The H and pseudo-H zones are distinct. In places the 43-nm periodicity of the myosin cross-bridges (*open arrow*), and the M line bridges (*arrow*) are visible. The section comprises several layers of thick filaments and the spacing and contrast varies, depending on the orientation of the crystal lattice (see Fig. 20). Bar, 1 μm

The thin filaments of adjacent sarcomeres insert into the Z discs, which in longitudinal sections appear as Z lines. Sometimes the Z line shows a zig-zag structure, but usually only a fuzzy dense line is seen. The Z line tends to be wider and less straight in slow than in fast contracting muscle fibres (see Sect. C.I.15, and Chaps. D, E, and F).

Longitudinal sections reveal two additional features of thick filaments. Over the last 150 nm of their length they taper at both ends, and between thick and thin filaments there are repeating transverse bridges about 10 nm in length (HUXLEY 1957) (Figs. 22, 23). The bridges are integral parts of the thick filaments, and are best seen in rigor (glycerinated or iodo-acetate treated) muscle fibres. Cross-bridges are lacking within the central section of the A band, on both sides of the M line. This zone appears lighter than the H zone in electron micrographs; it has been termed pseudo-H zone (also M zone). The cross bridges between thick and thin filaments seem to be arranged in pairs at regular intervals of about 43 nm. There are about 18 pairs of bridges to be seen on each half of the thick filament, i.e. 36 pairs in all. Because each thick filament is encircled by six thin filaments there must be three pairs of bridges for each 43 nm, i.e. each bridge pair one sees in a longitudinal section stands for six bridges. This brings the total number of bridges per thick filament to 216 (HUXLEY 1972). Recently, evidence has been presented that the bridges are in triplets rather than in pairs. These findings will be discussed below (Sect. C.I.8c).

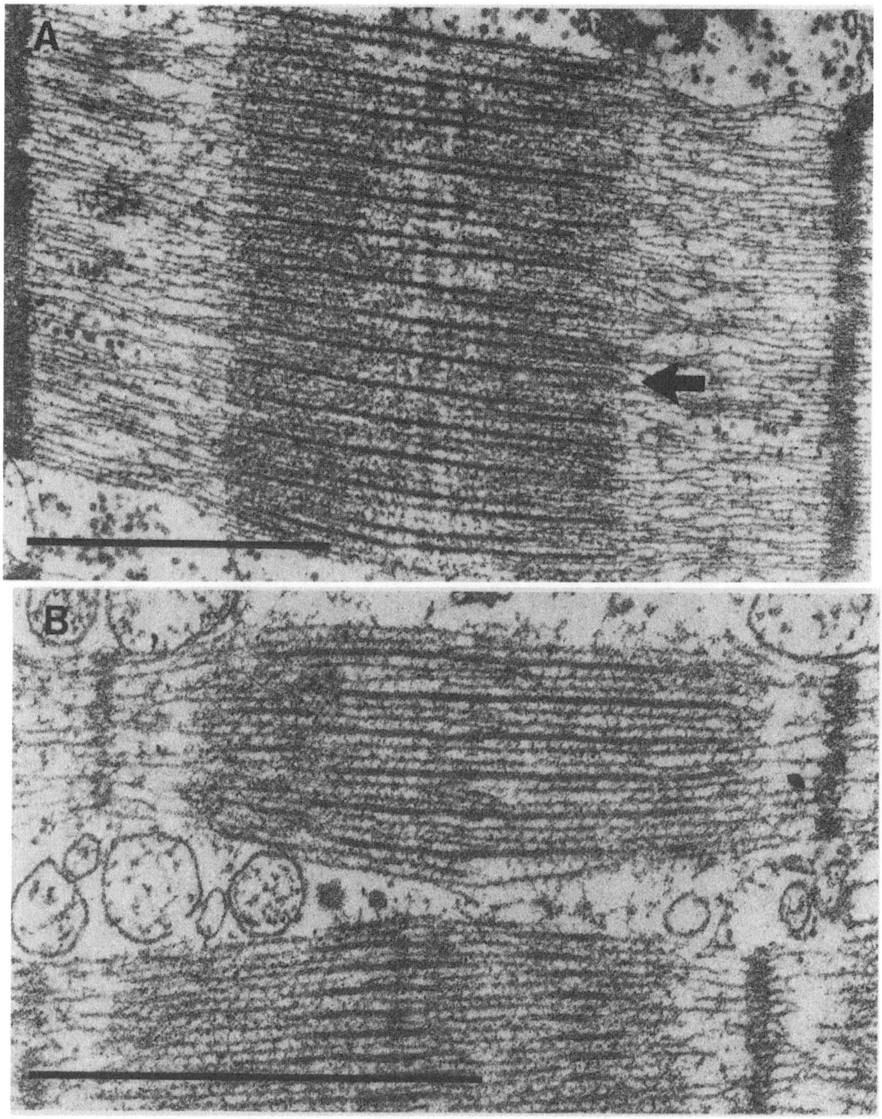

Fig. 22. A Human medial vastus muscle, relaxed. The section is thin and the overlap of thick and thin filaments in the lateral parts of the A band is visible. Only few thick filaments (*arrow*) are exactly longitudinally sectioned. Sometimes two and sometimes four thin filaments are between two thick filaments (for explanation, see legend to Fig. 20). **B** Rat extensor digitorum longus muscle in rigor (glycerinated). The sarcomere is short and the thin filaments of both I bands overlap in the middle of the A band. In the middle of the micrograph, two thin filaments are between two thick filaments (1,0 plane); the myosin cross bridges are visible. The myofibril below is cut in the 1,1 plane and there are alternating thick and thin filaments; the plane of sectioning partly passes between the thick filament layers and in places only thin filaments are seen. Both sections have been cut parallel to the fibre axes and have been compressed by 10%–25%, i.e. they are unsuited for measuring filament lengths. Bars, 1 μm

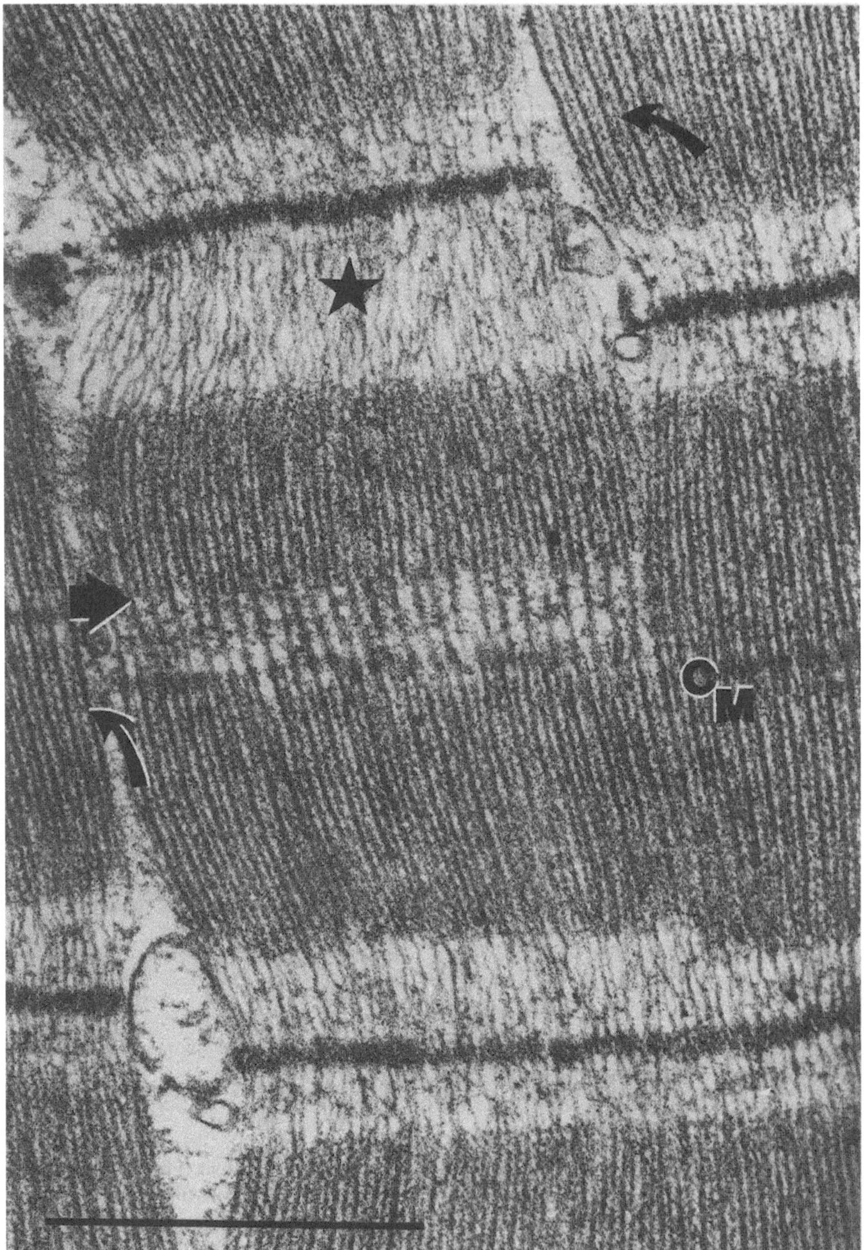

Fig. 23. Rat extensor digitorum longus muscle in rigor (glycerinated). Branching myofibril. The sarcomeres are shortened and the thin filaments overlap to form C_M bands (*C_M*). One-half of a sarcomere is stretched, the thin filaments have retracted, a half H zone has been formed (*straight arrow*), and the corresponding I band is wider than in other places (*star*). The section is thicker than those shown in Fig. 22; the filament array has been cut in the 1,1 plane (alternating thick and thin filaments). The cross-bridge periodicity (*bent arrows*) is identifiable. Bar, 1 μm

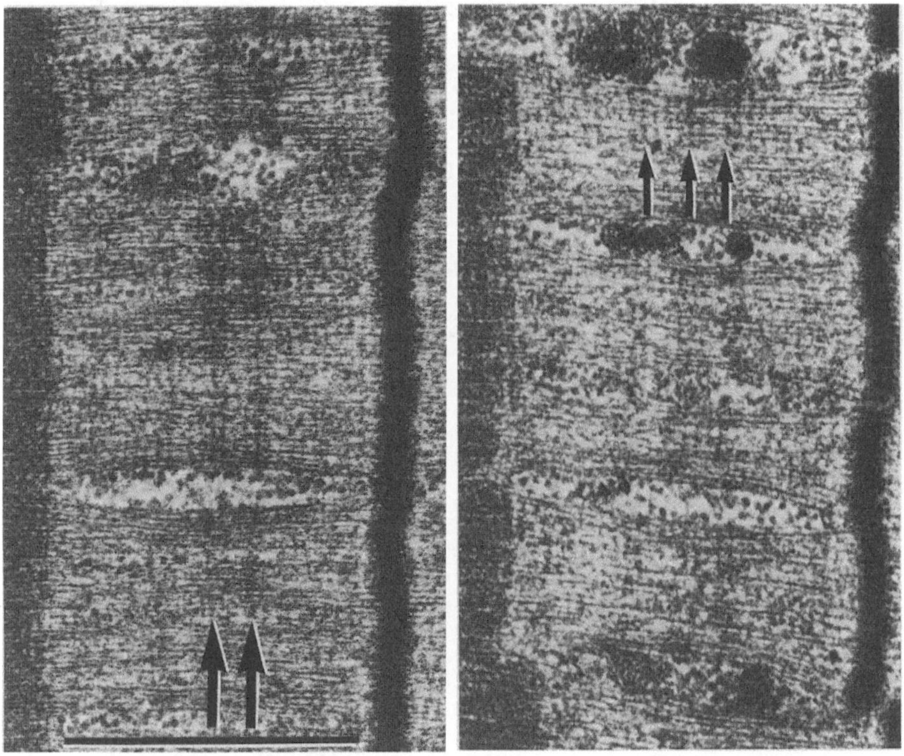

Fig. 24. I bands of human brachial biceps muscle fibres, about 3.5-μm sarcomere length. The Z lines are to the right and the edge of the A bands to the left. *Left*: Distinct N_1- and N_2-lines (*arrows*). *Right*: Distinct troponin periodicity of 38.5 nm (*arrows*). The same narrow band pattern is seen in the left micrograph, although it is less clear. The troponin-striation in the I band is a reliable calibration and allows, in this micrograph, estimation of the minimum length of the I filaments in human muscle. The width of the Z line corresponds to 2.5 of the narrow repeats, and the distance between the edge of the Z line and the edge of the A band is 30–31 repeats. Therefore, the I band width from A band to A band across the Z line is 62.5–64.5 × 38.5 nm = 2.40 to 2.48 μm. This is about 20% more than the I-filament length in frog muscles (PAGE 1968). The thin filaments have not yet pulled out of the thick filament array and the filaments may be even longer. Bar, 1 μm

Occasionally, the I band shows faint cross striations with a periodicity of about 40 nm (HALL et al. 1946; DRAPER and HODGE 1949; CARLSEN et al. 1961; PAGE and HUXLEY 1963; FRANZINI-ARMSTRONG 1970b) (Fig. 24). These striations are attributed to the presence of troponin within the thin filaments (OHTSUKI et al. 1967; EBASHI and ENDO 1968); they are most distinct in thick sections tilted in a high voltage electron microscope until the bands are precisely superimposed (ISHIKAWA 1980). The repeat is 38.5 nm (HUXLEY and BROWN 1967); originally PAGE and HUXLEY (1963) found a 41-nm repeat. Because the troponin periodicity (see below) is constant, the number of periods may be used to determine the length of the thin filaments. It varies in different species of vertebrates.

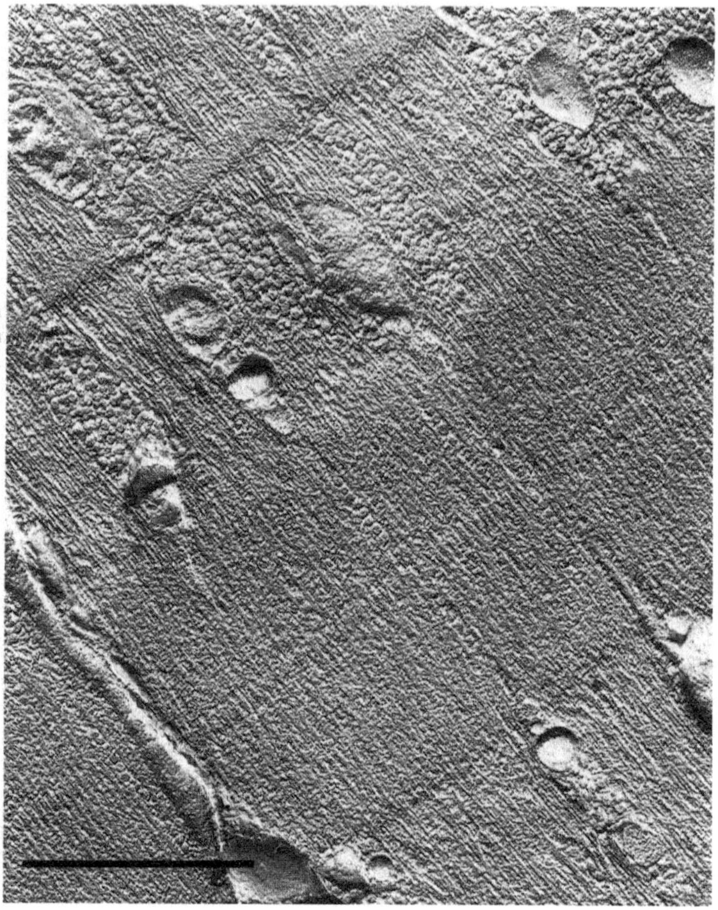

Fig. 25. Human lateral vastus muscle. Freeze-fracture replica showing a longitudinal view of one sarcomere. Overlapping thin and thick filaments, the *Z* and *M* lines, and the pseudo-H zone within which the thick filaments are devoid of cross bridges, may be seen. Clusters of glycogen granules at the I band level are distinct. The membrane fragments belong to mitochondria and elements of the sarcoplasmic reticulum. Bar, 1 μm

In frog, 24 periods are on either side of the Z line; in leg muscles of rat, 27 periods, and in human muscles, "at least 30" have been counted (PAGE and HUXLEY 1963). Tortoise muscles show 29–32 periods (PAGE 1968). This indicates that the length of the thin filaments is species dependent; in human muscles it is clearly larger than the 0.98 μm found in frog muscles (PAGE and HUXLEY 1963) (Fig. 24).

The I band may display one or two wider bands which are faint and only visible in stretched muscle fibres (PAGE 1968) (Fig. 24); they bind lead deposits (GILLIS and PAGE 1967) or ferritin (FRANZINI-ARMSTRONG 1970b) and become more distinct after myosin extraction (WALCOTT and RIDGWAY 1967). FRANZINI-ARMSTRONG (1970b) finds that one of these lines (N_1) is close to the Z line, where the thin filaments still are in an orthogonal array; and the N_2 line is

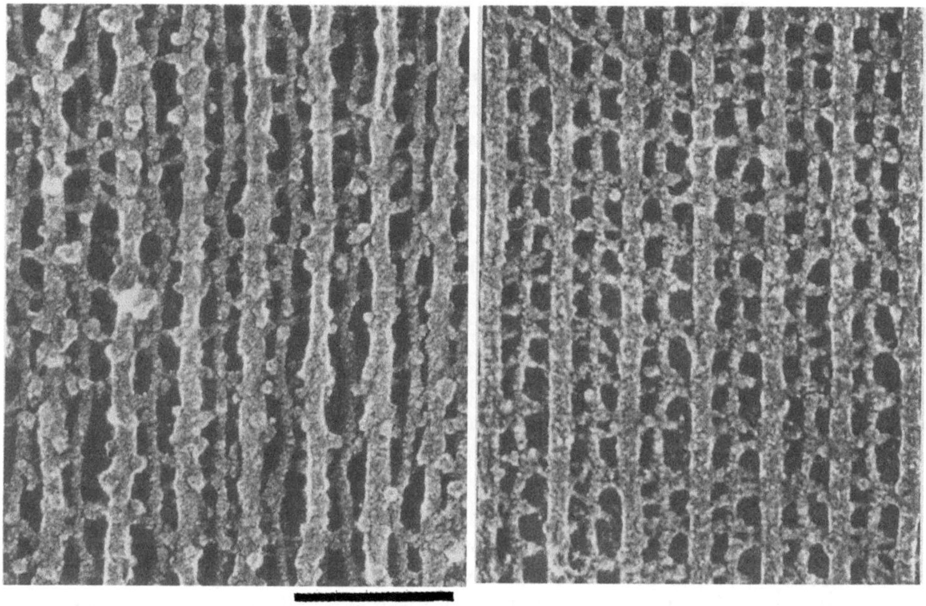

Fig. 26. Insect muscle, A band. Deep-etched rotary-shadowed freeze fractures of unfixed glycerinated and quick-frozen specimens. Direct comparison of ATP-relaxed (*left*) and rigor (*right*) muscle fibres. Note attached cross-bridges in the rigor as opposed to the relaxed muscle fibre. The absence of cross-bridges extending from the thick filaments in the relaxed muscle contrasts with the results of X-ray studies. Comparable micrographs of vertebrate muscles have not yet been obtained. Bar, 0.1 μm. (From HEUSER 1983, with copyright permission of Academic Press)

0.2–0.4 μm from the Z line, where the thin filaments are irregularly disposed. The distances between the Z line and N_1 line, and between the Z line and N_2 line, increase with increasing sarcomere lengths (PAGE 1968; FRANZINI-ARMSTRONG 1970b; LOCKER and LEET 1976b). This suggests that the thin filaments slide through the structures forming the N lines. The I band in stretched fibres is constricted at the N_2 line, closer to the A band the distance between the thin filaments increases. It appears as if material constituting the N_2 line links the thin filaments to each other. The N lines one sees in electron micrographs possibly mark the localization of cytoskeletal filaments (see Sect. C.II).

RAYNS (1972) tried to elucidate the sarcomere structure by freeze fracture. It was possible in insect muscles to see thick and thin filaments, and also cross bridges. Also freeze fractures of mammalian muscle fibres may show the filaments (Fig. 25). A clear view of the filament structure has first been obtained in deep-etched and rotary shadowed freeze-fracture specimens (HEUSER and COOKE 1983; HEUSER 1983) (Fig. 26). Unexpectedly, the cross-bridges are present in rigor muscles only; relaxed muscles showed irregular bumps on the thick filaments rather than periodic unattached bridges, as one should expect from the X-ray diffractograms of resting fibres (CRAIG 1983) (see below). Negatively stained frozen sections (SJÖSTRÖM and SQUIRE 1977b) represent another promising technique to surpass the artefacts undoubtedly connected with fixation and embedding.

4. The Localization of the Contractile Proteins

The two principal contractile proteins of skeletal muscle cells are actin and myosin. When easily soluble proteins are extracted from cross-striated muscle, the remainder consists of about 54% myosin and about 18% actin (HASSELBACH and SCHNEIDER 1951; HANSON and HUXLEY 1957; PERRY and CORSI 1958; WEBER et al. 1969). Myosin is extracted by strong salt solutions, and the A bands are dissolved (HASSELBACH 1953). Light microscopy shows ghost myofibrils consisting of Z lines and I bands only. The distance between the Z line and the border of the former H zone is about 1 μm; this corresponds to the length of the thin filaments. Interference microscopy in connection with densitometry allows determination of the amount of protein within a defined region of a specimen. The gap between the I bands, the original H zone, appears empty on inspection, but a small amount of material is left. The findings before and after myosin extraction from individual myofibrils are in agreement with large-scale biochemical analysis of muscle proteins, and indicate that practically all myosin is localized in the A band. The substance left in the I band makes up about 34% of the total myofibrillar protein. Because actin amounts to only 18% of the myofibrillar proteins, it is obvious that the I band contains additional proteins (HUXLEY and HANSON 1957). Prolonged extraction at low ionic strength dissolves Z lines and I bands and leaves mainly the A bands behind. The extract contains mainly actin and tropomyosin, the A band residues contain myosin and retain the ATPase activity of the actomyosin complex (PERRY and CORSI 1958; CORSI et al. 1967).

Additional evidence for the localization of the proteins is obtained by applying antibodies against the individual proteins to myofibrils or sections. The localization of the bound antibody may be inferred from a decreased solubility or an increased phase density of certain parts of the sarcomeres (SZENT-GYÖRGYI and HOLTZER 1960), or from the localization of fluorescent markers attached to the antibody (PEPE 1966, 1967a, b). Anti-myosin is bound to the A bands and anti-actin to that part of the sarcomere containing thin filaments (I band and lateral part of the A band).

When muscle fibres are minced in a medium that prevents interaction of actin and myosin, and are negatively stained and studied using electron microscopy, one sees thick and thin filaments resembling those seen in sections. The thin filaments, which often are still attached to fragments of Z discs, consist of a double helix of two intertwined chains of globular subunits. Each subunit is about 5 nm in diameter. The same double helix is found when purified actin is combined to filaments. This suggests that the globular subunits are the G actin monomers (HANSON and LOWY 1963; HUXLEY 1963). Also purified myosin aggregates under suitable conditions into filaments. These filaments resemble thick filaments isolated from muscle fibres. Both synthetic and natural thick filaments have tapered ends and bridges projecting from both sides at regular intervals; the central part of the filament is devoid of bridges. In contrast to natural thick filaments, artificial myosin filaments vary in length (HUXLEY 1963). "Synthetic" thick filaments of normal length are obtained when myosin is added to muscle cells from which myosin has been extracted. Under suitable conditions, these cells restore a normal sarcomere pattern and the myofibrils become contractile again (TANIGUCHI and ISHIKAWA 1982).

5. The Cross-Banding Pattern at Different Fibre Lengths

In living muscle fibres it is readily seen that the sarcomere length, i.e. the Z line interval, changes as the muscle fibre is stretched or shortens (Fig. 15). The sarcomere length may without difficulties be measured using light microscopy or, with better than 5-nm resolution, using laser diffraction (CLEWORTH and EDMAN 1972). With the use of laser diffraction, length changes during contraction have been recorded with a time resolution of 260 μs, limited only by the electronic data collecting system (LIEBER and BASKIN 1983).

The change of the band widths within a sarcomere is crucial for understanding the interaction of the two sets of filaments. The possible changes in vertebrate muscles are close to the resolution of the light microscope; the cylindrical shape of the fibre introduces optical artefacts. Most light microscopists found that both the A and the I band length decreased with shortening of the sarcomere. One of the few exceptions was SCHMIDT (1937), who noted that the A band length did not change. This observation had no consequences because at that time nothing was known about the fine structure of the contractile system.

In 1954, in the same issue of *Nature*, the results of two independent light microscopic studies were published, which initiated a new era of muscle research. H.E. HUXLEY and HANSON (1954) isolated fragments of myofibrils from glycerinated rabbit psoas muscles and stretched them by micromanipulation or made them contract by adding ATP. The thickness of the specimens was only 2 μm, and the cross-striation as seen by phase contrast or interference microscopy was very clear. The banding pattern could be measured with an accuracy of 5%–10%. A.F. HUXLEY and NIEDERGERKE (1954, 1958) used whole living muscle fibres of frog, and avoided the optical artefacts by immersing the fibre into a medium with a similar diffraction index. This made the fibre invisible by ordinary bright field light microscopy. Thin optical "sections" were obtained using an objective lens with a small depth of focus together with phase contrast. In both studies, over a wide range, length changes of the sarcomeres were exclusively due to length changes of the I bands. Living frog muscle fibres that were rapidly stretched attained the new I band length in less than 2 ms. The length of the A band was invariably 1.5 μm in resting, shortened, stretched, or isometrically contracting fibres. The cross-striation became faint at sarcomere lengths below 2 μm, and at 1.8 μm sarcomere length, a narrow dense band (C_M line) became visible in the centre of the A band. Additional bands occurred midway between M line and Z line at 1.7 μm sarcomere length (C_Z lines). The H zone disappeared in living fibres at 85% rest length, and it was lacking in isolated myofibrils that were allowed to shorten to equilibrium length before glycerination (about 80% rest length). The A bands and the C_Z lines disappeared when myosin was extracted from shortened myofibrils, but the C_M lines persisted. The myofibrils from which myosin had been extracted were not able to contract when ATP was added.

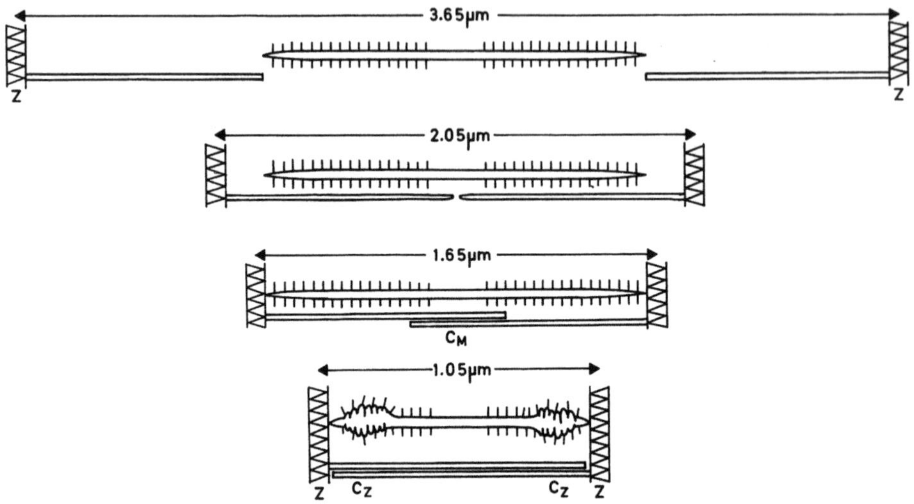

Fig. 27. Schematic illustration of the increasing filament overlap during sarcomere shortening (data for frog muscle, HUXLEY and PAGE 1963; all lengths from centre to centre Z line). The maximum sarcomere length without disrupting the filament array is 3.65 µm; at 2.05-µm sarcomere length the thin filaments meet at the M line, and the H zone disappears. The thin filaments overlap and give rise to the C_M line. Continued shortening increases the width of the C_M line until the thick filaments at 1.65-µm sarcomere length touch the Z lines. The fibre may continue to shorten and the thick filaments crumple up and produce C_Z lines (see also Figs. 15, 16). (Adapted from HUXLEY 1974)

The findings of both these studies suggested that the changes in sarcomere length reflect different degrees of overlap within the lateral parts of the A band, and that the two types of filaments slide past each other during stretch or contraction (Fig. 27).

Measuring the exact lengths of the filaments during different states, however, was, and remains, the problem. It can only be done using electron microscopy which, however, introduces new difficulties. The orientation of the fibre when the filament lengths are to be measured is of paramount importance (HUXLEY 1957); compression of the section during cutting and plastic flow of the embedding medium are other sources of error. Some authors reported changes of the length of filaments in the order of 10%–20% during isometric contractions (CARLSEN et al. 1961); in other studies the variation was very small and was explained by uncontrollable events during processing for electron microscopy (PAGE and HUXLEY 1963; PAGE 1964, 1966). One error often neglected has been the inaccuracy of the magnification of the electron microscope; in early microscopes without controlled lens temperature – some even without lens cooling – the magnification changed continuously.

Muscles that are fixed during electrical stimulation while they maximally shorten against load develop a C_M line; it represents a double array of thin filaments in the centre of the A band. This was interpreted to mean that the thin filaments meet at the centre of the sarcomere at a sarcomere length of

2 µm, and by-pass each other during continued shortening. Below 1.5 µm sarcomere length, when the ends of the thick filaments reach the Z lines, the thick filaments crumple up. The crumpled filaments form contraction bands close to the Z line; these are the C_Z lines. The thin filaments keep on sliding, also below 1.5-µm sarcomere length, and the region of double overlap in the middle of the sarcomere increases (HUXLEY 1972) (Fig. 27).

Mammalian muscles in rigor are stretched when enough force is applied. In places this causes lengthening of the sarcomeres without change in filament length; in other regions the sarcomere length is unchanged, possibly because the muscle yields outside the region studied (HEGARTY et al. 1973). By contrast, SUZUKI and SUGI (1983) stretched glycerinated rigor muscle fibres by 10% and found that the thin-thick filament overlap was unchanged although the sarcomere length was increased by 10%. This was attributed to lengthening of the I filaments outside the A band, and also to stretch of the H zone of the A filaments. The lengthening of the filaments was found to be reversible; it became irreversible when the preparation was stretched more than 20%. The authors conclude that the myofilament compliance contributes to the series-elastic component of the muscle fibre.

Stretch of a tetanized frog muscle fibre increases the sarcomere lengths, as in a resting fibre, without any change in the A band widths (HILL 1977).

6. The Sliding Filament Model

The energy for muscle contraction is derived from the splitting off of the terminal phosphate of ATP. It was shown by ENGELHARDT and LJUBIMOVA (1939) that the myosin molecule has ATPase activity, and by SZENT-GYÖRGYI (1942) that threads spun from actomyosin (a complex of myosin and actin) (WEBER 1934a, b) contract when immersed in a solution of ATP and suitable ions (WEBER 1951; HASSELBACH and WEBER 1955). These findings, taken together with the fact that actin and myosin are in different filaments which probably maintain their length independently of the sarcomere length, suggest that shortening of a muscle fibre involves sliding of the filaments past each other, and that ATP splitting and interaction of the two sorts of filaments generates the force which causes the filaments to slide. This brings into focus the multiple links between the two sorts of filaments, the cross-bridges occurring at regular intervals. Hence, in a mechanical sense, the bridges are involved in generating force, and in a biochemical sense they are involved in ATP splitting. This implies that the cross-bridges represent the site of the ATPase activity of the myosin molecule.

The bridges are short compared with the presumed range of filament displacement, but they might function repetitively in such a way that each one successively binds to different sites on the actin filaments. If the binding sites on the actin filaments were not in register with the origin of the cross-bridges, binding and shortening of some bridges would stretch and release others which then might bind to other sites, shorten, and stretch the first group of bridges, and so on. These cycles of binding-shortening-release of asynchronously attached bridges could maintain a steady force and at the same time cause successive sliding of the filaments past each other. The validity of this hypothesis for the action of the cross-bridges depends on the molecular structure and the arrangement of the molecules within the filaments.

From the length of the filaments, it is obvious that above 3.5-μm sarcomere length the thin filaments are pulled out of the A band which precludes the interaction of actin and myosin; hence, no force should be developed at this sarcomere length. This corresponds to the abrupt drop in extra-tension produced by contraction when a frog muscle fibre is stretched to about 1.6 times its equilibrium length (BUCHTHAL 1942) (Figs. 2, 3). Nevertheless, even fibres stretched to 180% equilibrium length and with more than 3.5-μm sarcomere length still develop a small fraction of force (CARLSEN et al. 1961). This is explained by the fact that the sarcomeres at the tendon ends are shorter than in the middle of the fibre, and therefore are still able to contract (HUXLEY and PEACHEY 1961). Stimulated frog muscle fibres shorten readily and reversibly to a sarcomere length of 1.3 μm at which the thick filaments already crumple and form C_Z bands (GORDON et al. 1966). Hence, the functional range of the sarcomere lengths is 1.3–3.5 μm.

7. X-Ray Diffraction of Muscle

The almost crystalline array of filaments in muscle fibres makes them a unique object for X-ray studies. Already before electron microscopy had elucidated the array of the contractile filaments, it was known that muscle produced a distinct X-ray diffractogram. Together with electron microscopy, X-ray studies have given insight into the molecular structure of the filaments, and have to a great extent confirmed the sliding filament mechanism (for review, see MILLER and TREGEAR 1971; WRAY and HOLMES 1981). The main advantage of X-ray diffraction is that it can be applied to living muscle fibres; a disadvantage is that the reflexes are rather weak. Initially, photographic exposures of hours or days were necessary. This exceeds the time isolated muscle fibres can be kept alive, and has hampered dynamic studies. The problem has recently been overcome by using a synchrotron as powerful X-ray source, and fast electronic signal detectors instead of photographic film. The useful time resolution is now in the millisecond range (HUXLEY et al. 1980, 1981, 1983).

Very low-angle diffractograms with long camera length are suited to reveal widespread periodicities; moderately low-angle diffractograms with relatively short camera length reveal narrow spacings. The first order reflection of a wide spacing is closer to the centre of a diffractogram than the first order reflections of a narrow spacing; the distance between the reflections and the centre is inversely proportional to the distance between the regular structures in the muscle. A diffractogram shows first order reflections of narrow spacings intermingled with higher order reflections of wider spacings. The low-order reflections of wide spacings may be hidden by the central blackening of the diffractogram. (For the principles of X-ray diffraction of muscle, see SQUIRE 1981.)

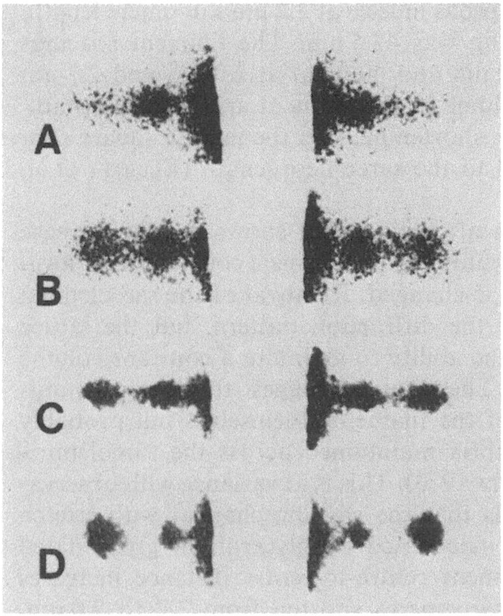

Fig. 28. **A** to **D** are a series of equatorial X-ray diffractograms (fibre axis vertical) from glycerinated (rigor) rabbit psoas muscle at decreasing sarcomere lengths (s): **A** $s=3.1$ µm, $d=36.5$ nm; **B** $s=2.85$ µm, $d=38.6$ nm; **C** $s=2.75$ µm, $d=3.96$ nm; **D** $s=2.3$ µm, $d=40.2$ nm (all measurements to $\pm 4\%$) (d: thick filament spacing in the 1,1 projection). Note change in the relative intensities of the outer 1,1 and the inner 1,0 reflections depending on the sarcomere length. (From Elliott et al. 1963, with permission of the author and Academic Press)

The reflection signals of a live muscle fibre at rest can be grouped into three classes:

1. Equatorial reflections originating from the regular lateral spacing of the filaments
2. Meridional reflections originating from the arrangement of molecules in the direction of the fibre axis
3. Off-meridional (close-to-meridional) reflections due to helical patterns within longitudinally running structures

a) Equatorial Reflections

The equatorial diffractogram of living frog and rabbit muscles shows two reflections which can be explained by the hexagonal lattice of the thick filaments. One of the signals indicates the spacings in the 1,0 and the other the spacings in the 1,1 orientation of the crystal lattice (Figs. 20, 28, 34). The 1,0 reflection is closer to the meridian than the 1,1 reflection. The hexagonal filament lattice spacing may be calculated from the 1,0 or 1,1 spacings. Data are given either as 1,0 spacing or as computed filament spacing. The filament spacing measured in electron micrographs is 35–45 nm, depending on the sarcomere length; it is subject to embedding artefacts (Carlsen et al. 1961; Brandt et al. 1967).

Elliott et al. (1963) found in living frog semitendinosus muscles at 2.7-µm sarcomere length a 1,0 spacing of 33.3 nm corresponding to a filament spacing of 38.4 nm. The 1,0 spacing in rabbit psoas muscle fibres was 40.2 nm at a sarcomere length of 2.3 µm, and it decreased to 36.5 nm when the fibres were stretched to a sarcomere length of 3.1 µm (Fig. 28). Huxley (1968) measured a filament spacing of 46.9 nm in rabbit psoas muscle at 2.2 μm sarcomere length, at 2.6 μm sarcomere length the spacing was 42.5 nm. The filament spacings in the frog sartorius muscle were 43.0 nm and 40.4 nm at 2.0-μm and 2.3-μm sarcomere length, respectively. The volume of the filament array was constant, independent of the degree of stretch or shortening, and the inverse square root of the 1,0 spacing was linearly related to the sarcomere length (ELLIOTT et al. 1963).

A frog muscle fibre stored for 4 h after dissection shows a 24% increase of the filament lattice due to swelling, but it still maintains a constant myofibrillar volume when the sarcomere length is changed. Removal of the sarcolemma ("skinning") does not basically alter the diffraction pattern, but the lattice volume increases by up to 100% and the ability to maintain a constant volume at different sarcomere lengths is lost. These results suggest that the constant-volume behaviour is not a property of the filaments themselves but probably is due to Donnan and osmotic equilibria maintained across the sarcolemma (MATSUBARA and ELLIOTT 1972; ELLIOTT 1973). This is at variance with observations by ROME (1967, 1968) who finds that the spacing changes with stretch even when the plasma membrane is inactivated by glycerol. In glycerinated rabbit psoas muscle, the myosin filament centre-to-centre distance increases from 24.0 nm to 31.0 nm when the sarcomeres shorten from 3.2 to 2.0 μm. This author attributes the constant-volume behaviour to long-range electrostatic forces between the filaments.

Frog muscle fibres at equilibrium length produce equally intense 1,0 and 1,1 reflections. The 1,0 reflection becomes more intense than the 1,1 reflection when the sarcomere length increases; with decreasing sarcomere length the 1,1 reflection becomes more intense than the 1,0 reflection. The same change in relative reflection intensity is seen in rabbit muscles; the increase in the intensity of the 1,1 reflection at short sarcomere length is less pronounced than in frog (ELLIOTT et al. 1963; HUXLEY 1968). The interpretation of these findings relates to the filament array (Fig. 20). At full overlap, the actin filaments are between the thick filaments. In 1,1 projection, the thin filaments project onto the thick filaments and thus enhance the contrast of the diffraction pattern; in the 1,0 plane, they fill the gaps betwen the projection planes of the thick filaments and thereby decrease the signal contrast. The thin filaments are pulled out of the thick filament array when the fibre is stretched, and the mass density between the thick filament planes in a 1,0 projection decreases. Therefore the 1,0 diffraction signal becomes more distinct than at short sarcomere length, whereas the contrast in 1,1 decreases.

The relative intensities of the 1,1 and 1,0 reflections also change when a live muscle develops rigor or when it contracts (HANSON and HUXLEY 1955; HUXLEY 1968) (Fig. 34). These findings shall be described below (Sect. C.I.10).

A third equatorial reflection has been found midway between the 1,0 and 1,1 reflections of the hexagonal lattice. This reflection is weak, and its intensity does not vary with the sarcomere length. Light diffraction of electron micrographs suggests that the third reflection is due to the structure of the Z line (YU et al. 1977).

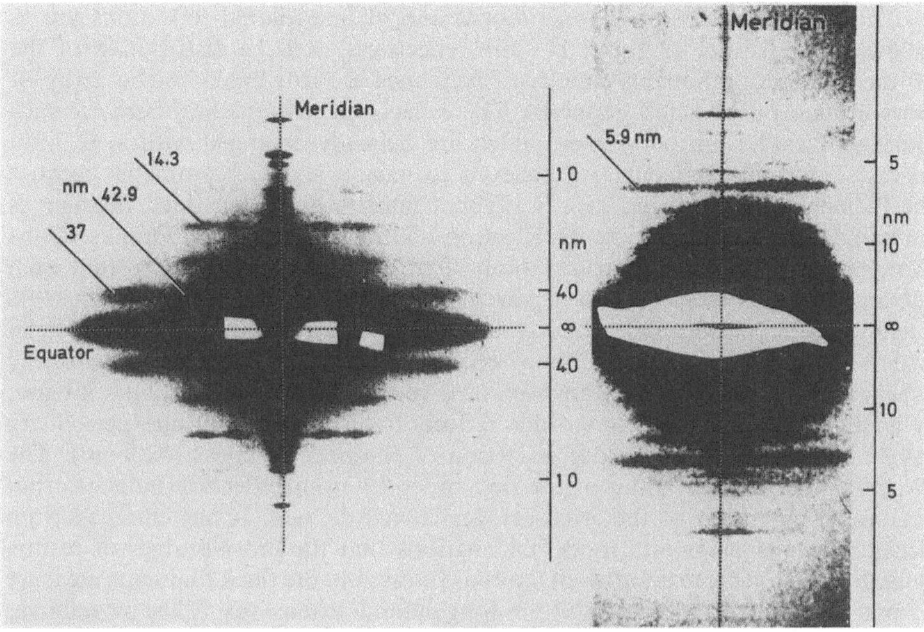

Fig. 29. Meridional X-ray diffractograms of living frog muscle fibres. The approximate positions of the reflections are indicated. The left diffractogram was recorded with longer camera length than the right one and shows wider reflections. The 42.9-nm layer lines originate from the cross-bridge array, the 37-nm reflection is believed to originate from the thin-filament helix, and the 5.9 nm reflection indicates the helix repeat of individual actin monomers within the thin filaments (see Fig. 33). For technical reasons it is not possible to demonstrate all reflections in the same print, the equatorial reflections (Fig. 28) are hidden by the central blackening. The blank central areas are projections of the stop for the zero beam. [Figure prepared by using diffractograms in HUXLEY 1968 (see also Fig. 36). With permission of the author and Academic Press]

b) Meridional and Off-Meridional Reflections

Low-angle long-camera length diffractograms from relaxed muscles show a series of prominent and regularly spaced layer lines which cross the projection of the fibre axis. Their repeat corresponds to a periodicity of 42.9 nm. These lines consist of both meridional and off-meridional components (WORTHINGTON 1959; ELLIOTT 1964) (Figs. 29, 36). The first layer line is stronger than the second one, and the third one which would represent the first order reflection of a 14.3 nm periodicity is exceptionally strong and has an intensity of the same order of magnitude as that of all the other reflections added together. In particular the first layer line has strong off-meridional components. The third layer line has a very strong and narrow meridional reflection; outside the meridian, the intensity of this component fades abruptly. The off-meridional reflections on the third layer line are sampled and give this line a dotted appearance. There are numerous weaker meridional reflections occurring between these layer lines.

The interpretations of the meridional and off-meridional reflections are as follows. The distinct 42.9 and 14.3 nm reflections, and the distribution of the off-meridional components on these layer lines are attributed to the array of cross-bridges of the thick filaments. The reflections suggest that there are subunits with a 14.3-nm periodicity which are helically arranged with a 42.9-nm repeat. The thick filaments are believed to have a six-fold symmetry because the filaments are arranged in a hexagonal fashion and each thick filament is surrounded by six thin filaments. Electron micrographs of longitudinal sections show bridge pairs at intervals of about 40 nm. It has been proposed that each subunit repeat corresponds to a pair of cross-bridges which in relation to its neighbours is rotated 120° so that three pairs of cross-bridges form a helix with 42.9-nm repeat. Three subunit repeats of 14.3-nm periodicity constituting a helical repeat of 42.9 nm can explain why the third layer line is so intense. It is simultaneously the first order reflection of the cross-bridge periodicity (14.3 nm), and the third order reflection of the helical repeat (42.9 nm). The sharp off-meridional fading of the first order 14.3-nm reflection indicates that the lateral extension of the cross-bridges is well defined. It has emerged from densitometer tracings and model calculations that the cross-bridges of resting muscle reach out from a radius of 6 nm to 13 nm, i.e. the thick filaments measure 12 nm in diameter and carry 7-nm-long lateral extensions. This harmonizes with the electron microscopic image. The 14.3-nm and 42.9-nm layer lines in vertebrate muscles disappear when myosin is selectively extracted. The spacing of the layer lines is unaltered when a muscle fibre is stretched, only the sampling of the off-meridional components becomes blurred. This is attributed to the less perfect lateral registration of the filaments in stretched muscle fibres (HUXLEY and BROWN 1967).

Moderately low-angle diffractograms best suited to reveal rather narrow periodicities show prominent meridional reflections of 2.73-nm repeat which represent the repeat of the G-actin monomers. Electron micrographs of negatively stained filaments show globules arranged in a twisted double strand; the two strands are displaced longitudinally relative to each other by half the subunit period, i.e. the repeat of each actin monomer within a helix is 2×2.73 nm = 5.46 nm (HANSON and LOWY 1963).

Other reflections (among others 37 nm, 5.9 nm, and 5.1 nm off-meridional) are ascribed to the pitch of the actin helix and to regularly spaced non-myosin and non-actin molecules (HUXLEY and BROWN 1967).

In insect muscles the array and the position of the cross-bridges have been studied in thin sections for electron microscopy (REEDY 1967, 1968; REEDY et al. 1965) and compared with the findings in X-ray diffractograms. The possible changes in configuration during the fixation and embedding procedures may be controlled by dynamic (see below) X-ray diffraction (REEDY et al. 1983).

8. The Thick Filament

a) The Myosin Molecule

Vertebrate thick filaments consist almost exclusively of myosin (rabbit >90%, MORIMOTO and HARRINGTON 1973), but thick filaments of invertebrate muscles may contain sizable amounts of specific backbone proteins (HANSON and LOWY 1965).

Individual myosin molecules have been visualized, using electron microscopy, in negatively stained or metal-shadowed preparations (RICE 1961 a, b; ZOBEL and CARLSON 1963; HUXLEY 1963; SLAYTER and LOWEY 1967) (Fig. 30). The total length of the molecule is 140–170 nm; it consists of a rod-shaped part about 2 nm thick and 150 nm long, and a globular part 5 nm thick and 10–20 nm long which is composed of two subunits (LOWEY et al. 1969, 1979; MOORE et al. 1970; COHEN et al. 1970). Length and shape of the molecule as determined by morphology are in harmony with early data obtained by measuring the light scattering in myosin solutions (MOMMAERTS 1951). The molecular weight is about 470,000 daltons. Tryptic digestion breaks the molecule into two pieces with different molecular weights. These pieces have been termed heavy and light meromyosin (HMM and LMM), respectively. HMM carries the ATPase activity of myosin, and LMM retains the ability of the myosin molecules to assemble themselves at suitable ionic strength into filaments (SZENT-GYÖRGYI 1953, 1968). LMM appears on electron micrographs as simple rod-shaped structures about 96 nm long; HMM has a double-headed tadpole-like structure consisting of two globules sharing a short tail 40–50 nm in length (black in Fig. 30). In the intact myosin molecule, LMM and HMM are connected end-to-end with a trypsin-sensitive region joining them. Papain cleaves HMM into a globular and a rod-shaped portion (LOWEY et al. 1969). The globular portion consists of two subunits, each with a molecular weight of 120,000 daltons, and each with ATPase activity. These globules are the "S1 subfragments", whereas the rod-shaped tail of HMM is the "S2 subfragment". The widely accepted model for the entire myosin molecule implies that the molecule consists of two identical polypeptide chains (the heavy chains) wound around each other forming a two-chain α-helix. The helix corresponds to the long rod-shaped part of the molecule. The ends of both chains fold up and form the two S1 subfragments (SZENT-GYÖRGYI et al. 1960; HUXLEY 1963; SLAYTER and LOWEY 1967; LOWEY et al. 1969, 1979). Each S1 subfragment contains two myosin light chains which may be different (see Sect. D.III.2).

b) The Packing Pattern of the Myosin Molecules

LMM forms, under suitable ionic conditions, paracrystals with an axial periodicity of 43.0 nm (PHILPOTT and SZENT-GYÖRGYI 1954; SZENT-GYÖRGYI et al. 1960), which is close to the 42.9-nm helical periodicity of natural filaments. It is reasonable to assume that LMM is oriented along the axis of the thick filament and constitutes its backbone; HMM projects out sideways and forms the cross-bridges. Only HMM reacts with actin. The cross-bridge array of the intact filament must be related to the packing pattern of the molecules.

Synthetic filaments made from the entire myosin molecule vary in length; they have tapered ends and carry lateral projections resembling cross-bridges on natural filaments. Like in natural filaments, a 150- to 200-nm-wide zone in the middle of the synthetic filaments is devoid of lateral projections and corresponds to the pseudo-H zone. The length of this bare zone is the same in short and long filaments. This appearance supports a model in which the

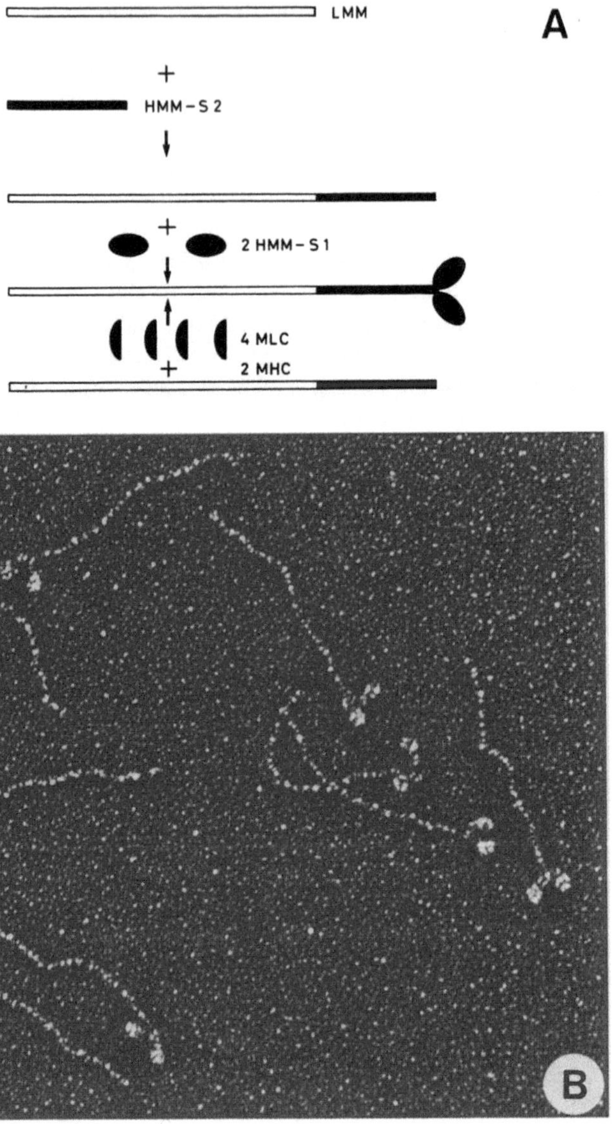

Fig. 30. A Schematic demonstration of the composition of the myosin molecule and the nomenclature of the myosin fragments. The entire molecule has a double-headed tadpole structure. The tail consists of the rod-shaped light meromysin (LMM) and the rod-shaped part of heavy meromyosin, the S2 subunit (*HMM-S2*). The two S1 subunits (*HMM-S1*) constitute the globular head region of the molecule. The entire molecule consists of two myosin heavy chains (*MHC*), which in a coiled coil constitute the tail and also part of the head region. Four myosin light chains (*MLC*) are attached to the head region; they are different, and the types of light chains characterize the different myosin isoenzymes (see also Fig. 80). **B** Rotary shadowed myosin molecules (magnification approx. ×200,000). The two-headed tadpole structure is beautifully shown. (Micrograph courtesy of Dr. S. Lowey, St. Louis, Missouri)

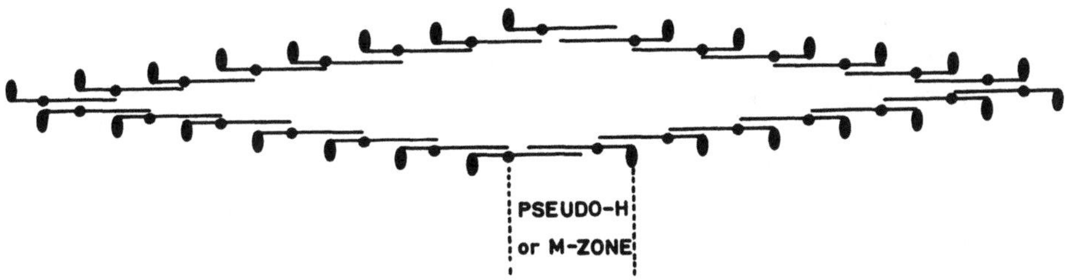

Fig. 31. Diagram illustrating the antiparallel packing of the myosin molecules within a thick filament. The tails constitute the shaft of the filaments, but the S2 subunits of HMM are not bound to the filament shaft. The S1 subunits stick out at regular intervals and form the cross-bridges. The hinge between LMM and S2 is indicated. The M zone of the filament is bare and does not carry bridges. The scheme is two-dimensional, and the tapering of the filament towards its ends is exaggerated. Whether the cross-bridges are arranged in pairs, each subsequent one being rotated by 120° such that three bridge pairs complete one turn, or whether single bridges form three or more helical strands, is undecided

myosin molecules within the filaments are arranged in an antiparallel fashion with the rod-shaped portions pointing towards the middle of the filaments. Then, rod-shaped LMM alone constitutes the bare central part of the filament at which the opposite sets of molecules overlap (HUXLEY 1963) (Fig. 31). If the myosin molecules were packed in pairs with the HMM part sticking out on either side, and if the successive pairs were staggered by 14.3 nm and rotated 120°, they would produce a two-stranded 42.9-nm helix. If one assumes that a thick filament is 1.5 μm long, that each myosin molecule has a 150-nm long tail, and that the myosin molecules are staggered in pairs by 14.3 nm, one should expect that a cross-section through the filament passes through about 20 myosin molecules. The staggered packing implies that there are fewer molecules in cross-sections through the ends of a filament; this is in harmony with the fact that the ends of synthetic and natural thick filaments taper off. The length of the cross-bridges in resting muscle fibres is only 7 nm, suggesting that only the globular S1 subfragments stick out, whereas the rod-shaped S2 subfragments are oriented alongside the LMM backbone.

This model, the principles of which are accepted by most workers, harmonizes with the assumption that opposite forces develop within a sarcomere such that the two sets of thin filaments are pulled into the array of thick filaments.

PEPE (1966, 1967a, b) performed antibody staining of myofibrils from chicken muscle to elucidate the packing of the myosin molecules, i.e. the structure of the shaft of the thick filament. Only certain regions of the sarcomere reacted with anti-myosin, or with antibodies against LMM and HMM, and he concluded that this reflected differences in the accessibility of the respective protein sites. The protein fragments were assumed to be inaccessible when they are buried in the depth of the filament shaft, or are bound to actin. In cross-sections through the pseudo-H zone of fish muscle fibres, the thick filaments were triangular; this shape became even more distinct in relatively thick sections. This was interpreted to mean that within the filaments the rod-shaped portions of the myosin molecules are arranged in a straight fashion, and not twisted.

If the rod-shaped tails were arranged parallel to the filament axis, for a given subunit periodicity, the degree of overlap of abutting tails of molecules in the middle of the filament would determine the degree of tapering of the filament ends. This again would determine the accessibility of the molecules to antibodies. Assuming that each cross-bridge represents one myosin molecule, PEPE (1967a, 1975) proposed a model for the packing of the molecules which was in concordance with his experimental data, and at the same time was in agreement with the then undisputed two-strand helical array of bridges. Each cross-section through a filament outside the pseudo-H zone and the tapered ends was assumed to contain 18 molecule tails being arranged in a triangular fashion with blunted apexes. The model rested on several assumptions, but it was indeed possible to visualize, in thick-filament cross-sections of chicken and fish muscles, substructures approximately arranged in the predicted fashion (PEPE and DRUCKER 1972; PEPE 1975; PEPE and DOWBEN 1977). Computer processing and optical diffraction of electron micrographs confirmed that the myosin filament is composed of subfilaments. Twelve subunits are arranged in a hexagonal lattice with 3.5–4-nm spacing. When the section is tilted, the subunit array changes corresponding to a straight course of the subfilaments within the thick filament. Spacing and number of the subfilaments suggest that they represent myosin dimers rather than individual molecules (PEPE 1979; STEWART et al. 1981). PEPE'S original model was based on a two-stranded cross-bridge array, but the subfilament pattern described would comply with a three-stranded arrangement of cross-bridges as well (STEWART et al. 1981). The non-homogeneous staining of the A filaments by myosin antibodies has been attributed to inadvertent reactions with non-myosin molecules, rather than to differences in the accessibility of the antigenic sites (OFFER 1976); in a recent study from Pepe's laboratory (WACHSBERGER et al. 1983), the selective reactibility of the tapered ends and the borders of the bare central zone has been confirmed with a monoclonal antibody against non-ATPase sequences of the S1 subfragment.

The structure of the backbone of the thick filament is difficult to assess using X-ray diffraction because only weak reflections can be expected. In some invertebrate muscles the diffractogram suggests that the rod-shaped parts of the myosin molecules are twisted into intermediate subfilaments, which are then arranged in a ring to form a hollow tubular backbone (WRAY 1979; WRAY and HOLMES 1981). This agrees with the electron microscopic appearance of these muscles, but there is no evidence for a similar array in vertebrate muscles.

c) The Number of Myosin Molecules and of Cross-Bridges per Subunit Repeat

A crucial question for the understanding of the packing pattern is how many myosin molecules contribute to a cross-bridge. If there are pairs of cross-bridges with an axial distance (subunit repeat) of 14.3 nm, there must be 216 cross-bridges for each thick filament. HUXLEY (1972) estimated from the number of thick filaments per cm^3 muscle fibre, the myosin content of wet muscle, and the molecular weight of myosin, that there are 384 myosin molecules per thick filament. This means that each cross-bridge represents either one or two myosin molecules, because it is unlikely that the number of molecules per cross-bridge is not an integer.

The fibrillar flight muscle of the water bug *Lethocerus cordofanus* has been studied extensively using various methods; it is particularly suited for X-ray studies because of its almost crystalline structure. CHAPLAIN and TREGEAR (1966) made the same calculation for this muscle and arrived at three myosin molecules per cross-bridge. This result was based on two cross-bridges per subunit repeat; REEDY et al. (1965) and REEDY (1967) found that also in this muscle the cross-bridges are in pairs. Nevertheless, in a subsequent study, REEDY (1968) found four cross-bridges for every 14.4 nm.

An alternative approach to determine the number of myosin molecules per subunit repeat was made by TREGEAR and SQUIRE (1973). They measured the relative amounts of one of the chains of HMM and of actin, and calculated the number of myosin molecules per subunit repeat from the relative lengths of thick and thin filaments, and the periodicity of the actin monomers. Thus, they circumvented the difficulty of estimating the number of thick filaments per unit volume muscle. The number of myosin molecules per subunit repeat was three in rabbit psoas muscle and six in *Lethocerus* flight muscle. Based on these results, the authors disputed the two-stranded cross-bridge helix and instead assumed that there are three strands of cross-bridges in vertebrate muscles, and six strands in *Lethocerus* muscle. LAMVIK (1978), in isolated thick filaments of vertebrate muscles, measured the mass per unit length using scanning transmission electron microscopy. The mass was 87 kdaltons/nm, i.e. 1,244 kdalton/14.3 nm. Because the molecular weight of myosin is 470,000 daltons, the results were interpreted as supporting the view that there are three myosin molecules per subunit repeat.

Recently also morphological evidence for a three-stranded model has been presented. Frayed thick filaments of rat psoas muscle, studied in negatively stained preparations, appeared three-stranded (MAW and ROWE 1980). When the filaments split into two subfilaments, one was thicker than the other; three subfilaments were of equal thickness. No evidence for coiling of the subfilaments was found, indicating that the single strands are arranged in a straight fashion as predicted by PEPE's model (see above). KENSLER and STEWART (1983) succeeded in isolating thick filaments from frog muscle with the cross-bridge array intact. They obtained optical diffractograms of electron micrographs which resembled X-ray diffractograms of living muscle fibres. This ascertained that the filament structure was intact. Computer image analysis revealed a three-stranded array of the bridges, but did not elucidate whether the tails forming the filament backbone were arranged in straight or helical fashion. MAEDA (1983) obtained X-ray diffractograms from crab muscle, suggesting that in this case there is a fourfold rotational symmetry of the cross-bridges.

Whatever the final outcome of this dispute, it is obvious that the original two-stranded model, which appears in most textbooks, has come under fire from several sides.

d) C Protein

Not only in invertebrate muscles, but also in vertebrate muscles, the thick filaments contain proteins other than myosin. The most prominent one is C protein (OFFER et al. 1973; MORIMOTO and HARRINGTON 1973). It has a molecular weight of 140,000 daltons and consists of a single polypeptide chain which is not coiled. The length of the molecule is about 20 nm. C protein amounts to 2% of the myofibril. Labelled antibodies against C protein produced nine stripes on both sides of the M line, indicating that the C-protein molecules of adjacent thick filaments are in register. The 2 × 9 lines were grouped in the centre of the sarcomere adjacent to the pseudo-H zone (OFFER et al. 1973). It is now known that the innermost two of the stripes were stained by unspecific antibodies; therefore, each filament has 2 × 7 sites with C protein (PEPE and DRUCKER 1975; CRAIG and OFFER 1976b). There are about 37 C-protein molecules per thick filament (OFFER et al. 1973), i.e. two to three at each reactive site (MORIMOTO and HARRINGTON 1973; CRAIG and OFFER 1976b). The C protein is at least partly on the surface of the filament (CRAIG and OFFER 1976b). Also negatively stained frozen sections for electron microscopy show the C-protein stripes (SQUIRE 1973; SJÖSTRÖM and SQUIRE 1977a; SQUIRE et al. 1982b) (Fig. 32).

ROME et al. (1973a) labelled the C protein in rabbit muscles by antibodies, and in X-ray diffractograms obtained strong 44.2-nm reflections; the corresponding reflections were rather weak in unlabelled fibres. This does not mean that the C-protein repeat is 44.2 nm. Sections for electron microscopy show a 43.0-nm periodicity of the labelled sites. The authors demonstrated that the 44.2-nm reflection was the resultant of interference of the 43.0-nm periodicity of the C-protein sites, and the 665-nm periodicity of the grouped C-protein lines in each half A band. The 43.0-nm C-protein periodicity seems to be identical with the 42.9-nm myosin periodicity (CRAIG and MEGERMAN 1979), but SQUIRE et al. (1982a, b) found a 43.4-nm repeat for the C-protein sites, and conclude that C protein and the myosin cross-bridges are not in register.

The C-protein stripes have an axial extent of 7 nm, i.e. less than the length of the molecule (20 nm) (CRAIG and OFFER 1976b). The molecules do not seem to extend radially, and it is assumed that they are wrapped circumferentially around the shaft of the thick filament.

The function of C protein is enigmatic. Three possibilities are discussed (MOOS et al. 1978):

1. The length of the C-protein period is slightly different from the myosin period, and some sort of Vernier effect determines the length of the thick filament (SQUIRE et al. 1976; SQUIRE 1981; SQUIRE et al. 1982b), which is constant in vivo but variable when pure myosin filaments are assembled in vitro (HUXLEY 1963). Some authors, however, find no difference in the axial repeats of myosin and C protein (CRAIG and OFFER 1976a, b).

2. C protein inhibits the activity of actin-activated myosin ATPase, but not the ATPase activity of pure myosin. One C-protein molecule binds to 3–5 G-actin monomers and competes with HMM for binding to actin. It is possible that C protein modifies the kinetics of the actin-myosin interaction (MOOS et al. 1978).

3. The C-protein molecule would be long enough to bridge the gap between thick and thin filaments. If, in resting muscles, C protein established contacts between thick and thin filaments, it could account for mechanical measurements indicating that even in resting muscles there is a weak interaction between the

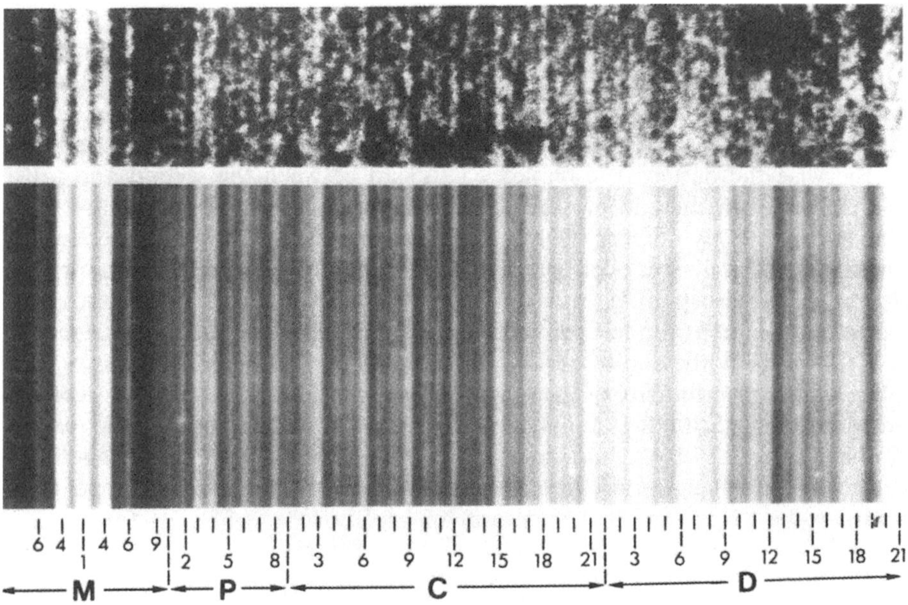

Fig. 32. *Top:* Negatively stained frozen sections for electron microscopy showing one-half of an A band. All structures that appear dark in sections of embedded material with this technique appear white. The fibre axis is horizontal; the M line is to the left and the edge of the A band to the right. A series of lines are visible throughout the A band. *Middle:* Laterally averaged image of the micrograph shown above; the line pattern is distinct. *Bottom:* Denotation of the different zones and lines within the A band. The M zone displays three strong and two moderately strong lines representing the five-bridge system of the M line. Each line throughout the A band corresponds to a myosin cross-bridge repeat. The lines beyond D17 seem to merge, but D19 is clearly missing, indicating a gap in the cross-bridge array. The seven strong lines in the C zone reflect the localization of C-protein molecules within the A filaments. (From SQUIRE 1981, with permission of the author and Plenum Press)

filaments (HILL 1968). C protein added to rabbit myofibrils binds strongly to the thin filaments; this binding is Ca^{2+} regulated, which may suggest that actin-C-protein interaction is physiologically significant (MOOS 1981).

e) The Periodicities of the A Band of Vertebrate Muscle

Each half of the A band has been subdivided into zones (SJÖSTRÖM and SQUIRE 1977a): the M zone (with M line and pseudo-H) and the bridge zone. The width of the M zone is 160 nm. The bridge zone comprises the P zone (for proximal, adjacent to the M zone), the C zone (for C protein) with the seven bands of C protein, and the D zone (for distal). Also the P and D zones contain non-myosin proteins; the various proteins forming the M band are localized in the M zone. The array of cross-bridges, of C-protein sites, and of M bridges and unidentified proteins results in an elaborate system of periodicities and lines which have consecutively been numbered within the zones

(Fig. 32). The fact that there are stripes in negatively stained frozen sections indicates that adjacent myosin filaments are in precise register, and are rotated relative to one another in a precisely determined fashion (super lattice, HUXLEY and BROWN 1967; SQUIRE 1981). There is always one gap in the 14.3-nm cross-bridge array near each end of the filament such that band D19 is lacking, but otherwise the pattern is complete. This gap between the second and third cross-bridges from the end of the filament indicates that one particular cross-bridge is consistently missing (CRAIG and OFFER 1976a). This seems incompatible with PEPE'S (1975) model for the thick filament, which does not predict any gap in the cross-bridge array. For exhaustive review and discussion of the different filament models for vertebrate and invertebrate muscles, SQUIRE'S book (1981) should be consulted.

The 14.3-nm periodicity of the cross-bridges (the subunit repeat) has recently been questioned. SQUIRE et al. (1982a, b) found, in human muscle, a regular cross-bridge repeat of 42.9 nm, but periodic perturbations of the cross-bridge distribution. The lengths of the three subunits within each helical repeat are reported to be 15–16 nm, 11–13 nm, and 14–16.5 nm.

f) Myosin ATPase and Cytochemistry

Some authors attempted to demonstrate the ATPase activity of the myosin molecule in thin sections by adopting the histochemical method for electron microscopy, and found lead precipitates in the cross-bridge region exactly where one would expect the enzyme to be localized (SOMOGYI and SOTONYI 1969; JAGENDORF-ELFVIN 1970). Nevertheless, TICE and BARRNETT (1962), GILLIS and PAGE (1967), and TUNIK (1971) showed that the same "staining" may be obtained without substrate; this suggests that differential lead binding of the protein filaments rather than the enzyme localization determines the non-uniform grain distribution.

9. The Thin Filament

a) The Array of Actin Monomers

Actin isolated from skeletal muscle appears in a globular form (G actin) which under suitable ionic conditions assembles itself into filamentous F actin. The formation of synthetic actin filaments was first demonstrated using electron microscopy by JAKUS and HALL (1947). The monomers are arranged in two helically wound strands (Fig. 33). The subunit repeat is 5.46 nm; in X-ray diffractograms it is 2.73 nm, because the two strands are displaced relative to each other, by half the subunit period. The double helix structure was inferred from X-ray diffractograms and has since been visualized in negatively stained preparations (SELBY and BEAR 1956; HANSON and LOWY 1963, 1965; HUXLEY 1963).

The actin monomers are slightly elongated. Their dimensions are 5–7 × 4 × 3 nm (AEBI et al. 1981). The orientation of the molecules within the

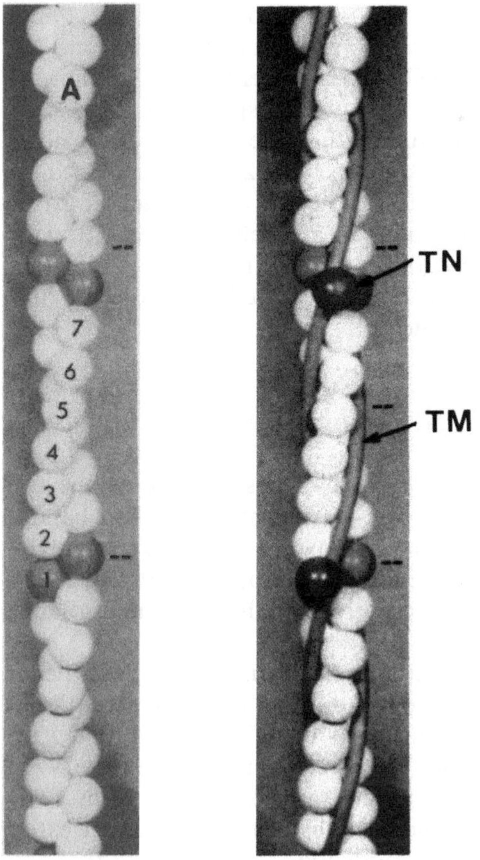

Fig. 33. Models of the structure of F actin (*left*) and a thin filament (*right*). The almost globular G actin monomers (*A*) form a double-stranded helix within the F-actin filament. According to EBASHI's (1969) model, thread-like tropomyosin molecules (*TM*) lie in the grooves of the helix and at regular intervals of about seven actin monomers, globular troponin molecules (*TN*) are attached to the system. The half-pitch of the helix is about 37 nm; also the individual monomers constitute a helix which gives rise to the 5.9-nm reflection in the X-ray diffractogram (Fig. 29). (From SQUIRE 1981, with permission of the author and Plenum Press)

double helix determines the diameter of the thin filaments. If they were oriented with their long axis perpendicular to the filament axis, the filament diameter would be at least 9.5 nm, and if the orientation were parallel to the filament axis, the filament diameter would be 6.0–8.0 nm. It is not yet clear which one of these two models is correct. Thin sections and negatively stained specimens provide evidence that the filament diameter is below 8 nm; freeze fractures (HEUSER and KIRSCHNER 1980) and X-ray diffractograms of live muscles (EGELMAN and PADRON 1984) indicate that the filament diameter is close to 10 nm. Nevertheless, FOWLER and AEBI (1983) measured also in frozen rotary-shadowed preparations a diameter of 7–8 nm and attributed the measurements of HEUSER and KIRSCHNER (1980) to thick platinum shadowing.

b) The Pitch of the Actin Helix

The pitch of the actin helix is still uncertain (WRAY and HOLMES 1981); the exact value is of interest for the question how HMM is attached to actin and whether the actin helix is integral or not. HANSON and LOWY (1963, 1965) found, in negatively stained specimens, a cross-over distance (half pitch) of 34.9 nm and, in X-ray diffractograms, a helix repeat of either 2×31.5 nm or 2×40.6 nm with 13 or 15 actin monomers per turn, respectively. MILLMAN et al. (1967) concluded from their X-ray studies that the actin helix is non-integral, and that it has about 15 subunits and a repeat of 2×40.0 nm. HUXLEY and BROWN (1967) measured 2×36.0 to 2×37.0 nm with a non-integral array of 6–7 actin monomers per half turn. This result is now widely accepted (HUXLEY 1972; WRAY and HOLMES 1981; HUXLEY 1982). The X-ray reflections from which shorter or longer helical repeats had been calculated are probably due to the array of backbone and regulatory proteins.

c) Tropomyosin and Troponin

The A bands of myofibrils of vertebrate muscle consist almost exclusively of myosin, but there is a sizable amount of non-actin proteins in the I bands. Besides myosin and actin, myofibrils contain tropomyosin. "Native tropomyosin" isolated from muscle is always "contaminated" by troponin (EBASHI 1963; EBASHI and EBASHI 1964). Both tropomyosin and troponin have by immunfluorescence been located in the I band.

Tropomyosin is a thread-like molecule about 45 nm in length. It is polar and forms linear aggregates or paracrystals, depending on the ionic conditions. In all states, it shows a periodicity of around 40 nm (COHEN and LONGLEY 1966; CASPAR et al. 1969; HIGASHI and OOI 1968; EBASHI and NONOMURA 1973; KATAYAMA and NONOMURA 1979a, b). Antibodies against tropomyosin stain the I bands homogeneously, and electron micrographs of thin filaments treated with tropomyosin antibody do not reveal periodicities. This suggests that the tropomyosin molecules form a continuous structure; they are probably arranged head-to-tail in two strands (CASPAR et al. 1969). If this is correct, seven actin monomers would correspond to one tropomyosin molecule period. In a three-dimensional reconstruction from electron micrographs of negatively stained thin filaments, two strands of protein molecules have been found in the grooves of the double helix of actin subunits; these strands probably consist of tropomyosin. When the same technique was applied to F-actin filaments assembled from purified G actin, only the two helical chains were found (MOORE et al. 1970).

The troponin molecule is globular and it always binds to a defined site of the thread-like tropomyosin molecule. Thin filaments of chicken or rabbit muscle stained with antibodies against troponin or fragments of troponin reveal a periodicity of 38 nm (OHTSUKI et al. 1967; EBASHI et al. 1972; OHTSUKI 1975, 1979). Myofibrils of rabbit muscles labelled with antibodies against troponin produce 38.3-nm reflections in the X-ray diffractogram, which are 3–5 times more intense than those of unlabelled myofibrils. The reflection is commonly referred to as "38.5-nm reflection", the true repeat is somewhat smaller and seems to fit with seven times the actin subunit repeat (7×5.46 nm = 38.22 nm) (ROME et al. 1973b).

d) A Thin Filament Model

The following model for the thin filament structure is now widely accepted (EBASHI et al. 1969) (Fig. 33). The actin monomers form a non-integral double helix with a repeat of 2 × 36 nm, with 2 × 6–7 G-actin molecules per half-turn. Two strands of tropomyosin molecules, each 40 nm in length, are within the grooves of the helix such that each tropomyosin molecule interacts with seven actin molecules (7 × 5.5 nm = 38.5 nm). The ends of the individual tropomyosin molecules probably overlap by 1.5 nm to accommodate to the 38.5-nm repeat. At regular intervals of about 38 nm, globular troponin molecules are attached to the system. It is probably of functional significance that the pitches and subunit repeats are different from those of the thick filaments.

Actin, tropomyosin, and troponin in vitro form unique complex paracrystals which may directly reflect the molecular structure of the thin filaments. Paracrystals of pure actin consist of parallely arranged F-actin filaments; the helices of the filaments are in register. The actin filaments are arranged in the same fashion within complex paracrystals of actin, tropomyosin, and troponin; end-to-end filaments of tropomyosin lie in the grooves of the actin helices, and every 38-nm troponin is attached to the filaments. Depending on the relative concentration of tropomyosin and troponin, the paracrystals occur in two modifications: either the actin helices are in register and the sites of troponin attachment are not, or the troponin sites are in register and the actin helices are not (OHTSUKI and WAKABAYASHI 1972). Troponin causes a distinct 38-nm band pattern in the latter type of crystal which is reminiscent of the cross-striations within the I band of muscle fibres (HALL et al. 1946; DRAPER and HODGE 1949; CARLSEN et al. 1961; PAGE and HUXLEY 1963; PAGE 1964; FRANZINI-ARMSTRONG 1970b) (Fig. 24).

The length of the thick filaments in vivo is uniform, but the internal structure of the thin filament apparently allows appositional growth. TRAEGER and GOLDSTEIN (1983) found a considerable variation in length in rat soleus muscle. In adult rats, the length varied from 0.18–1.2 μm, with 25% of the filaments less than 0.7 μm in length. In 3-day-old rats, the length was more uniform but there were 25% fewer thin filaments per unit cross-sectional area compared with adults.

e) Binding Sites for the Cross-Bridges

The actin monomer is an ellipsoid consisting of two domains. It must have specific sites which can interact with the S1 myosin subunits, the cross-bridges of the thick filaments. The system would be most efficient if not only the myosin molecules of the thick filaments but also the actin sites were arranged in an antiparallel fashion. The thin filaments reveal, in fact, a polarity. Electron micrographs show characteristic arrowhead formations when isolated actin filaments are allowed to combine with HMM or S1 subfragments. All myosin subfragments insert at an angle of about 45° and all point in the same direction away from the Z lines (HUXLEY 1963). The formation of "arrowhead complexes" with HMM or S1 subfragments is a property exploited to identify actin filaments in homogenates and sections of muscle and non-muscle cells (ISHIKAWA et al.

1969). There is evidence that each S1 subfragment of HMM binds not to one, but to two, actin monomers simultaneously, and that pairs of actin monomers act as functional subunits for specific binding of a myosin head (MORNET et al. 1981; AMOS et al. 1982).

10. Morphological Changes of the Thick Filament Structure During Rigor and Contraction

The structure of resting muscle fibres studied using electron microscopy and X-ray diffraction at different degrees of stretch, suggests that the filaments slide along each other when the sarcomeres change their lengths (HUXLEY and NIEDERGERKE 1954; HUXLEY and HANSON 1954). Additional information, in particular with respect to the question of how the system works, might be expected from muscles investigated during active contraction.

Muscle fibres treated with iodo-acetate or glycerol develop rigor, which may be understood as complete and irreversible "contraction". The loss of extensibility is associated with characteristic changes of the X-ray diffractogram assumed to indicate attachment of all cross-bridges to the thin filaments: (a) The equatorial 1,1 reflection becomes stronger than the 1,0 reflection (Fig. 34). (b) The helical 42.9-nm periodicity of the cross-bridges disappears. (c) The 14.3-nm reflections of the cross-bridge subunits decrease in intensity. (d) A new pattern of layer lines indicates that the cross-bridges now form a 36- to 37-nm helix (HUXLEY and BROWN 1967) (Fig. 36).

The explanation for these changes is as follows (Fig. 35). The S1 subfragments of HMM, the cross-bridges, in resting muscle fibres reach out for only 7 nm from the surface of the thick filaments because the S2 subfragments of HMM lie alongside the backbone of the thick filament. The actin-to-myosin centre-to-centre distance is 19–26 nm (ELLIOTT et al. 1963) and the distance from the surface of a thick to the surface of a thin filament is at least 10 nm. The S1 subfragments, therefore, have to move for about 3 nm to become attached to the actin subunits, and a certain amount of mass moves from thick to thin filaments. This increases the mass around the thin filaments. In the 1,0 orientation, the thin filaments lie between the thick filament planes and a rigor-induced increase in mass will blur the diffractogram so that the 1,0 equatorial reflection becomes relatively weaker (Fig. 34). The thin filaments in electron micrographs of cross-sections of rigor muscles appear thicker than those of resting muscles, and in rigor the two sets of filaments differ less in diameter than they do at rest. When a rigor muscle fiber is "relaxed" by removing calcium and adding ATP, the difference in filament diameter is re-established. The shift in mass from thick to thin filaments in rigor has been demonstrated in Fourier reconstructions based on the intensity shifts of the 1,0 and 1,1 reflections (HUXLEY 1968). The fact that the helical 42.9-nm periodicity is lost during rigor indicates that the actin attachment sites for the cross-bridges do not have the same helical repeat as the array of bridges on the thick filament. The new cross-bridge order closely resembles the helical pitch of the actin monomers of the thin filaments. The change in helical order requires a certain

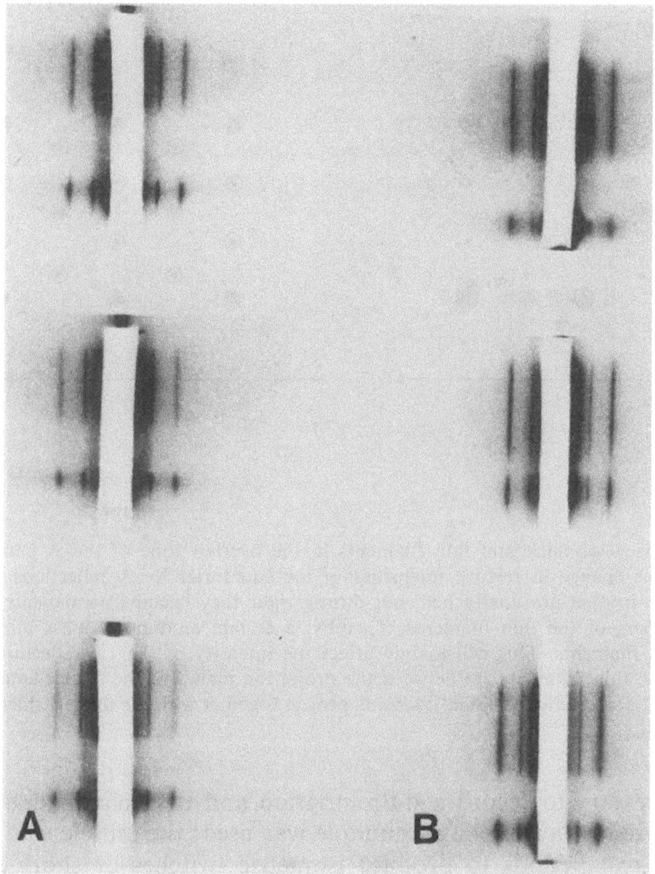

Fig. 34. Equatorial X-ray diffractograms from live (**A**) and rigor (**B**) rabbit psoas muscles showing the 1,0 and 1,1 reflections from the hexagonal lattice. A combined slit (*upper part* of the diagrams) and pin-hole (*lower part* of the diagrams) type diagram has been recorded in each case. Fibre axes are vertical. All fibres were at equilibrium length, 2.20 μm ± 0.05 μm. Note the pronounced increase of the intensity of the (*outer*) 1,1 reflection compared with the (*inner*) 1,0 reflection during rigor. (From HUXLEY and BROWN, 1967, with permission of the authors and Academic Press)

amount of longitudinal movement; this explains why the 14.3-nm meridional reflections become weaker (Fig. 36).

There is also less sampling on the layer lines (they are less "dotted") in rigor than in resting muscles, which indicates that the transverse orientation of the filaments is less perfect in rigor than at rest (HUXLEY and BROWN 1967). The orientation of the cross-bridges in resting frog muscle fibres and in muscles stretched beyond filament overlap (4.2 μm sarcomere length) is random; the cross-bridges are attached at a 45° angle to the actin filaments in rigor muscles at full-overlap sarcomere length (2.3 μm) (POULSEN and LOWY 1983).

Rigor is a special, non-physiological form of contraction. The investigation of actively contracting muscles using X-ray diffraction was hampered by the

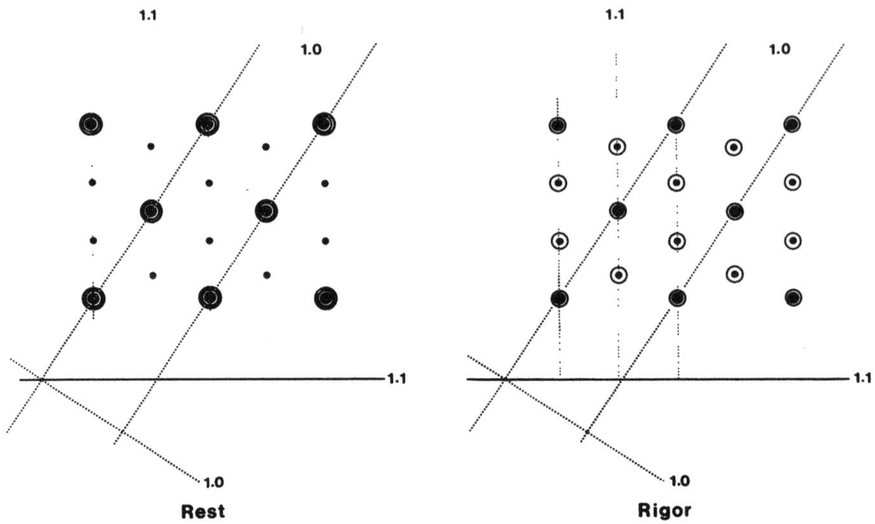

Fig. 35. End-on view of thick and thin filaments in the overlap zone of the A band. Schematic explanation of the change in relative intensities of the equatorial X-ray reflections during rigor. At rest, the cross-bridges are unattached, but during rigor they become permanently attached to the actin monomers of the thin filaments. Thereby, a certain amount of mass moves from the thick to the thin filaments. This will mainly affect the intensity of the 1,0 reflection, because in this projection the thin filaments are between the projection planes of the thick filaments, whereas in the 1,1 projection the thin filaments project together with the thick filaments

long time necessary to record a diffractogram and the unavoidable fatigue of the muscle. Initially, a sampling technique was used: the muscle was stimulated with suitable rest periods to produce isometric twitches or short tetani, and the diffractogram was recorded only when the muscle had reached a certain amount of tension. These experiments were performed at low temperature to slow the contractile response and to improve survival of the muscle fibres (Elliott et al. 1965; Huxley and Brown 1967; Haselgrove and Huxley 1973). Like in rigor muscles, in contracting muscles the intensity of the equatorial 1,1 reflection increases and that of the 1,0 reflection decreases. The 14.3-nm periodicity of the cross-bridges is preserved, and the 42.9-nm helical repeat disappears. In contrast to rigor muscles, no new helical order is established. This has been interpreted to mean that the bridges are attached for a short time only in such a manner that their orientation varies over a wide range of angles. The actin filament reflections become more distinct than at rest, possibly because the orientation of the filaments is more regular and straight.

These results were confirmed when very intense X-rax sources (synchrotron radiation) and fast electronic detectors became available. Now, the changes in the diffractogram may be assessed within a very short time window of a twitch, and only relatively few twitches are necessary to average a sufficiently strong signal. The changes in the diffractogram of a twitching muscle commence shortly after stimulation. The intensity changes of both the equatorial reflections and the off-meridional reflections lead, by a few ms, the change in tension

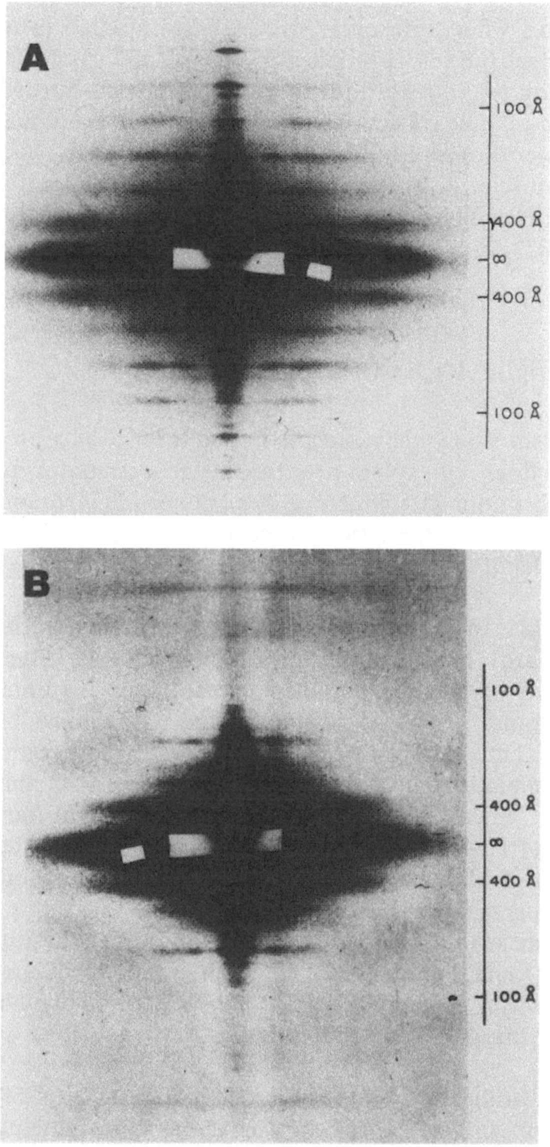

Fig. 36 A, B. Change in the meridional X-ray diffractogram during rigor. **A** Diffraction pattern from live frog sartorius muscle; fibre axis is vertical. The pattern of layer lines corresponds to a repeat of 42.9 nm and is believed to arise from the helical arrangement of cross-bridges on the myosin filaments. Note the strong first and third layer lines, and sampling of the third layer line (see Fig. 29). **B** Diffraction pattern from frog sartorius muscle in iodo-acetate rigor. Fibre axis is vertical. The strong 14.3-nm reflection (*third layer* line in **A**) is still present, but the system of layer lines based on 42.9 nm has disappeared. Instead, there is a strong but diffuse layer line at about 37 nm, also the 5.9-nm actin reflection can be seen (not recorded in **A**). (From HUXLEY 1968, with permission of the author and Academic Press)

during the rise phase of the twitch (frog sartorius muscle, 10° C). Stimulation does not induce any change when the muscle is stretched beyond filament overlap. The cross-bridges do not move outwards unless there is an actin filament alongside (MATSUBARA and YAGI 1978; HUXLEY 1979; HUXLEY et al. 1980, 1981, 1983; YAGI and MATSUBARA 1980).

The time relationship between contraction and change of the X-ray diffractogram was first studied in insect muscles (ARMITAGE et al. 1975), which are easier to maintain in vivo than are vertebrate muscles. The fibrillar flight muscle of *Lethocerus cordofanus* oscillates when sinusoidal stretches are applied in a solution containing Ca^{2+}. An electronic X-ray detector was placed over the appropriate region of the diffractogram to record the intensity variations during the contraction cycles. The actin-related reflections were stable, but both the equatorial reflections and the meridional 14.5-nm reflections oscillated with the mechanical output of the muscle. The intensity changes compared with those in rigor, when probably all cross-bridges are attached, indicate that at a given point in time, between 10% and 20% of the cross-bridges are attached to actin filaments.

11. Swinging Cross-Bridges

The sliding filament mechanism was widely accepted soon after it was presented. Nevertheless, this process does not explain how force is generated during contraction. The idea that asynchronous attachment cycles between HMM and actin produce movements which are large in relation to the length of the individual cross-bridges has led to a model of cycling and swinging cross-bridges (HUXLEY 1969) (Fig. 37a).

The S1 subfragments of HMM can only reach the thin filaments when the S2 subfragments are tilted outward. The S1 subfragment must also be tilted in relation to the S2 subfragment to become attached to the actin subunits. The cross-bridge and actin-monomer arrays are different, and there must be a certain amount of circumferential movement as well. The X-ray diffractograms of resting muscles change during rigor and contraction in harmony with this prediction (HUXLEY 1982). The "hinge" mechanism correlates with the fact that the rod-shaped S2 subunit of HMM does not aggregate with LMM, and hence is not tightly bound to the LMM backbone of the thick filament, and that the rod section of the molecule is flexible (KING and YOUNG 1972).

Tropomyosin, in combination with troponin and calcium ions, enhances the affinity of myosin for actin (EBASHI et al. 1972). HUXLEY (1982) speculates that the cross-bridges might always undergo a certain amount of Brownian movement, and in an activated muscle become attached if they get close to actin.

It is not clear how the cross-bridges generate force. H.E. HUXLEY's (1969) hypothesis implies that the S1 subunits of HMM approach the thin filaments always at the same angle, regardless of the interfilament distance, and then

Fig. 37 A–D. Several models for the cross-bridge action. **A** Model proposed by HUXLEY (1969). He suggested that the force originates from a tendency of the myosin head (S1) to rotate relative to the thin filament, and is transmitted to the thick filament by the S2 portion of the myosin molecule acting as an inextensible link. Flexible points at each end of S2 permit S1 to rotate, and allow for variations in the separation between the filaments. **B** Development from **A** proposed by HUXLEY and SIMMONS (1971) to incorporate the elastic and stepwise-shortening elements for which they gave evidence derived from tension transients. The strength of the attached sites is higher in position 2 than in position 1, and in position 3 than in position 2. During isometric contractions, the myosin head oscillates rapidly between its three stable positions. The myosin head can be detached from

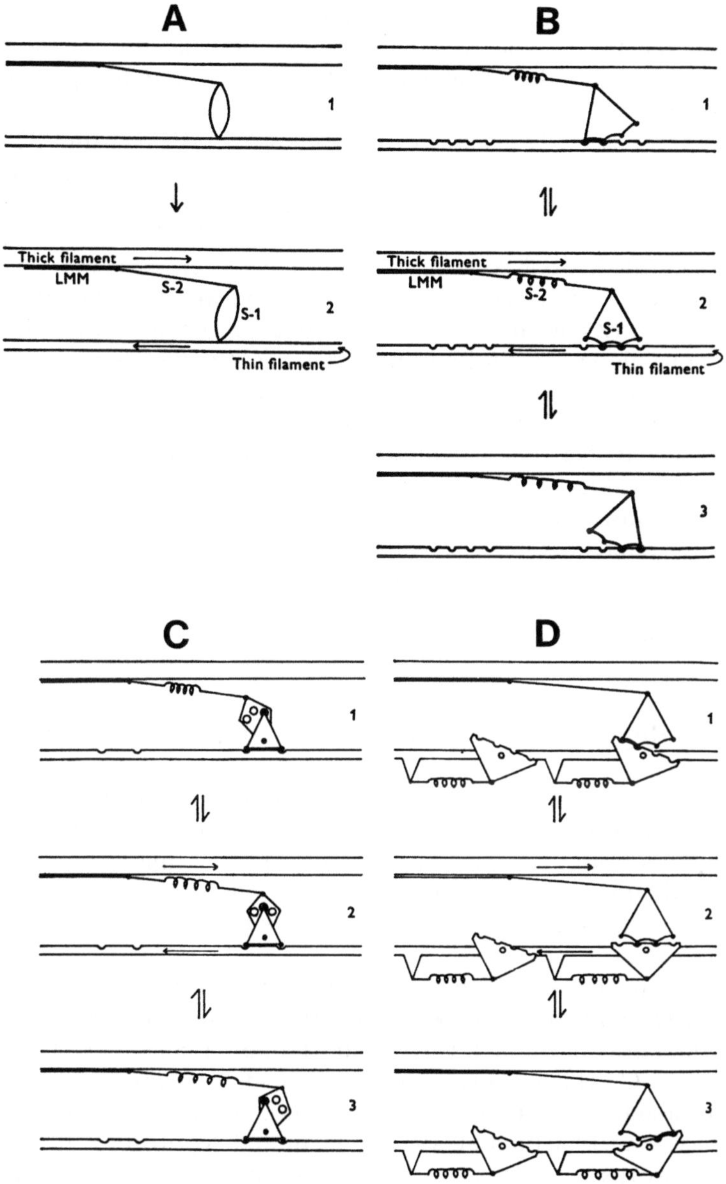

position 3 with the utilization of a molecule of ATP; this is the predominant process during shortening. During stretch, the myosin head can dissociate from position 1 without utilization of ATP. **C** Alternative to **B** as a structural basis for HUXLEY and SIMMONS' (1971) elastic and stepwise components. The attachment to the thin filament is rigid, and the stepwise movement takes place within the myosin head. The elastic element is in S2 as in **B**. **D** Another alternative to **B**. The stepwise movement is produced in the same way as in **B**, but the elastic element is provided by rotation, relative to the rest of the thin filament, of the piece to which the myosin head attaches. (Diagrams and explanations composed from two figures in HUXLEY 1974, with permission of the author and the Physiological Society) bend. This requires a hinge mechanism, not only between LMM and HMM, but also between the S2 and S1 subunits of HMM. Bending of the attached S1 subunit in relation to the S2 subunit would move the actin filament in relation to the thick filament.

Whether force is really generated in the hinge between S1 and S2 subfragments is unknown. A.F. HUXLEY and SIMMONS (1971) and A.F. HUXLEY (1974) proposed that during the active bending, a negative elastic force is exerted, which swings the S1 subunit back after detachment (Fig. 37b–d). The quantitative aspects of the cross-bridge movements have been reviewed by HUXLEY (1974), HARRINGTON et al. (1979), and TREGEAR and MARSTON (1979).

Recently, SHEETZ and SPUDICH (1983) were able to demonstrate that HMM alone can move along F-actin filaments. The alga *Nitella* has giant cells which contain thick cables of actin filaments with unidirectional polarity. Rabbit HMM was attached to fluorescent beads, which were applied to the isolated actin cables of the alga cell. The myosin-coated beads moved in one direction along the actin cables with a velocity of about 5 μm/s. This is surprisingly close to the intrinsic speed of shortening of mammalian muscle (10–60 μm/s/sarcomere; for references, see CLOSE 1972), in particular when the low temperature of the in vitro system is taken into account. Also the movement of individual F-actin filaments interacting with solubilized myosin fragments has been demonstrated (YANAGIDA et al. 1984).

12. The Regulatory Proteins and the Action of ATP and Ca^{2+}

Ca^{2+} ions activate attachment of the cross-bridges, whereas splitting of ATP provides the chemical energy which in some way is transmitted into mechanical work. The cross-bridges are not attached in a resting muscle containing ATP and rather little free Ca^{2+}. Myofibrils and also isolated actin-myosin complexes get into a contracted state (rigor) in the absence of Ca^{2+} and ATP. Also glycerination causes rigor.

Tropomyosin (formerly tropomyosin B in contrast to tropomyosin A, now paramyosin) (EBASHI and NONOMURA 1973) has long and slender molecules, whereas troponin is more or less globular. Troponin consists of three components: troponin-C (which is involved in Ca^{2+} binding), troponin-I (which inhibits the myosin-actin interaction), and troponin-T (which binds to tropomyosin) (for references, see EBASHI et al. 1972; EBASHI and NOMOMURA 1973; OHTSUKI 1979). Ca^{2+} works as a de-repressor by blocking troponin-I and enables the cross-bridges to attach to actin. One troponin molecule inhibits the binding of myosin to 6–7 actin monomers, which roughly corresponds to the proportion of troponin to actin molecules in the thin filaments. The depression of the troponin effect by Ca^{2+} becomes permanent (is "frozen") when glutaraldehyde is added at an appropriate concentration (MIKAWA 1979). The regulation of muscle contraction by Ca^{2+} and the role of the non-myosin and non-actin proteins has been reviewed by EBASHI (1980).

Ca^{2+} stimulation does not produce outward movement of the myosin cross-bridges in vertebrate muscles stretched beyond filament overlap (YAGI and MAT-

SUBARA 1980), because the regulatory system lies in the thin filaments. In striated molluscan muscles, however, Ca^{2+} acts directly on the myosin of the thick filaments, and in fact many invertebrate muscles contain a myosin-linked regulatory system (KENDRICK-JONES et al. 1970; WALLIMANN and SZENT-GYÖRGYI 1981 a, b). Striated muscle of lobster resembles vertebrate muscles, in that muscle contraction is regulated by an actin-linked troponin-tropomyosin system; whereas, in insect muscles a myosin-linked regulatory system coexists with an actin-linked troponin-tropomyosin system (for references, see WEBER and MURRAY 1973; REGENSTEIN and SZENT-GYÖRGYI 1975; EBASHI 1980; KENDRICK-JONES and SCHOLEY 1981).

Release of attached cross-bridges (i.e. of S1 subfragments) is an energy-consuming process and requires ATP. ATP causes dissociation of actin and myosin in vitro, and glycerinated muscles "relax" when ATP is added and Ca^{2+} is removed. The continuous breaking of cross-bridges during the attachment cycles of contraction is asociated with the splitting of ATP. Attachment of S1 subfragments to actin greatly enhances the activity of the Mg^{2+}-dependent myosin ATPase. The cross-bridges are broken without energy consumption when a contracted muscle is passively stretched. H.E. HUXLEY's (1969) swinging cross-bridge model requires a conformational change to occur within the myosin molecule before or during bridge attachment because the bridges must swing back when detached. Some workers believe that two ATP molecules per cross-bridge cycle are split, one to drive the conformational changes before attachment, and the other to detach the myosin head again (TONOMURA et al. 1961; EBASHI et al. 1969; EBASHI and NONOMURA 1973). By contrast, A.F. HUXLEY and SIMMONS (1971) and A.F. HUXLEY (1974) designed a model according to which one ATP molecule would be enough to drive the system. If during tilting of the attached myosin head energy is stored, this elastic energy would bring about the backward swing of the myosin head after detachment.

13. Alternative Contraction Theories

Before 1954, it was generally believed that actomyosin formed continuous filaments which during contraction shortened by some form of folding or coiling. This theory was based on the fact that continuous actomyosin threads (WEBER 1934a, b) can be made to contract or relax by various means (for review, see HASSELBACH and WEBER 1955). Most authors abandoned the folding or coiling filaments when the sliding filament model was put forward. It is noteworthy that SCHMITT et al. (1947), in one of the first electron microscopic studies on muscle, emphasized that they failed to see filament coiling in contracted myofibrils.

Some authors regarded sliding of the thick and thin filaments as an epiphenomenon not directly related to force generation. Extra filaments were described, which connected the ends of thick and thin filaments (SJÖSTRAND and ANDERSSON-CEDERGREN 1957; SJÖSTRAND 1962, 1964; GALEY 1964; JAGENDORF-ELFVIN 1967; SJÖSTRAND and JAGENDORF-ELFVIN 1967), or the Z lines across the entire sarcomere (HOYLE et al. 1965; HOYLE 1969), or the ends of thick filaments and the Z lines (GUBA et al. 1968). It was supposed that coiling of these filaments caused the sliding mechanism. These theories were not developed to a point where they could be tested quantitatively. Several of the mechanisms in which extra filaments are involved were based on early electron micrographs which did not allow visibility of the structures claimed to be present (for review, see HUXLEY 1964).

Filamentous connections between the ends of the I filaments crossing the H band (S filaments) and possibly also between A filaments and Z discs ("connecting filaments") have been described in insect muscles (ULLRICK et al. 1977a; BULLARD et al. 1977), but there is no evidence for filaments of that type in vertebrate muscles. Highly stretched vertebrate muscles may show filaments within the gap between A and I band ("gap filaments") (LOCKER and LEET 1975, 1976a, b; LOCKER et al. 1976). It has never been shown beyond doubt that these filaments connect to thick or thin filaments, and one might speculate whether the "gap filaments" belong to the ubiquitous cytoskeletal system of the muscle fibre (see Sect. C.II).

The following experiment speaks against sliding as a secondary phenomenon: Individual myofibrils from glycerinated rabbit muscles were locally injured by a fine light beam of high energy; the injured sarcomere failed to contract when ATP was added, provided the A band had been completely destroyed. When only two-thirds of an A band were destroyed and one edge of it remained intact, half of the sarcomere contracted. After disruption of the I band close to the Z line, the released I filaments slid into the A band (STEPHENS 1965).

A special class of theories claims that the force is generated transversely, and that axial shortening of the muscle fibre is due to the constant volume behaviour of the system. ULLRICK (1967) assumes that the Z discs expand laterally, and SPENCER and WORTHINGTON (1960) and ELLIOTT et al. (1970) suggest that repulsive forces between the thick and thin filaments due to electrostatic charges cause the lateral expansion of the fibre. These theories are incompatible with the fact that skinned muscle fibres are still contractile, and that in skinned fibres the interfilament distances probably do not increase during shortening (MATSUBARA and ELLIOTT 1972).

Two types of theories for the force generation in a sliding filament system are distinguished (HUXLEY 1974): "theories of maximization of the number of interaction sites" and "theories of independent force generators". The swinging-cross bridge model belongs to the second type. The maximization theory (HUXLEY and HANSON 1954) implies that the filaments only move so as to increase the number of sites at which chemical interaction between actin and myosin can exist, which would be at a sarcomere length of about 1.6 μm. Nevertheless, the sarcomeres continue to shorten also after this sarcomere length is reached. In a live frog muscle fibre shortening without restraint, the I filaments finally overlap in the middle of the sarcomere and form the C_M band (HUXLEY and NIEDERGERKE 1954); the fibre continues to shorten to 1.3-μm sarcomere length until crumpled A filaments form C_Z bands (GORDON et al. 1966). Passive shortening by axial compression renders the myofibrils wavy, but the thin filaments do not pass across the M line, and no C_M and C_Z bands are formed (BROWN et al. 1970; HUXLEY 1974). Invertebrate muscles may shorten even more, and the filaments may pass straight through the Z line into the next sarcomere. This is uncommon for vertebrate muscles, but in the tongue muscle of the chameleon the same phenomenon has been seen. In this muscle, the sarcomere length may vary by a factor of six although the length of the filaments is the same as in other vertebrate muscle (RICE 1973). These observations, together with the fact that individual HMM molecules move along actin filaments (SHEETZ and SPUDICH 1983), invalidate the theory of "maximization of the number of interaction sites".

Within the frame of the sliding filament mechanism, it has been proposed that the force is generated not by attachment and release of the cross-bridges, but by generation of electric charges of opposite sign (INGELS and THOMPSON 1966; YU et al. 1970), or by rearrangement of protein molecules or of water molecules within the filaments (WARNER 1970).

PROCHNIEWICZ-NAKAYAMA et al. (1983) observed changes in the polarization of light during active contraction of muscles in which the actin filaments had been labelled with fluorescent phalloidin. The authors conclude that conformational changes in F actin are involved in the process of tension development.

Many of the alternative contraction theories appear to be entirely out of line with the experimental evidence available, or they rest on now-obsolete data. Nevertheless, the fact that the sliding filament theory has dominated the discussion for 30 years, and has proved to be a powerfully effective tool for planning and interpreting experiments, entails the danger that divergent findings are neglected. POLLACK (1983) reviewed many conflicting studies, in particular those which report changes in filament length, and emphasized the inherent danger of "Bellman's principle": what I tell you three times is true (see also HUXLEY 1974).

14. The M Line

The M line marks the middle of the sarcomere; under favourable conditions it is visible using light microscopy. The M line is about 75 nm wide and on both sides is bordered by the low-opacity pseudo-H zone. KNAPPEIS and CARLSEN (1968) demonstrated that the M line represents a system of 3–5 transverse bridges with an axial distance of 20 nm. These bridges connect each A filament with its six neighbours (Fig. 38). In addition, each set of bridges is reported to be linked by short M filaments running parallel to and midway between the A filaments (Fig. 39). Stretch of the muscle fibre does not alter this array, but at high degrees of stretch the M line becomes indistinct due to staggering of the thick filaments. The major features of this model have been confirmed in recent studies using frozen sections for electron microscopy (LUTHER and SQUIRE 1978), or image reconstruction techniques (Fig. 40), which suggests that there are additional bridges cross-linking the M filaments directly (LUTHER and CROWTHER 1984; CROWTHER and LUTHER 1984). The M line has been suggested by many authors to maintain the orientation of adjacent thick filaments in relation to each other. In muscles which do not have M lines, the thick filaments tend to be slightly out of register (see Chap. E).

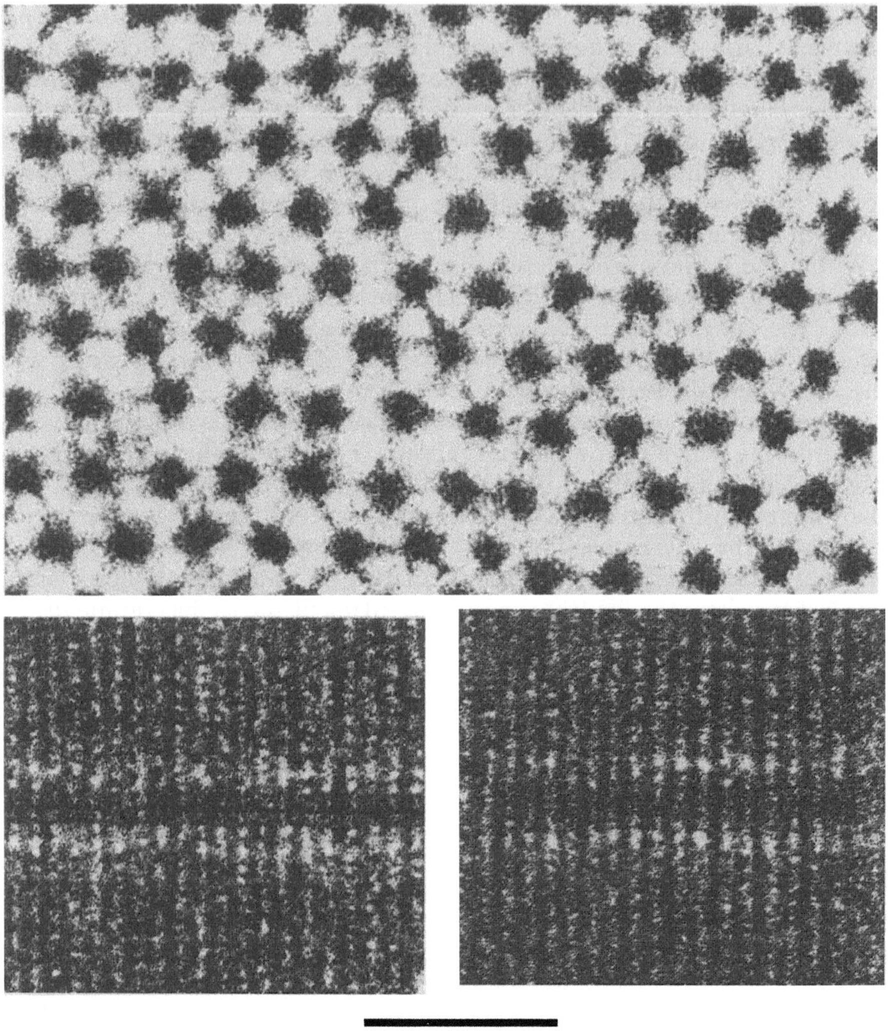

Fig. 38. Human brachial biceps muscle. M line in cross- and longitudinal sections. The thick filaments are arranged in an hexagonal array and each is connected to its six neighbours by six M bridges. The longitudinal sections are from fibres poor (*left*) and rich (*right*) in mitochondria, which presumably were fast- and slow-twitch fibres, respectively. The M lines consist of three and five systems of bridges which are best seen when viewed at a shallow angle along the M lines (with respect to the differences in M line structure in fast- and slow-twitch muscle fibres, see Sect. D.IX). Bar, 0.1 μm (cross-section), 0.25 μm (longitudinal sections)

Specific M-line proteins have been demonstrated by antibody staining of intact myofibrils (PEPE 1966) and by extraction of the protein forming the M line (KUNDRAT and PEPE 1971). The M lines are restored when the extracted proteins and the rest of the myofibrils are allowed to interact (STROMER et al. 1969). Several M-line proteins have been characterized (MASAKI et al. 1968; MORIMOTO and HARRINGTON 1972; CHOWRASHI and PEPE 1979; WALLIMAN et al. 1979;

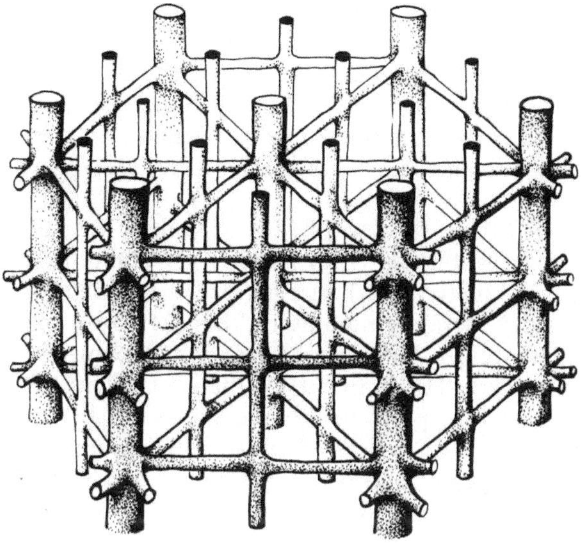

Fig. 39. M line model as proposed by KNAPPEIS and CARLSEN (1968). Each thick filament is linked to its six neighbours by 3–5 M bridges which are interlinked by short longitudinal M filaments

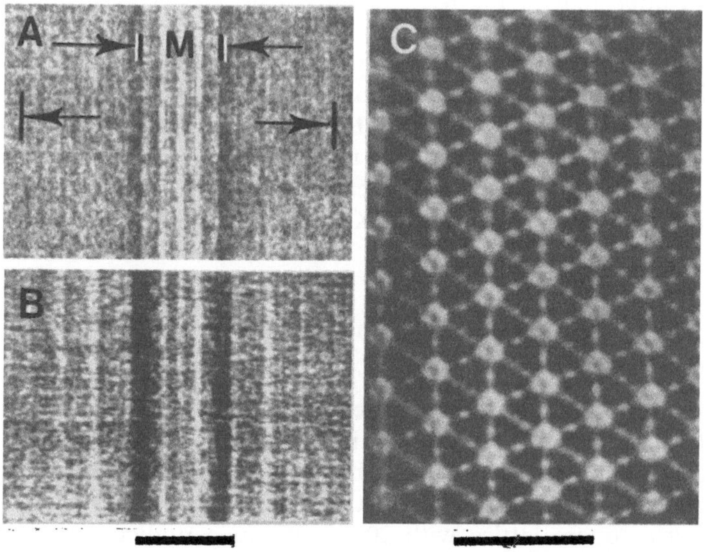

Fig. 40 A, B. Negatively stained frozen sections of the M zones of human (**A**) and frog (**B**) muscle fibres. The M lines (*M*) consist of five- and three-bridge systems, respectively. Note that the bridges with this technique appear white. **C** Cross-section through the M line of a fish muscle. The micrograph has been optically filtered and printed as negative (compare with Fig. 38). The M bridges and also the M filaments (Fig. 39) are distinct. In places, additional bridges linking the M filaments directly can be detected. Bars, 0.1 μm. (**A**, **B** from SQUIRE et al. 1982, with permission of the author and the Royal Microscopical Society. **C** from SQUIRE 1981, with permission of the author and Plenum Press)

GROVE et al. 1984). A globular protein with a mol. w. of about 90,000 isolated from M line extracts causes the aggregation of myosin filaments (MORIMOTO and HARRINGTON 1972). Creatinkinase has been shown to be a structural protein of the M line as well. To date, three proteins have definitely been associated with the M line bridges: MM-CK (M-creatinkinase), a 165-kdalton protein (M protein), and a 185-kdalton protein (myomesin) (for references, see GROVE et al. 1984).

Several bridges and bands are distinguished within the M zone. These are best seen in negatively stained frozen sections (Figs. 32, 40). The middle of the sarcomere is occupied by the M1 line (nomenclature according to SJÖSTRÖM and SQUIRE 1977a) and on both sides there are nine other lines (M2–M9) plus some weak "sublines". These lines occupy the entire M zone which comprises the M line and the bare region (pseudo-H) of the thick filaments. The M1, M4, (+M4′), and M6 (+M6′) lines correspond to the five transverse bridges of the M line. The other lines (and "sublines") are assumed to reflect the packing pattern of the tails of the myosin molecules and possibly also the localization of unidentified non-myosin molecules (SJÖSTRÖM and SQUIRE 1977b; SQUIRE et al. 1982).

It has long been known that in slow (tonic) muscle fibres of amphibia and birds, and of extraocular muscles of mammals, the M line is either indistinct or lacking (HESS and PILAR 1963; PEACHEY and HUXLEY 1962; PAGE 1965, 1969; PEACHEY 1968). SJÖSTRÖM and SQUIRE (1977b) describe differences in the relative intensities of the M1, M4, and M6 lines of fast- and slow-twitch fibres and of fibres of different species. The authors state that "the M-region appearance is a characteristic of the species, muscle and fibre type from which the section has been taken". This statement refers to the M bridges only, whereas the array of weak lines believed to reflect the myosin packing is similar in all fibres. SJÖSTRÖM et al. (1982a, b) use the M line structure as seen in routine sections for electron microscopy as a criterion for distinguishing human muscle fibres of different types (see Sect. D.IX).

15. The Z Disc

Whereas there is little disagreement with respect to the structure of the M line, the structure of the Z disc is still a matter of controversy.

The thin filaments within the A band are in a hexagonal array, dictated by the array of the thick filaments; they are without an order in the I band, and display a square lattice with 20- to 25-nm periodicity as they approach the Z disc (Fig. 19). The squares are displaced on both sides of the Z disc by half the lattice period in both *x*- and *y*-axes. Thus, each thin filament is positioned opposite the centre of a square formed by the ends of four thin filaments from the next sarcomere. The width of the Z disc, as seen in longitudinal sections, ranges from 50 to 150 nm depending on the fibre type and the animal species. The Z line often appears amorphous; the approaching thin filaments seem to thicken before they merge with the Z disc itself (Figs. 41–44). The Z line has no really defined border and measuring its width entails an element of arbitrariness.

The first model for the Z disc was proposed by KNAPPEIS and CARLSEN (1962), who studied frog muscle fibres. They found that each thin filament inserts at four short Z filaments which run obliquely through the Z disc, deviating 10° from the direction of the fibre axis, and connect to four thin filaments from the opposite side (Fig. 42 A). In cross-sections, the array of Z filaments is tilted 45° with respect to the squares formed by the I filaments. Hence, cross-sections through the Z disc show an angled tetragonal lattice with 16-nm periodicity (Fig. 43). Stretch does not change the structure of the Z disc. The variable appearance of the Z lines in different micrographs of the same muscle fibres was attributed to variations in orientation and section thickness. KATCHBURIAN et al. (1973) confirmed this assumption by tilting sections of rat muscle fibres in the electron microscope. FRANZINI-ARMSTRONG and PORTER (1964c) arrived at a similar model for a variety of invertebrate and vertebrate muscles. They proposed that the structures connecting the thin filaments of adjacent sarcomeres are not oblique Z filaments, but that the Z disc is composed of a lamina stretched into opposite directions by the inserting thin filaments. This would give rise to the same zig-zag appearance in longitudinal sections as an array of oblique Z filaments, and also produce the angled tetragonal lattice in cross-sections. The difficulty of explaining how an actin filament representing a double helix can split into four subunits would thus be circumvented. REEDY (1964a, b) demonstrated in cross-sections a weave-like appearance of the Z filaments, and modified the KNAPPEIS and CARLSEN (1962) model, in that he assumed that each thin filament untwists and frays into four Z filaments, the sense of twist being the same in all filaments approaching the Z disc from one side, and opposite in those of the opposite sarcomere (Fig. 42 B). That model deviates only in details from the Knappeis and Carlsen model, but explains the large "woven" pattern sometimes seen in cross-sections. KELLY (1967, 1969) proposed a radically new model to explain the paradox that a two-stranded helix gives off four subfilaments. He assumed that from each I filament only two subfilaments arise which run through the Z line, interlink with the subfilaments of opposite thin filaments, then return to the side of origin, and again enter the double helix of a thin filament. At each thin filament tip one would see two originating Z filaments and both the "afferent" and the "efferent" parts of a loop, i.e. four Z filaments in all.

The early "non-looping" models were based on investigations for which the muscles had been fixed with osmium tetroxide (KNAPPEIS and CARLSEN 1962; FRANZINI-ARMSTRONG and PORTER 1964c) or glutaraldehyde (REEDY 1964a, b; KATCHBURIAN et al. 1973). LANDON (1970a) and MACDONALD and ENGEL (1971), however, found that the image depended on the method of fixation. Glutaraldehyde should produce an "11-nm square lattice pattern". According to MACDONALD and ENGEL (1971), only a "looping" model is compatible with the image after glutaraldehyde fixation, whereas the "non-looping" KNAPPEIS and CARLSEN (1962) model only accounts for the osmium tetroxide image. LANDON (1970a, 1982) explained the "11-nm square lattice" pattern using a non-looping model: the thin filaments of adjacent sarcomeres overlap

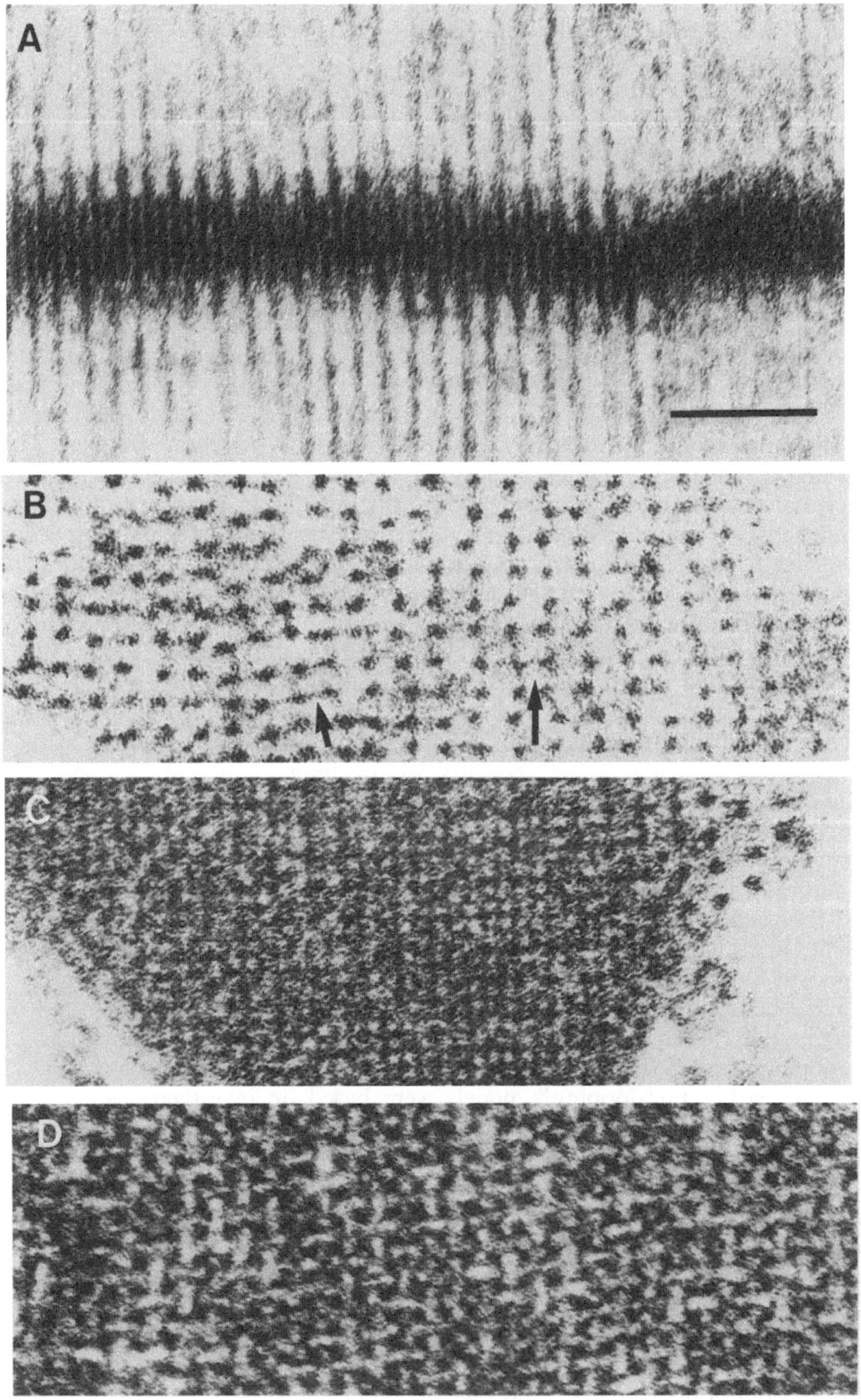
A
B
C
D

each other and their ends are cross-linked by transverse filaments. Depending on the orientation, longitudinal sections would then either show interdigitating thin filaments within the Z line, or thin filaments that seem to run continuously across the Z line. At variance with KNAPPEIS and CARLSEN (1962), LANDON (1970b) found different Z-line configurations in resting and contracted fibres. Resting muscle fibres of rat showed an "11-nm square lattice", whereas tetanized fibres in places showed a coarse "woven lattice", also when they had been fixed with glutaraldehyde.

ROWE (1971) proposed a model in which four subfilaments (two actin strands and two tropomyosin strands) arise from each thin filament, return after looping, and insert at the same side of the Z disc in other thin filaments. The different images seen in sections were explained by different levels of sectioning and by mismatches between the plane of sectioning and the square lattice. In "red" fibres of rat muscles, the Z disc is thicker than in "white" fibres. ROWE (1973) claimed that his four-stranded looping model applies to all types of Z lines, and that the different widths of the Z line are due to the fact that in "white" fibres all loops from one side of the Z line are in the same plane, in "red" fibres they are in three planes, and in intermediate fibres in two planes.

FRANZINI-ARMSTRONG (1973b) confirmed the KNAPPEIS and CARLSEN (1962) model for the very distinct Z line of a fish muscle, and found no proof for the claim that the type of fixation changes the image seen in sections. ULLRICK et al. (1977b) studied a variety of muscles, including human muscle, and concluded that the "11-nm square lattice" is a fixation artefact. The authors found only the "woven lattice" pattern, and explained this by a model in which each thin filament, as it enters the Z line, is continuous with three curved strands which, after looping, unite with other I filaments of the same sarcomere.

Thus, in addition to the still viable KNAPPEIS and CARLSEN (1962) model with four non-looping Z filaments connecting to each thin filament, there are various looping models with two, three, or four Z filaments connecting to each I filament. LANDON'S (1970a) model with transverse bridges is a special form of a non-looping model.

Z discs of insect muscles are basically different from those of vertebrate muscles. Cross-sections show a hexagonal rather than tetragonal array; also tubular structures have been found (AUBER and COUTEAUX 1963). Furthermore, some authors find connecting filaments between A filaments and Z line which, according to others, are non-existent (ASHBURST 1967; TROMBITAS and TIGYI-SEBES 1975; ULLRICK et al. 1977a).

The problem with the Z discs of vertebrate muscles is that cross-sections show at least three different images (Fig. 43): (a) an angled "large woven pat-

Fig. 41 A–D. Human brachial biceps muscle, Z disc. **A** Longitudinal section. The thin filaments of two adjacent sarcomeres become thicker, enter the Z line, and seemingly interdigitate. They are linked by oblique and transverse structures. **B** Cross-section through the I band, immediately adjacent to the Z disc. The thin filaments show an orthogonal array. In places amorphous material adheres to the filaments and also connects the filament tips (*arrows*). **C, D** Cross-sections through the Z disc showing the "small square lattice" (**C**) or the "large woven lattice" (**D**). Note that the magnification is the same in all four micrographs. Bar, 0.1 μm

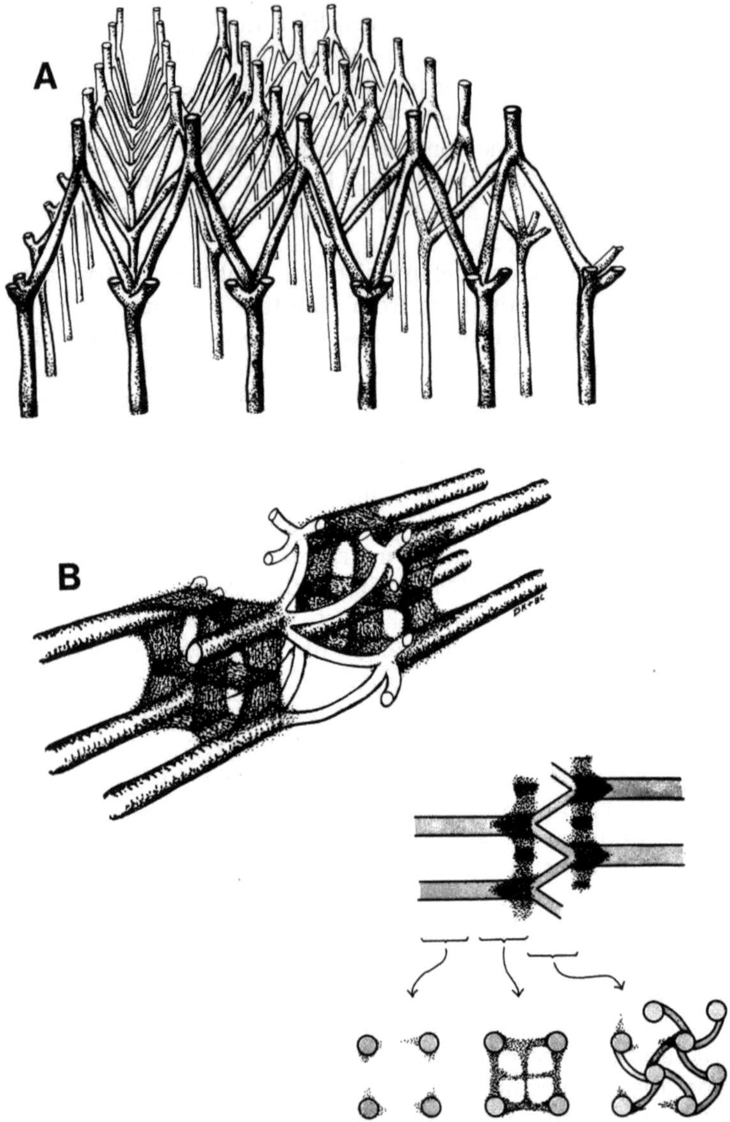

Fig. 42 A, B. The non-looping Z-disc models of Knappeis and Carlsen (1962) and Kelly and Cahill (1972). **A** According to Knappeis and Carlsen (1962), the thin filaments give rise to four obliquely running Z filaments which are linked to the thin filaments of the next sarcomere (adapted from Knappeis and Carlsen 1962). **B** Kelly and Cahill (1972) keep the filament array as in **A**, although it is modified according to Reedy (1964a, b) to account for the "large woven pattern" (*bottom right*). In addition, it is assumed that matrix material is deposited in a squared array between the actin filament tips in a way that would produce the "small squared pattern" (*bottom middle*). For the sake of clarity, the matrix substance has been excluded from the centre of the Z-disc diagrams. The lateral view of the model has been oriented to illustrate the different longitudinal appearances in sections. The model does not assume that the filament tips interdigitate (Fig. 41 A), but that shelf-like deposits of matrix in the centre of the large squares project as continuation of a filament from the opposite sarcomere. (**B** from Kelly and Cahill 1972, with permission of the authors and the Wistar Institute Press)

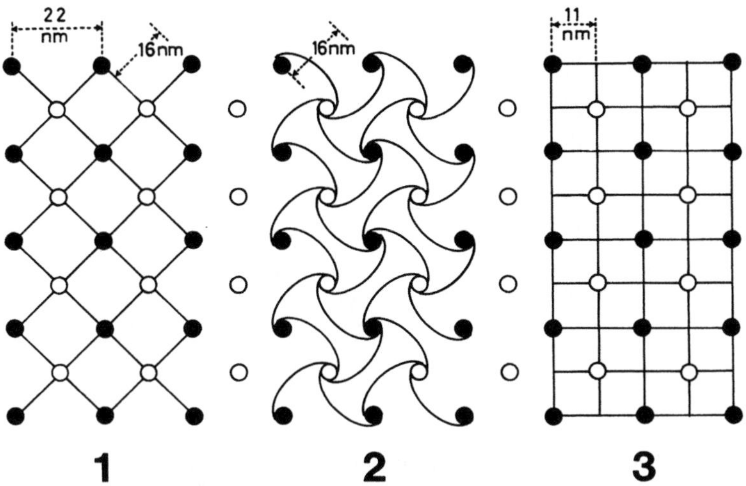

Fig. 43. Schematic illustration of the three most frequently seen images in cross-sections through the Z disc. *Open* and *filled circles* represent the tips of the thin filaments of two adjacent sarcomeres. The filaments are in an orthogonal array with 22-nm lattice and those of the two sarcomeres are off-set in both axes by half the lattice period. The large "square lattice" (*1*) is tilted 45° in relation to the filament lattice, and its periodicity is 16 nm. It is explained by diagonally running Z filaments projecting into the cross-sectional plane. The "large woven lattice" (*2*) is very similar, and it is assumed that each thin filament untwists to give rise to 4 Z filaments, and that the screw sense is the same in all filaments of one sarcomere and opposite in all filaments of the next sarcomere. The lattice period of the "large woven lattice" is also 16 nm. The "small square lattice" (*3*) has half the period of the actin filament lattice (11 nm) and it is oriented parallel to the filament lattice. In addition, a "parallel large square lattice" may be found in sections through the edge of the Z disc. It is probably an incomplete "small square lattice". Whether one or the other image is found, according to KELLY and CAHILL (1972), depends on the level at which the Z disc is cut, on the conditions of fixation, and on the animal species. All three images have been explained by the model depicted in Fig. 42B

tern"; (b) an angled "large square pattern", each with 16-nm periodicity; and (c) a "small square lattice" with 11-nm periodicity. The large woven or square patterns arise from two sets of 22-nm squares connected by bent or straight Z filaments running diagonally from the I filaments of one side of the Z disc to the I filaments of the other side of the Z disc. The lattice is oriented at 45° angle with respect to the rows of dots representing the ends of actin filaments. The "small square lattice" is oriented parallel to the nearby rows of aligned actin filament dots. Longitudinal sections show either seemingly continuous I filaments, or I filaments interdigitating in the Z line, or I filaments inserting in a zig-zag structure.

KELLY and CAHILL (1972) emphasize the possibility that components of the Z discs may be solubilized during the fixation and embedding procedures. They studied, in serial sections, fast muscle fibres of frogs and newts fixed using different methods. The components of the Z disc were successively digested. Cross-sections not only showed the angled large 16-nm square or woven lattices, and the small 11-nm square lattice, but also a large 24-nm square

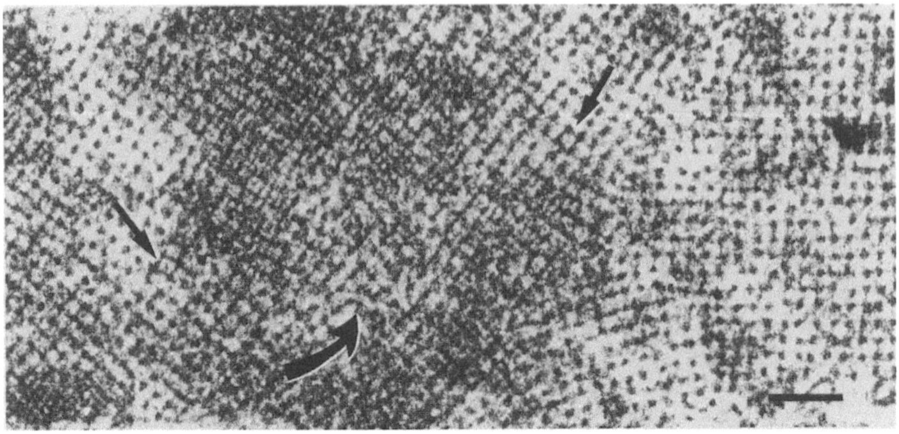

Fig. 44. Human brachial biceps muscle. Slightly oblique section through the Z disc. The approaching thin filaments are seen on the right and left sides. The matrix material deposited between the tips sometimes forms "parallel large squares" with 22-nm side length (*straight arrows*). Most of the section through the Z disc shows the "small square lattice" with 11-nm side length. In the middle of the Z disc (*bent arrow*), no regular pattern is seen. Note the variability of the orientation of the squares formed by the filament tips. Bar, 0.1 µm

lattice with sides parallel to the dots representing the actin filaments (Fig. 44). The different configurations were found in the same section. This is at variance with the claim that the different images are typical for a certain procedure of fixation or relate to the state of contraction. KELLY and CAHILL (1972) suggest that the Z disc consists of oblique Z filaments and a matrix; the filament pattern is to a large extent independent of fixation, whereas the matrix may be partly or completely extracted under conditions of primary osmium fixation. The tendency of the matrix to become extracted varies in different species. Cross-sections passing through the region where the I filaments approach the Z disc show a parallel lattice with either 24-nm or 12-nm periodicity. A large angled (often woven) lattice is seen in sections passing through the Z disc proper. Sometimes the Z disc consists of a dense meshwork within which no lattice can be detected, indicating that the matrix is preserved. The authors arrive at a model that includes the original KNAPPEIS and CARLSEN (1962) Z filament pattern (Fig. 42); in addition, matrix material is deposited in a squared array between the actin filament tips as proposed by LANDON (1970a). Apparent interdigitation of the filament tips is attributed to a shelf-like deposition of matrix in the same plane as the actin filaments; therefore, the tips of the interdigitating filaments seen in longitudinal sections are in reality projections of shelf-like deposited matrix lamellae (Fig. 41 A). KELLY and CAHILL (1972) believe that also a looping model could explain the oblique filaments in the centre of the Z disc, giving rise to the large angled lattice; they do not exclude the possibility that the matrix strands giving rise to the parallel large and small square lattices are precipitation products of fixation, i.e. artefacts. This study is the most complete account of the problem. Although there is no evidence

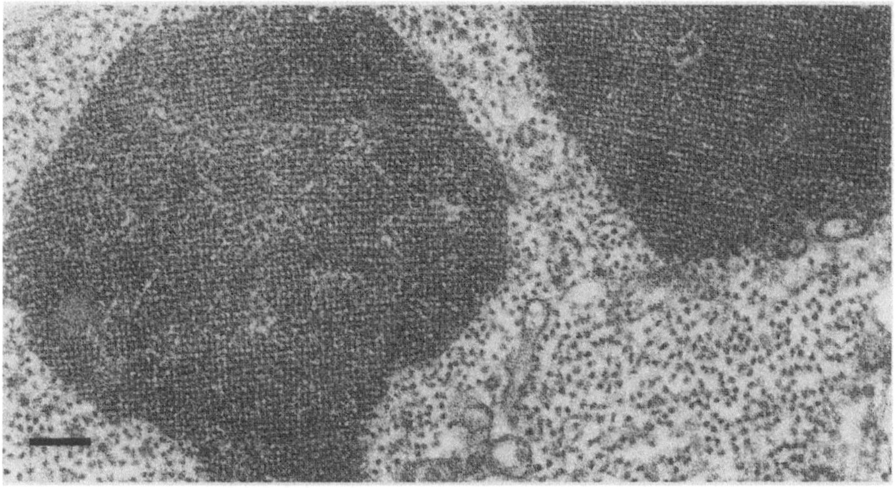

Fig. 45. Cross-section of a fibre from the medial vastus muscle of a child with congenital nemaline myopathy. The micrograph shows two nemaline bodies between cross-cut thin filaments. The nemaline bodies show a "small square lattice" comparable with that in Fig. 44. They represent α-actinin-actin complexes. Bar, 0.1 μm

against looping Z filaments, there is very little direct evidence for such loops. The advantage of the KELLY and CAHILL (1972) model is its simplicity and the fact that it explains all observations in electron micrographs.

Originally, tropomyosin was suspected to be the constituent of the Z discs because pure protein crystals showed a lattice similar to that in native Z discs (HUXLEY 1963; EBASHI and KODOMA 1966; RASH et al. 1968). STROMER et al. (1969), however, demonstrated that the crystal lattice of tropomyosin differs from the Z-disc lattice. Crude muscle extracts not containing tropomyosin are able to reconstitute solubilized Z lines (STROMER et al. 1967, 1969); the extract obtained by solubilizing the Z lines contains α-actinin (BRISKEY et al. 1967; GOLL et al. 1969), a protein that cross-links actin polymers (EBASHI and EBASHI 1965). α-Actinin has subsequently been shown to amount to about 50% of the mass of the Z disc (SUZUKI et al. 1976). A similar protein is present in non-muscle cells in the electron-dense zones (dense bodies) connecting the cytoplasmic actin filaments. When actin filaments of muscle or non-muscle cells are "decorated" with HMM-S1 subfragments, they always point away from the Z lines or the dense zones, respectively (ISHIKAWA et al. 1969). It is conceivable that α-actinin is involved in the organization of the actin filaments, and that it is responsible for the polarization of the thin filaments within the sarcomeres of muscle fibres (HUXLEY 1963; GOLL et al. 1972).

The interaction of actin and α-actinin is illustrated by the composition of nemaline or rod bodies, peculiar structures formed by muscle fibres under pathological conditions. These bodies are rather electron dense, develop in association with the Z lines, and display a crystalline structure. In one plane they resemble cross-sections through Z lines showing the "small square lattice" (Fig. 45), in others they reveal parallel filaments with a transverse periodicity of 12–20 nm (ENGEL 1966; ENGEL and GOMEZ 1967). HMM-S1 binds to the filaments within the rods indicating that they contain actin (YAMAGUCHI et al. 1978), and antibody staining reveals that the rods contain α-actinin as well. A Ca^{2+}-activated protease which is known to dissolve the Z lines of muscle fibres almost selectively (BUSCH et al. 1972) dissolves the matrix of the rods but leaves the actin filament lattice intact (STROMER et al. 1976).

16. The Turnover Rates of Myofibrillar Proteins

The various proteins of the myofibrils have different turnover rates. In rabbit muscles the following life times have been measured: actin 75 days, myosin 29 days, α-actinin 24 days, tropomyosin 22 days, troponin 12 days, M protein 11 days (KOIZUMI 1974).

17. Helicoidal Sarcomeres

The cross-striation pattern of sarcomeres, as seen using light microscopy, in isolated fibres and in sections may be irregular, to the extent that a particular sarcomere plane does not traverse the entire fibre but appears incomplete. This gives rise to wedge-shaped "sphenoden" and Vernier-type shifts (HEIDENHAIN 1919). Along a given distance, one myofibril has one sarcomere more than an adjacent one. These configurations have been interpreted by some authors to indicate that the sarcomeres form a screw, and that the Vernier shifts arise when the section plane passes through the axis of the screw (for references, see HÄGGQUIST 1956; RUSKA and EDWARDS 1957). ENGELHARDT and POPP (1953) and ENGELHARDT (1955) rotated individual muscle fibres under the microscope, and claimed to have seen that the sarcomeres of the whole fibre always form a left-handed screw; they related this to the asymmetry of living organisms. HUXLEY and TAYLOR (1958) saw Vernier-type shifts in living frog muscle fibres when they studied the spread of local activation across the muscle fibre (see Sect. C.III.1). The local contraction of one sarcomere plane, at the site of the shift, spread to the two sarcomeres in continuation of the excited sarcomere. This assured that the Vernier shifts are not artefacts produced by distortion of the fibre, as claimed by BUCHTHAL and KNAPPEIS (1940). SCHMALBRUCH (1968) distinguished two types of Vernier shifts: one was due to the fact that adjacent myofibril bundles were cut at slightly different angles, so that two series of sarcomeres projected at different lengths; the other one was apparently due to branching of a sarcomere plane. In the latter type, the T systems branched in the same fashion as one would expect from the observations of HUXLEY and TAYLOR (1958). This type corresponded to HEIDENHAIN'S (1919) wedge-shaped "sphenoden". SCHMALBRUCH (1968) speculated that the branching sarcomere planes were due to sarcomere "splitting" and represented sites of longitudinal growth (see Sect. G.V.2).

The problem of the helicoidal structure of muscle fibres has been reinvestigated by PEACHEY and EISENBERG (1978). They filled and labelled the T system of frog muscle fibres with peroxidase, and studied thick serial cross-sections using high voltage electron microscopy. According to this study, adjacent sarcomeres are connected by sloping ramps, much the way successive floors of a multilevel garage are connected by ramps. These ramps constitute helicoids with right- and left-hand screw direction when seen along the fibre axis. Up to eight helicoids with both directions have been found in the same fibre cross-section. All of the 66 helicoids assessed were completed in the length of one sarcomere. These observations invalidate the hypothesis that the sarcomeres form a continuous screw running the entire fibre length, but are compatible with the observations by HUXLEY and TAYLOR (1958), and also with SCHMALBRUCH'S (1968) speculation that incomplete sarcomere planes (the "ramps") arise by "splitting" of the sarcomeres.

II. Cytoskeletal Elements

The cytoskeletal elements of skeletal muscle fibres are apparently scarce in routine preparations, and microtubules and 10-nm filaments have only occasionally been observed (SANDBORN et al. 1967). Nevertheless, there is indirect evidence that a muscle fibre is not just a tube filled with myofibrils and aqueous cytoplasm. Several recent studies describe an elaborate sarcomere-related system which is believed to link the sarcolemma, and, by means of fibronectin (see Sect. G.IV.5), also the interstitial connective tissue to the myofibrils. Adjacent sarcomeres and even the sarcomeres of adjacent fibres, at different degrees of stretch, remain in register (BUCHTHAL and KNAPPEIS 1940). A fibre is still able to transmit a large fraction of force after the chain of contractile elements has been interrupted (STREET and RAMSEY 1965) (see Sect. B.IV.1). This indicates that force is transmitted not only at the myotendinous junctions, but along the entire length of the fibre. In invertebrate muscle fibres with few myofibrils and much sarcoplasm, light microscopy of longitudinal sections occasionally shows dense lines at the Z-line level that pass through the sarcoplasm (HÄGGQUIST 1956). These "lines" suggest that there exist Z-disc-related structures which are not part of the myofibrils.

Intermediate filaments have been found at the Z-line level (KELLY 1969; PAGE 1969; PIEROBON-BORMIOLI 1981; THORNELL et al. 1980, 1983) (Fig. 46). NUNZI and FRANZINI-ARMSTRONG (1980) distinguish two sets of transverse filaments: short radially oriented ones connecting the Z discs to sarcoplasmic reticulum and transverse tubular system membranes, and long filaments running tangentially to the myofibrils and connecting the fibrils with each other and to the plasma membrane. The link between plasma membrane and external lamina of the sarcolemma may then be established by the periodic filaments within the glycocalyx of the plasma membrane that have been described by BONILLA (1983) (see Sect. C.IV).

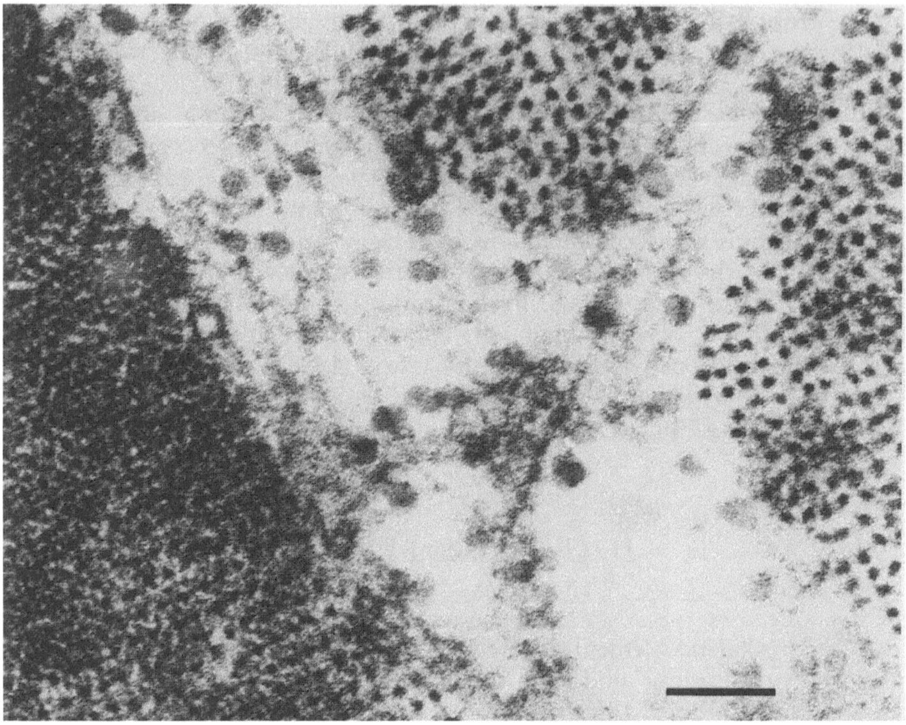

Fig. 46. Human brachial biceps muscle, cross-section at the level of the Z disc. Note transversely arranged filaments apparently connecting the myofibrils. Glycogen granules are visible as well. Bar, 0.1 μm

The "intermediate filaments" of the cytoskeleton do not represent a homogeneous class of proteins. At least five different proteins are now distinguished (for review, see LAZARIDES 1980, 1982). When smooth muscle cells are treated with appropriate salt solutions to extract myosin and actin, a filament network connecting to the dense bodies, the analogues of the Z discs of striated muscle, is left behind. In addition to the α-actinin of the dense bodies, this filament system contains cytoplasmic actin and desmin, a 55,000-dalton protein (COOKE 1976; HUBBARD and LAZARIDES 1979). Labelled antibodies against desmin stain

Fig. 47 A–D. Myofibrils of rabbit muscles which have been extracted with KI to show the array of cytoskeletal filaments. **A**, **B** Scanning electron micrographs showing transverse bridges at the level of Z and M lines (*TZ*, *TM*) connected by longitudinally running filaments (*LZ*), which seem to ensheathe the sarcomere. **C** Transmission electron micrograph taken at low magnification. The cytoskeletal systems of two myofibrils are shown. The Z discs are surrounded by doublet rings of filaments (*DZ*). The myofibrils are transversed at the Z-disc level by internal filaments (*IZ*) which have been related to the N_1 lines of the I bands. The longitudinal filaments (*LZ*) and the connecting systems at the Z and M levels (*TZ*, *TM*) are visible (for comparison see **A** and **B**). **D** A highly schematic diagram of the sarcomere-related cytoskeletal system. (From WANG and RAMIREZ-MITCHELL 1983, with permission of the authors and the Rockefeller University Press.) (Original plates kindly provided by Dr. WANG, Austin, Texas)

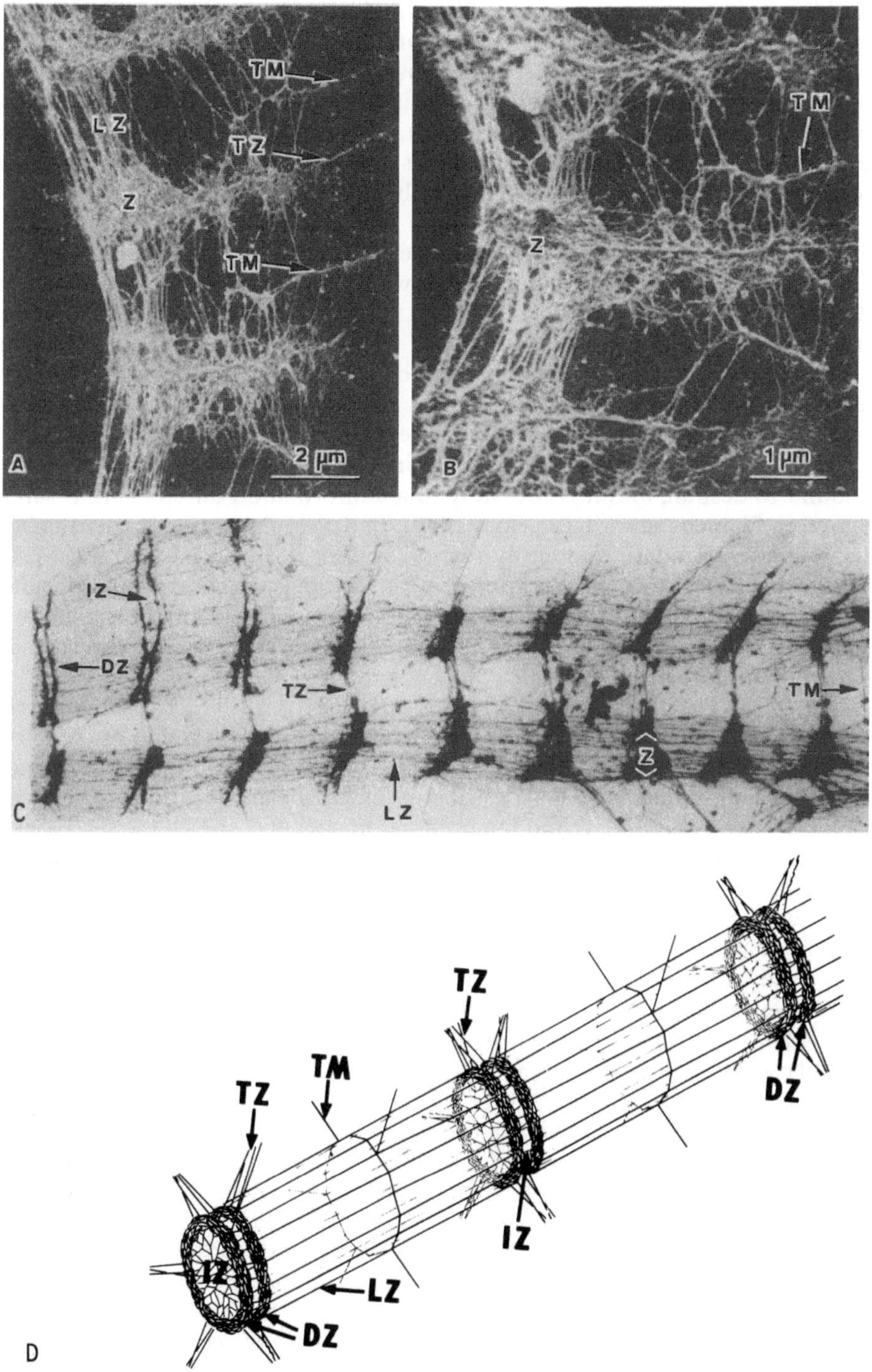
TM
LZ
TZ
Z
TM
A
2 μm
TM
Z
B
1 μm
IZ
DZ
TZ
TM
Z
LZ
C
TZ
TM
TZ
DZ
IZ
LZ
DZ
IZ
D

the Z lines of cross-striated muscle fibres and also the sites where the Z discs are close to the plasma membrane (LAZARIDES and HUBBARD 1976; LAZARIDES and GRANGER 1978; GARD and LAZARIDES 1979). Honeycomb-like sheets of Z discs are obtained when myosin and actin are extracted from chicken muscle fibres and the fibres are then blended (GRANGER and LAZARIDES 1978). These sheets are well suited to study the spatial distribution of the different cytoskeletal proteins using immunofluorescence. Antibodies against actin stain the entire Z plane; antibodies against α-actinin stain only the individual Z discs, whereas antibodies against desmin produce a fluorescent pattern complimentary to that of α-actinin. The same complimentary network surrounding each Z disc is seen after antibody staining for vimentin (GRANGER and LAZARIDES 1979), and for synemin (GRANGER and LAZARIDES 1980), which also are intermediate filament proteins. Synemin is a 230,000-dalton protein closely associated with desmin and vimentin filaments. Thus, the Z discs are composed of two molecular domains, a peripheral one which contains desmin, vimentin, synemin, and actin, and a central one which contains α-actinin and actin. The actin of the Z disc is bound to desmin, and it is probably in a conformation different from sarcomeric thin filament actin (LAZARIDES 1980). In developing muscles, intermediate filaments are abundant and easily seen in routine electron micrographs (ISHIKAWA et al. 1968); they exist scattered in the cytoplasm and are not associated with the Z discs (BENNETT et al. 1979). The rearrangement of the cytoskeletal filaments is one of the main events of myotube maturation during development (see Sect. G.III.2b).

WANG and RAMIREZ-MITCHELL (1983) treated myofibrils from rabbit muscles with potassium iodide, which dissolves actin and myosin filaments. Electron microscopy of the remnants reveals a highly ordered three-dimensional system of cytoskeletal filaments (Fig. 47). Short bundles of filaments connect adjacent Z and M lines, respectively, and longitudinal filaments ensheath the sarcomeres. Each Z disc is surrounded by a pair of rings formed by tightly interwoven filaments. The distance between the two rings is 0.1 μm. The filaments connecting adjacent Z discs insert in these rings, which, again, connect to a fine network of filaments traversing the myofibril. Thus, the cytoskeletal system extends even into the myofibrils. These authors speculate on whether the intrafibrillar networks give rise to the N_1 lines within the I bands (see Sect. C.I.3). The intermediate filament systems within the myofibrils contain only little desmin, but a rather large amount of titin. Titin is the name of a group of more than 500,000-dalton proteins, found, using immunofluorescence, predominantly at the Z- and M-line level of the sarcomeres (WANG et al. 1979). Another large-molecule protein occurring together with titin is "band 3" protein which is at the N_2 line (WANG and WILLIAMSON 1980). This suggests that both N lines of the I band are related to the cytoskeletal system.

In many non-muscle cells, actin and other cytoplasmic proteins are linked to the plasma membrane by the proteins spectrin and vinculin (for references, see BRANTON et al. 1981; WILKINS and LIN 1982). Recently, both spectrin and vinculin have been demonstrated, using immunofluorescence, in skeletal muscles of chicken (REPASKY et al. 1982; PARDO et al. 1983). The reactive sites form a zebra-patterned lattice consisting of rib-like transverse bands on the fibre surface ("costameres", PARDO et al. 1983), which in doublets flank the Z lines and encircle the muscle fibre. The "costameres" probably represent the sites at which the cytoskeletal system of the Z discs is attached to the sarcolemma.

III. Sarcoplasmic Reticulum and T System

1. Historical Background

The endoplasmic reticulum was one of the first cellular structures identified using electron microscopy and characterized as a system of membrane-bounded compartments. In skeletal muscle fibres it was found between the myofibrils (BENNETT and PORTER 1953). At that time it was already well established that an action potential of the plasma membrane triggered a contractile response of the fibre; how these two events were linked was unknown. SANDOW (1952) introduced the term "excitation-contraction coupling" for the unsolved problem. HILL (1949) demonstrated that the size of a muscle fibre and the fast onset of contraction following the action potential were incompatible with the assumption that a diffusible transmitter connected the electrical reaction of the plasma membrane and the contractile system. Independently, two groups of workers suggested that not only the sarcolemma, but also the membranes of the endo-(sarco-)plasmic reticulum were polarized, and that the stimulus was conducted by depolarization of the internal membranes (EDWARDS et al. 1956; RUSKA et al. 1958; RUSKA 1958; PORTER 1956; PORTER and PALADE 1957; PEACHEY and PORTER 1959).

Several light microscopists had seen reticular intracellular structures in muscle fibres, and the Golgi study by VERATTI (1902) is regarded as the most comprehensive description (SMITH 1961) of them. VERATTI (1902) had no suggestions with respect to the function of the reticulum; previously, GERLACH (1877) had described an intracellular network in gold-stained specimens, and had assumed that this was an intrasarcolemmal nerve plexus connecting the nerve fibre directly with the isotropic substance of the muscle fibre. This speculation, erroneous as it may appear from a modern point of view, indicates that the physiological problem had already been recognized. Nevertheless, the histological technique was unreliable, and others (KRAUSE 1877; FISCHER 1877), after gold staining, found granules instead of a network. The illustrations suggest that sometimes mitochondria and sometimes the sarcoplasmic reticulum had been stained. In my opinion, the same applies to all light microscopic descriptions of "reticular structures" in muscle fibres, whether they have been stained by gold or silver (for references, see HÄGGQUIST 1931).

Biochemists had isolated a "factor" from muscle that inhibited the ATPase activity of myosin, and relaxed contracted muscle fibres or "contracted" actomyosin models ("relaxing factor") (MARSH 1951; BENDALL 1952; GOODALL and SZENT-GYÖRGYI 1953; GERGELY 1959; BRIGGS and FUCHS 1960; FUCHS and BRIGGS 1961). The "relaxing factor" was in a granule fraction which was identical with the lipid-rich microsome fraction with magnesium-activated ATP-ase activity isolated by KIELLY and MEYERHOF (1948). The granules were in reality vesicles derived from the membranes of the sarcoplasmic reticulum (NAGAI et al. 1960; MUSCATELLO et al. 1961; EBASHI and LIPMANN 1962), and were able to accumulate Ca^{2+} at the expense of ATP, and to lower the Ca^{2+} content of the surrounding medium. Since it had just been shown that very small amounts of Ca^{2+} were indispensible for skeletal muscle contraction – amounts that occur as contamination in any solution and hence had to be captured by a chelating agent to demonstrate the Ca^{2+} effect – it was proposed that the "relaxing factor" was not an inhibiting agent but rather the ability of the sarcoplasmic reticulum to capture free Ca^{2+} (WEBER 1959; HASSELBACH 1960; EBASHI 1960, 1961 a, b; NAGAI et al. 1960; WEBER and WINICÜR 1961; HASSELBACH and MAKINOSE 1961; MUSCATELLO et al. 1961; EBASHI and LIPMANN 1962; EBASHI and EBASHI 1962; EBASHI et al. 1962).

The observations that (a) Ca^{2+} is essential for the interaction of myosin and actin, (b) the granule fraction relaxes contracted muscle fibres or actomyosin models, (c) the granule fraction accumulates Ca^{2+}, and (d) the granules in reality are vesicles derived from the sarcoplasmic reticulum, resulted in a theory for the intracellular regulation of muscle activity by Ca^{2+}. Several independent laboratories were involved, and the question of who was the first to have the idea has since been discussed (EBASHI and LIPMANN 1962; HASSELBACH 1963; HASSELBACH and WEBER 1965; EBASHI 1980). The role of Ca^{2+} for contraction of a living muscle cell had already been shown by NIEDERGERKE (1955) who produced local contractions by intracellular injection of Ca^{2+}. HUXLEY (1964) emphasizes that, as early as 1942, BAILEY stated the theory that in resting muscle fibres Ca^{2+}, which activates the myosin-ATPase, by some mechanism is separated from the enzyme, and that after excitation both are brought together to make the muscle fibre contract.

In 1959, ANDERSSON-CEDERGREN subdivided the "endoplasmic reticulum" of skeletal muscle fibres into two different membrane systems. One system consists of cisterns wrapped around the myofibrils and connected by longitudinal tubules; the other one forms branching tubules running transversely across the muscle fibre, either at the level of the Z disc (amphibia) or at the A-I boundary (mammals). The two systems were named sarcoplasmic reticulum (SR) (also sarcotubular system or longitudinal system) and transverse tubular system (T system), respectively. Subsequently, it was shown that the tubules of the T system are invaginations of the plasma membrane. The two membrane systems frequently touch each other at structures called triadic junctions, but there is no obvious connection between the two membrane systems. Each triadic junction consists of two elements of the SR plus one interposed T tubule. The ability to conduct the excitation across the fibre was attributed to the T system, and control of contraction and relaxation by release and recapturing of Ca^{2+} was attributed to the SR. Already before the T system was known as a separate entity, HUXLEY and TAYLOR (1958) performed the decisive experiments showing that each sarcomere had its own intracellular conduction system. They applied subthreshold stimuli through a microelectrode to the surface of a frog muscle fibre and filmed the contractile response. The two adjacent half-sarcomeres contracted when the tip of the electrode was at the Z-line level, but no response was seen when the electrode was positioned opposite an A band. By moving the electrode circumferentially, distinct sensitive spots were found at the Z-line level. The spots were about 5 μm apart, and they probably correspond to the openings of the T tubules. Development in this field was rapid, and already in 1961 a supplement volume of the *Journal of Biophysical and Biochemical Cytology,* covering the role of the internal membrane systems of skeletal muscle fibres was published (PORTER 1961).

2. The Sarcoplasmic Reticulum

a) Array

The sarcoplasmic reticulum (SR) forms tubules and cisterns running transversely and longitudinally in the narrow spaces that delineate the myofibrils. The basic array in mammalian muscle fibres is as follows. Pairs of elongated cisterns are arranged transversely at the level of the I band and the outer part of the A band, such that there are four cisterns per sarcomere. Between the two cisterns of a pair is a transversely running T tubule. Across the A band, the cisterns directed towards the M line are connected by narrow tubules running parallel to the myofilaments. These tubules, in the middle of the sarcomere, unite and branch in an elaborate fashion (Fig. 48). They form a flat central cistern in fibres rich in SR; only few confluent and branching tubules are seen in fibres poor in SR. The cisterns pointing towards the Z line connect to narrow branching tubules which cross the Z-line level and connect to the corresponding cisterns of the next sarcomere. These tubules are less distinct and less ordered than those at the A-band level. In longitudinal sections, one usually sees only the T tubules and the adjacent SR elements forming triadic junctions. Grazing sections rarely expose the entire system (Fig. 49); a more complete view of the organization of the SR may be obtained in freeze fractures (KELLY and KUDA 1979) (Figs. 50, 51). The following nomenclature for the different components is now in general use: Adjacent to the T tubule are two terminal cisterns or terminal sacs, connected across the A band by longitudinal tubules and across the Z line by the I-band SR. The system of branching and reuniting longitudinal tubules at the M-line level forms the central cistern or the fenestrated collar. When a fenestrated collar is distinct, numerous 15- to 20-nm-wide pores are seen. These pores are not true discontinuities of the membrane of the SR but channels penetrating the central cistern. The terminal cistern of mammalian muscle fibres may have a strand of similar pores where the longitudinal tubules arise (Figs. 49, 50). The array described depends on the presence of two T tubules per sarcomere, which is the case in mammalian but also in some avian (PAGE 1969; HIKIDA 1972), fish (FAWCETT and REVEL 1961; KILARSKI 1964, 1965, 1966, 1967; GOLDSTEIN 1969a), and invertebrate muscles, but usually not in muscle of amphibia (cf. SMITH 1966; HOYLE 1969). Tortoise muscles, however, have two T tubules per sarcomere (PAGE 1968). The T tubules in twitch muscle fibres of frog are at the Z-line level. There is only one transverse system

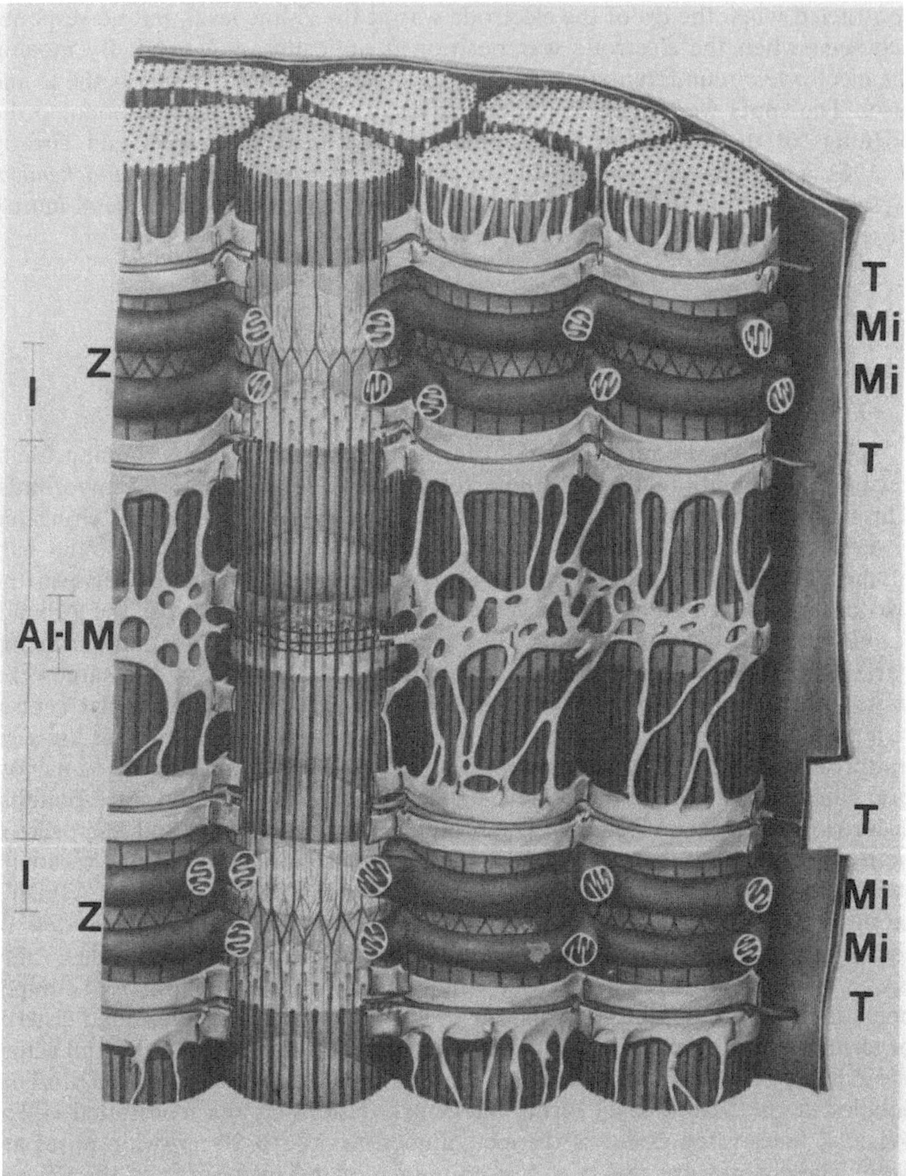

Fig. 48. Diagram illustrating the sarcomeric array of the membrane systems. The myofibrils are at both sides of the Z lines encircled by transversely running mitochondria (*Mi*) which are part of the mitochondrial grids. At the A–I junction are pairs of transversely arranged terminal cisterns of the sarcoplasmic reticulum (*SR*) accompanying T tubules (*T*), which are invaginations of the plasma membrane of the sarcolemma (*right*). The SR cisterns pointing towards the middle of the sarcomere connect to narrow longitudinal tubules which close to the M line branch and re-unite to form a "fenestrated collar". The discrete I band SR linking the outer terminal cisterns across the Z line is not shown. (Drawn on the basis of electron micrographs of sections of muscle fibres of the rat diaphragm). (From SCHMALBRUCH 1970)

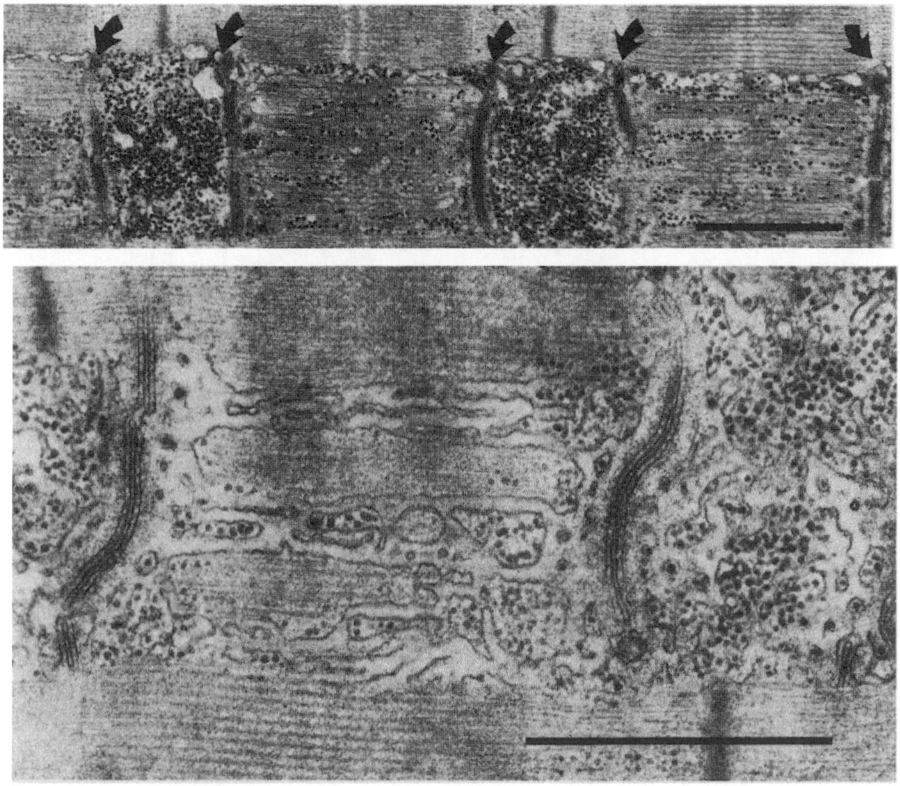

Fig. 49. Anterior tibial muscle of rat. Longitudinal sections showing the sarcoplasmic reticulum and the T system forming triadic junctions (*arrows*) at the A–I junction. The junctional gaps between T tubules and the terminal cisterns contain periodic densities ("connecting feet", Figs. 54, 55). Note high glycogen concentration at the I-band level. Bars, 1 μm

and only two terminal cisterns per sarcomere; a flat intermediate cistern connects the terminal cisterns and the longitudinal tubules. The I-band SR is lacking in amphibian muscles and there are only few connections across the Z line (PEACHEY 1965).

The content of the longitudinal tubules, the fenestrated collar, and the I-band SR, is electron-translucent; whereas, the terminal cisterns or sacs contain a moderately electron-dense granular substance. They are readily detected in appropriate cross-sections showing the terminal cisterns meandering across the mass of contractile filaments which thus subdivides into myofibrils. The terminal cisterns do not encircle but rather have a tendency to pass around two sides of a myofibril (BIRKS 1965). The terminal cisterns of a given sarcomere plane are continuous across the fibre.

YASUMURA and SCALES (1982) stained rabbit muscle fibres using a modified Golgi technique and in some cases the entire SR and T system were heavily impregnated. Sections about 1 μm thick were studied using electron microscopy, and in stereoscopic pictures a three-dimensional view of the SR was obtained.

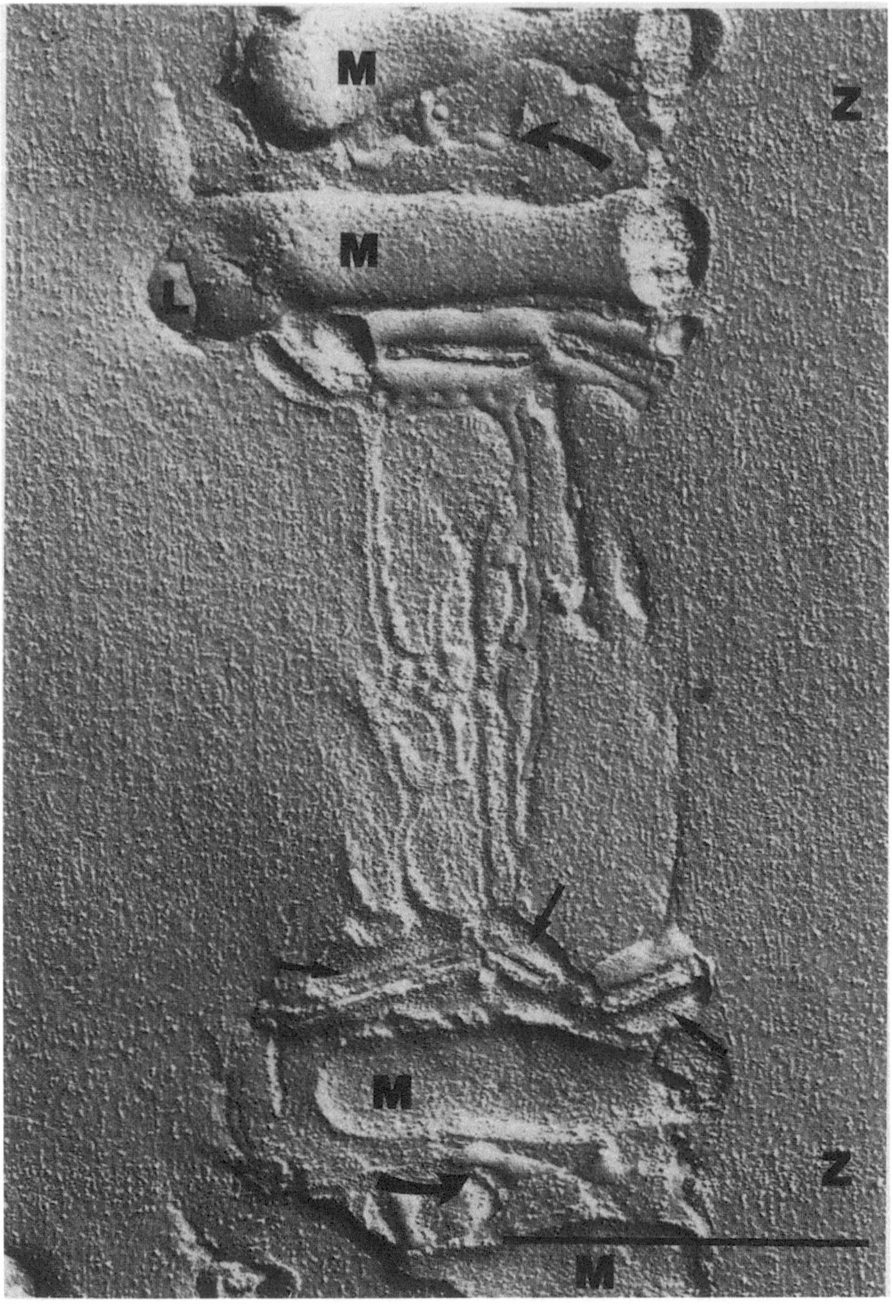

Fig. 50. Cat gastrocnemius muscle. Freeze fracture demonstrating the organization of the membrane systems around the sarcomere. (Magnification and orientation identical to Fig. 48). The position of the Z lines is indicated (*Z*). The four terminal cisterns of one sarcomere are fractured to expose the E faces (*top*) and the P faces (*bottom*). The P faces are covered with numerous intramembrane particles (*straight arrow*), whereas the E faces appear rather smooth. At the Z line level, fragments of the I band SR are visible (*bent arrows*). Four mitochondrial branches (*M*) are exposed, the upper ones displaying the E face and the lower ones the P faces of the outer mitochondrial membrane. *L*, lipid droplet; bar, 1 μm

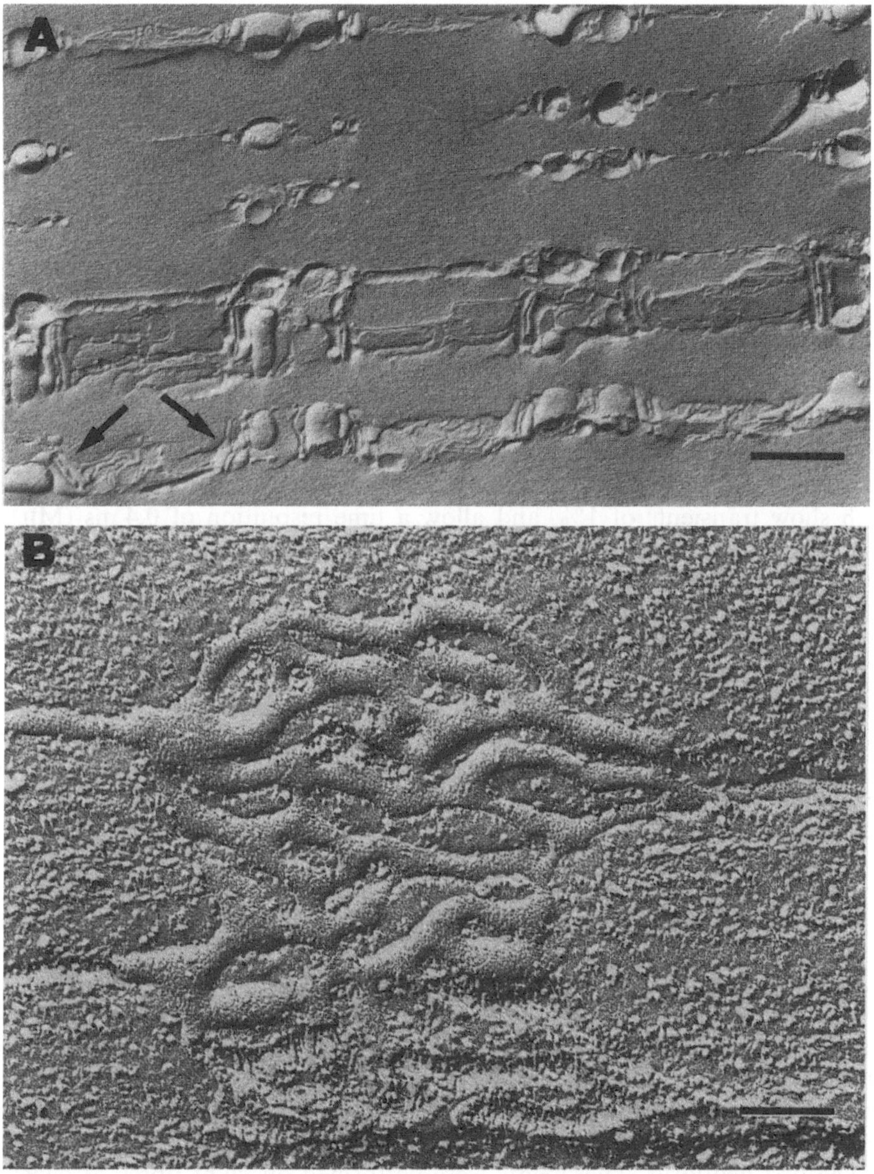

Fig. 51 A, B. Longitudinal freeze fractures demonstrating the array of the sarcoplasmic reticulum. **A** Cat gastrocnemius muscle. Three sarcomeres are shown. Two T-tubules and their accompanying terminal cisterns run obliquely (*arrows*) and possibly accommodate to a Vernier-type shift of the cross-striation (see Fig. 128). Bar, 1 μm. **B** Rat anterior tibial muscle, fenestrated collar. The smooth E face and the particle-carrying P face of the SR membrane are exposed. Bar, 0.1 μm

The authors compared the array at different sarcomere lengths and found that the configurational changes were most pronounced in the A-band SR. They conclude that there are cytoskeletal filaments connecting the Z and M lines and the adjacent SR elements.

b) Morphological Methods for the Study of Ca^{2+} Movements and Internal Membrane Changes During Contraction

The localization of Ca^{2+} and in particular its movements have been visualized using precipitation as oxalate (HASSELBACH 1964), autoradiography (WINEGRAD 1965, 1968), and with the help of the bioluminiscent dye, aequorin, or other Ca^{2+} probes. Aequorin emits blue light in the presence of even low Ca^{2+} concentrations (ASHLEY and RIDGEWAY 1970); it has a slow time resolution and is difficult to obtain. Murexide (JOBSIS and O'CONNOR 1966) is another Ca^{2+} monitor which, however, is rather insensitive. Recently, mainly the metallochromic indicator dyes, arsenazo III and antipyrylazo III, have been used which show transients of 1%, and allow a time resolution of 0.4 ms (MILEDI et al. 1977, 1982; KOVACS et al. 1983). The advantages and disadvantages of the various intracellular Ca^{2+} probes have been reviewed by BLINKS et al. (1982).

The oxalate method for visualizing Ca^{2+} is suitable for electron microscopy, but is hampered by the fact that the sarcolemma is not permeable to oxalate. This problem has been overcome by destroying the plasma membrane with glycerol for a period after which the internal membranes are still intact, or by skinning the muscle fibre mechanically. Both types of experiments suggest that most of the Ca^{2+} is localized in the terminal cisterns (HASSELBACH 1964; PEASE et al. 1965; COSTANTIN et al. 1967; COSTANTIN and PODOLSKY 1967; PODOLSKY et al. 1970).

When resting frog muscle fibres are soaked for 5 min in ^{45}Ca Ringer's solution activity of ^{45}Ca is found exclusively above the terminal cisterns. After 5 h, the activity also occurs in the intermediate cisterns, and in the A band portion of the myofibrils. The exchange between the initially labelled terminal cisterns and the two other sites is accelerated by electrical stimulation, according to WINEGRAD (1965, 1968), who interprets his findings as indicating that, during rest, extracellular Ca^{2+} enters the T tubules and is taken up by the terminal cisterns from where it is released during stimulation. Re-uptake during relaxation is by the intermediate cisterns and longitudinal tubules with subsequent binding within the terminal cisterns (for diverging results, see below).

The first aequorin experiments were performed on giant fibres of barnacle muscles (ASHLEY and RIDGEWAY 1970; HOYLE 1970). Aequorin does not emit light in resting frog muscle fibres, but when electrically stimulated, an aequorin-injected fibre emits flashes of light. The onset of the flash during a twitch contraction precedes the onset of force production, and the intensity of light reaches its peak before the peak force, and coincides with the maximum rate of force development. The light intensity declines, and falls to zero shortly after peak force has developed. Repetitive stimulation at high frequency produces a steady increase in luminiscence, i.e. of the sarcoplasmic Ca^{2+} concentration. There is a persisting increase in light intensity after tetanus (TAYLOR et al. 1975;

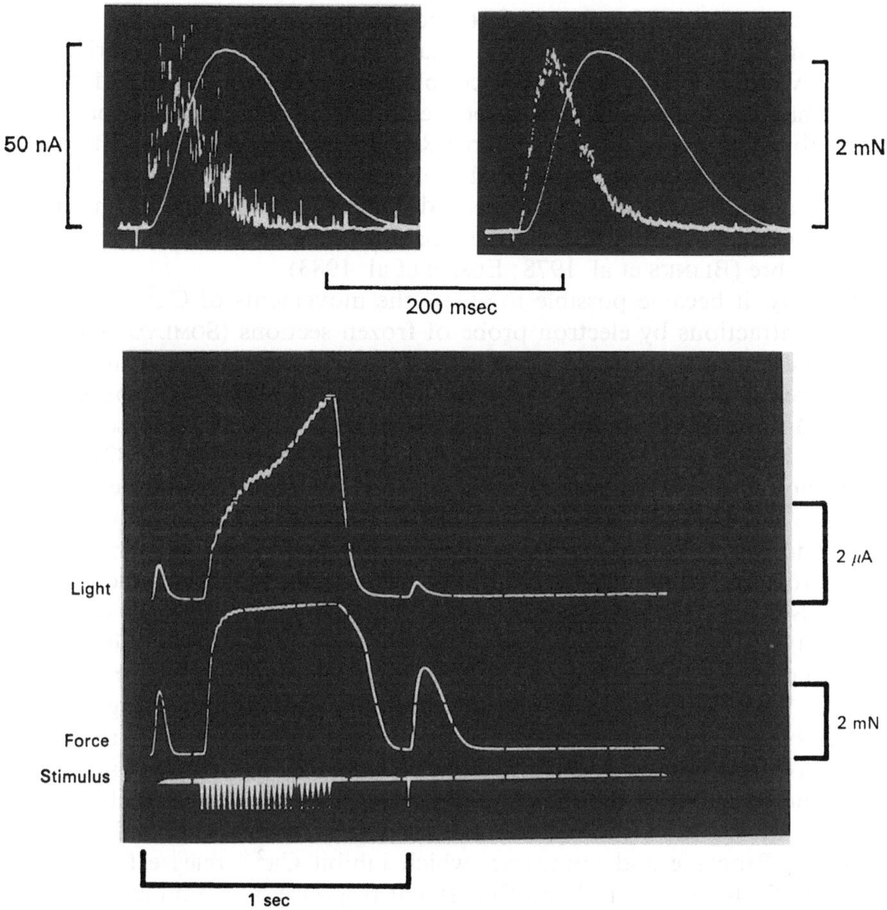

Fig. 52. Luminescent and mechanical responses of amphibian twitch muscle fibres loaded with aequorin. *Top:* Single fibres from the anterior tibial muscle of *Rana temporaria*. Striation spacing 2.3 μm, temperature 10° C. *Left panel* shows records of light (noisy trace) and force from a single rested state contraction. In the *right panel*, seven twitches have been averaged to reduce photomultipler shot noise. The time of the stimulus is indicated by the *vertical mark* below the baseline. Note early onset and decline of the light response in comparison with the mechanical response. *Bottom:* Post-tetanic response. Isometric contractions of a single fibre from the iliofibularis muscle of *Xenopus laevis*. Striation spacing 2.5 μm, temperature 15° C. The first twitch is a rested state contraction, this was followed by 50-Hz tetanus, and post-tetanic twitch. Note the steady increase of the light response during tetanus, and the potentiation of the subsequent twitch contraction. The output of the photomultiplier is given as current (nA and μA); this is a measure for the sarcoplasmic Ca^{2+} concentration, but Ca^{2+} concentration and light intensity are not linearly related. (From Blinks et al. 1978, with permission of the authors and the Physiological Society)

Blinks et al. 1978), which may explain why twitch contractions after repetitive stimulation often are more powerful (potentiated) than when the fibre is stimulated after a period of rest (Fig. 52). The arsenazo III response to the Ca^{2+} transient in a fibre stimulated by two action potentials is smaller in the second than in the first response, for stimulus intervals of up to several seconds. Recovery is in two phases, the first is fast and reflects re-entry of Ca^{2+} from the sarcoplasm into the SR, the second phase is slow and possibly arises from the inactivation of the T-tubule-SR coupling process (MILEDI et al. 1983a, b). The mechanical and luminescent responses in twitches of aequorin-loaded frog muscle fibres are unaltered in Ca^{2+}-free Ringer's solution, indicating that Ca^{2+} necessary for contraction is liberated from intracellular stores only. There is, however, a gradual decrease in force and overall luminescence during long or repeated tetani, probably because the stores become depleted and Ca^{2+} is lost from the fibre (BLINKS et al. 1978; EUSEBI et al. 1983).

Recently, it became possible to assess the movements of Ca^{2+} ions during tetanic contractions by electron probe of frozen sections (SOMLYO et al. 1981). Half of the calcium content of the sarcoplasmic reticulum of frog semitendinosus fibres is released during a 1.2 s tetanus. The sarcoplasmic Ca^{2+} concentration rises by 1 mM which compares well with the 0.6–0.7 mM increase estimated from the arsenazo III response of living fibres (MILEDI et al. 1982). Na^+ or Cl^- do not enter the SR, which argues against a large and sustained potential difference across the SR membrane. The Ca^{2+} release is associated with uptake of K^+ and Mg^{2+} into the terminal cisterns, but there remains a charge deficit which probably is compensated for by the movements of organic ions. The A-band SR does not show selective Ca^{2+} uptake; this contrasts with WINEGRAD's (1965, 1968) isotope studies. The Ca^{2+} concentration within the mitochondria does not change during contraction, indicating that the intramitochondrial Ca^{2+} stores are not involved in excitation-contraction coupling.

Nile blue is another intracellular fluorescent dye probe that responds to stimulation. It is not Ca^{2+} sensitive and probably indicates changes in the transmembrane potential of the SR. The twitch responses in frog muscle fibres are similar to those seen with aequorin, but there is no additive effect during trains of stimuli. Procaine and tetracaine, which inhibit Ca^{2+} release from the SR, abolish both the mechanical and fluorescent response of a stimulated fibre (BEZANILLA and HOROWICZ 1975; CAPUTO et al. 1979). NAKAJIMA and GILAI (1980) introduced a new potential-sensitive dye, NK 2367, which allows recording of potential changes of the plasma and T-tubule membranes.

c) Other Ca^{2+}-Binding Systems

In addition to the Ca^{2+}-binding capacity of the SR, there are other mechanisms that could regulate the concentration of free Ca^{2+} ions; Ca^{2+}-binding sites in mitochondria and in nuclei, a Ca^{2+}-transport mechanism across the sarcolemma, and several cytoplasmic Ca^{2+}-binding proteins (for review, see CARAFOLI and CROMPTON 1978; MARTONOSI 1982a) (Fig. 53). The significance of these systems for the contraction of twitch fibres is questionable, but they may play a role in slow muscle fibres and in cardiac and smooth muscles. One of the Ca^{2+}-binding proteins, parvalbumin (GILLIS et al. 1979), was, using immunohistochemistry, shown to be present in fast-twitch rat muscles only; it was lacking in slow-twitch fibres (CELIO and HEIZMANN 1982). The authors suggest that parvalbumin might be concerned with rapid muscle relaxation. Nevertheless, RANATUNGA and WYLIE (1983) studied the temperature depen-

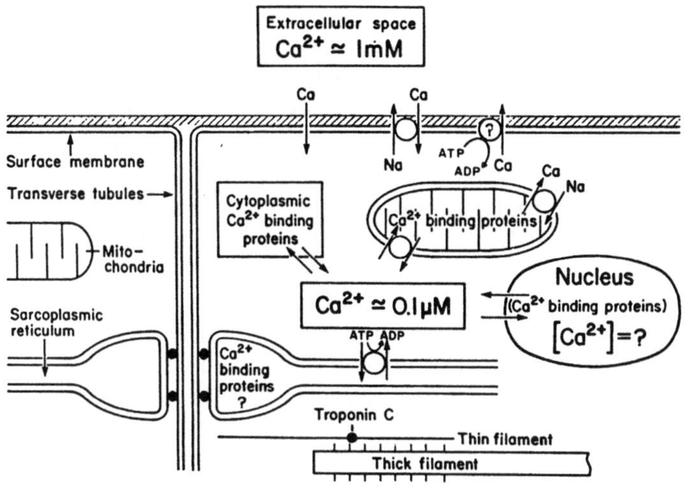

Fig. 53. Schematic illustration of the different Ca^{2+} binding and transporting systems in a skeletal muscle fibre. (From MARTONOSI 1982a, with permission of the author and John Wiley and Sons)

dence of tension relaxation in fast- and slow-twitch rat muscles, and found no evidence for a special relaxation mechanism in fast-twitch fibres.

d) The SR Membrane

The SR is a specialized smooth endoplasmic reticulum. During development, it is derived from the rough endoplasmic reticulum (EZERMAN and ISHIKAWA 1967; for review, see MARTONOSI 1982b); it never carries ribosomes. When the membrane is fragmented and isolated, it reseals and forms small vesicles. Vesicle fractions have been studied biochemically and morphologically. The ATPase which is able to pump Ca^{2+} across the membrane into the interior of the system and to maintain a gradient of at least 1:1,000 (HASSELBACH 1979; YAMAMOTO et al. 1979) is an integral membrane protein. Calsequestrin, another protein of the SR, has a strong capacity to bind Ca^{2+}; it is water soluble and not membrane bound. The localization of the ATPase and of calsequestrin has been demonstrated using immunofluorescence (JORGENSEN et al. 1979) and immunocytochemistry (JORGENSEN et al. 1982, 1983); the ATPase has also been shown using cytochemistry (GAUTHIER and PADYKULA 1965; ZEBE 1965; SCHULZE and WOLLENBERGER 1967; PODOLSKY 1968; VYE et al. 1969; AGOSTINI and HASSELBACH 1971; DE MEIS et al. 1974). Calsequestrin is restricted to the terminal cisterns, and probably corresponds to the dense granular content seen in electron micrographs (MACLENNAN and WONG 1971; MACLENNAN and HOLLAND 1975) (Fig. 54), whereas the Ca^{2+}-transport ATPase is distributed throughout the I band and within the fenestrated collar around the M line. It is absent from the junctional face of the terminal cistern.

The crude vesicular preparation of SR obtained by mincing and centrifugation is heterogeneous. Different fractions may be separated by gradient centrifugation. The two extremes, the "heavy" and "light" fractions, consist of terminal cisterns and longitudinal tubules, respectively (MEISSNER 1975; CHEVALLIER et al. 1977b). The Ca^{2+} transport ATPase is uniformly distributed among the various fractions; in rabbit muscle, it amounts to 50%–70% of all membrane proteins (MEISSNER and FLEISCHER 1971; SARZALA et al. 1975). The ATPase requires Mg^{2+} (or other divalent ions) and is regulated by the Ca^{2+} concentration in the medium. Under optimal conditions uptake of one mole Ca^{2+} requires hydrolysis of 0.4 mole ATP (MARTONOSI and FERETOS 1964) (for review, see TADA et al. 1978; MARTONOSI 1982a; IKEMOTO 1982).

The molecules of the enzyme are concentrated in the outer half of the membrane. Thin sections (HASSELBACH and ELFVIN 1967), and X-ray diffraction (WORTHINGTON and LIU 1973) of the membrane show marked asymmetry. The P face of freeze-fracture replicas carries numerous 8 to 8.5-nm particles (DEAMER and BASKIN 1969). These particles do not leave pits on the E face of the SR membrane. On the outer surface of the SR membrane there are 4-nm-high particles with a distance of 4–7 nm between them as seen in negatively stained and also in freeze-dried and shadowed vesicle preparations. The 4-nm surface particles represent extramembranous tails of the intramembranous particles. The array of the tails and also of the P-face particles is random (FRANZINI-ARMSTRONG 1963; SCALES and YASUMARA 1983). This is in harmony with X-ray diffraction, which does not indicate a regular array of the protein molecules (WORTHINGTON and LIU 1973). Vandate "stabilizes" the Ca^{2+} transport ATPase. SR vesicles treated with vandate in the absence of Ca^{2+} show that the previously 4-nm-high tails extend very little above the membrane surface, the P face displays a regular two-dimensional lattice of particles, and the E face now displays corresponding deep pits (SCALES and YASUMURA 1983). This is interpreted to mean that the Ca^{2+}-ATPase-vandate complexes have moved perpendicularly to the plane of the membrane, and that this mimics a step in calcium translocation across the SR membrane. Treatment of SR preparations from rabbit muscle with desoxycholate yields non-aggregated particles 10–20 nm in diameter. The particles represent 90% of the membrane proteins and phospholipids, and have a very high ATPase activity (SELINGER et al. 1969). The unit membrane structure of the vesicles remains intact when lecithin is extracted by phospholipase C, but the ATPase activity and the ability to accumulate Ca^{2+} are lost. Both are restored when synthetic phospholipids are added (MARTONOSI 1964). Hence, both the enzyme protein and a fraction of the membrane lipids are essential for Ca^{2+} transport across the membrane.

About 16,000 surface particles ("tails")/μm^2 membrane are seen in negatively stained samples (JILKA et al. 1975); freeze fractures show only 4,000 P face particles/μm^2 (SCALES and INESI 1976). This discrepancy and the large size of the intramembrane particles suggest that they represent clusters of probably four molecules.

IKEMOTO et al. (1971) found that the 4-nm surface particles, when visualized in negatively stained preparations, are connected to the membrane by 2-nm-wide stalks. Trypsin treatment removed both particles and stalks, but did not alter the ATPase activity and the uptake of Ca^{2+}. The authors conclude that "the surface spheres and stalks have little or no role to play in Ca^{2+} transport".

This is not generally accepted, and MARTONOSI (1982a) maintains that the 4-nm particles represent a major portion of the ATPase molecule which projects from the membrane into the water phase of the sarcoplasm.

The activity of the Ca^{2+}-transport ATPase relates to the incidence of intramembrane particles. The enzyme activity is very low in immature muscles (FANBURG et al. 1968) and also the incidence of 8.5-nm particles in freeze fractures is only 10% of that in adult muscles (TILLACK et al. 1974; for reviews, see MARTONOSI 1982a, b). The ATPase content of the SR is smaller in slow-twitch muscle fibres than in fast-twitch muscle fibres, and also the number of 8.5-nm particles is smaller than in fast-twitch fibres (BERINGER 1976; BRAY and RAYNS 1976; BRIGGS et al. 1977). The difference in particle density between the fibre types is not very pronounced in mice (3,000 vs 2,600/μm^2 P-face SR) (BERINGER 1976), but in chicken there are more than twice as many particles in the fast-twitch posterior latissimus dorsi muscle than in the slow (tonic) anterior latissimus dorsi muscle (1,500 vs 3,100/μm^2 P-face SR; RYAN and SHAFIQ 1980).

e) The Effect of Various Drugs and of Electrical Stimulation on Ca^{2+} Release from the SR

The interplay of Ca^{2+}, the SR, and the contractile apparatus has provided a basis for the understanding of the action of various drugs on muscle. Caffeine induces contracture which is due to the release of Ca^{2+} from the SR; the local anaesthetics, procaine and tetracaine, prevent both contracture and Ca^{2+} release (FEINSTEIN 1963; CARVALHO and LEO 1967; CARVALHO 1968b; WEBER and HERZ 1968; BONDANI and KARLER 1970; BORYS and KARLER 1971). Procaine and tetracaine do not influence the contractile system directly, and contractures elicited by intracellular application of Ca^{2+}, or by application of Ca^{2+} to skinned fibres, are not inhibited (ALMERS and BEST 1976). Divalent ions known to potentiate the twitch force of a muscle depress the active binding of Ca^{2+} and thereby increase the sarcoplasmic Ca^{2+} concentration (CARVALHO 1968a).

Dantrolene causes relaxation by reversibly depressing excitation-contraction coupling; dantrolene does not interfere with the neuromuscular transmission or the conduction of the action potential. It abolishes the Ca^{2+} response in fibres loaded with aequorin, and $^{45}Ca^{2+}$-kinetic studies show that dantrolene selectively inhibits Ca^{2+} release from the SR into the sarcoplasm. The selective action of dantrolene on skeletal muscle is of great clinical interest. This drug has a life-saving effect during attacks of malignant hyperthermia, a disorder in which anaesthesia causes massive Ca^{2+} release from the SR with generalized hyperactivity of skeletal muscle, rise in temperature, and mostly fatal outcome (DESMEDT and HAINAUT 1977; WOOD et al. 1982). Catecholamines enhance the twitch force in frog muscles. This is attributed to an increase of the activity of the Ca^{2+} pump in the SR such that extra Ca^{2+} is available when the fibre is stimulated (GONZALES-SERRATOS et al. 1981).

Skinned muscle fibres (without plasma membranes but preserved SR) are by electrical stimulation induced to contract (COSTANTIN and PODOLSKY 1967), i.e. electrical signals may, at least experimentally, release Ca^{2+} from the SR.

3. The T System

a) Array

The T system consists of branching tubules running transversely across the fibre, either at the level of the A-I junction (mammals) or at the Z-line level (most amphibia) (for species differences, see SMITH 1966; HOYLE 1969; FRANZINI-ARMSTRONG 1973c). They form the central elements of the triadic junctions and are placed between two terminal cisterns. The T tubules are flattened and measure 20–30 nm in the direction of the fibre axis and 80–100 nm in a direction perpendicular to the fibre axis (Fig. 54).

It was easy to show that the T tubules of insect muscles and of heart muscle cells are invaginations of the plasma membrane, in particular because in heart muscle cells the basal lamina penetrates into the opening of the tubules. The T tubules of vertebrate skeletal muscle fibres, in places where they are not part of triadic junctions, disintegrate during primary osmium tetroxide fixation into rows of vesicles. Glutaraldehyde preserves the tubules better than does osmium tetroxide, and in fish (FRANZINI-ARMSTRONG and PORTER 1964a, b; JASPER 1967) and frog muscles (FRANZINI-ARMSTRONG et al. 1975) openings have been demonstrated. After leaving the last triadic junction, the tubules take a tortuous course, which explains why openings rarely appear in thin sections (FRANZINI-ARMSTRONG et al. 1975). The entire T system also of mammalian muscle fibres is accessible to extracellular tracer molecules (H.E. HUXLEY 1964; HILL 1964; ENDO 1964, 1966).

The name T system implies that its tubules are running transversely. This is not always so. In laryngeal muscles of bat, the T tubules sometimes are duplicated such that one sarcomere shows not two but four T tubules which form pentads (each with two T tubules and three terminal cisterns), rather than triads. This indicates longitudinal branching, and has been related to the fact that these muscles are very fast (REVEL 1962). Nevertheless, pentadic junctions and also longitudinally oriented T tubules are frequent in developing fibres (WALKER and SCHRODT 1967; LUFF and ATWOOD 1971; KELLY 1980) and in denervated adult fibres (GORI 1972); both contract slowly. Therefore, an excess of T tubules does not necessarily indicate that contraction is fast. About 3% of the T tubules of adult frog muscle fibres run longitudinally and connect the transverse networks (EISENBERG 1972); 10%–30% of the T tubule membrane belongs to short longitudinal extensions (PEACHEY and SCHILD 1968; EISENBERG and EISENBERG 1968).

The transverse array of the T system has been investigated in frog using high voltage electron microscopy of thick cross-sections of fibres loaded with exogenous peroxidase (EISENBERG and PEACHEY 1975). The tubules form meshes of about 1.6-μm diameter, and most branchings are threefold. The circumferential distance between points at which the tubules approach the plasma membrane is 1.2 μm, which is less than the circumferential distance between the sensitive spots on the surface of frog muscle fibres (about 5 μm; HUXLEY and TAYLOR 1958) (see Sect. C.III.1).

b) The T-Tubule Membrane

The membrane of the T system contributes to the total surface area of the muscle fibre, and hence the "gross" electrical properties of the plasma membrane differ from what one would expect from a membrane covering a cylinder with the length and diameter of the fibre. The surface area of the T tubules in frog muscle fibres is 4.5–7 times larger than that of the plasma membrane (PEACHEY 1965; PEACHEY and SCHILD 1968; MOBLEY and EISENBERG 1975). The specific membrane capacity of the fibre surface appears much larger than in nerve fibres (FALK and FATT 1964). In an intact fibre, it is 6.10 $\mu F/cm^2$ and drops to 2.25 $\mu F/cm^2$ when the T tubules are selectively destroyed ("detubulated") (EISENBERG and GAGE 1967; GAGE and EISENBERG 1967, 1969a, b). This can be achieved by bathing the fibre in glycerol Ringer's solution and then returning it to normal Ringer's solution (HOWELL and JENDEN 1967; HOWELL 1969; KROLENKO 1969). The mechanism probably involves the hypertonic solution entering the tubules and causing disruption when the muscle fibre is returned to normotonic Ringer's solution. The remnants of the T system become inaccessible to tracer molecules (NAKAJIMA et al. 1973). The glycerol Ringer's treatment works well in amphibian muscles, but it may be less appropriate for mammalian muscles because the SR is also heavily damaged (DAVEY et al. 1980).

It is generally agreed that the T system is the pathway of the internal spread of excitation. This was surmised on the basis of the local activation experiments by HUXLEY and TAYLOR (1958), and was ascertained by "detubulation". Detubulated fibres react to caffeine with contracture, indicating that the release of Ca^{2+} from the SR is unimpaired. The surface membrane still generates and conducts an action potential which, however, does not elicit contraction because excitation and contraction are uncoupled (GAGE and EISENBERG 1967, 1969b).

The properties of the T-tubule membrane can only be assessed indirectly. There is a reversible increase in the diameter of the T tubules of frog and crayfish muscle fibres when the fibres are bathed in hypertonic or low-chloride Ringer's solutions. The selective movement of water has been assumed to indicate different ionic conductances of plasma membrane and T-tubule membrane (FREYGANG et al. 1964; GIRARDIER et al. 1963; BRANDT et al. 1968). Nevertheless, also isotonic sucrose Ringer's solution causes T-tubule swelling, suggesting that there are fixed charges within the tubules (RAPOPORT et al. 1969). EISENBERG and GAGE (1968) studied the membrane properties of frog muscle fibres with intact and destroyed T tubules and interpreted their results to mean that either T tubules and plasma membrane have different ion conductances, or the tubules contain a material with negative fixed charges. The interpretation of the electrophysiological data strongly depends on the knowledge of the size and the configuration of the T-tubule system; the matter would be further complicated if there existed small ionic conductances at the triadic junctions (see below) (for references, see EISENBERG and EISENBERG 1982).

The T tubules of vertebrate muscles appear empty in electron micrographs, but tannic acid and ruthenium red stains the T-tubule content. CULLEN et al.

(1984), however, observed a particulate and sometimes laminar substance in the lumen of the T tubules of rat muscles without special staining. Possibly, the T tubules contain a special pool of proteins with anionic binding sites (GOLDSTEIN 1969b; BONILLA 1977). On the other hand, HOWELL (1974) and FORBES and SPERELAKIS (1979) found that ruthenium red stained the terminal cisterns but not the T tubules; alternatively, the intermediate cisterns were stained when the fixation method was modified. These discrepancies emphasize the sensitivity of the morphological technique, and it apparently remains unsettled whether the content of the T tubules really differs from that of the extracellular space.

HUXLEY and TAYLOR (1958) found a graded inward spread of the contraction after local subthreshold stimuli. They assumed that the membrane of the T tubules was not regenerative and not able to propagate a signal, and that conduction was due to passive electrotonic spread of the depolarization of the plasma membrane. It then became apparent that also the T-tubule membrane is regenerative and that intracellular conduction is not by passive electrotonic spread alone (GONZALES-SERRATOS 1966, 1971; ADRIAN et al. 1969; STRICKHOLM 1962, 1966; COSTANTIN and TAYLOR 1973). SUGI and OCHI (1967) and SUGI (1974) observed in frog muscle fibres a phasic spread of excitation, but the safety factor was low. The contraction after weak local depolarization in frog muscle fibres comprised only the peripheral part of the fibre. Surprisingly, contraction started in the centre of the fibre when the area of depolarization was large. The response showed a considerable decrement, and only rarely spread across the entire fibre. The phasic response was not always sensitive to tetrodotoxin, as should be expected when the membrane properties were identical with those of the plasma membrane. By contrast, COSTANTIN (1975) observed in frog muscle fibres a phasic inward spread which became passive electrotonic under tetrodotoxin. Like in SUGI's (1974) experiments, the contractile response was most pronounced in the axial part of the fibre. The author believed that the tubules nearest the surface were voltage-clamped at the applied depolarizing step.

4. Triadic Junctions

The excitation conducting membrane system (plasma membrane, T tubules) and the Ca^{2+}-accumulating membrane system (SR) are separated from each other, and the signal to release Ca^{2+} is transmitted at morphologically distinct junctions. These typically consist of two terminal cisterns with one interposed T tubule (triadic junction, triad). Sometimes there is only one terminal cistern

Fig. 54. Various aspects of SR-membrane couplings in rat (**A–C**) and human (**D**) muscle fibres. *T*, T tubule; *P*, plasma membrane. **A** Longitudinal section through a triadic junction. The membranes of the terminal cisterns are scalloping, connecting feet are barely visible. Note electron-dense content of the SR compared with the T-tubule lumen. **B** Transversely sectioned junction with plate-like density midway between the membranes (*arrows*). **C** Peripheral coupling between a cistern of the SR and the plasma membrane. Attached to the inner face of the SR membrane is amorphous material (*bent arrow*). **D** Cross-sectioned triadic junction showing four rows of connecting feet which rest on the membrane of the terminal cisterns, but do not reach out to the T tubule membrane. Bars, 0.1 μm

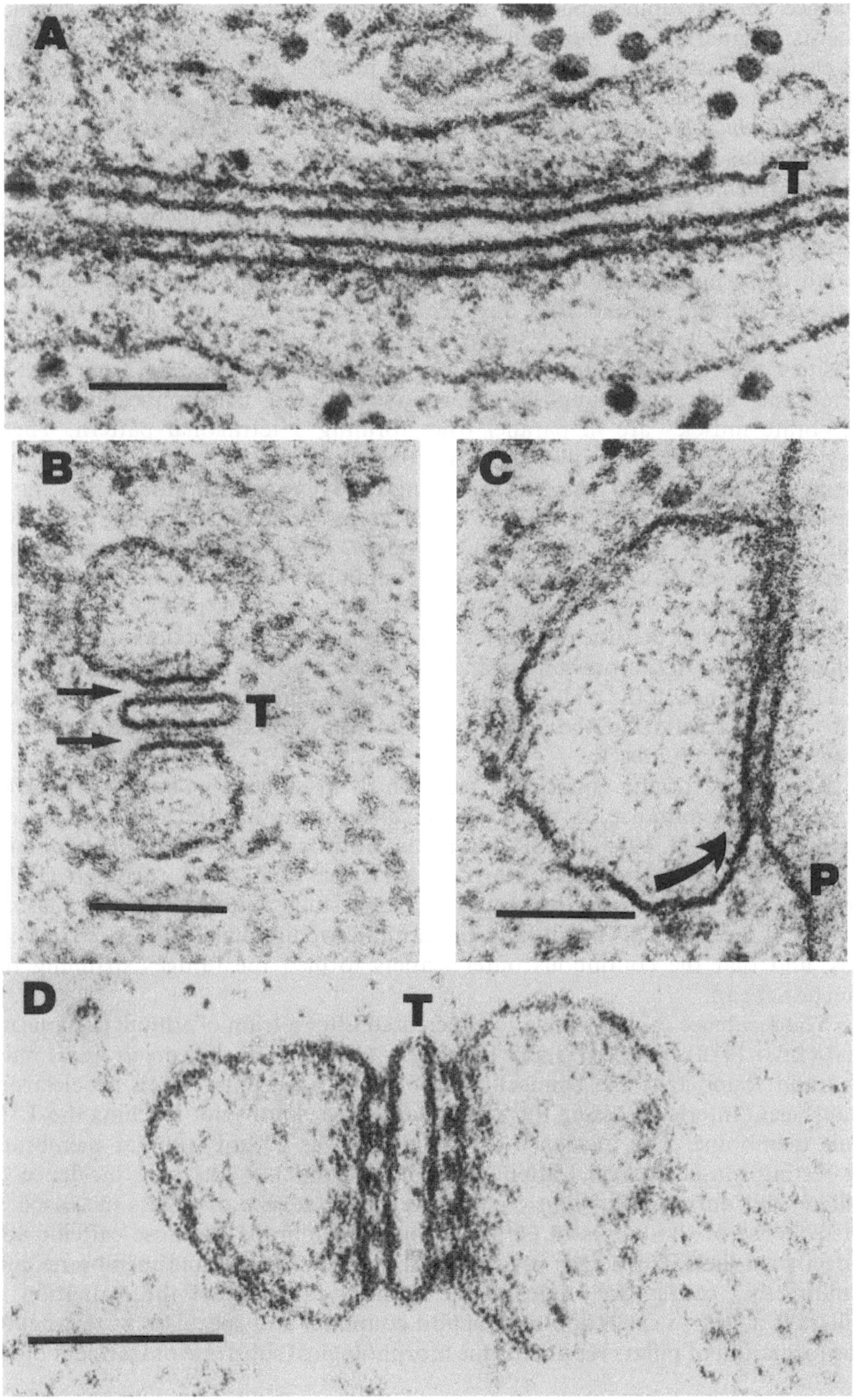
A
T
B
T
C
P
D
T

attached to a T tubule (diad) or there are two T tubules and three terminal cisterns (pentad); even more elaborate systems of alternating T tubules and SR elements have been seen. In immature muscles one also finds peripheral couplings between the plasma membrane and the SR (Fig. 54 C). This is the rule in *Amphioxus* muscles which have no T system and no typical SR (PEACHEY 1961). In this species, the Ca^{2+}-accumulating system occurs as subsarcolemmal vesicles which form many peripheral couplings at the I-band level (FLOOD 1977). Peripheral couplings are numerous at the myotendinous junction of skeletal muscles of the lamprey (NAKAO 1975).

All these junctions have common features. The opposing membranes are flattened and run parallel at a distance of 10–13 nm leaving a junctional gap between them. At intervals of 30 nm, the junctional gap is crossed by periodic densities, the connecting feet. These consist of scallops projecting from the SR membrane, and small lumps of amorphous material. The connecting feet in vertebrate muscles are in parallel rows forming a tetragonal pattern. There may be two or four parallel rows of connecting feet on the SR membrane facing a T tubule (or, in peripheral couplings, the plasma membrane) (FRANZINI-ARMSTRONG 1970c, 1973c, 1975, 1984; KELLY and KUDA 1979) (Figs. 54 C, 55). This picture is not seen in all instances. Sometimes the junctional gap contains a plate of amorphous material running midway between and parallel to the two apposed membranes (Fig. 54 B). CULLEN et al. (1984) did not find differences in the densities of connecting feet in triadic junctions of the fast-twitch extensor digitorum longus and slow-twitch soleus muscles of rat. There was no difference in the width of the T tubules of fibres from these two muscles. The volume fraction of the entire T system was 50% larger in the fast- than in the slow-twitch muscle.

The nature of the connecting feet, which in some way may be involved in signal transmittance, is unknown. In experimentally "uncoupled" muscle fibres (WOOD et al. 1975) the SR and the myofibrils remained remarkably intact but the connecting feet disappeared (EASTWOOD et al. 1979). SOMLYO (1979) presented micrographs of frog muscles, suggesting that the two cytoplasmic leaflets of the T and SR membrane, but not the two luminal leaflets, are continuous, and that the continuous outer leaflets formed the bridges spanning the junctional gap.

The incidence of connecting feet increased after a train of stimuli (EISENBERG and GILAI 1979). EISENBERG and EISENBERG (1982) studied *Xenopus laevis* muscles and found that the connecting feet either were pillars with an electron-translucent interior crossing the junctional gap, or knobs not reaching the T tubule membrane. The incidence of the pillars was $39/\mu m^2$ tubular membrane in resting muscle fibres. Caffeine contracture did not alter the incidence of pillars, but during potassium contracture the incidence of pillars increased at the expense of the knobs to $66/\mu m^2$ tubular membrane. Because caffeine acts directly on the SR, whereas potassium depolarizes the plasma membrane, and "indirectly" causes Ca^{2+} release, the authors assume that the formation of pillars is a step in excitation-contraction coupling, and speculate as to whether the formation of pillars represents the morphological substrate of transient open-

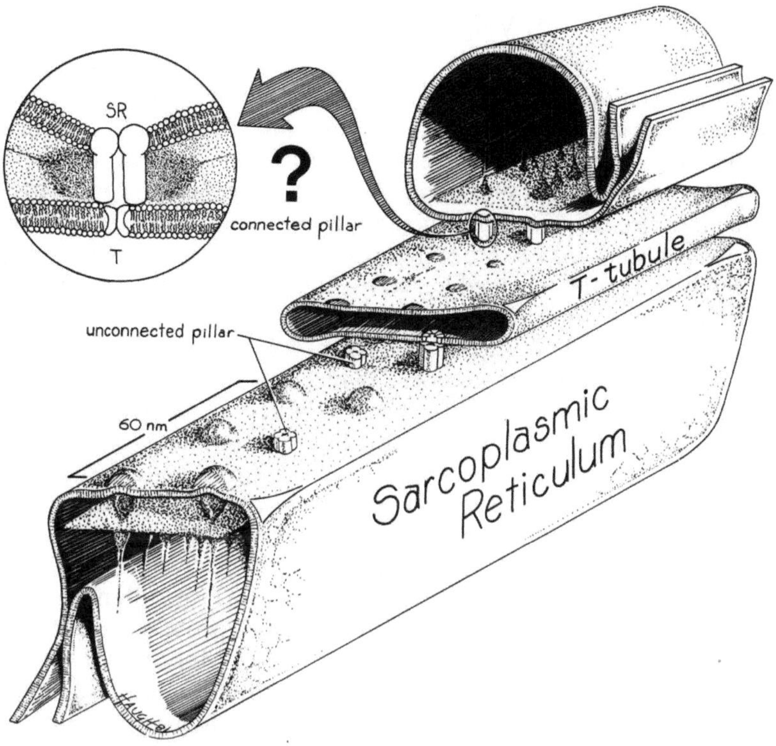

Fig. 55. Diagram showing the structure of the triadic junction and a highly speculative proposal for the coupling mechanism. The connecting feet (pillars) are in two rows (amphibia; four rows in human muscles as shown in Fig. 54D); the unconnected feet have been drawn as bumps of the reticulum membrane. The amorphous material adhering to the inner face of the junctional reticulum membrane (calsequestrin?) is shown as well. The diagram for the function assumes that the connecting feet are involved in pore formation between the interior of the T tubule and the sarcoplasmic reticulum. (From EISENBERG and EISENBERG 1982; with permission of the authors and the Rockefeller University Press; original plate kindly provided by Dr. B.R. EISENBERG, Chicago, Ill.)

ings between T tubules and SR. Ca^{2+} release from the SR is triggered by Ca^{2+} so that the release process itself is regenerative; hence, a small amount of Ca^{2+} entering the sarcoplasm might act as "transmitter" at the triadic junction (PODOLSKY 1975).

The accepted view is that T tubules are permanently open to the extracellular space, but that the SR is not in open connection with the T tubules or the extracellular space. Nevertheless, there are observations indicating that also the SR is accessible from the extracellular space. The SR may change its volume in relation to changes in the osmotic pressure in the bath solution (BIRKS and DAVEY 1969), and peroxidase, which is an extracellular marker, may occur in the SR (RUBIO and SPERLAKIS 1972). Also, connections between mitochondria and the SR have been reported (WALKER and SCHRODT 1966). The significance of these findings is not clear.

5. Very Fast Muscles

Muscle fibres with outstanding morphological and functional properties have received particular attention, because these fibres allow testing of the hypothesis that T tubules and SR membranes are involved in excitation-contraction coupling. FAWCETT and REVEL (1961) found in the very fast (SKOGLUND 1961) swim-bladder muscles of sound-producing fish particularly well developed SR and triadic junctions at the A-I boundaries. The myofibrils appear in cross-sections as narrow, about 0.3-μm wide, band-shaped structures; hence, the diffusion distance between SR and myofilaments is short. REVEL (1962) studied, in bats, the fast laryngeal muscles involved in the generation of ultrasound; the SR is abundant and each sarcomere plane is supplied by four T tubules. Nevertheless, the number of T systems per sarcomere is not necessarily related to speed of contraction. Tortoise muscles have two systems per sarcomere, but are slower than frog muscles which have only one system (PAGE 1968). The second antenna of the lobster, *Homarus americanus,* contains a muscle bundle which at 16° C has twitch contraction times of 2.5–3 ms; the total duration of the twitch is <10 ms. The fibres can mechanically follow stimulation frequences of >100 Hz. Only one-quarter of the muscle fibre volume is occupied by myofibrils; the rest is filled with masses of SR, which probably are responsible for the rapid decline of tension (ROSENBLUTH 1969; MENDELSON 1969). Still higher mechanical frequencies (up to 1,000 Hz) are found in indirect flight muscles of some insects. Nevertheless, these muscles are asynchronous, and respond with a series of oscillations to a single nerve stimulus. The individual contractions are not controlled by Ca^{2+} shuttling between the SR and the contractile filaments; the SR is reduced and the myofibrils are large (SMITH 1966; HOYLE 1969). SMITH (1983) states that a frequency of 100 Hz probably represents the upper limit of myoneuronal synchrony. This view has recently been contested by YOUNG and JOSEPHSON (1984). Tymbal muscles of cicada may be asynchronous or synchronous. In nine species with synchronous muscles, the authors found cycle periods (frequency^{-1}) and twitch durations of 4.5–25 ms. The cross-sectional areas of the myofibrils ranged from 0.4 to 0.8 μm^2; they were linearly related to the twitch duration in ms, and they were inversely related to the quantity of SR.

6. Quantitative Approaches to the Internal Membrane Systems

The amount of SR, as determined using morphometry (for review, see EISENBERG 1983), is smaller in large (cat, man) than in small mammals (mouse, rat, guinea pig) (Table 3). This may relate to the fact that twitch contractions in small animals have a shorter duration than in large animals (see Sect. D.VIII). T tubules and SR are more abundant in fast- than in slow-twitch fibres within the same species (see Sect. D.IX). This is part of the explanation for why the twitches differ in time course. In addition, the Ca^{2+}-pump capacity of the SR

Table 3. Morphometric data for T-system and sarcoplasmic reticulum (*SR*) in different muscles

Species	Muscle/Fibre type[a]	Myofibrils (Vol.%)	T-System (Vol.%)	T-System (Area per area fibre surface)	SR (Vol.%)	Term. Cistern (Vol.%)	References
Frog	Sartorius/F	83	0.32	5.5[b]	9.1	4.1	MOBLEY and EISENBERG (1975)
Guinea pig	"White" vastus/F	82	0.27	1.8[c]	4.8	1.6	EISENBERG and KUDA (1975)
	"Red" vastus/F	80	0.28	1.9[c]	3.3	1.2	EISENBERG and KUDA (1976)
	Soleus/S	87	0.14	0.8[c]	3.2	0.9	EISENBERG et al. (1974)
Rat	Ext. digit. long./F				9.3		DAVEY and O'BRIEN (1978)
	Soleus/S				3.3–3.5		DAVEY and WONG (1980)
Mouse	Ext. digit. long./F	75	0.4	3.1[d]	5.5		LUFF and ATWOOD (1971)
	Soleus/S	81	0.22	1.8[d]	2.9		
Cat	Gastrocnemius/F				1.3	1.0[e]	SCHMALBRUCH (1979a)
	Gastrocnemius/S				0.7	0.5[e]	
	Soleus/S				0.6	0.5[e]	
Man	Quadriceps fem./F		0.28	10.8[c]	1.9	1.2	EISENBERG (1983)
	Quadriceps fem./S		0.13	5.5[c]	1.2	0.7	

[a] *F*, fast-twitch; *S*, slow-twitch
[b] Calculated for a fibre diameter of 100 μm
[c] Calculated for a fibre diameter of 50 μm
[d] Calculated for a fibre diameter of 30 μm
[e] Including I-band SR and T-system

is higher in fast-twitch than in slow-twitch muscle fibres (see Sect. C.III.2d), and fast-twitch fibres have a higher myosin ATPase activity than do slow-twitch fibres (see Sect. D.III). It thus appears as if – at least in mature muscle fibres – contractile proteins, amount of internal membrane systems, and specific Ca^{2+}-transport capacity are mutually related.

The fast- and slow-twitch fibres of the gastrocnemius and soleus muscles of cat differ with respect to the volume fraction of the fibre occupied by terminal cisterns. Interestingly, slow-twitch fibres of the gastrocnemius contain as many longitudinal tubules as the fast-twitch fibres do, but in slow-twitch soleus fibres longitudinal tubules are rare. This difference between slow-twitch gastrocnemius and soleus fibres (SCHMALBRUCH 1979a) may tentatively be connected with functional differences between these two types of slow-twitch fibres. Slow-twitch gastrocnemius fibres contract faster than slow-twitch soleus fibres, and a twitch is potentiated by preceding tetanus, which is not the case for soleus fibres (BURKE et al. 1973, 1974).

IV. Sarcolemma

1. Historical Background

The sarcolemma envelops the muscle fibre; it separates the interior of the muscle fibre from the connective tissue space and is part of the neuromuscular and myotendinous junctions. When a muscle fibre is segmentally injured, the sarcolemma, at least its "basement membrane" (see below), may be seen as an empty tube connecting the retracting fibre fragments ("retraction clots") (see Sect. B.IV, and Fig. 118). The sarcolemmal tube was first observed and also depicted by SCHWANN (1839). BOWMAN (1840), who is usually credited with the detection of the sarcolemma, described it as a "tubular membranaceous sheath of the most exquisite delicacy investing every fasciculus from end to end". For more than a century, this remained the best and most complete characterization of the sarcolemma. Some authors disputed its very existence, whereas others became engaged in speculations about the "nature" of the sarcolemma. The speculations published during the pre-electron microscope era have been reviewed by CLARA (1961).

LONG (1947) succeeded in separating the sarcolemma into two layers; the inner one was amorphous and the outer consisted of fine filaments. The first electron microscopic studies confirmed that BOWMAN'S (1840) "membranaceous sheath" was multilayered (REED and RUDALL 1948; DRAPER and HODGE 1949; SITARAMAYYA 1951; RUSKA 1954; ROBERTSON 1956; PORTER and PALADE 1957; FAWCETT and SELBY 1958; MAURO and ADAMS 1961). Three layers have been identified: a plasma membrane, a basal lamina, and a network of collagen fibrils.

2. The Non-Membrane Components

For descriptive purposes, the sarcolemma is divided into two main layers: the plasma membrane, and the "basement membrane" consisting of the basal lamina and the reticular lamina formed by collagen fibrils (Fig. 56).

The organization and the mechanical role of the collagen fibrils for the muscle fibre have been discussed together with the connective tissue of muscle (Sect. B.IV).

Isolated sarcolemmal tubes consist of protein (mostly collagen of type IV and V; DUANCE et al. 1977; BAILEY et al. 1979), carbohydrates (GOLDSTEIN 1959), and lipids. Neither collagenase treatment nor lipid extraction alone destroys the tubular structure of isolated sarcolemma, but successive application of both treatments produces an amorphous precipitate (KONO et al. 1964).

The plasma membrane and the basal lamina are separated by a 20-nm-wide gap, which appears empty in electron micrographs. The content of this gap binds colloidal iron or thorium (ZACKS et al. 1973), and corresponds to the cell coat or glycocalyx of other cells (ITO 1965; RAMBOURG et al. 1966; RAMBOURG and LEBLOND 1967).

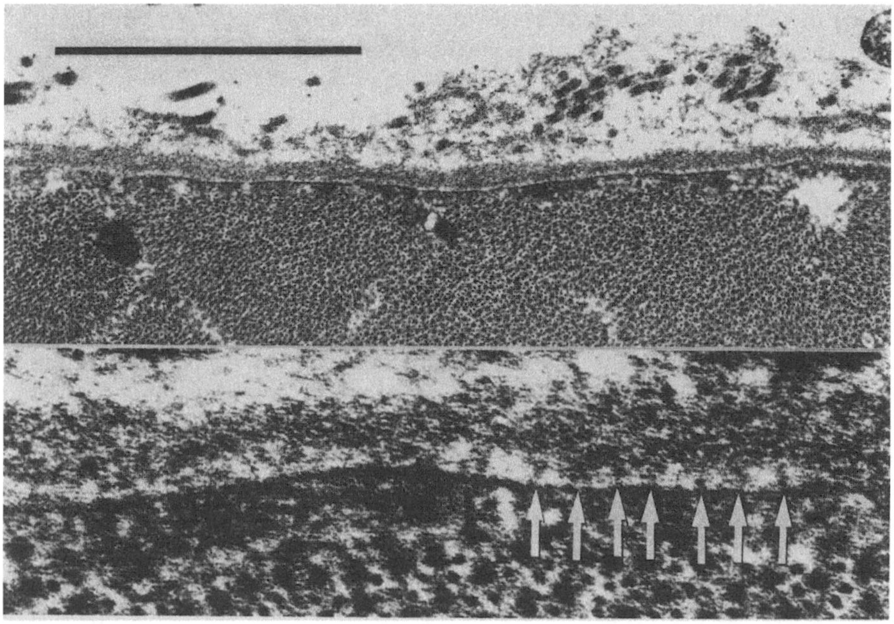

Fig. 56. Human brachial biceps muscle. Cross-section showing the layers of the sarcolemma. The following structures are shown (*upper figure*, from bottom): myofilaments, the plasma membrane, an empty gap representing the glycolcalyx of the membrane, the basal lamina consisting of amorphous or fine-filamentous material which extends into the external lamina consisting of rather few collagen fibrils. At higher magnification (*bottom*) one sees that the apparently empty gap between basal lamina and plasma membrane is traversed by structures (*arrows*) which probably are identical to the periodic structures found by BONILLA (1983) in freeze fractures (see text). Bar, 1 μm and 0.25 μm

Deep-etched freeze fractures show fine filaments crossing the glycocalyx at regular intervals (BONILLA 1983). Similar periodicities may be seen in thin sections heavily stained with uranyl acetate and lead citrate (Fig. 56). This extracellular filament system is reminiscent of the periodic filaments which at the myotendinous junction cross the plasma membrane and link the terminal I filaments to the collagen of the tendon (TROTTER et al. 1981, 1983) (see Sect. C.IV.3d). It is conceivable that the cytoskeleton of the muscle fibre links the Z lines of the myofibrils to the plasma membrane (see Sect. C.II), and that the filaments described by BONILLA (1983) represent the link between plasma membrane and the basal lamina and ultimately the endomysium. The sarcomeres of adjacent fibres in living muscles tend to remain in register (BUCHTHAL and KNAPPEIS 1940), indicating that the myofibrils are not freely moving inside the fibre. Segments of the sarcolemmal tube can only be isolated when the Z-band material is solubilized. The expulsion of the fibre content through the open ends is then simply brought about by osmotic vesiculation of the internal membrane systems (BERINGER and KOENIG 1975).

Beyond the glycocalyx is the basal lamina which has a strong affinity to ruthenium red. This dye also stains material extending from the basal lamina between the innermost collagen fibrils of the reticular layer; ZACKS et al. (1973) named the entire ruthenium red-positive layer of the sarcolemma "external lamina". This layer is rich in glycoproteins.

The basal lamina encloses the entire muscle fibre; it is continuous with the basal lamina of the Schwann cells of the motor axon, and also extends into the cleft between the pre- and postsynaptic membranes of the neuromuscular junction (Figs. 57, 58). The basal lamina is permeable for ferritin which has a diameter of 12 nm, but not for a colloidal gold-horseradish peroxidase complex with a diameter of 20–25 nm. Hence, the basal lamina may act as a diffusion barrier for large macromolecules (OLDFORS and FARDEAU 1983).

The chemical composition of extrajunctional and junctional basal lamina is different. The junctional basal lamina binds divalent ions more intensely than the extrajunctional basal lamina. This property has been utilized to stain the endplates for light microscopy (SAVAY and CSILLIK 1959; NAKAMURA et al. 1967).

The basal lamina persists when a muscle fibre necrotizes and provides a mechanical scaffolding for the regenerating myoblasts. This is probably not the only role the basal lamina has during muscle repair. A denervated muscle fibre is more apt to be reinnervated by a sprouting motor axon at the site of the former endplate than at the extrajunctional sarcolemma. Originally, it was surmised that either the persisting Schwann cells guide the axon (IWAYAMA 1969) or the specific membrane properties of the former postsynaptic membrane attract the growing axon (BENNETT et al. 1973). Now it appears as if specific properties of the junctional basal lamina both attract the axon and determine the specialization of plasma membrane of the muscle fibres. SANES et al. (1978) damaged the cutaneous pectoris muscle of frog, and at the same time interrupted the motor axons by crushing. Muscle regeneration was then blocked by X-irradiation, which did not prevent axonal outgrowth. The regenerated axons terminated at the empty basal lamina tubes at the precise site of the former endplates. Also BURDEN et al. (1979) denervated and damaged the muscle, but they prevented reinnervation instead of muscle regeneration. The plasma membranes of the new myofibres showed a high concentration of acetylcholine receptors at the site of the former endplate, although there was no attached terminal axon. In both studies, the site of the former neuromuscular junction was identified by staining for cholinesterase, which persists within the basal lamina material for up to several weeks (MCMAHAN et al. 1978). It is likely that specific molecules within the junctional basal lamina attract the axon and are also responsible for the differentiation of the muscle cell membrane (SANES et al. 1978).

SANES and HALL (1979) and SANES (1982) stained rat skeletal muscle fibres with labelled antibodies to locate the various components of the muscle fibre basement membrane and found laminin, fibronectin, and collagen type IV in the basal lamina of the junctional and extrajunctional region. Collagen type V and a specific soluble collagenous protein (high-salt-soluble protein, HSP) occur in the extrajunctional region only. HSP and collagen type V are in the reticular lamina, i.e. in the outer part of the basement membrane. The endplates have been selectively stained by antibodies against acetylcholinesterase, by rhodamin-labelled α-bungarotoxin, by antibodies against cytoplasmic actin (HALL et al.

1981), and by fluorescein-labelled lectin (Sanes and Cheney 1982). α-Bungarotoxin binds to acetylcholine receptors which, like acetylcholinesterase, are necessary for neuromuscular transmission (see below). Cytoplasmic actin may be involved in clustering or stabilization of acetylcholine receptors in the postjunctional membrane. The binding of lectin indicates that there are synapse-specific carbohydrates in the basal lamina, which may be involved in the targeting of the axons and recognition of synaptic sites. The lectin receptors persist after denervation or muscle fibre necrosis (Sanes and Cheney 1982). By contrast, Pena et al. (1981) and Watanabe et al. (1981) using the same type of lectin (agglutinin I from the legume *Doliches biflorus*, DBA) found, in human and mouse skeletal muscles, that it was only bound to the myonuclei or that it was not bound at all.

3. The Plasma Membrane

a) Functional Differences Between Junctional and Extrajunctional Membrane

The plasma membrane of a muscle fibre contains specific proteins which are unevenly distributed and which determine the properties of the different parts of the membrane (for reviews, see Almers et al. 1982; Stefani and Chiarandini 1982). The Na^+ channels of the extrajunctional membrane are involved in the generation of the action potential; they are blocked by tetrodotoxin. The endplate region contains acetylcholine receptors (AChRs) and cholinergic channels which open under the influence of acetylcholine (ACh). ACh induces junctional potentials and may elicit an action potential propagating along the extrajunctional membrane. Acetylcholinesterase (AChE) is present in the endplate region to destroy ACh after it has been liberated from the nerve terminal.

That ACh is a neurotransmitter not only in the autonomic nervous system but also in skeletal muscles became evident when Dale et al. (1936) found that blood from electrically stimulated muscles contained an increased amount of ACh, and when Brown et al. (1936), by injecting ACh into the artery of a cat muscle, produced contractions comparable to those after electrical stimulation. Buchthal and Lindhard (1937, 1942) mimicked the presumed action of the nerve by applying a small amount of ACh to the endplates of individual muscle fibres which then twitched; the extrajunctional plasma membrane did not respond. Peper and McMahan (1972) found that the junctional ACh sensitivity in frog muscle fibres fell steeply over a distance of only 5–10 µm. According to Kuffler and Yoshikami (1975), the ACh sensitivity is at least 50 times lower 2 µm outside the endplate than at the postsynaptic membrane; in distant parts of the extrajunctional membrane, it is only 1% of the junctional ACh sensitivity.

In mammalian muscles, the drop in sensitivity outside the endplate region is most pronounced in fast-twitch fibres. Slow-twitch fibres show a weak sensitivity also along the extrajunctional membrane; towards the myotendinous junctions, the sensitivity increases again (Katz and Miledi 1964; Miledi and Zelena 1966; Albuquerque and Thesleff 1968). The fibre-type-specific pattern of ACh sensitivity is governed by the motoneuron; slow-twitch fibres cross-innervated by a "fast" nerve lose the sensitivity along the extrajunctional plasma membrane and at the myotendinous junction, and fast-twitch fibres cross-innervated by the nerve to a slow-twitch muscle develop ACh sensitivity outside the synaptic region (MILEDI et al. 1968).

b) The Structure of the Neuromuscular Junction

Before the advent of electron microscopy, it was an open question among morphologists whether there was a cytoplasmic continuity between nerve fibre and muscle cell or not; physiological and pharmacological findings had clearly indicated that the two were separated by membranes. In 1954, this dispute was settled by the results of three independent studies (REGER 1954; ROBERTSON 1954; PALADE 1954). When the terminal axon approaches the muscle fibre it loses its myelin sheath, and the axolemma becomes apposed to the plasma membrane of the muscle fibre leaving a 50-nm-wide gap. This gap contains amorphous basal lamina material. The presynaptic membrane of the axon terminal is more or less straight, and less electron dense than the postsynaptic membrane. The postsynaptic membrane is usually folded, the main orientation of the folds being at right angles to the fibre axis (Figs. 57, 58).

The axon terminal, with its covering Schwann cells, lies in a depression of the muscle fibre surface known as synaptic gutter or trough. The gap between axon and muscle fibre is the primary synaptic cleft from which the fissures between the postsynaptic folds, the secondary synaptic clefts, originate (for reviews, see COËRS 1967; BOWDEN and DUCHEN 1976). The bottom of the secondary clefts sometimes shows rather many T-tubule openings (rat diaphragm and soleus muscle, KORNELIUSSEN 1977).

The depth, shape, and size of the gutters and folds vary considerably in different species and also in different muscle fibre types of the same animal. Within the same species, the size of the endplate region increases with the diameter of the muscle fibre (ANZENBACHER and ZENKER 1963; NYSTRÖM 1968a; SALPETER and ELDEFRAWI 1973; KORDYLEWSKI 1979b; SLACK et al. 1983). The shape of the neuromuscular junction, and the terminal arborization of the axon provided the basis for the distinction of muscle fibres with en plaque and en grappe endings (see Chap. E) (for references, see KRÜGER 1952; BOWDEN and DUCHEN 1976). The relation between physiological properties and endplate configuration is not as distinct as it was once believed. Tonic (slow) fibres have en grappe endings but not all twitch fibres have typical en plaque endings. There is a tendency for the gutters to be deeper in mammalian muscle than in muscles of amphibia, reptilia, fish, or birds. Fish muscles have no postsynaptic folds. Synaptic folds are lacking in slow frog or snake muscle fibres (PAGE 1965; HESS 1965), but are present in frog twitch muscle fibres. ANDERSSON-CEDERGREN (1959) demonstrated a regular pattern of parallel folds and clefts in mouse; in guinea pig muscles, FARDEAU (1973) described a superficial cribriform layer with holes and a deeper part with continuous intercommunicating spaces. The postsynaptic membrane as seen in longitudinal sections of human muscle fibres is 10 times longer than the presynaptic membrane (SANTA et al. 1972; ENGEL and SANTA 1973).

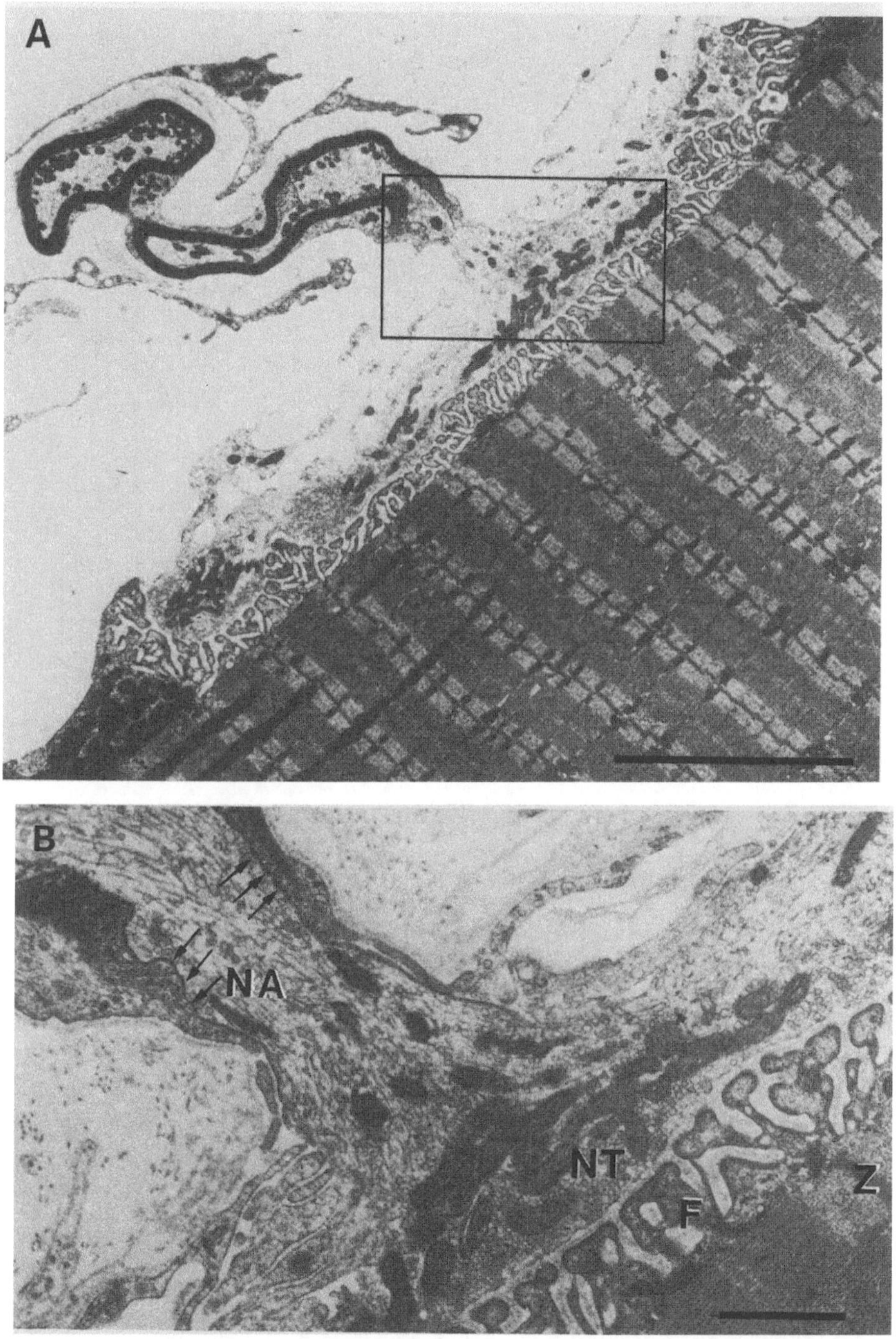

Fig. 57 A, B. Soleus muscle rat, endplate. **A** Low-power view. The terminal axon loses its myelin sheath and ends in a flattened nerve terminal embedded in the synaptic gutter of the muscle fibre. The bottom of the synaptic gutter consists of numerous, rather regular indentations, oriented transversely to the axis of the muscle fibre. These identations represent the secondary synaptic clefts, the primary synaptic cleft is the gap between the presynaptic membrane of the nerve terminal and the bottom of the gutter. **B** Serial section of the area inscribed in **A** photographed at higher magnification. After the terminal axon (*NA*) has left its myelin sheath, which ends with terminal loops (*arrows*), it extends into the flat nerve terminal (*NT*) containing mitochondria and synaptic vesicles. Primary and secondary synaptic clefts contain basal lamina material. The junctional folds of the postsynaptic membrane (*F*) are, on their crests, covered by a thickened plasma membrane. Bars, 5 μm and 1 μm. (From Ellisman et al. 1976, with permission of the authors and the Rockefeller University Press)

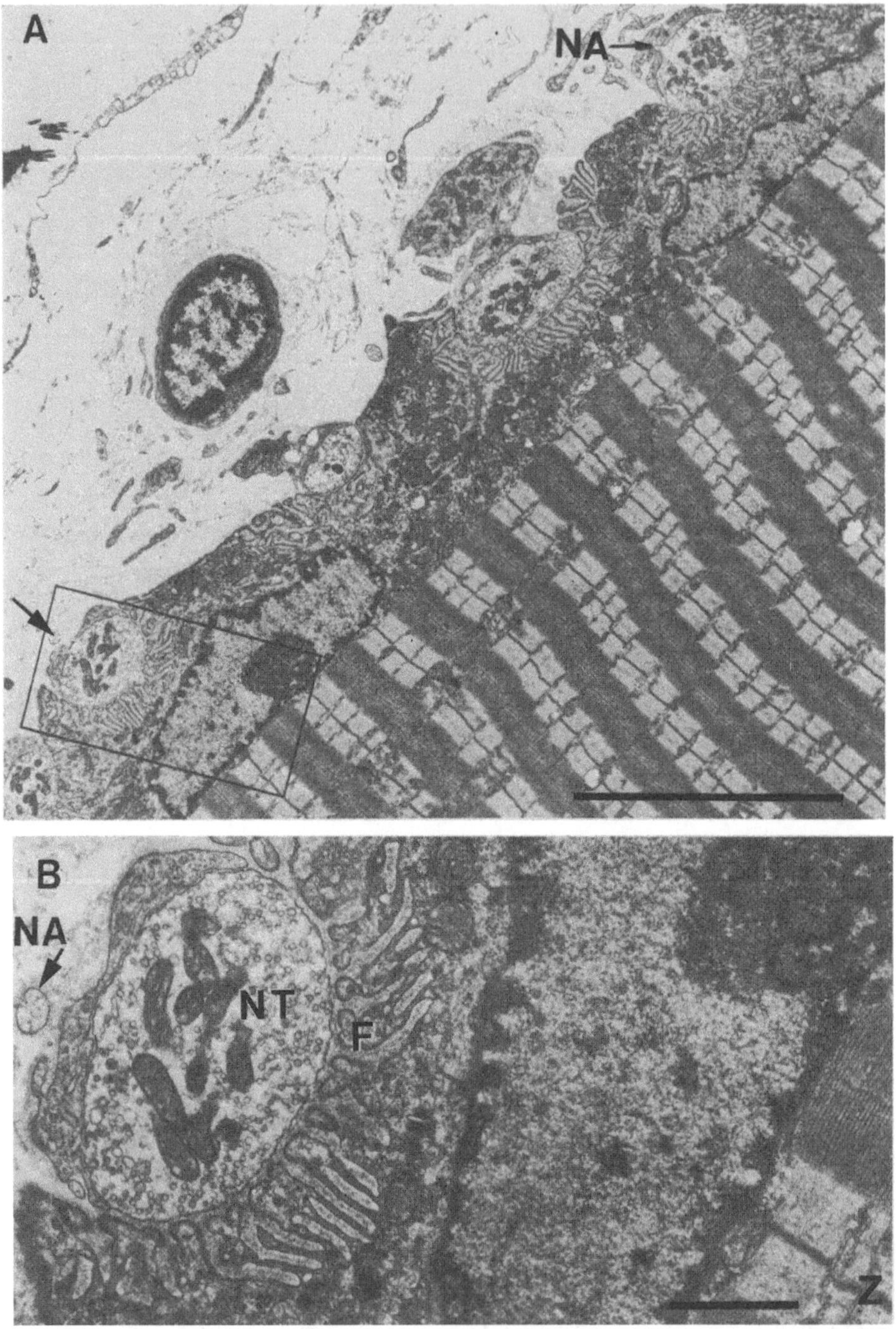

Fig. 58 A, B. Extensor digitorum longus muscle of rat, endplate. **A** Low-power view. The terminal axon (*NA*) has numerous bulb-like extensions which lie in deep synaptic gutters with regular synaptic folds. **B** Higher magnification of area inscribed in **A**. The bulb-like nerve terminals (*NT*) are filled with synaptic vesicles. The junctional folds (*F*) appear deeper than in the soleus muscle (Fig. 57). Bars, 5 μm and 1 μm. (From Ellisman et al. 1976, with permission of the authors and the Rockefeller University Press)

Morphological differences between endplates of fibres of different types in man, rat, and mouse have been described by MURATA and OGATA (1969), PADYKULA and GAUTHIER (1970), DUCHEN (1971), and SANTA and ENGEL (1973). In general, there are more and deeper clefts in the postsynaptic membranes in "white" or fast- than in "red" or slow-twitch fibres.

PADYKULA and GAUTHIER (1970) found, in the diaphragm of rat, that in fibres rich in mitochondria the secondary junctional clefts are 0.5 μm deep, as opposed to 1.0 μm in fibres poor in mitochondria. In "red" fibres the folds are sparser and more branched than in "white" fibres. In "red" fibres the axonal profiles are small and elliptical in outline (2.8–4.0 μm by 1.1–1.7 μm; in "white" fibres they are flat and long (4.6–8.4 μm by 1 μm). Intermediate fibres have deep secondary clefts (1.2 μm) and large axonal endings. The difficulty with these results is that many "red" fibres of rat muscles are fast (Sect. D.IV.2). DUCHEN (1971) compared muscle fibres of the mouse soleus, which contains both fast- and slow-twitch fibres, with fibres of the fast-twitch gastrocnemius muscle, and found in general deeper clefts in the gastrocnemius (0.8 μm) than in the soleus muscle (0.5 μm). The folds are twice as numerous in the gastrocnemius than in the soleus muscle. SANTA and ENGEL (1973) studied rat soleus and gastrocnemius fibres and report that the nerve terminals in "red" soleus fibres have more and larger synaptic vesicles than in "white" fibres. They do not find consistent differences in the shape of the nerve terminals or the postjunctional structures, but considerable size variations within the same fibre type have been observed. Only the ratio between post- and presynaptic membrane lengths as seen in longitudinal sections is larger in "white" (10.5) than in "red" or intermediate fibres (8.1). By contrast, ELLISMAN et al. (1976) describe distinct differences between soleus (slow-twitch) and extensor digitorum longus (fast-twitch) fibres of rat. The nerve terminals of rat soleus fibres resemble "flattened plaques with vesicle-rich expansions interconnected by filamentous neuroplasm"; whereas, in the extensor digitorum longus muscle, there are "numerous expansions arising from ramifications of the terminal axon". The soleus endplates measure 30–50 μm by 20–30 μm, the extensor digitorum longus endplates are 75–100 μm long. The depth of the secondary clefts is 0.75 μm and 1.0 μm, respectively (Figs. 57, 58). Acetylcholinesterase staining shows that, in the soleus, the reactive area of the endplates is more lobulated than in the extensor digitorum longus (Fig. 59). It was consistently found that the endplates of the soleus muscle are longer; this possibly relates to the larger fibre diameter (SCHMALBRUCH, unpublished), and contrasts with the results of ELLISMAN et al. (1976). KORNELIUSSEN and WAERHAUG (1973) in the diaphragm of rat distinguish, using light microscopy, three different types of endplates after intravital methylene blue staining. Similar differences were found by KEMPLAY and STOLKIN (1980) in rat sternocostalis muscle after zinc-iodide-osmium impregnation. It is obscure how these endplate "types" relate to the physiological fibre types.

The shape of the endplate may not only reflect the fibre type; FAHIM and ROBBINS (1982) found that in mice the postsynaptic area becomes more complex with age. CARDASIS and PADYKULA (1981) observed degenerative changes in the endplates of soleus muscles of rats 3–5 months old, which they assume indicates reorganization either during growth or during the physiological postna-

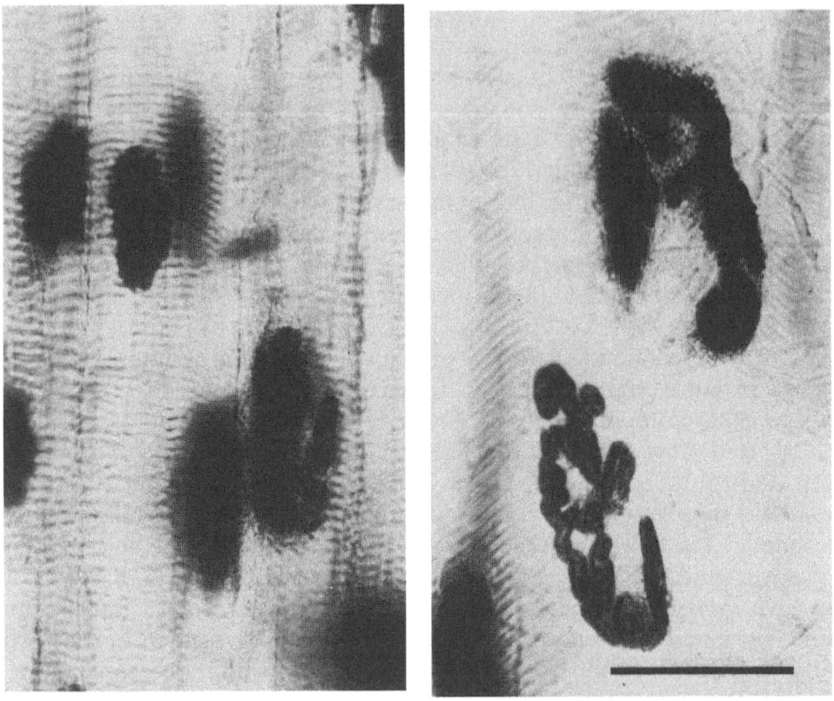

Fig. 59. Rat extensor digitorum longus (*left*) and soleus (*right*) muscles. Whole mount stained for acetylcholinesterase to show the endplates. The reactive areas were consistently longer in the soleus than in the extensor digitorum longus muscle. The shape of the endplates appears more lobulated in the soleus muscle, which apparently contrasts with the electron micrographs shown in Figs. 57 and 58. Bar, 50 µm

tal motor unit conversion (KUGELBERG 1973a, 1976) (see Sect. D.VI.1). BANKER et al. (1983) found in the extensor digitorum longus muscles of mice 2–3 years of age, postsynaptic membranes without attached axon terminals, as well as decreased nerve terminal areas. The changes are mainly presynaptic; neuromuscular transmission is unimpaired.

α) The Presynaptic Membrane

The axon terminal is filled with clear vesicles 45–52 nm in diameter, which tend to cluster at thickenings of the presynaptic membrane (the "active sites") (HUBBARD and KWANBUNBUMPEN 1968; HUBBARD 1973). The proximal part of the terminal axon contains many mitochondria; the lateral faces of the axon terminal are covered by Schwann cells and the axon membrane reveals coated pits and vesicles.

The synaptic vesicles disappear when the nerve is stimulated for minutes or hours (KORNELIUSSEN 1972; HEUSER and REESE 1973); concomitantly the total amount of presynaptic membrane increases, suggesting that exocytosis of the vesicle contents has taken place. HEUSER et al. (1974) found in freeze fractures of endplates rapidly frozen during stimulation, an increased number of pits and dimples in the presynaptic membrane representing vesicles during exocytosis (Fig. 60).

The number of coated vesicles and of intra-axonal membranes which form cisterns instead of vesicles, increases. This has been taken as evidence that the vesicle membranes are recycled (HEUSER and REESE 1973). CECCARELLI et al. (1979b) found no increase in the number of coated vesicles during recovery from stimulation, and observed endocytosis only close to the site of exocytosis. MILLER and HEUSER (1984) distinguish two types of endocytosis: a rapid one commencing seconds after stimulation, which without coated vesicle formation occurs both near and far from the active zones, and a slow one commencing 1 s after stimulation and continuing for 90 s. The second type involves coated vesicles and does not occur close to the active zones.

When a frog nerve-muscle preparation is stimulated (for 20 min) and during recovery, when the presynaptic membrane material is recycled, incubated, with labelled AChE, the label occurs inside the synaptic vesicles of the axon terminal. Transmission is blocked and remains so, even when AChE inhibitors are added. The explanation for this finding is that the vesicle membrane during recycling has taken up AChE, which then hydrolyzes ACh within the vesicles, and thereby blocks transmission (POLITOFF et al. 1975).

The dense presynaptic membrane areas ("active zones") in freeze fractures of frog muscles appear as series of transverse ridges. The ridges are bordered by two parallel rows of relatively large intramembrane particles (Fig. 60). These particles may constitute the calcium channels mediating the calcium influx necessary for vesicle exocytosis. In stimulated terminals, the ridges carry dimples representing synaptic vesicles during exocytosis (HEUSER et al. 1974, 1979; PEPER et al. 1974; CECCARELLI et al. 1979b). The active zones become disorganized in the absence of Ca^{2+} and the large particles are scattered over the P face of the presynaptic membrane; pharmacologically induced exocytosis of synaptic vesicles then takes place close to these particles (CECCARELLI et al. 1979a). RASH and ELLISMAN (1974), ELLISMAN et al. (1976), and RASH et al. (1981) find in rat and human muscles that the active zones of the presynaptic membrane consist of four short parallel rows, each row being formed by 6–10 particles. the particles are in a square lattice with 20-nm distance between the outer and inner row, and 40-nm distance between the two inner rows. The vesicle openings that have occurred during stimulation are between the two inner rows (Fig. 61).

The thickenings of the presynaptic membrane, corresponding to the active zones or rows of intramembrane particles, in frog muscles appear in sections opposite the folds of the postsynaptic membrane (BIRKS et al. 1960); in rat, only 60% of the postsynaptic folds are opposite a presynaptic thickening (HUBBARD and KWANBUNBUMPEN 1968). In freeze fractures, the different orientation of the active zones is obvious: in frog, they are long and overlie the largest axis of the clefts between the postsynaptic folds (HEUSER et al. 1974); in rat, the active zones are short and span the clefts between the postsynaptic folds (ELLISMAN et al. 1976).

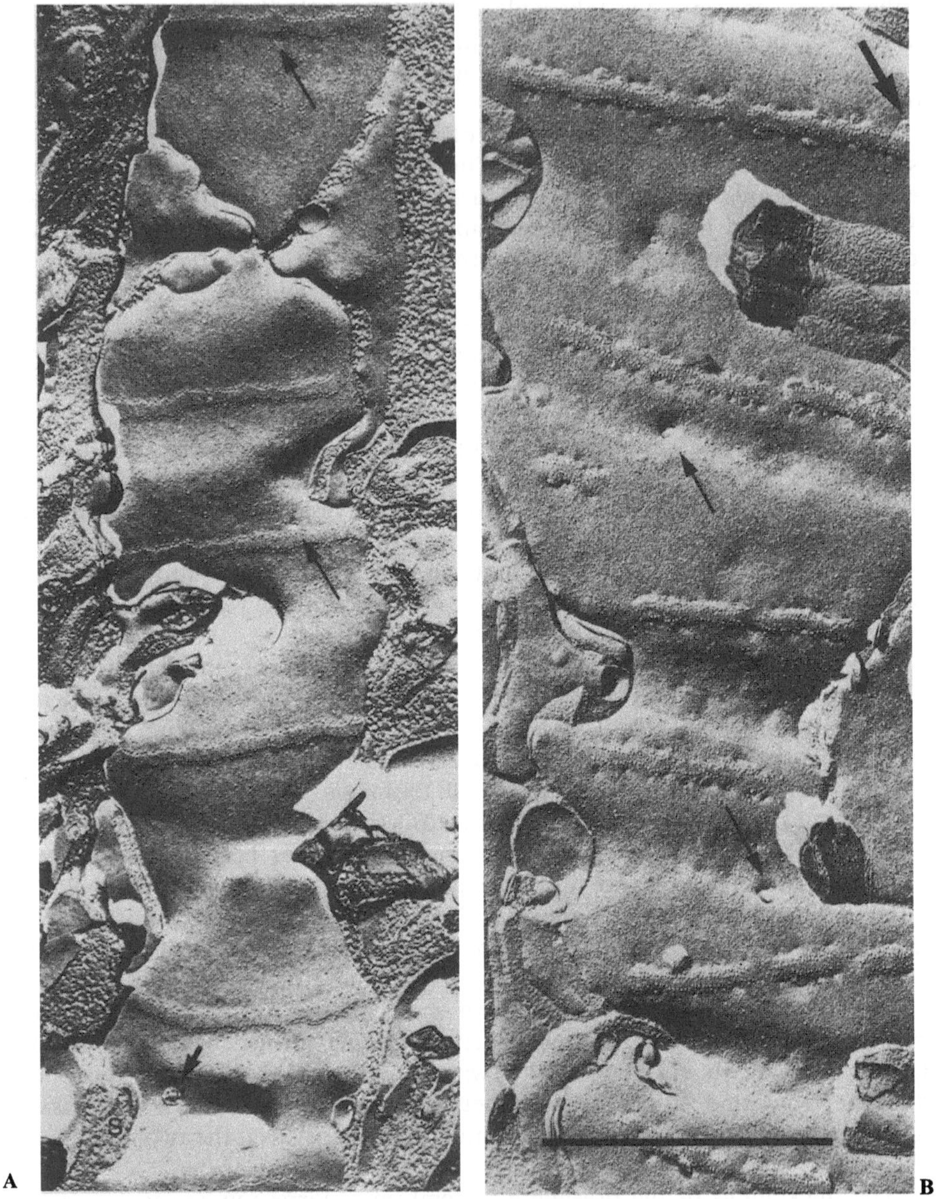

Fig. 60 A, B. Freeze fractures of the presynaptic membrane in frog endplates, in both micrographs the P face is shown. The fibre axis is vertical. **A** At rest. Prominent transverse ridges (*arrows*) run parallel to the junctional folds of the postsynaptic membrane. The ridges are bordered by rows of intramembrane particles. **B** After electrical stimulation. Numerous dimples have appeared (*arrows*), both in connection with the ridges and between them. The small dimples at the ridges probably represent sites of exocytosis of synaptic vesicles, and the large dimples are probably sites of endocytosis to retrieve membrane material. Bar, 1 μm. (From HEUSER et al. 1974, with copyright permission of Chapman and Hall)

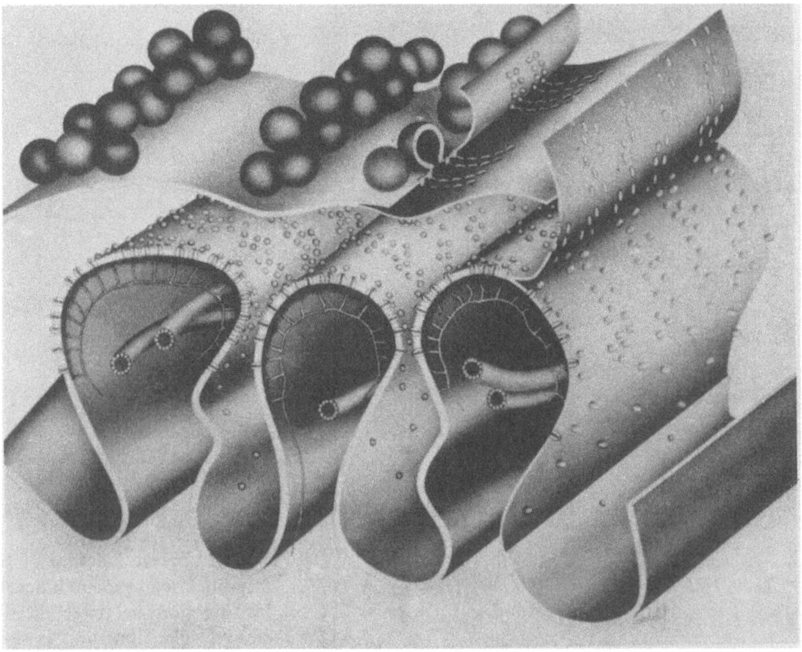

Fig. 61. Highly schematic diagram of a mammalian neuromuscular junction (note difference to amphibian neuromuscular junction shown in Fig. 60). From top: Clusters of synaptic vesicles containing acetylcholine are close to the presynaptic membrane which displays short ridges bordered by double rows of intramembrane particles. The orientation of the ridges is perpendicular to the junctional folds of the postsynaptic membrane. The synaptic vesicles open between the double rows of particles to release acetylcholine into the synaptic cleft. The junctional folds carry acetylcholine receptors (shown as large particles) on their crests. Also depicted is the filamentous network beneath the postsynaptic membrane, and microtubules running within the folds. (From ELLISMAN et al. 1976, with permission of the authors and the Rockefeller University Press)

β) The Postsynaptic Membrane

The postsynaptic membrane is less electron dense in the depth than at the crest of the folds (REGER 1954, 1958, 1959; ROBERTSON 1956; DE ROBERTIS 1958; ANDERSSON-CEDERGREN 1959; DE HARVEN and COËRS 1959; BIRKS et al. 1960; ZACKS and BLUMBERG 1961; COUTEAUX 1963; ZENKER and ANZENBACHER 1964; VON DÜRING 1967). Freeze fractures show that the crests of the folds are paved with clusters of large (11–14 mm) particles adhering to the P face. The P face also shows orthogonal arrays (or "assemblies") of 6-nm particles with corresponding pits on the E face (HEUSER et al. 1974; RASH and ELLISMAN 1974; ELLISMAN et al. 1976) (Fig. 62). These formations shall be discussed together with the extrajunctional sarcolemma.

γ) Acetylcholinesterase

The postsynaptic membrane is associated with acetylcholinesterase (AChE). No AChE is found 2 μm outside the endplate (KUFFLER and YOSHIKAMI 1975).

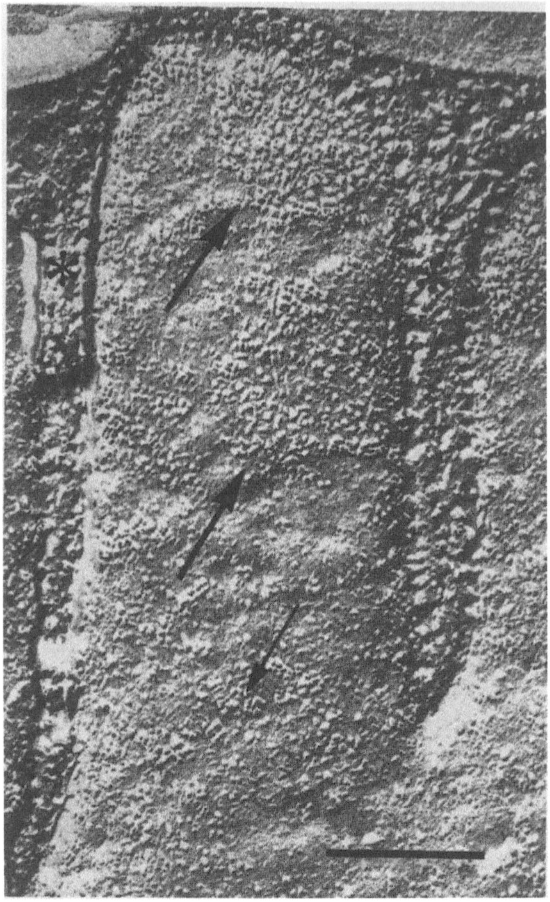

Fig. 62. Freeze fracture of the postsynaptic membrane of a neuromuscular junction in frog; the P face is shown. The fibre axis is horizontal. The crest of a junctional fold is bordered by openings of secondary clefts. The fracture face of the membrane carries large angular particles representing acetylcholine receptors (*large arrows*) and few 6-nm particles in an orthogonal array ("square arrays") (*small arrow*). Bar, 0.5 µm. (From HEUSER et al. 1974, with copyright permission of Chapman and Hall)

AChE is a rather stable enzyme and it is conveniently shown using histochemistry (KOELLE and FRIEDENWALD 1949; COËRS 1953; COUTEAUX 1955) (Fig. 59). In slow-twitch fibres there is also some AChE activity at the myotendinous junctions (COUTEAUX 1953; GEREBTZOFF 1956, 1957; SCHWARZACHER 1960a, c, d, 1961), corresponding to the acetylcholine sensitivity at this site (MILEDI and ZELENA 1966). Staining for AChE has become a routine procedure which can be applied to macroscopic specimens to show the course of the endplate zone, but also to samples for electron microscopy.

The basic principle of the classical histochemical method for AChE is that thiocholine is offered as substrate. Many modifications for light and electron microscopy have been described (for review, see FRIEDENBERG and SELIGMAN 1972). The major problems with staining for electron microscopy are diffusion, and attachment of the reaction product to membranes. This limits the spatial resolution. Attempts to localize the enzyme with a modified Koelle technique gave inconsistent results (ZACKS and BLUMBERG 1961; BARRNETT 1962; MILEDI

1964; IWAYAMA 1966; TERÄVÄINEN 1967; DAVIS and KOELLE 1967), but eventually it has been assured that the reaction product is in the primary and secondary synaptic clefts (DAVIS et al. 1972).

Various enzyme inhibitors serve to distinguish AChE, pseudo-cholinesterase, and unspecific esterases. One of the inhibitors, di-isopropylfluorophosphate (DFP), strongly binds to the molecule; as ^{3}H-DFP or ^{32}P-DFP it is used as an in situ label for cholinesterase. The complex is then shown using autoradiography (SALPETER 1969; SALPETER et al. 1972), or it serves to quantify the amount of cholinesterase by liquid-scintillation counting; 35% of the DFP-binding sites at the endplates represent AChE (BARNARD and ROGERS 1967). In a more direct approach, the localization of the enzyme has been demonstrated using fluorescent antibodies against the AChE molecule (GREENBERG et al. 1977; WEINBERG et al. 1981).

AChE is synthesized by the muscle cell; a labile intracellular fraction which rapidly disappears after inhibition of protein synthesis, and a stable extracellular fraction which persists after inhibition of synthesis, can be distinguished (GOLDER et al. 1977). The extracellular fraction is bound to the basal lamina of the synaptic region (WEINBERG et al. 1981). When living muscle fibres are treated with appropriate concentrations of collagenase, trypsin, or protease, the basal lamina and the extracellular activity of AChE disappears, whereas the pre- and postsynaptic membranes remain functionally intact (HALL and KELLY 1971; BETZ and SAKMANN 1973). Different molecular forms of AChE occur in different relative amounts in fast- and slow-twitch muscles (BACOU et al. 1982). The number of AChE molecules per endplate is 0.4×10^{7} to 3×10^{7} (WASER and RELLER 1965; BARNARD and ROGERS 1967; ROGERS et al. 1969; ROGERS and BARNARD 1969). According to SALPETER (1969) and SALPETER et al. (1972), the DFP-binding sites are mostly in the depth of the postjunctional clefts; there the density is 12,000 molecules per μm^{2} postjunctional membrane area; 3,000–4,000 of these DFP-binding sites represent AChE.

δ) Acetylcholine Receptors

It was of advantage for the work on acetylcholine receptors (AChRs) that large amounts can be isolated from the cholinergic electric organ of *Torpedo* (FELDBERG and FESSARD 1942; BENNETT 1970; WHITTAKER 1977), and that the snake venom α-bungarotoxin irreversibly binds to the receptor (CHANG and LEE 1963; BERG et al. 1972). Thus, AChRs can be labelled for light and electron microscopy with α-bungarotoxin. To visualize the label, it may be conjugated with horseradish peroxidase (LENTZ et al. 1977) or fluorescent dyes such as rhodamine (ANDERSON and COHEN 1974; RAVDIN and AXELROD 1977), or it may be marked by ^{131}I (MILEDI and POTTER 1971) or ^{125}I (FAMBROUGH and HARTZELL 1972; HARTZELL and FAMBROUGH 1972). Unmarked α-bungarotoxin bound to AChRs can also be visualized using indirect immunocytochemistry with a peroxidase-labelled antibody (DANIELS and VOGEL 1975; BENDER et al. 1976; RINGEL et al. 1975, 1976). It is also feasible to omit the α-bungarotoxin step and to label the AChR molecule with fluorescent antibodies (WEINBERG and HALL 1979). In rat diaphragm, the concentration of the AChRs at the endplate is at least 2,500 times higher than at the extrajunctional membrane (HARTZELL and FAMBROUGH 1972). In fast- and slow-twitch cat muscle fibres, STEINBACH (1981) found no binding of α-bungarotoxin in the distant extrajunctional region. Junctional and extrajunctional AChRs differ immunologically and also with respect to their functional properties (KATZ and MILEDI 1972; NEHER and SAKMANN 1976; FAMBROUGH 1979; WEINBERG and HALL 1979; REINESS and HALL 1981; FAMBROUGH et al. 1982) (see also Sect. G.III.2dα).

The secondary peaks of acetylcholine sensitivity at the myotendinous junctions (frog, MILEDI 1960; rat, MILEDI and ZELENA 1966; man, CULL-CANDY et al. 1982) are due to the presence of AChRs as well. In frog muscles, these AChRs have similar functional properties to those at the neuromuscular junctions. The function of the myotendinous AChRs remains obscure (MILEDI et al. 1984).

From *Torpedo* postsynaptic membrane, FROEHNER et al. (1981) isolated, as well as AChRs, an unknown 43,000-dalton protein, which using immunohistochemistry was shown to be present in the postsynaptic membrane of rat diaphragm as well. The protein persisted after denervation; it was different from AChRs. Because it has been "developmentally conserved", the authors suspect that it has a role in neuromuscular transmission.

In cultured non-innervated muscle fibres, AChRs appear in clusters along the entire length of the cell (VOGEL et al. 1972; SYTKOWSKI et al. 1973; FISCHBACH and COHEN 1973; VOGEL and DANIELS 1976). Freeze fractures show at the same sites large, angular P-face particles (PENG and NAKAJIMA 1977; YEE et al. 1978; COHEN and PUMPLIN 1979). These particles resemble the large particles covering the crests of the postsynaptic folds; it is reasonable to assume that in both cases they represent AChRs (Figs. 62, 63).

A discrepancy exists between the density of AChRs in the postsynaptic membrane, as determined using autoradiography (25,000 to 30,000/μm²), and the number of large intramembrane particles at the corresponding sites (1,800 to 6,000/μm²; RASH and ELLISMAN 1974; HEUSER et al. 1974). If the particles represent the morphological correlate of AChR, it would seem that each particle corresponds to a complex of several molecules.

ε) *Quantitative Aspects*

The total number of AChRs per endplate is $0.9–4.7 \times 10^7$ (MILEDI and POTTER 1971; BERG et al. 1972; FAMBROUGH and HARTZELL 1972), the amount of ACh liberated per endplate and stimulus is 3×10^6 molecules (POTTER 1970), and the rate of ACh hydrolysis by AChE is 2.7×10^8 molecules/ms/endplate (NAMBA and GROB 1968). The size of the postsynaptic membrane is 3,000 μm² (mouse) to 7,000 μm² (rat), and the cleft volume is 450 μm³ (rat). The density of AChRs within the postsynaptic membrane is 7,200/μm², both in mouse and rat, because the total number of AChRs is larger in rat than in mouse. The AChRs occupy 10%–15% of the postsynaptic membrane area; the concentration within a 10-nm-thick membrane is 10^{-3} M. If ACh is uniformly distributed after release, the concentration would be 10^{-5} M, i.e. 100 times less than the concentration of AChRs. The activity of AChE would suffice to clear the cleft

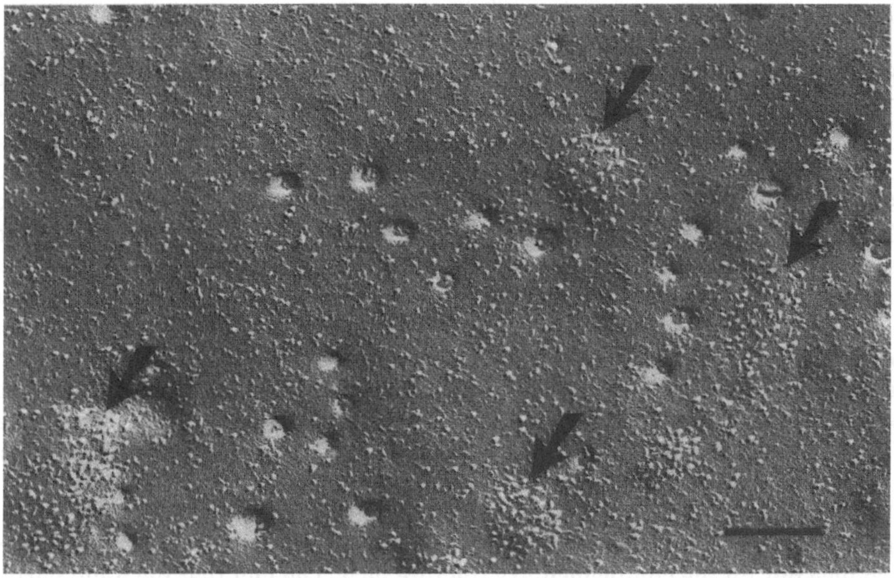

Fig. 63. Freeze fracture of the extrajunctional plasma membrane of an immature muscle fibre of rat (neonatal flexor digitorum brevis muscle). The P face is exposed. Among the 10-nm particles are larger particles which are clustered *(arrows)*. The number of clusters is unusually high in the area depicted. The particles probably represent extrajunctional acetylcholine receptors. Bar, 0,2 μm

within 1 ms (SALPETER and ELDEFRAWI 1973). These calculations are based on the assumption that the molecules involved are evenly distributed. Recent autoradiographic results suggest that this is not the case.

FERTUCK and SALPETER (1976) and MATTHEWS-BELLINGER and SALPETER (1978) labelled AChRs in endplates of the mouse sternomastoid and frog cutaneous pectoris muscles with ^{125}I-bungarotoxin. The binding sites were found only above the thickened membranes covering the crests of the postjunctional folds. These thickened membrane areas ("receptive area") comprise 25% and 50% of the postsynaptic membrane in mouse and frog, respectively. The AChR densities at these sites are 30,000 (mouse) and 26,000/μm^2 membrane (frog). The density of AChRs in the secondary clefts half-way down (0.4 μm) from the top of the junctional fold is only 3% of its peak subsynaptic value (mouse). There is also a sharp drop in AChR concentration outside the junction; 7 μm (mouse) and 17 μm (frog) from the edge of the axon terminal it is only 0.2% of the subsynaptic value. Even when the higher concentration of the AChRs at the receptive sites is taken into account, only 4% (mouse) to 8% (frog) of the ACRs would be saturated by ACh after a stimulus (MATTHEWS-BELLINGER and SALPETER 1978).

c) The Extrajunctional Plasma Membrane

By far the largest part of the plasma membrane of a muscle fibre is extrajunctional. Along this membrane the action potential is conducted, the T tubules which lead the excitation across the fibre originate at regular intervals, and all substances that go into or come out of the muscle fibre must pass through the sarcolemma, either by endo- or exocytosis or through membrane channels. The plasma membrane appears in electron micrographs of sections as a thin line with rather many open caveolae. Some of these caveolae may connect to convoluted T tubules. In longitudinal sections the course of the membrane is straight, in shortened fibres it may be festooned with inward bends at each Z-line level. Freeze fractures reveal more details (Figs. 63–65). The membrane is more or less plane, but there are folds, and openings of caveolae or T tubules. Intramembrane particles are evenly distributed (RAYNS et al. 1968). The particles measure 9–10 nm in diameter (BERTAUD et al. 1970) and are more frequent on P than on E faces. In addition to these many "10-nm particles" there are few "6-nm particles" which are always part of orthogonal arrays comprising 6–100 individual particles (RASH et al. 1974; HEUSER et al. 1974; SMITH et al. 1975; ELLISMAN et al. 1976; SHAFIQ et al. 1979; SCHMALBRUCH 1979c). These orthogonal or square arrays ("assemblies", HEUSER et al. 1974) are only on the P face, but they leave corresponding pits on the E face. This distinguishes them from scattered 10-nm particles which even in "perfect" fractures rarely produce pits on the complementary fracture face. The fact that the 6-nm particle square arrays always leave pits has been used to assess the degree of contamination during replication because, ideally, particles and pits should have the same size (RASH et al. 1979).

α) Folds and Caveolae

The volume of a muscle fibre remains constant at different lengths; hence, the surface must increase when the fibre is stretched. DULHUNTY and FRANZINI-ARMSTRONG (1975) studied this problem in freeze fractures of frog muscle fibres. At a sarcomere length of 2.8 µm, the membrane is folded perpendicularly to the longitudinal axis of the fibre; 37 caveolae per µm² fibre surface have been found. At 2.4-µm sarcomere length, 80% of the surface (without T tubules) is in the caveolae. They are distributed in two circumferential bands on either side of the Z-line level. At a sarcomere length of 3.0 µm, the folds have stretched out but the number of caveolae is unchanged. For a stretch beyond 3-µm sarcomere length, additional membrane is provided by the opening of caveolae, and at 8-µm(!) sarcomere length all caveolae are open. The membrane ruptures with further stretch. The authors conclude that the surface membrane is relatively inelastic, and that the increase in fibre surface area is compensated for by an increasing contribution of the membrane in folds and caveolae. As seen in sections, 47% of the plasma membrane of frog muscle fibres (without T tubules) are in the caveolae (MOBLEY and EISENBERG 1975).

From 12 to 16 (SCHMALBRUCH 1979c) to 19 (BONILLA et al. 1981) caveolae/µm² surface area have been found in human muscle fibres. BONILLA et al. (1981) stated that the caveolae, like in frog muscles (DULHUNTY and FRANZINI-ARMSTRONG 1975), are preferentially localized above the I band; SCHMALBRUCH (1982a), however, found only patchy density variations without relation to the sarcomere pattern. The number of caveolae in rat muscle fibres decreases after

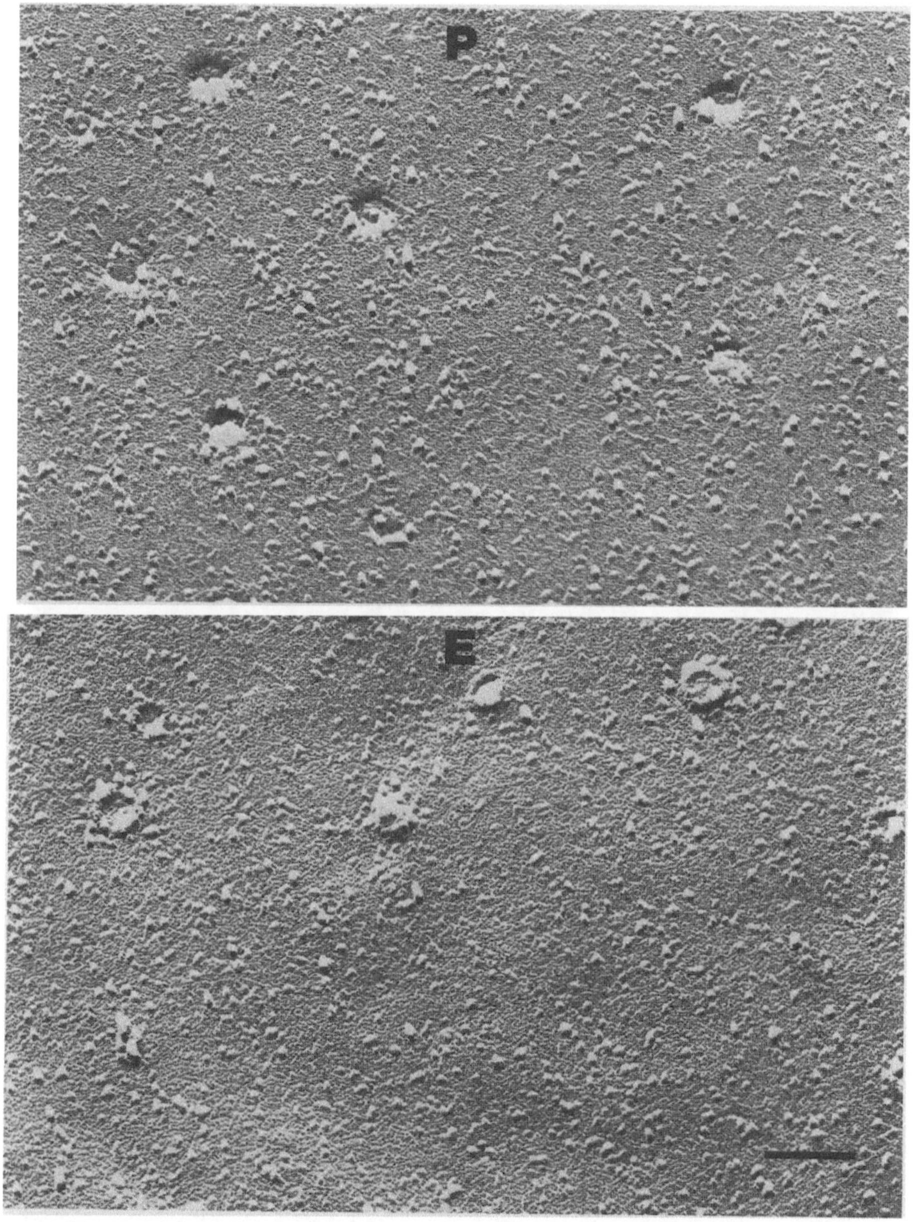

Fig. 64. Freeze fractures of fetal human muscle, exposing the *P* and *E* faces of the plasma membrane. The caveolae of the membrane appear as dimples and elevations, respectively; 10-nm particles are more frequent on the P face than on the E face. The density is lower than in adult muscle fibres, and square arrays are missing (see Fig. 66). Bar, 0.1 μm

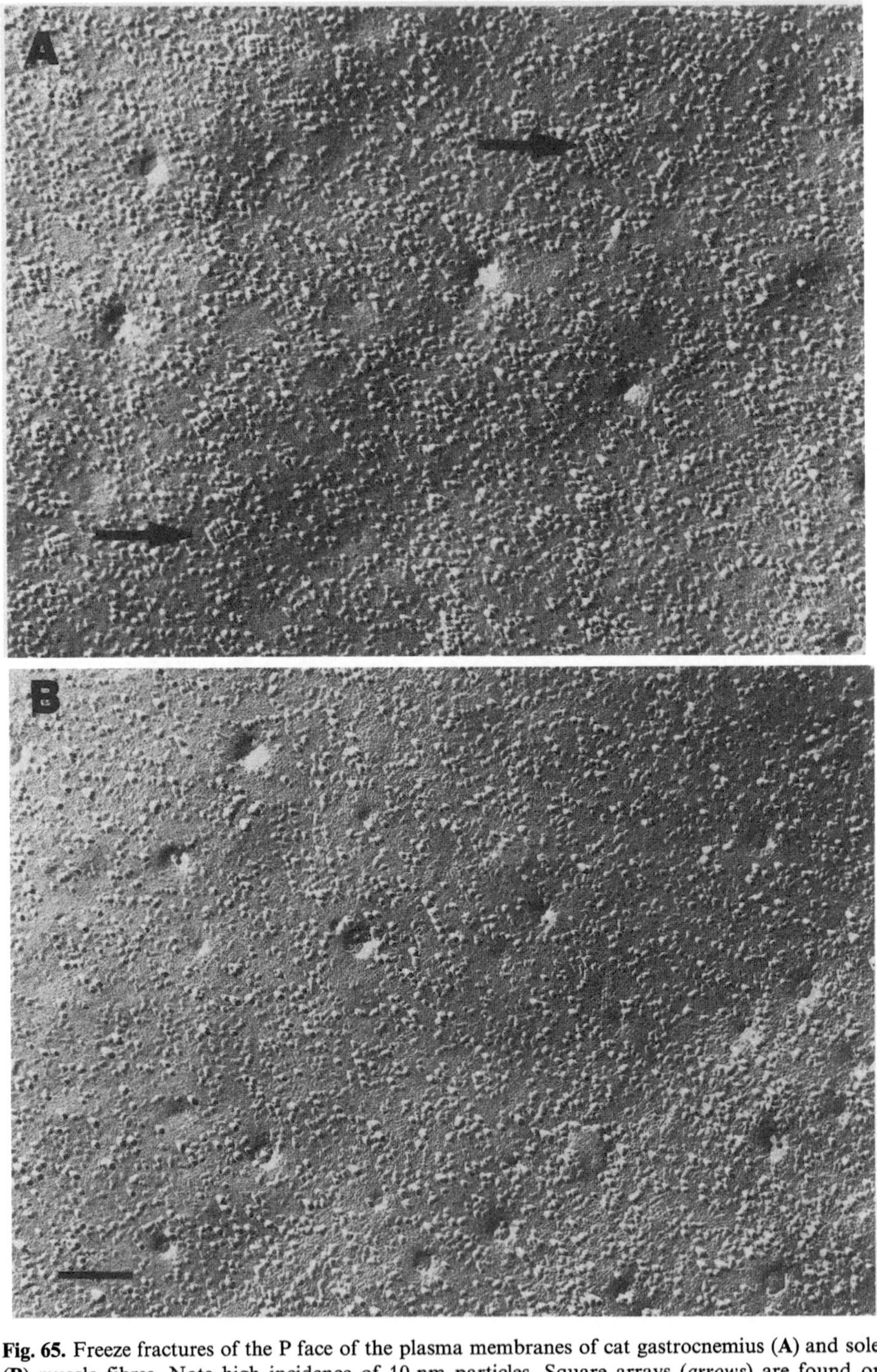

Fig. 65. Freeze fractures of the P face of the plasma membranes of cat gastrocnemius (**A**) and soleus (**B**) muscle fibres. Note high incidence of 10-nm particles. Square arrays (*arrows*) are found only on the presumably fast-twitch gastrocnemius fibre, but not on the slow-twitch soleus fibre. Bar, 0.1 μm

30–60 min hypoxia; in immature (fetal or regenerating) muscles, fewer caveolae than in mature muscles are found. The density is only 7/μm² surface area in human fetal muscle fibres (SCHMALBRUCH 1979c, 1980b, 1982a). LEE et al. (1983) investigated the influence of fixation on size and array of the sarcolemmal caveolae in rat muscles. Unfixed rapidly frozen fibres have few small-diameter, randomly scattered caveolae. The number of caveolae increases after perfusion fixation, but they are still small and randomly distributed. Immersion-fixed muscle fibres, however, in rat and also in man, show large caveolae, which in linear arrays are oriented over the I bands (see above). This change in pattern and size is most pronounced in fibres taken from the central part of the specimen, i.e. in those fibres which are the last to get into contact with the fixing solution. This finding underlines the extent to which the freeze-fracture image is preparation-dependent.

β) 10-nm Particles

The vast majority of intramembrane particles belong to the 10-nm class; they are believed to represent integral membrane proteins (MACLENNAN et al. 1971; MARCHESI et al. 1972). It is not possible morphologically to distinguish different types of 10-nm intramembrane particles, although there are various proteins within a membrane.

AChRs which have been identified in the extrajunctional sarcolemma of myotubes as angular particles with a size of 10–20 nm (Fig. 63) do not occur in mature muscle fibres (ELLISMAN et al. 1976; COHEN and PUMPLIN 1979).

PUMPLIN and FAMBROUGH (1983) labelled cultured chick myotubes with a monoclonal antibody against $(Na^+ + K^+)$ATPase. The binding sites could be induced to cluster, and samples studied by freeze fracture showed that a large number of intramembrane particles had clustered as well. Antibodies against other surface proteins applied after the binding sites had been induced to cluster did not bind in a patchy fashion. Clustered and non-clustered particles did not differ in size. The authors conclude that more than 50% of the intramembrane particles, i.e. about 400/μm² P-face area, each represented one $(Na^+ + K^+)$ ATPase molecule.

The observation that muscular dystrophy at a very early stage affects the muscle cell membrane incited interest in freeze fractures of the plasma membrane of diseased human muscle. Several workers have quantitated the particle density. The results are inconsistent (for references, see SCHMALBRUCH 1982b; SHOTTON 1982). Nevertheless, even for normal muscles variable counts have been reported for the number of 10-nm particles per μm² P-face area: 240 (KETELSEN 1974, 1975), 168 (YOSHIOKA and OKUDA 1977), 262 (SCHOTLAND et al. 1977), 550 (KETELSEN 1980), 1,564 (SCHOTLAND et al. 1981). OSAME et al. (1981) found 1,621 particles/μm² P face in cultured human muscle fibres, i.e. in non-innervated myotubes.

In hypoxic rat muscle fibres, the particles are clustered, and band-like particle-free areas are formed. The number of particles is initially unchanged, but decreases by more than 50% after 60 min hypoxia (SCHMALBRUCH 1980b). Occasionally one sees band-like particle-free areas in human muscle; the unavoidable

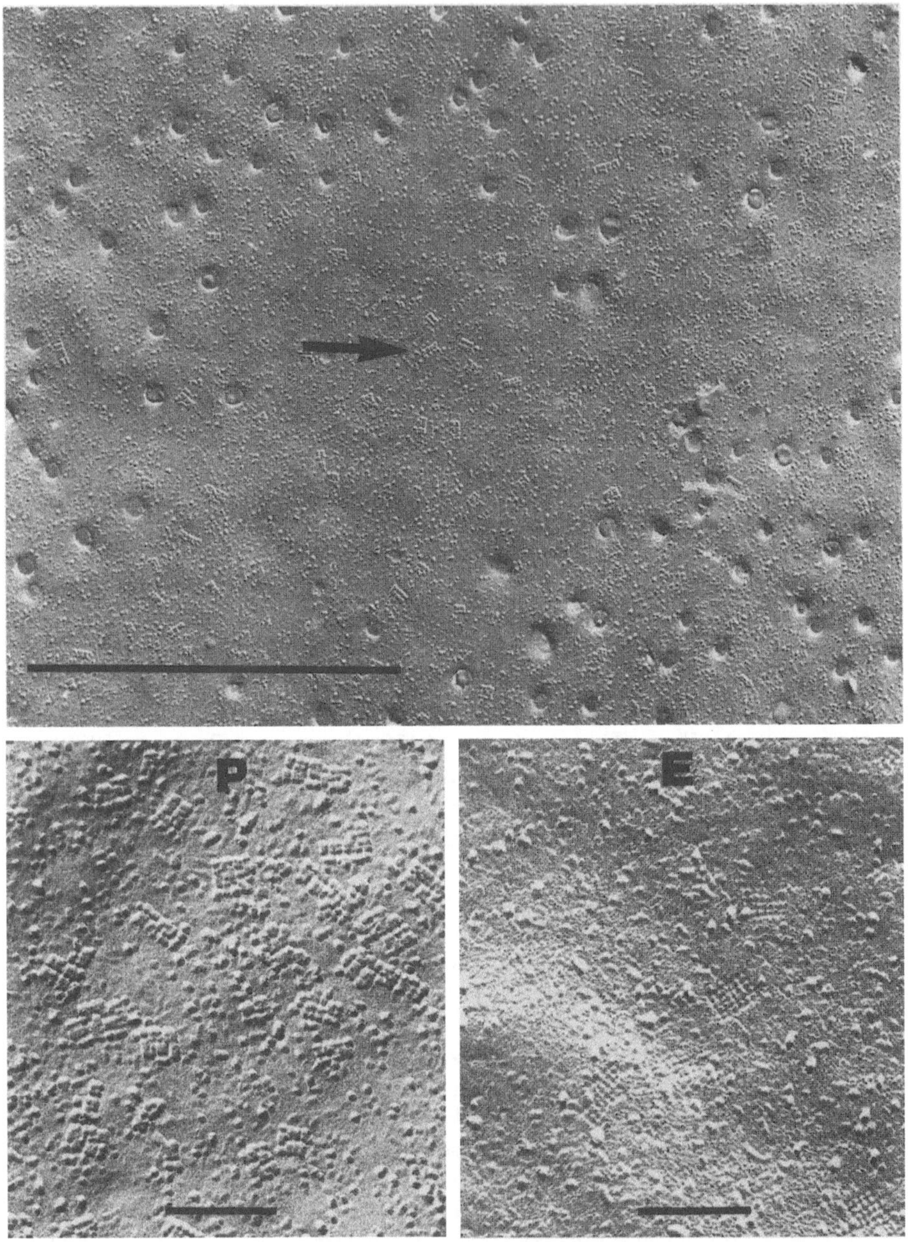

Fig. 66. Freeze fractures of the plasma membrane of an adult human medial vastus muscle fibre. *Top:* P face showing openings of the caveolae, numerous 10-nm particles, and rather many square arrrays (*arrow*). Bar, 1 μm. *Bottom: P* and *E* face of the plasma membrane at higher magnification. The square arrays are attached to the P face and leave distinct pits on the E face. Bars, 0.1 μm

hypoxia during the preparation of biopsy samples may explain the varying particle densities.

γ) 6-nm Particles in Square Arrays

The P face of the plasma membrane carries rectangular arrays of particles, each about 6 nm in diameter. Usually, several arrays with different orientation are clustered (Figs. 65, 66). Similar particle arrays occur in other cell types as well (astrocytes, DERMIETZEL 1973; LANDIS and REESE 1974; hepatocytes, KREUTZIGER 1968; intestinal epithelial cells, STAEHELIN 1972).

RASH and ELLISMAN (1974) and ELLISMAN et al. (1976), found, in rat muscle fibres, that the density is a function of the distance from the endplate, and that it varies in different fibre types. Few square arrays are present in slow-twitch soleus fibres at and near the endplate; more than 2 mm from the endplate none are found. None are present at and around the endplates of fast-twitch fibres, but square arrays are abundant in the extrajunctional plasma membrane. The density at 0.5 mm from the endplate is up to 100 arrays/μm^2. Denervation does not alter the distribution of square arrays (ELLISMAN and RASH 1977), but soleus muscle fibres cross-innervated by the nerve to the fast extensor digitorum longus muscle show the same density and distribution as fast-twitch fibres (ELLISMAN et al. 1978). Also slow- and fast-twitch fibres of cat differ with respect to the incidence of sarcolemmal square arrays (SCHMALBRUCH 1979a) (Fig. 65).

The density of square arrays in human muscle varies between individual fibres, and in the vastus muscle consisting of about 50% type I (slow-twitch) and 50% type II fibres (fast-twitch), the densities form a two-peak histogram. It appears justified to conclude that the slow-twitch fibres are represented in the low-density peak (1.3 arrays/μm^2) and the fast-twitch fibres in the high-density peak (5.9 arrays/μm^2). No square arrays have been found in fetal human muscle (Fig. 64) (SCHMALBRUCH 1979c). SHAFIQ et al. (1979) were able to distinguish "slow oxidative type I fibres" and "fast glycolytic type II fibres" in freeze fractures of human muscles; they found 0–2 square arrays per μm^2 plasma membrane in type I and up to 70 per μm^2 in type II fibres. KETELSEN (1980) counted 4/μm^2 in fast fibres, and SCHOTLAND et al. (1981) counted 13.2/μm^2 (fibre type not specified). The discrepancies probably relate to the difficulty in deciding whether a cluster of arrays is to be counted as one or as several arrays.

The function and composition of these square arrays are enigmatic. Initially, they were believed to represent specialized gap junctions (STAEHELIN 1972), but this view was invalidated by the same author (RASH et al. 1974). Because square arrays are abundant on astrocyte projections rich in ($Na^+ + K^+$)ATPase, it has been suggested that the number of arrays might be related to the activity of the enzyme in the muscle-fibre membrane (ELLISMAN et al. 1976); nevertheless, the activity of ($Na^+ + K^+$)ATPase is the same in plasma membranes of fast- and slow-twitch fibres (FESTOFF et al. 1977). One could relate the square arrays to a particular type of sodium channel, because only in mature, and not in immature, muscle fibres the action potentials are blocked by tetrodotoxin

(HARRIS and MARSHALL 1973). One might also consider the distribution of extrajunctional AChRs which show an inverse relation to the distribution of square arrays (MILEDI and ZELENA 1966). Cross-innervation converts the distribution of extrajunctional AChRs (MILEDI et al. 1968) (see above) and also of square arrays (ELLISMAN et al. 1978). Nevertheless, in denervated fibres the density of extrajunctional AChRs increases (see below) but the density of square arrays remains unaltered (ELLISMAN and RASH 1977).

Finally, the possibility cannot be excluded that the square arrays do not represent proteins but lipids, in particular because the square arrays unlike other intramembrane particles always leave complementary pits. Freeze fractures of protein-free lipid bilayers consisting of different types of lipids may reveal intramembrane particles with distinct pits in the opposite fracture face. These particles represent inverted micelles with hydrophilic cores and hydrophobic edges. Uniform particles of 4.5- to 17.8-nm diameter, depending on the lipid, have been produced; sometimes the particles are arranged in a square pattern (VERKLEIJ et al. 1979; CULLIS and DE KRUIFF 1979; SEN et al. 1981; HUI and BONI 1982). The different density of square arrays in plasma membranes of fast- and slow-twitch fibres might reflect the different lipid composition of the membranes. Rat muscle fibres treated with digitonin, which binds to cholesterol, in freeze fractures reveal patches of tubular complexes within the plasma membranes. These complexes take up a greater proportion of the surface area in slow-twitch than in fast-twitch fibres, in harmony with the fact that the plasma membrane of slow-twitch fibres contains relatively more cholesterol than that of fast-twitch fibres (FISCHBECK et al. 1982).

δ) The Extrajunctional Plasma Membrane and the Motor Nerve

The extrajunctional plasma membrane of adult slow-twitch fibres has a low but distinct ACh sensitivity; this indicates that AChRs are present. Large particles believed to represent AChRs are frequent in developing non-innervated fibres (see Sect. C.IV.3 b δ).

The number of extrajunctional AChRs increases in denervated muscle fibres (AXELSSON and THESLEFF 1959; MILEDI 1960; ALBUQUERQUE and THESLEFF 1968) from less than 5/μm^2 to about 1,500/μm^2 (rat diaphragm, HARTZELL and FAMBROUGH 1972). Nevertheless, it is not possible to identify corresponding particles in freeze fractures. The action potentials become tetrodotoxin-resistant after denervation, probably because the immature type of sodium channels reappears. Also this does not change the freeze-fracture image of the extrajunctional sarcolemma (TACHIKAWA and CLEMENTI 1979). Only MALHOTRA and TIPNIS (1978) claim to have found in denervated rat muscles clustered 15-nm P-face particles, which they believe to be newly synthesized AChRs. The illustrations accompanying the paper do not show such particles.

The following physiological properties of the extrajunctional plasma membrane after denervation are used to assess the so-called trophic effect of the nerve; the increase of extrajunctional ACh sensitivity, the occurrence of tetrodotoxin-resistant action potentials, and a drop in resting membrane potential. It was assumed that these changes are at least in part due to lack of one or several specific neurotrophic substances produced by the motoneuron (MILEDI 1963; GUTH 1968; HOFMANN and THESLEFF 1972; HOFMANN and PEACOCK 1973; GILLIATT et al. 1978; GUTH and ALBUQUERQUE 1978, 1979; DESHPANDE et al. 1980). There is now increasing evidence that the normal organization of the plasma membrane is preserved or restored when the denervated muscle is electrically stimulated (LØMO and ROSENTHAL 1972; LØMO and WESTGAARD 1975), and that in the innervated muscle fibre quantal and non-quantal release of ACh from the axon terminal suffices to maintain the integrity of the membrane (DRACHMAN and JOHNSTON 1975; MATHERS and THESLEFF 1978; BRAY et al. 1982; PESTRONK et al. 1980).

The extrajunctional plasma membrane of an innervated fibre rejects any approaching motor axon and thereby prevents double innervation. Denervated fibres will preferentially become reinnervated at the former endplate region (see Sect. C.IV.2), but ectopic reinnervation is possible if no axon reaches the endplate (FRANK et al. 1975). The original endplate site eventually rejects approaching axons; then the ectopic endplate becomes permanent. Under certain experimental conditions, lasting re-innervation by more than one axon has been achieved (KUFFLER et al. 1980). Electrical stimulation of a denervated muscle not only prevents the "denervation" changes of the extrajunctional plasma membrane, but also prevents the formation of ectopic neuromuscular junctions (LØMO and SLATER 1978).

ε) Birefringence Changes of the Plasma Membrane During Excitation

Muscle fibres act as a birefringent medium when light passes through the ordered myofilaments. A decrease in birefringence during isometric twitches and tetanuses was observed by several authors (for references, see EBERSTEIN and ROSENFALCK 1963). The onset of these changes precedes the onset of tension but the decline follows the decline of tension, i.e. they are slow and hence are not related to the action potential at the membrane.

COHEN et al. (1971) described a rapid decrease in birefringence of squid giant axons in the order of 10^{-5} in relation to resting intensity. They attribute this change in birefringence to the potential changes across the surface membrane. An early component of the birefringence signal of muscle fibres following the action potential (CARNAY and BARRY 1969) is in the order of 2×10^{-4}, 20 times larger than in axons. This may be due to the large amount of intracellular membranes. BAYLOR and OETLIKER (1975, 1977) distinguish two early components during the twitch of a single frog muscle fibre. The size of the first component is $1–2 \times 10^{-5}$ of the resting intensity; it coincides with the surface action potential suggesting that the signal is due to the depolarization of the plasma membrane alone. The second component has a peak size of 5×10^{-4}, it is propagated with the speed of the action potential, and is attributed to potential changes across the internal membranes.

d) The Myotendinous Junction

The question of whether or not myofibrils are continuous with collagen fibrils was discussed for more than a century (for references, see HÄGGQUIST

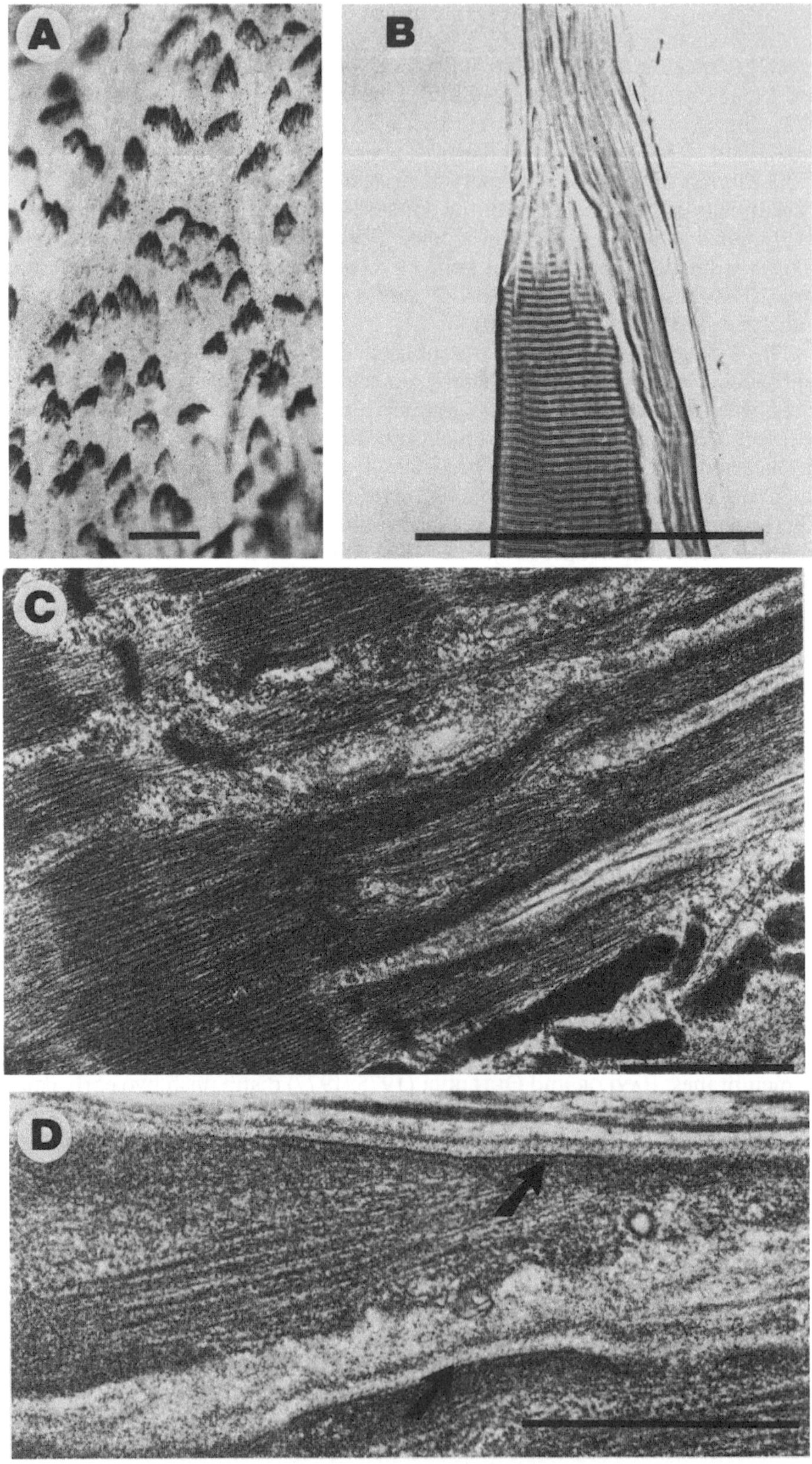
A
B
C
D

1931, 1956; SCHWARZACHER 1960b). PORTER (1954), RUSKA (1954), EDWARDS et al. (1956), COUTEAUX (1959), GELBER et al. (1960), and SCHWARZACHER (1960a, b, c) demonstrated that the fibre end is covered by a plasma membrane with basal lamina. The membrane at the fibre end forms finger-like processes and crypts. The sarcomeres end with I bands, and the thin filaments insert at varying levels in cytoplasmic thickenings of the longitudinally oriented parts of the plasma membrane. The collagen fibrils enter the crypts and folds of the membrane from the extracellular side. Staining of the myotendinous junctions for AChE (COUTEAUX 1953; GEREBTZOFF 1956, 1957) allows identification of the array of processes and invaginations using light microscopy (SCHWARZACHER 1960d); the activity of AChE is about 20% of that at the neuromuscular junction of the same fibre (SCHWARZACHER 1960a, 1961) (Fig. 67). The collagen fibrils within the myotendinous junction do not insert at the plasma membrane but at the basal lamina (REISSIG and SCHIPPEL 1966; MACKAY et al. 1969). KORNELIUSSEN (1973) describes in *Myxine* 6-nm-wide spine-like profiles arranged in transverse circular ridges between plasma membrane and basal lamina. The spines are 15 nm apart. Similar, but less regular, spines are found in myotendinous junctions of rat muscles. The author suggests that the spines connect plasma membrane and basal lamina ("lamina densa") across the "gap" ("lamina lucida", or glycocalyx, see above) and thereby transmit force from muscle fibre to tendon. TROTTER et al. (1981) extracted the membranes of mouse muscles using non-ionic detergents. The muscle fibres still contracted when ATP was added, and force transmission was unimpaired. Electron microscopy revealed that the plasma membrane had vanished, and showed fine filaments between the cytoplasmic densities, into which the I filaments inserted, and the basal lamina. This suggests that in intact fibres, filaments pass through the hydrophobic compartment of the plasma membrane, and that the plasma membrane itself does not play a significant role in the transmission of tension. TROTTER et al. (1983) describe a dense 0.05- to 0.1-µm-thick intracellular layer (internal lamina) of cytoplasmic actin filaments oriented parallel to the plasma membrane and to the principal vector of contractile force. These actin filaments are continuous with the thin filaments of the last sarcomere, and are apparently interlinked by an unknown protein. They are not decorated by the S1 subfragment of heavy meromyosin (see Chap. C I 9e). The centre-to-centre spacing of these actin filaments is only 12 nm. The internal lamina is connected to 2- to 8-nm-thick filaments (connecting domain) passing perpendicularly to the vector of force through the plasma membrane and the gap (lamina lucida) to merge

Fig. 67 A–D. Rat soleus muscle, myotendinous junction. **A** Whole mount stained for acetylcholinesterase. The fibre ends are covered by a reactive cuff and terminate in a blunt fashion. The insertions of the fibres are staggered (see Sect. B.IV.2). Bar, 100 µm. **B** Semithin plastic section stained with p-phenylenediamine. A fibre inserts obliquely into the tendon sheet which is rotated 90%, compared with **A.** Deep indentations of the fibre end are visible. Bar, 100 µm. **C** Electron micrograph showing three indentations of the fibre end and the insertion of the terminal I band into a fine filamentous network beneath the plasma membrane. Bar, 1 µm. **D** Higher magnification showing the submembraneous system of cytoplasmic filaments. The gap between plasma membrane and basal lamina displays ill-defined periodic structures (*arrows*). Collagen fibrils are attached to the outer face of the basal lamina. Bar, 1 µm with the basal lamina (lamina densa). These filaments are subjected to shearing rather than tensile force.

There is little doubt that the connecting domain of TROTTER et al. (1981, 1983) is identical with KORNELIUSSEN'S (1973) periodic spines (see above). The nature of these transmembrane filaments is obscure. Future studies should take into account the role of the stress fibres in myotubes during myofibrillogenesis (Sect. G.III.2a), the periodic structures shown by BONILLA (1983) to connect plasma membrane and basal lamina outside the myotendinous junctions (Sect. C.IV.2) (see also Fig. 56), and the interesting role of fibronectin, which across the cell membrane may link cytoplasmic actin to collagen (Sect. G.III.3).

In extraocular muscles of bird (MAYR et al. 1967), rat (TERÄVÄINEN 1969a), and cat (FLOYD 1970) AChE-positive junctions between longitudinally connected muscle fibres have been found. Electron microscopy revealed a remote similarity with intercalated discs between heart muscle cells, but there were no membrane junctions. The gap between the membranes of the linked cells was 70–160-nm wide and contained amorphous material and collagen fibrils. The cytoplasmic face of both plasma membranes showed attachment sites of thin filaments, as in myotendinous junctions (MAYR et al. 1967). These junctions, like myotendinous junctions, react for acetylcholinesterase. "Myomuscular" junctions have been found in regenerated (SCHMALBRUCH 1976b) and reinnervated (PIEROBON BORMIOLI and SCHIAFFINO 1977) rat skeletal muscles as well.

V. Metabolic Systems

The energy necessary for maintenance and function of the muscle fibre is obtained by anaerobic and aerobic metabolization of glucose and by oxidation of free fatty acids (DELBRÜCK et al. 1959; HOLLOSZY and BOOTH 1976). The subcellular structures related to energy metabolism are mitochondria, lipid droplets, and glycogen granules. Histochemistry also demonstrates the enzymes of anaerobic glycolysis contained in the aqueous sarcoplasm. Flight and leg muscles of locusts show gross differences in the number of mitochondria, and these muscles were the first in which the different enzyme patterns and the metabolic specialization of muscles were thoroughly investigated (DELBRÜCK et al. 1959). The enzyme activities differ also in mammalian muscles and muscle fibres with different functional properties (WACHSTEIN and MEISEL 1955); this observation ultimately led to the establishment of types of muscle fibres and motor units (see Chap. D).

1. Mitochondria

a) The Array of Muscle Mitochondria

The mitochondria of mammalian skeletal muscle fibres usually lie at the I-band level; the early light microscopists named them "I granules". In some insect muscles, and also in mammalian heart muscle cells, the array of mitochon-

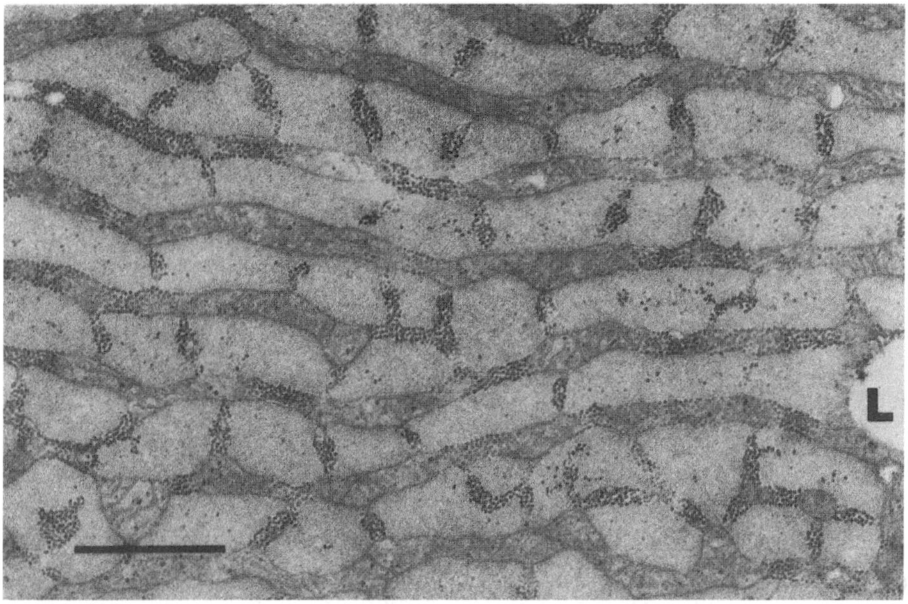

Fig. 68. Human medial vastus muscle. Cross-section through the I band. The mitochondria form long branches separating the myofibrils. The electron-dense granules are glycogen. *L*, triglyceride droplet. Bar, 1 μm

dria does not reflect the sarcomere pattern; in these muscle cells the mitochondria were described as "A (or Q) granules". The I-band mitochondria apparently formed transverse threads, which, when they projected into the cross-striation pattern, gave rise to the light microscopical N bands (HÄGGQUIST 1956).

Electron micrographs of longitudinal sections of mammalian skeletal muscle fibres show pairs of mitochondrial profiles at the I-band level on both sides of the Z line (Figs. 16A, 48, 88). They have occasional longitudinal projections connecting the I-band mitochondria across one or several sarcomere planes, in addition, there may be clusters of rather large mitochondria beneath the sarcolemma. The clustered subsarcolemmal mitochondria are always close to capillaries (Fig. 87) (ANDERSSON-CEDERGREN 1959).

BUBENZER (1964, 1966) and GAUTHIER and PADYKULA (1966) were the first to describe the three-dimensional array of the mitochondria in different types of muscle fibres of the diaphragm. BUBENZER (1964, 1966) distinguished "thin" and "thick" fibres in rats. The subsarcolemmal mitochondria of "thin" fibres have slender extensions running transversely across the fibre. These bars of mitochondria undulate and branch, but remain in the plane of the I band, and separate the mass of myofilaments into irregularly band-shaped myofibrils. At low magnification, the transverse bars form a duckboard or grate pattern. The grates are longitudinally connected across several sarcomeres. Hence, the "thin" fibres in reality contain giant mitochondria stretching throughout the fibre. Subsarcolemmal mitochondria are inconspicuous in "thick" fibres; also in these fibres transversely arranged mitochondria run at both sides of the

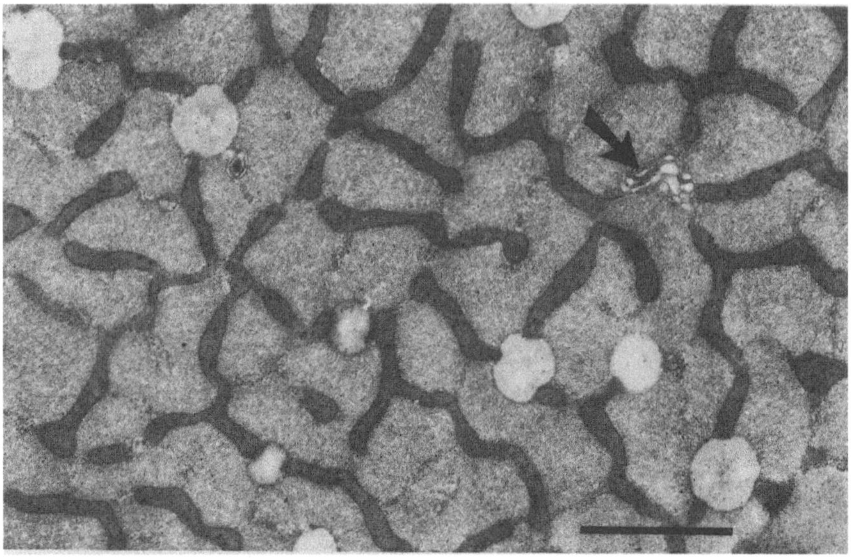

Fig. 69. Cat gastrocnemius muscle. Cross-section through the I band showing the array of mitochondria and lipid droplets. *Arrow*, small Golgi zone; bar, 1 μm

Z disc, but they are thinner, and branch more extensively than in "thin" fibres. Longitudinal connections are lacking. Appropriately oriented cross-sections show mitochondrial grids through which circular or polygonal myofibrils pass.

GAUTHIER and PADYKULA (1966) compared the diaphragm of a multitude of mammalian species ranging from *Myotis mucifugus* (5 g body weight) to *Bos taurus* (500 kg body weight). The mitochondrial content is inversely related to body size, and species of <65 g weight have only "red" fibres corresponding to BUBENZER'S (1966) "thin" fibres. Longitudinal sections of "red" fibres show either ribbon-like mitochondria or pairs of circular profiles at the I band. This reflects the parallel orientation of the transverse bars. MILEDI and SLATER (1968) describe the mitochondria of rat diaphragm as worm-like, long and branched, and surrounding the myofibrils at both sides of the Z line. It is noteworthy that the transverse mitochondrial threads disintegrated rapidly after denervation. RAMBOURG and SEGRETAIN (1980) studied the "red" fibres of the diaphragm of rat by electron microscopy using a new staining method for thick sections. Their findings conform to those by BUBENZER (1964, 1966); longitudinal sections show longitudinal profiles with transverse branches embracing the myofibrils. Grids of mitochondria more or less completely encircling each myofibril were also demonstrated in limb muscles (GAUTHIER 1969; SCHMALBRUCH 1971). The mitochondrial network is difficult to visualize in thin sections (Figs. 68, 69); if the cross-striation pattern is out of register, cross-sections show only "worm-like" mitochondria. Parts of the framework are readily seen in freeze fractures (KAMIENIECKA and SCHMALBRUCH 1980) (Fig. 70).

The extraocular muscles and some laryngeal muscles of mammals have a mitochondrial array comparable to that in heart muscle cells. All mitochondria

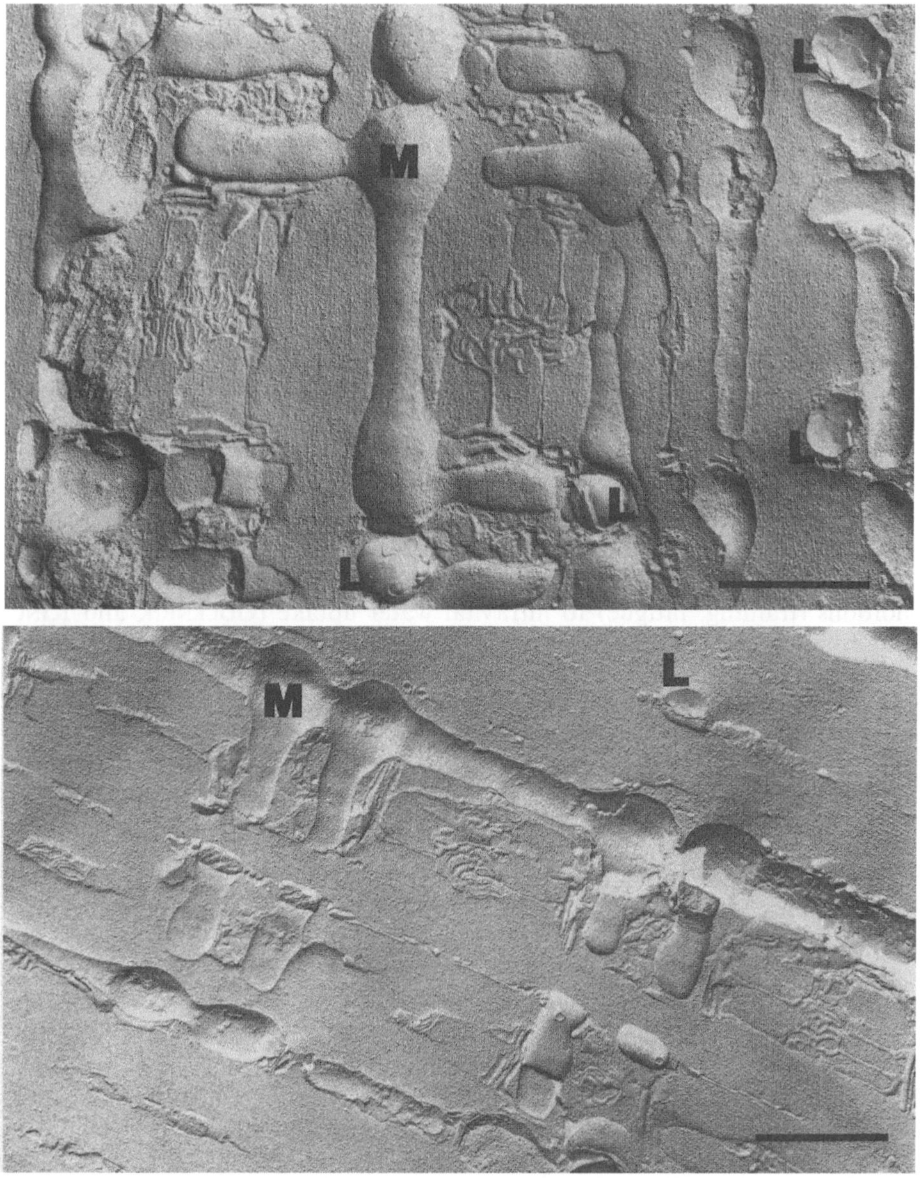

Fig. 70. Longitudinally oriented freeze fractures of fibres of the anterior tibial muscle of rat, showing the array of mitochondria (*M*). In both micrographs sarcoplasmic reticulum and T system indicate the sarcomeric pattern. The E face of the outer mitochondrial membrane appears convex, and the P face appears concave. The mitochondria form doublets running transversely on both sides of the Z disc; they are interlinked by branches which may stretch over several sarcomere planes (*below*). The fracture faces of triglyceride droplets are smooth (*L*), occasionally they appear layered (see text). Bars, 1 μm

are ovoid bodies distributed throughout the cell without relation to the cross-banding pattern (SCHMALBRUCH 1971; PEACHEY et al. 1974). These muscles will be discussed separately (Chap. F).

The relative volume of the muscle fibre occupied by mitochondria ranges from about 1% (frog sartorius muscle, MOBLEY and EISENBERG 1975; fast-twitch fibres of human quadriceps muscle, EISENBERG 1983) to 8%–10% (red vastus muscle of guinea pig, EISENBERG and KUDA 1976; rat soleus muscle, STONNINGTON and ENGEL 1973; DAVEY and WONG 1980). For oxidative fibres of mouse muscles, values of up to 20% have been reported (SILVERMAN and ATWOOD 1980).

The quantitation of muscle mitochondria using morphometry is hampered by several sources of error. The spatial inhomogeneities and the specific array in relation to the cross-striation have to be taken into account (HOPPELER et al. 1973; JAMES and MEEK 1979; EISENBERG 1983). It is obvious that mitochondrial "counts" do not give the number of mitochondria but only the number of profiles in a section. Swelling of mitochondria during the preparation process decisively influences the results of morphometry. Whenever possible, the muscles should be fixed by perfusion, which at least renders the artefacts reproducible. One might be tempted to attribute mitochondrial swelling under experimental conditions (training, fatigue) to intravital water uptake (GOLLNICK and KING 1969), but it is not possible to distinguish this from artefactual swelling during specimen preparation. Muscle mitochondria always contain electron-dense matrix granules (Figs. 68, 69, 75). The absence of matrix granules usually indicates improper handling of the specimen before or during fixation.

b) Isolated Mitochondria

Mitochondria are routinely isolated from muscle tissue for the study of their metabolism. A mitochondrial fraction seen by electron microscopy consists of spherical bodies usually <1 µm in diameter. During the isolation procedure, the mitochondrial framework or grids are disrupted and the fragments then reseal. This may explain some of the peculiarities of muscle mitochondria studied in vitro. Liver mitochondria change their ultrastructural configuration depending on their metabolic state (HACKENBROCK 1966); this is not the case for mitochondria from skeletal muscles (KUNER and BEYER 1970). HULSMANN et al. (1968) isolated two types of mitochondria from rat skeletal muscle; mild fragmentation yielded mitochondria with loosely coupled oxidative phosphorylation, whereas a more thorough procedure gave a mitochondrial fraction with tightly coupled phosphorylation. The authors assumed that the easily isolated fraction comprised subsarcolemmal mitochondria which had "aged". This is in harmony with the observation that the mitochondria of muscles of aging insects lose the tight coupling of oxidation and phosphorylation (TRIBE and ASHHURST 1972; SACKTOR and SHIMADA 1972).

c) Training and Hypoxia

Training increases the activity of mitochondrial enzymes; the enzyme activity decreases in inactive muscles. RIFENBERICK et al. (1973) found in mitochondrial fractions of rat muscles, immobilized by surgical pinning of the joint, a relative decrease in mitochondrial proteins together with a decrease in enzyme activity. This indicates that not only the enzyme activity decreases but that the mitochondrial system "atrophies". The increase in enzyme activity after training (HOLLOSZY and BOOTH 1976) is reflected by a corresponding increase in the volume fraction of the mitochondria, and also by an increase in the number of cristae. HOPPELER et al. (1973) found a volume increase in the lateral vastus muscle from 5.0% to 6.6% after training (orienteers) (fibre types not distinguished). The increasing volume fraction of mitochondria was related to the increasing maximum oxygen intake of the subject. The surface of the mitochondrial cristae was 1.6 times larger in trained than in untrained muscles. SCHÖN (1978) measured a more than twofold higher volume fraction of mitochondria in "red" fibres of the vastus muscle of marathon runners compared with untrained subjects; "white", i.e. glycolytic, fibres in long-distance runners showed no change in mitochondrial volume. FRIDÉN et al. (1984) studied the lateral vastus muscles of 4 untrained males and 6 trained long-distance cross-country runners (mean age of the subjects 30 years, range 25–35 years). The samples were obtained by surgical biopsy, not by needle biopsy. The mitochondrial volume in the untrained subjects was 1.9%–2.9% of the muscle fibre volume; type I fibres (slow-twitch oxidative, see Chapt. D) contained most mitochondria. The mitochondrial volume in the trained subjects was 6.5% of the fibre volume in slow-twitch type I fibres and 5.4% in fast-twitch type II fibres.

SJÖSTRÖM et al. (1982b) performed a controlled training study in sedentary middle-aged men and found a pronounced increase of the relative mitochondrial volume in oxidative (presumably both fast- and slow-twitch, see Chap. D) fibres, but not in glycolytic fibres. The relative mitochondrial volume was larger than normal in patients with claudicatio intermittens. Concomitantly, the activity of mitochondrial enzymes increased (ÄNGQUIST and SJÖSTRÖM 1980). This may reflect an attempt of the muscle fibres to adjust to the decreased oxygen supply. HANZLIKOVA and SCHIAFFINO (1977) produced ischemia in the hindlimb muscles in rats by ligating the aorta. Subsarcolemmal giant mitochondria occurred in the soleus muscle within 6 h. These mitochondria contained stacks of cristae. In addition, crystalline inclusions were found between the inner and outer mitochondrial membranes.

d) Intramitochondrial Crystalloids

Intramitochondrial crystalloids are characteristic for a variety of muscular disorders, but they are by no means specific (for references, see KAMIENIECKA and SCHMALBRUCH 1980; SCARLATO and CERRI 1983) and may be encountered in normal muscles as well (HAMMERSEN et al. 1980). Nevertheless, HAMMERSEN'S material stems from patients operated on for inveterated knee lesions, and the biopsies were taken after an Esmarch tourniquet had been applied. Mitochondrial inclusions were encountered in mice more than 3 years of age (A. KELLY, personal communication), which suggests that they also may reflect a metabolic change related to aging.

Two types of intramitochondrial crystalloids have been distinguished, both of which lie between the outer and inner mitochondrial membranes. The "park-

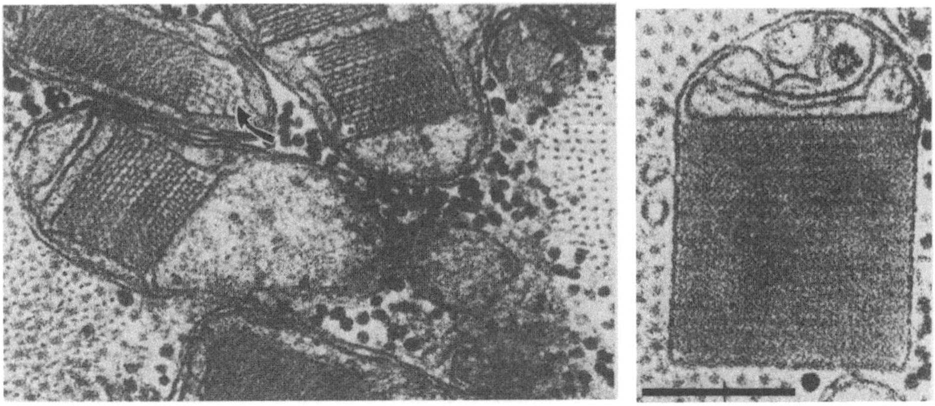

Fig. 71. Intramitochondrial crystalloids from the brachial biceps muscle of a patient with "mitochondrial" myopathy. The "parking-lot type" (*left*) consists of subunits each composed of four lamellae connected by densities with 16-nm repeat. The "rectangular type" (*right*) shows indistinct narrow spacings or a herring-bone pattern. Note that both types of crystalloids are present in the crista- or membrane-space of the mitochondrium. It is conceivable that the "types" are merely different projections of the same structure. Mitochondrial crystalloids occur in "normal" muscle as well: their nature is obscure. Bar, 0.25 μm

ing lot" type (Fig. 71) consists of 2–4 subunits, each formed by four lamellae 0.2–2.0 μm in length with 6-nm gaps between them. The lamellae are connected by periodic densities with a 16-nm repeat. The densities in the two outer gaps are in register, those in the inner gap are shifted half a period (Fig. 72). The periodicities within these subunits are less than the usual section thickness and several repeats are superimposed; different images may represent different directions of cutting. The appearance of the periodicities changes when the section is tilted in the microscope. The densities connecting the lamellae probably represent bridges. A plexiglass model under different angles projected onto photographic paper shows similar image shifts (SCHMALBRUCH 1983). The second type of intramitochondrial crystalloids has been described as "rectangular" (Fig. 71); it reveals narrow striations or a herring-bone pattern. Evidence that these are two crystalloid types and not different projections of the same structure is lacking.

The composition of the crystalloids is unknown (for references, see KAMIENIECKA and SCHMALBRUCH 1980). LANDON (1982) believes that the "parking-lot" type resembles cytochrome oxidase membrane (MANILOFF et al. 1973).

2. Glycogen

a) The Intracellular Localization

The glycogen of muscle fibres appears in electron micrographs as spherical particles 20–30 nm in diameter (β-glycogen) (REVEL 1964). These granules are found mostly at the I-band level (Fig. 73). In cross-sections, strands of granules

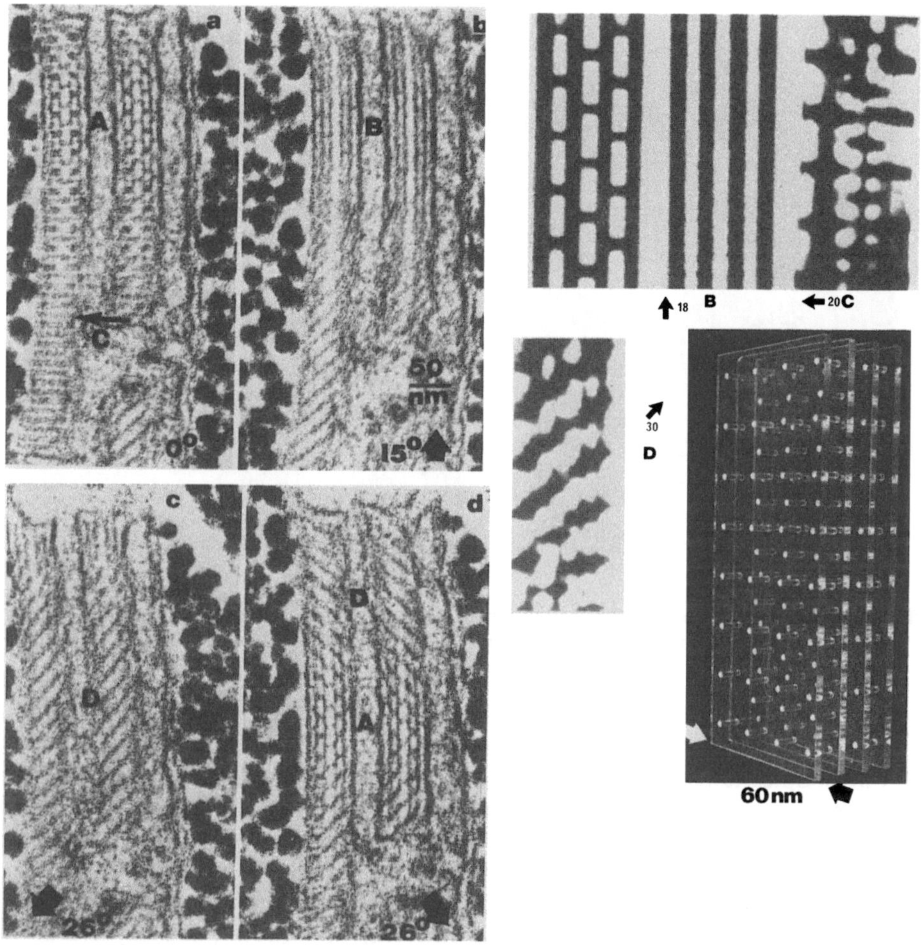

Fig. 72. Intramitochondrial crystalloid photographed in four different projections (*left*), and different projections of a Plexiglass model for the crystalloid structure (*right*). *Left*: The tilt is indicated (*thick arrows*). Four different images are seen. *A*, four lamellae connected by periodic densities with 16-nm repeat; *B*, four parallel lines, but no densities; *C*, periodic densities with 8-nm repeat, but no lamellae; *D*, oblique striations with alternating orientation depending on the tilt. *Right:* The model for the crystalloid structure consists of four plates connected by bridges with 16-nm repeat, such that the bridges in the two outer gaps are in register, whereas those in the middle gap are off-set by half the repeat on both axes. A 60-nm-thick section would comprise four bridge repeats. To test this model, it was projected onto photographic paper under different angles. The shadow image changed with the orientation because the mass density depends on which structures project onto each other; the images produced correspond to the images seen in electron micrographs at similar degrees of tilt. (From SCHMALBRUCH 1983, with permission of Piccin Medical Books)

may encircle the myofibrils together with mitochondria, T tubules, and terminal cisterns of the SR. PAS-staining shows this network in the light microscope. In the electron microscope, glycogen granules are best visible after lead staining; after staining with uranyl acetate they are inconspicuous. Only few glycogen granules are seen in muscle fibres which have been hypoxic before fixation,

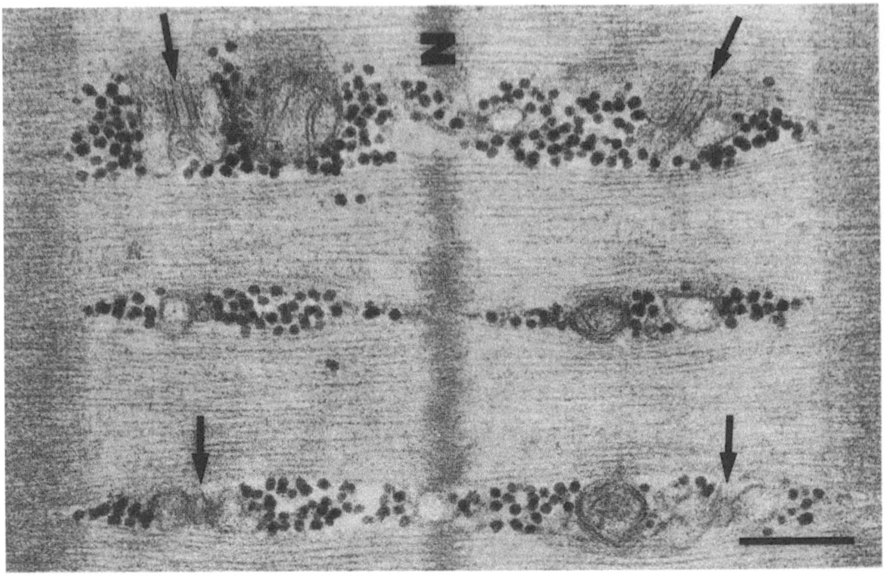

Fig. 73. Cat gastrocnemius muscle. Longitudinal section showing a *Z* line and the two adjacent I bands. Numerous electron-dense *β*-glycogen granules surround the myofibrils at the I-band level. *Arrows*, triadic junctions; bar, 0.25 μm

and in samples taken at autopsy (SCHMALBRUCH 1970). Glycolytic muscle fibre types contain more glycogen than do oxidative fibres (Sect. D.II.1), but in human muscles no difference has been found (PIEHL 1974). Glycogen is lost after lasting activity, more so from glycolytic than from oxidative fibres (Sect. D.II.2).

The subcellular localization of glycogen granules differs in fast- and slow-twitch fibres. In fast-twitch fibres of rat, cat, and rabbit, the myofibrils at the I-band level are clearly delineated by strands of glycogen granules. The delineation of myofibrils is less clear in slow-twitch muscle fibres, and scattered glycogen granules are seen between the thin filaments of the I band (SCHMALBRUCH 1971, 1979) (see Sect. D.IX, Fig. 88). The same applies to human muscles, but the difference is less clear than in animal muscles fixed by vascular perfusion (SCHMALBRUCH and KAMIENIECKA 1975). Subsarcolemmal aggregations of glycogen granules are common in diseased muscles but are scarcely found in normal muscles (own observation; CARPENTER and KARPATI 1984). Membrane-enclosed glycogen granules (lysosomes?) are probably a degenerative sign. They are never encountered in muscles of young adult laboratory animals.

b) The *β*-Glycogen Granule

Negatively stained preparations of isolated glycogen granules from rabbit skeletal muscle are 39.4 nm in diameter (mean); in sections the diameter is 27.3 nm. "Purified" glycogen granules contain about 3% protein (WANSON and DROCHMANS 1968). ROSATI (1967) treated thin sections of rat liver and muscle

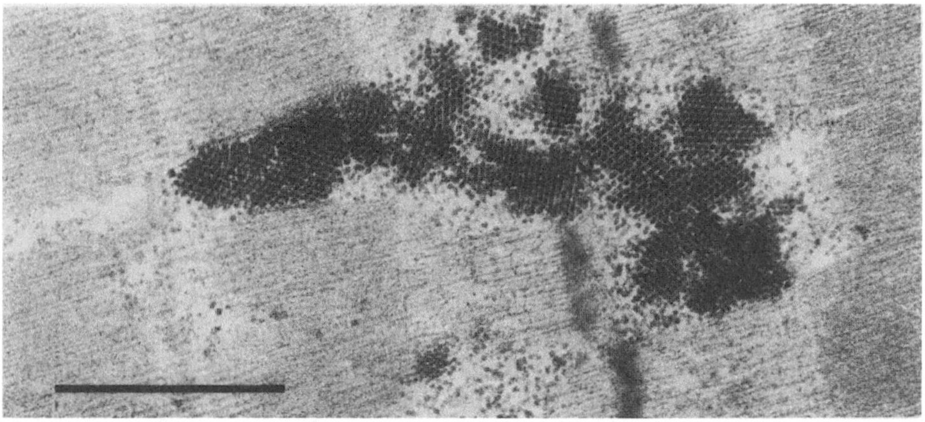

Fig. 74. Human brachial biceps muscle (male 35 years, sudden death in traffic accident, specimen taken 5 h after death). The sarcoplasm contains small glycogen granules many of which have aggregated to form a crystalloid with hexagonal and orthogonal packing pattern. Bar, 1 µm

with diastase; only the α-glycogen granules of liver cells, but not the β-glycogen granules of muscle, were dissolved. The glycogen granules increased in size when muscle tissue was incubated with glucose. Diastase was able to digest the "extra" glycogen until the granules had again reached their "normal" size. ROSATI concludes that in muscle fibres a "core" of glycogen is tightly bound to protein, and that diastase can digest only the "extra" glycogen.

According to RYBICKA (1981), the portion of the glycogen granule that is stained by lead does not contain glycogen, but protein. For the protein-glycogen complex she proposes the name "glycosome" instead of "β-granule of glycogen", and finds that the glycosomes – at least in heart muscle cells – are bound to the SR and to intermediate filaments.

c) "Glycogen Paracrystals"

The concept that β-glycogen granules consist of a protein-glycogen core with defined size, and that the size variation seen in sections is due to easily digested "extra" glycogen, conforms to the fact that glycogen granules under certain conditions form paracrystals. Normally, glycogen granules lack a specific array, although WANSON and DROCHMANS (1968) mentioned in rabbit muscles a "more or less regular" packing pattern. SCHMALBRUCH (1967a) observed, in autopsy samples from healthy subjects who had died in accidents, crystalline structures consisting of hexagonally and orthogonally arranged β-glycogen granules (Fig. 74).

There is no doubt that the glycogen molecules, because of their variable size, do not crystallize. Nevertheless, non-molecular elements may occur in a crystalline pattern if they vary little in size (virus crystals, LABAW and WYCKOFF 1957; KLUG et al. 1959; Latex particles, LUCK 1967). The array is usually "cubic close packed". One might speculate that postmortal digestion of glycogen renders the β-glycogen particles uniform in size and thus creates the condition for "crystallization". The fact that the paracrystals have repeatedly been found in a series of normal muscles taken at autopsy (SCHMALBRUCH 1967a), but never in several hundred normal and diseased muscles taken by biopsy (SCHMALBRUCH, unpublished), suggests that they are "sophisticated artefacts". It appears worthwhile to mention them because many authors have mistaken them for "virus crystals" (for review, see PALMUCCI et al. 1983).

3. Intracellular Triglycerides

Free fatty acids enter the oxidative metabolism and provide an important source of energy for the muscle fibres. Because combustion of fat is linked to mitochondrial enzymes, it makes sense that neutral lipids are predominantly in fibres with many mitochondria (GAUTHIER and PADYKULA 1966). Staining for lipids was the first histochemical procedure that allowed certain identification of different fibre types in an inhomogeneous skeletal muscle (BULLARD 1919).

Triglyceride droplets appear "black" in electron micrographs, and the contours are irregular, when the specimen has primarily been fixed with osmium tetroxide. After glutaraldehyde fixation, the droplets appear smooth contoured and are mostly electron translucent (Figs. 68, 69). Whether they are "white" or "gray" probably depends on the nature of the lipid, and on the time the sample has remained in the different media used during fixation and embedding. Lipid droplets have no surrounding membrane, and are always close to mitochondria. The preferred localization is the I band; throughout a cross section they are evenly distributed. Lipid droplets appear layered in freeze fractures, similar to fractured polar lipids (Fig. 70), although the triglyceride molecules are in random array. The layer structure does not imply that the droplets contain polar lipids, but it is probably an unavoidable freezing artefact (RUSKA and RUSKA 1971).

The slow-twitch soleus muscle of guinea pigs contains more than twice as many neutral lipids than the fast-twitch oxidative vastus muscle does. In rabbits, the neutral lipid content of the soleus muscle is five times larger than that of the fast-glycolytic semimembranous muscle (FIEHN and PETER 1973). The volume fraction of the fibres occupied by lipid droplets is 0.2%–0.4% in guinea pig (EISENBERG and KUDA 1976) and rat (DAVEY and WONG 1980) soleus muscles; it is 0.4% and 0.01% in the red and white vastus muscle, respectively (EISENBERG and KUDA 1976). The lipid droplets in human muscles occupy 0.2% and 0.5% of the volume of fast- and slow-twitch fibres (EISENBERG 1983).

MAUNSBACH and WIRSÉN (1966) mobilized fatty acids from the fat cells in dogs by continuously infusing noradrenaline. The free fatty acids entered the skeletal muscle fibres, and were esterified and eventually deposited as triglyceride droplets. HAYASHIDA and SCHMALBRUCH (1972) measured the change of the size of lipid droplets in fast-twitch oxidative fibres of the anterior tibial muscle in rat. After 24 h fasting, the droplets had a mean diameter of 0.8 µm; 2 h after fat feeding, it increased to 0.95 µm (computed from the size in electron micrographs of sections). This corresponded to an increase in volume of

60%–70%. The diaphragm of guinea pigs contains 2–3 times more Oil Red 0 positive material than the soleus muscle. The lipid content increases in both muscles after 48 h fasting because fatty acids are liberated from the fat stores of the body (ODUSOTE et al. 1981).

Endurance training increases the amount of mitochondria and the concentration of enzymes for β- and terminal oxidation in skeletal muscles, the total body maximal oxygen uptake, and also the contribution of fat to the energy metabolism during submaximal exercise (HOLLOSZY and BOOTH 1976; SALTIN et al. 1977). According to a hypothesis presented by GOLLNICK and SALTIN (1982), this shift in energy metabolism enhances the submaximal work capacity, because the greater enzyme concentration causes an increase in the metabolic rate of fatty acids, thereby sparing muscle glycogen stores. Hence, the "aim" of the increase in mitochondrial content of muscle is not to increase the oxygen uptake, but to enhance the fatty acid flux through the oxidative pathways during submaximal exercise. The maximum oxidative capacity of the muscle and the submaximal fatty acid utilization are interrelated; this has been demonstrated in subjects in whom only one leg had been trained (SALTIN et al. 1976).

VI. Myonuclei

The nuclei of mammalian skeletal muscle fibres are 10–20 μm long and plate-like; they are oriented parallel with and close to the plasma membrane. The shape of the myonuclei depends on the sarcomere length. In stretched fibres they are smooth contoured, but in shortened fibres they appear screw-like or coiled. Isolated myonuclei retain the coiled shape which in sections gives rise to deep indentations (FRANKE and SCHINKO 1969). It is conceivable that myonuclei spontaneously attain a coiled shape, and that they are linked to the cytoskeleton and are passively stretched together with the sarcomeres.

The myonuclei are randomly distributed around the fibre; in longitudinal sections they tend to occur alternatingly on both sides of the fibre. Cross-sections 10 μm thick, show 1–2 nuclei per fibre cross-section (SCHMALBRUCH and HELLHAMMER 1977), but more may be seen in extremely contracted fibres.

All myonuclei are diploid (LASH et al. 1957). The amount of DNA per nucleus is constant (rat, 6.2×10^{-6} g) and the number of nuclei may be determined biochemically; ENESCO and LEBLOND (1962) arrive at 5×10^4 nuclei per mg gastrocnemius muscle. They find that the total number of nuclei increases between 17 and 95 days of age by a factor 3.8, whereas the weight of the muscle increases by a factor 21.2. Thus, the muscle weight per nucleus increases 5.5-fold during this period. These counts include all nuclei, i.e. also those of connective tissue cells. The number of myonuclei proper increases three times in the rat soleus muscle during that period (ENESCO 1961). The total number of nuclei, as determined in rat gastrocnemius muscle using biochemistry, compares well with morphometrical results for the histochemically similar rat anterior tibial muscle. The number of nuclei per g muscle tissue is twice as high in the soleus muscle and three times as high in the diaphragm than in the "white" part

Table 4. Number of nuclei in rat and rabbit muscles as determined by light microscopy (Burleigh 1977), by light and electron microscopy (Schmalbruch and Hellhammer 1977), or by biochemistry and electron microscopy (Gibson and Schultz 1983). *EDL*, extensor digitorum longus muscle. The number of nuclei per mg muscle has been calculated from the data of Gibson and Schultz (1983)

Species and muscle	Nuclei per mg muscle	Myonuclei per mm fibre length	% Myonuclei of all nuclei	% Fibrocyte nuclei	% Endothelial cell nuclei	Length of myonuclei in μm	Sarcomere length in μm	References
Rat male (200–250 g)								
Soleus	10.5×10^4	76	44	16	18	13.4	2.8–2.9	Schmalbruch and Hellhammer (1977)
Anterior tibial. superfic. part	4.5×10^4	44	51	11	14	16.0	2.5–2.6	
Diaphragm	14.9×10^4	62	43	13	25	11.5	2.8	
Rat male								
1-month-old/soleus	29.3×10^4		44.7					Gibson and Schultz (1983)
1-month-old/EDL	22.3×10^4		49.5					
12-month-old/soleus	13.6×10^4		36.8					
12-month-old/EDL	7.8×10^4		49.9					
24-month-old/soleus	12.5×10^4		41.5					
24-month-old/EDL	9.3×10^4		50.3					
Rat female (247–265 g)								
Semitendinosus		86			Less than 8		2.16	Burleigh (1977)
Rat female (260–341 g)								
Semitendinosus		101					2.21	
Rabbit female (4.0–4.2 kg)								
Adductor magnus		118					1.89	

of the anterior tibial muscle. The number of nuclei per mm fibre length ranges from 44 in the rat anterior tibial to 118 in the rabbit adductor magnus muscle (Table 4). The fibres of the brachial biceps muscle of man contain about 30 myonuclei per mm fibre length (SCHMALBRUCH, unpublished). Hence, the short fibres of the soleus muscle of rat contain about 1,000 myonuclei, and a 10-cm-long human biceps fibre contains 3,000 myonuclei. (The implications of the large number of nuclei per cell are discussed in Sect. G.III.1.)

VII. The Lysosomal System

Primary lysosomes are membrane-bound organelles which contain a number of hydrolytic enzymes (DE DUVE and WATTIAUX 1966). The morphology of these organelles varies. They are involved in degradation of cellular constituents (autophagocytosis) but also of foreign material taken up by the cell (heterophagocytosis). The digestion of the material to be disposed of takes place within membrane-bound vacuoles which, like primary lysosomes, contain hydrolytic enzymes; they are named secondary lysosomes. Secondary lysosomes range in appearance from myeloid bodies or dense bodies to giant phagolysosomes or phagocytotic vacuoles containing membrane whorls and still identifiable cell organelles. Ultimately, the undigestible remnants are expelled by exocytosis, whereby the limiting membrane of the vacuole becomes part of the plasma membrane, or they are stored within the cell as residual body. Residual bodies are by definition devoid of hydrolytic enzyme activity. It is not possible to assure that structures resembling lysosomes really are lysosomes without the cytochemical demonstration of one of the hydrolytic enzymes, usually acid phosphatase. The hydrolytic enzymes are synthesized within the rough endoplasmic reticulum and are packed within the Golgi apparatus, which also is the usual source of the membrane limiting the phagolysosome. Also in cells not engaged in heterophagocytosis, the function of the lysosomal system is closely linked to endocytotic activity which serves to internalize elements of the surface membrane. Endocytosis may be demonstrated by extracellular tracers (e.g. horseradish peroxidase, ^{3}H-inulin), which are recovered in membrane-bound vacuoles within the cell. Protamine stimulates and colchicine or vinblastine inhibit endocytosis (for references, see LIBELIUS 1975, 1978; LIBELIUS et al. 1978, 1979).

The lysosomal system of skeletal muscle is poorly developed. Rarely small Golgi systems are found (Fig. 69). Dense bodies or multivesicular bodies, which usually are lysosomes in other cells, may occur in developing fibres (CARPENTER and KARPATI 1984); they are practically lacking in mature fibres. Only few authors have, using cytochemistry, demonstrated acid phosphatase in normal muscle (GORDON et al. 1967; TROUT et al. 1979; BIRD et al. 1980). This may not only reflect the scarcity of organelles containing the enzyme, but also the technical difficulty of demonstrating the lysosomal marker enzymes in muscle fibres. Even when (under pathological conditions) lysosomal structures are abundant, acid phosphatase is difficult to demonstrate in muscle fibres (CULLEN and MASTAGLIA 1982). The reason for this difficulty is obscure. Various hydro-

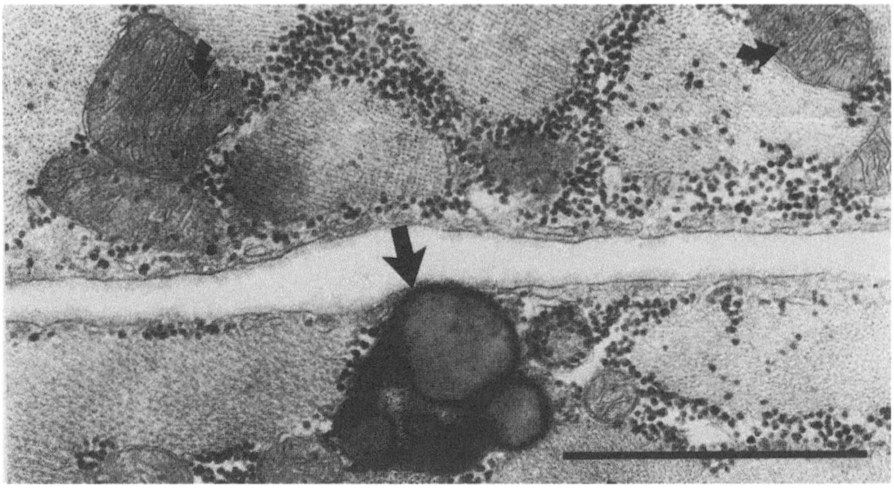

Fig. 75. Human medial vastus muscle (female, 25 years of age). Cross-section through two fibres separated by a narrow extracellular space. A lipofuscin granule (*straight arrow*) is seen beneath the plasma membrane of the fibre below. The sarcoplasm of both fibres contains many glycogen granules. Note distinct matrix granules in the mitochondria (*bent arrows*). Bar, 1 μm

lytic enzymes are present in homogenates of muscle (for review, see BIRD et al. 1980), but it is not always clear whether they are from muscle fibres or from cells of the connective tissue. Lipofuscin granules which represent secondary lysosomes or residual bodies are regularly found in old animals and also in human muscles (Fig. 75). This indicates that lysosomes are involved in the normal turnover of cellular constituents. Whether myofibrillar proteins are degradated by lysosomal or cytoplasmic proteases is a matter of controversy (for references, see LIBBY and GOLDBERG 1982).

Signs of lysosomal activity are prominent in some neuromuscular disorders and also in denervated muscles. Concomitantly, the activity of endocytosis and also of biochemically assessed lysosomal enzymes increases (POLLACK and BIRD 1968; SCHIAFFINO and HANZLIKOVA 1972b; LIBELIUS 1978; LIBELIUS et al. 1978, 1979; MRAK et al. 1982).

Chloroquine and several other compounds with similar physicochemical properties (amphiphilic drugs, LÜLLMANN et al. 1975) induce the formation of myeloid bodies and of autophagic vacuoles; in muscle fibres they cause a vacuolar myopathy and eventually fibre necrosis (MACDONALD and ENGEL 1970; DRENCKHAHN and LÜLLMANN-RAUCH 1976, 1979; SCHMALBRUCH 1978b, 1980a; TROUT et al. 1981; STAUBER et al. 1981a, b; BURSZTAIN and LIBBY 1981). The fact that (after chloroquine treatment) signs of phagocytosis and exocytosis are prominent, probably does not reflect an enhancement of autophagocytosis, but inhibition of the process (LIE and SCHOFIELD 1973; WIBO and POOLE 1974). According to BURSZTAIN and LIBBY (1981), chloroquine interferes with the fusion of lysosomes and coated vesicles originated by endocytosis.

Some authors demonstrated lysosomal enzymes in cisterns of the sarcoplasmic reticulum by use of cytochemical methods (TROUT et al. 1979). The origin

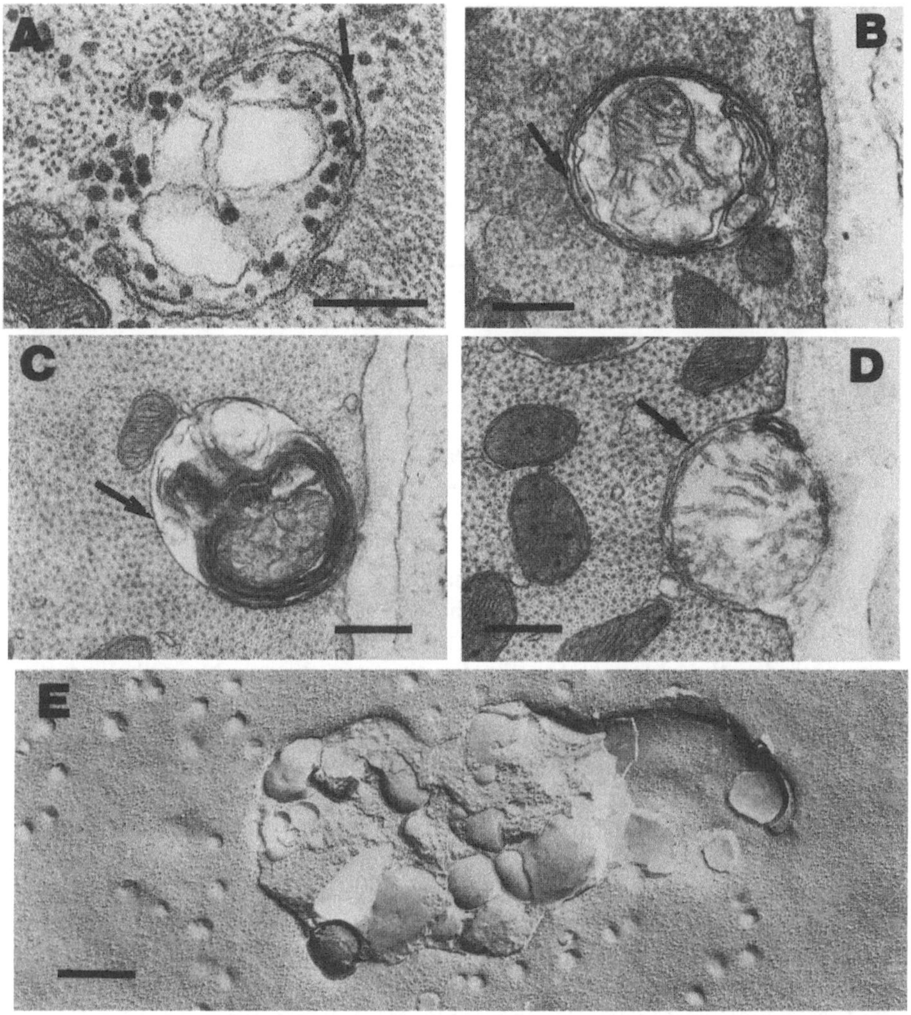

Fig. 76 A–E. Soleus muscle of a rat which has been treated for 5 days with chloroquine (25 mg/day/kg body weight) to demonstrate putative steps of autophagocytosis. **A** A cistern (*arrow*) of unknown origin (sarcoplasmic reticulum?) envelops sarcoplasmic elements (membrane remnants and glycogen granules). **B** A double-walled (*arrow*) vacuole encloses fragments of mitochondria. **C** The vacuole has become single-walled (*arrow*); it contains a mitochondrium and membrane remnants. The vacuole apparently touches the plasma membrane of the muscle fibre. **D** The wall of the vacuole (*arrow*) has become part of the plasma membrane, and its content, a mitochondrium, is expelled into the interstitium, but is still covered by the basal lamina of the sarcolemma. **E** Freeze-fracture of the plasma membrane (P face) showing a similar situation as depicted in **D**. One sees an excavation of the membrane still containing cell debris. The membranes of the cell debris are particle-free and consist of pure lipid bilayers, whereas the membrane forming the bottom of the open vacuole carries membrane particles. Bars, 0.25 μm

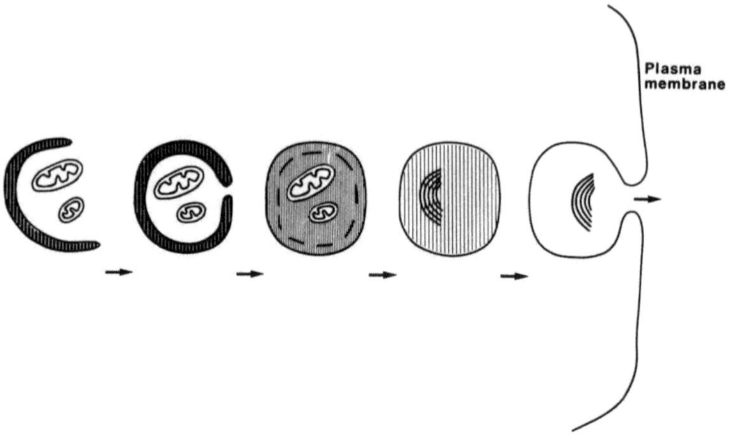

Fig. 77. Diagram summarizing the sequence of events during autophagocytosis in muscle fibres derived from the steps shown in Fig. 76. *Shaded area*, assumed localization of lysosomal enzymes

of the membranes enclosing the autophagolysosomes is not clear. Cell debris and also activity of acid phosphatase has been found in vacuoles connected to the T system; MacDonald and Engel (1970) believe that this membrane system provides the material for the limiting membrane. Nevertheless, their observation may indicate that the contents of a vacuole have been expelled into the T system, comparable to exocytosis through the plasma membrane. A more likely source for the limiting membrane would be the sarcoplasmic reticulum; the reticulum has the same developmental origin as the Golgi membranes, which in other cells form the vacuoles. There is morphological evidence that cisterns envelop the material to be sequestered and thus give rise to a double-walled limiting membrane (MacDonald and Engel 1970; Schmalbruch 1980a). Nevertheless, most autophagic vacuoles are limited by a single-layered membrane. It is conceivable that a cistern of the sarcoplasmic reticulum containing lysosomal enzymes sequesters a region of the sarcoplasm and that, when the vacuole is sealed off, the contents of the cistern are released into the vacuole by dissolving the inner membrane. This would conform to one of the pathways for hydrolytic enzymes, as proposed by Novikoff et al. (1964) (autophagic vacuole type 1, Novikoff 1976). These authors propose that the hydrolytic enzymes pass from the rough endoplasmic reticulum into the smooth endoplasmic reticulum, but instead of reaching the Golgi apparatus enter the double-layered wall of a forming autophagic vacuole, the limiting membrane of which is connected to the smooth endoplasmic reticulum (see also De Duve and Wattiaux 1966; Ericsson 1969) (Figs. 76, 77).

Chloroquine and related drugs damage red muscle fibres more severely than fibres poor in mitochondria (MacDonald and Engel 1970; Drenckhahn and Lüllmann-Rauch 1976, 1979; Schmalbruch 1978b, 1980a; Stauber et al. 1981b). Red muscle fibres normally have a higher activity of lysosomal enzymes than do white fibres and may therefore be more sensitive to blockage of enzyme activity (Stauber et al. 1981b).

D. Muscle Fibre Types in Mammalian Muscles

I. Historical Background

Human muscle tissue appears macroscopically more or less alike, but in small mammals the muscles often differ in colour. RANVIER (1873) observed that red muscles of rabbits contracted slowly and that they consisted of fibres with much sarcoplasm and distinct longitudinal striations; the white muscles were fast and the histological picture of the fibres was dominated by regular cross-striations.

A mixture of fibres of the "white" and "red" type was observed in many animal muscles and in all human muscles, but it remained unsettled whether one muscle consisted of functionally different fibres. DENNY-BROWN (1929) found no evidence for a functionally heterogeneous fibre population in the histologically heterogeneous cat gastrocnemius muscle. The difficulty was that physiological studies could only be done on whole muscles. The histological fibre "types" were not always distinct; first electron microscopy and histochemistry ascertained that muscle fibres were morphologically different. In the 1960s, it became possible to stimulate single anterior horn cells or motor axons and to study the contractile properties of individual motor units (ANDERSEN and SEARS 1964; DEVANANDAN et al. 1965), the final common pathways of motor activity (LIDDELL and SHERRINGTON 1925; SHERRINGTON 1930) (Chap. H.). It was found that the functional differences between motor units are parallelled by morphological differences between their muscle fibres.

The morphological differentiation of individual muscle fibres is usually assessed using light microscopical histochemistry; it is less tedious than electron microscopy, and in many cases the only feasible procedure. The combination of various staining methods has resulted in a multitude of classifications. Nevertheless, the point of reference is still RANVIER'S (1873) classical distinction of slow and well vascularized, as opposed to fast and poorly vascularized, fibres of red and white rabbit muscles. The designations "red" and "white" fibres are still in use, meaning rich and poor in mitochondria, although there is no colour difference between individual fast and slow fibres; fast and slow muscles may have the same colour, or macroscopically red muscles may contract even faster than white muscles.

II. Anaerobic and Aerobic Energy Metabolism of Muscle Fibres as Reflected by Morphology

1. Preferred Pathways of Metabolism

Chemical energy to be converted into mechanical work, in the form of ATP, is derived either from the anaerobic glycolysis of glucose stored as glycogen, or from oxidation of pyruvate or free fatty acids. The glycogen of the muscle fibres forms β-glycogen granules; fatty acids are bound as neutral fat which occurs as droplets between the myofibrils. Oxidative enzymes are localized inside mitochondria; the enzymes of anaerobic glycolysis are not membrane bound and are solubilized in the aqueous cytoplasm. Although the efficiency of anaerobic glycolysis is relatively small, it has the advantage that it is unimpaired by the lack of blood supply during sustained contractions. The amount of glycogen stored inside a muscle fibre is limited, and a muscle working exclusively anaerobically without supply of fuel will soon fatigue. The oxidative metabolism provides energy for a practically unlimited period of time as long as a sufficient blood supply is maintained. Each skeletal muscle fibre can, to a certain extent, shift between both metabolic pathways, but most fibres are specialized and differ in their content of enzymes and substrates of metabolism. Already EHRLICH (1885) observed that the oxidative capacity of muscles destined for permanent work was higher than that of muscles destined for rapid movements or short-lasting activity.

The speed of a muscle is not linked to its preferred pathway of energy metabolism, but slowly contracting muscles usually have a high oxidative capacity. The slow-twitch cat soleus muscle, at lower energy cost, maintains the same isometric tension as the fast-twitch gastrocnemius muscle. This difference is most pronounced for submaximal contractions. Anaerobic glycolysis in the soleus muscle provides 18%, and in the gastrocnemius muscle 74%, of the energy necessary to maintain 10% of maximum tetanic tension. At 75% maximum tetanic tension, anaerobic glycolysis accounts for 75% and 96% of the energy, respectively (SAWKA et al. 1981).

Lipid droplets, one of the substrates of oxidative metabolism, are readily shown using histochemistry. Staining by Oil Red or Sudan Black B distinguishes three classes of fibres with different lipid content (BULLARD 1919; GAUTHIER and PADYKULA 1966; MAN-I et al. 1967). Sudan Black B not only stains the triglyceride droplets but also the phosphatides within the mitochondrial membranes. Hence, staining for one of the enzymes of the oxidative pathway of metabolism or Sudan Black staining produces the same checkerboard pattern of weakly ("white"), intermediately, and intensely ("red") stained fibres (WACHSTEIN and MEISEL 1955; GAUTHIER and PADYKULA 1966). Staining for glycogen or for a glycolytic enzyme results in a checkerboard pattern of fibre types as well. In general, there is a tendency for the reactions for enzymes and substrates of oxidative and anaerobic metabolism to be reciprocal; fibres rich in oxidative enzymes are poor in anaerobic enzymes and vice versa (Fig. 78). This is, however, not the case for all fibres, and the three groups of fibres established by staining for oxidative enzymes are not identical with the three groups established by staining for glycolytic enzymes (BROOKE and KAISER 1970a). Unexpectedly, in several species (rat, rabbit, guinea pig, mouse) the fibres rich in mitochondrial enzymes ("red") have been found to contain 30% more glycogen than fibres poor in mitochondrial enzymes ("white") (GILLESPIE et al. 1970; PETER et al. 1972). In man, the intensity of the PAS stain is the same in fibres rich and poor in mitochondrial enzymes (GOLLNICK et al. 1974a, b) (see below).

2. The Glycogen Depletion Method

"White" muscle fibres empty their glycogen stores during prolonged activity; they are susceptible to fatigue. This fact has been exploited to mark the fibres of individual motor units. When one motor axon is electrically stimulated until the force developed by the muscle fibres of the unit declines, the previously active muscle fibres become PAS negative and thus can be histologically identified. The fibres may then be classified in serial sections using appropriate staining procedures. The "glycogen depletion method" has become an important tool for studying individual motor units. It was originally suggested by KRNJEVIC and MILEDI (1958) and was first used for rat muscles by EDSTRÖM and KUGELBERG (1968) and KUGELBERG and EDSTRÖM (1968), and for cat muscles by BRANDSTATER and LAMBERT (1969, 1973), DOYLE and MAYER (1969), MAYER and DOYLE (1970), and MAYER (1973). The method has proved that all fibres of a motor unit are of the same metabolic type, and that the speed of contraction, the fatiguability of a motor unit, and the histochemical properties of its muscle fibres are related (BURKE et al. 1971, 1973). For the first time, the number of fibres of individual motor units could be determined (see Chap. H). A major problem is that muscle fibres rich in oxidative enzymes are difficult to fatigue, and that most data are from motor units with fibres poor in oxidative enzymes. Fatigue-restistant units are only depleted in a reproducible fashion when the fibres are stimulated under hypoxia (KUGELBERG and LINDEGREN 1979).

The application of the glycogen depletion method to voluntary contractions in man and monkeys (GOLLNICK et al. 1974a, b; GILLESPIE et al. 1974) is problematic. Motor units of different types may be activated with different innervation rates, and oxidative units may have been active without being fatigued. Hence the degree of glycogen depletion does not necessarily reflect the role a given motor unit has played during a motor task. The relative contribution of the muscles during a complex movement (jumping, walking) is usually unknown. Oxidative fibres may have spared their glycogen stores. This could explain the paradox that some authors find more glycogen in "rested" oxidative than in "rested" glycolytic fibres (GILLESPIE et al. 1970; PETER et al. 1972). It is difficult to exclude uncontrolled activity before or during the preparation of the muscle sample.

3. Method-Related Problems of Fibre Type Histochemistry

Technical difficulties may affect the histochemical characterization of fibres with different metabolic pathways. Freezing of the muscle sample for histochemistry always produces ice crystals which disrupt the mitochondria; in "successfully" frozen fibres, the crystals are too small to be seen using light microscopy. Nevertheless, one should be aware that the granules of the reaction product that are seen at high magnification do not represent individual mitochondria, but only mark the approximate site of the enzyme.

A large and variable amount of glycogen or of soluble enzymes is lost when the section is incubated for a histochemical reaction. This may enhance or cloud differences between neighbouring fibres, or it may result in different staining intensities in different parts of the specimen. To overcome the problem of diffusion and enzyme loss, the substrate for the reaction has been offered in a gel film rather than as solution (SIGEL and PETTE 1969). This method is particularly useful for the quantitation of enzymes (NOLTE and PETTE 1972a, b), or for the subcellular localization of non-membrane-bound enzymes of the cytosol. It seems as if several of the glycolytic enzymes are restricted to the compartment between the myofilaments, or are even bound to the filaments, rather than being equally distributed throughout the sarcoplasm (ARNOLD et al. 1969, 1971; ARNOLD and PETTE 1968, 1970; BROSEMER 1972; PETTE 1975; DÖLKEN et al. 1975).

III. Fast and Slow Muscle Fibres and Their Histochemical Correlates

The first to present physiological evidence for the fact that muscles are inhomogeneous with respect to the contraction velocity of their fibres, were GORDON and PHILLIPS (1953); for the cat anterior tibial muscle they showed that the superficial and deep parts of the muscle had fast and slow contractions, respectively. ANDERSEN and SEARS (1964) and DEVANANDAN et al. (1965) stimulated single motor axons of the cat intercostal muscles; the isometric twitch contraction times of individual motor units of this muscle fell into two distinct classes. The attempts to identify fast- and slow-twitch muscle fibres using histology are based on the facts that myosin has ATPase activity (ENGELHARDT and LJUBIMOVA 1939), which over a wide range of species increases with the speed of shortening of the muscle (BARANY et al. 1965; BARANY 1967), and that the structure of the myosin molecule is different in fast and slow muscles.

1. Myosin ATPase and Fibre Typing

The activity of the histochemically demonstrated "myosin" ATPase activity is widely used to morphologically identify fast- and slow-twitch fibres (PADY-

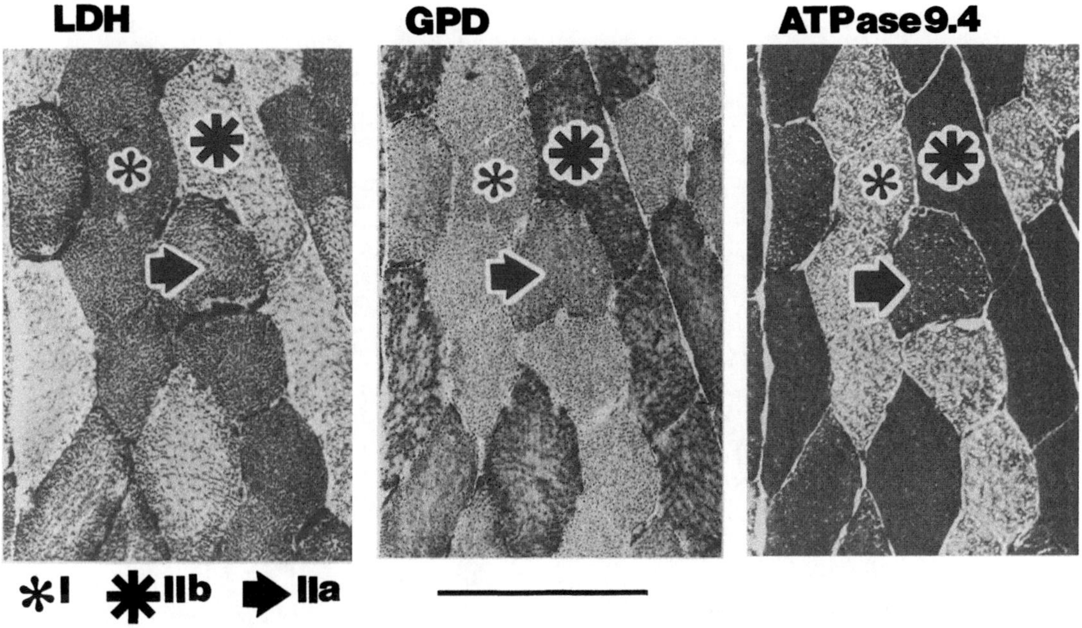

Fig. 78. Medial gastrocnemius muscle, cat. Serial cross-sections reacted for different enzymes. Identical fibres are marked by identical symbols. Type I fibres (*small stars*) stain intensely for the mitochondrial enzyme nicotinamide adenine dinucleotide-linked lactic dehydrogenase (*LDH*), weakly for the glycolytic enzyme menadione-linked alpha-glycerophosphate dehydrogenase (*GPD*), and weakly for adenosine triphosphatase at pH 9.4 (*ATPase 9.4*). IIB fibres (*large stars*) have an inverse staining pattern, and IIA fibres (*arrows*) stain intermediately for the mitochondrial and glycolytic enzymes, and intensely for ATPase at pH 9.4. Note that some fibres have circular and others oblong cross-sectional shapes. Bar, 100 μm. (From SCHMALBRUCH and KAMIENIECKA 1975, with permission of the Histochemical Society)

KULA and HERMAN 1955a, b; ENGEL 1962). The rationale for the staining method is that myosin ATPase is activated by Ca^{2+}, and that it has a high activity at an alkaline pH (9.4); other ATPases of the muscle cell are believed to be inactivated under these conditions. All muscle fibres of fast motor units, as identified by glycogen depletion, stain intensely for ATPase at pH 9.4 (type II fibres), and all fibres of slow-twitch motor units stain weakly (type I) (BURKE et al. 1971, 1973). One usually finds, in cross-sections of skeletal muscles, a checkerboard pattern of dark and lightly stained fibres. Alkaline preincubation enhances the contrast. Occasionally three gradations are visible because fast fibres rich in mitochondria may stain slightly less intensely than fast fibres poor in mitochondria (Fig. 78). Intermediately stained fibres are consistently found in masticatory muscles (RINGQVIST 1973; RINGQVIST et al. 1982; ERIKSSON 1982), and also in skeletal muscles when the ATPase method is modified (GOTTSCHALL et al. 1980a; STARON et al. 1983). The fact that the ATPase method sometimes reveals more than two fibre types with varying staining intensities should not be overstressed and not be taken as evidence that this reflects gradual differences in the myosin ATPase activity. SAMAHA and YUNIS (1973) showed

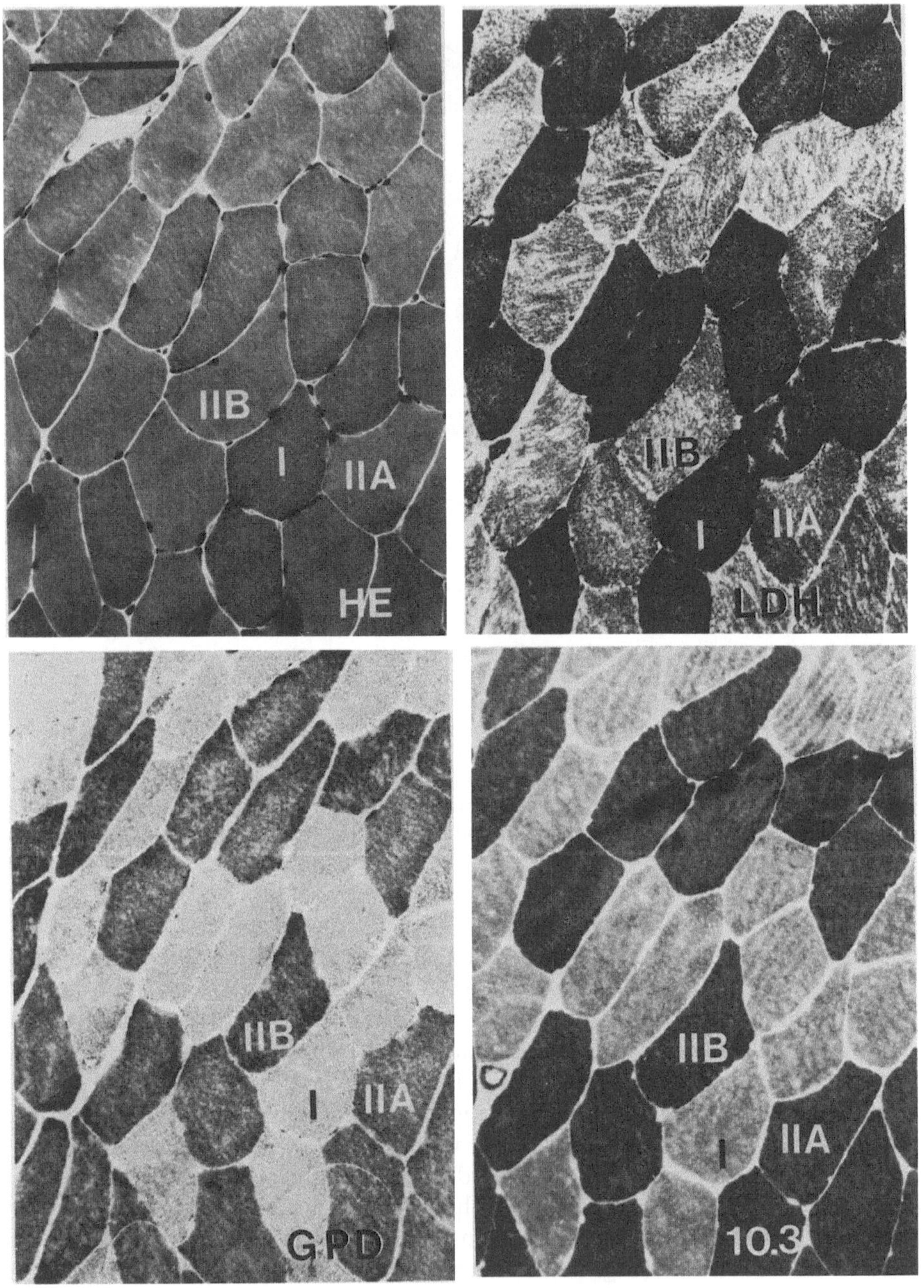

Fig 79

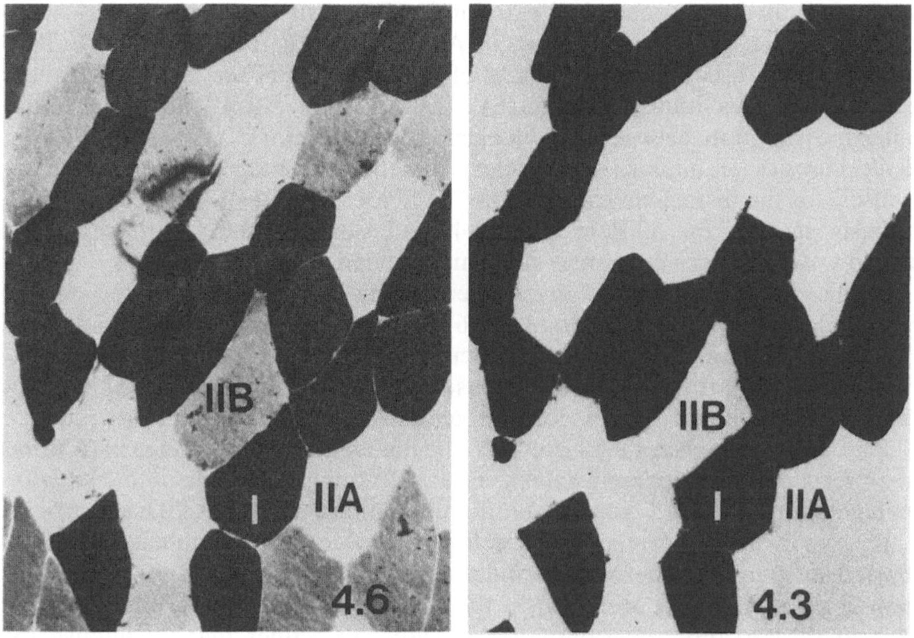

Fig. 79. Human medial vastus muscle. Serial cross-sections stained using various methods to demonstrate the histochemical profiles of the fibre types. The staining method is indicated (*HE*, haematoxylin eosin; *LDH*, nicotinamide adenine dinucleotide-linked lactic dehydrogenase; *GPD*, menadione-linked alpha-glycerophosphate dehydrogenase; *10.3*, *4.6*, and *4.3*, adenosine triphosphatase at pH 9.4 after preincubation at pH 10.3, 4.6, and 4.3). Three fibres of type I, IIA, and IIB have been marked throughout. Type I fibres react intensely for the mitochondrial enzyme LDH and weakly for the glycolytic enzyme GPD; the ATPase reaction is weak after alkaline preincubation and strong after acid preincubation. IIA and IIB fibres are clearly distinguishable by ATPase staining at pH 9.4 after acid preincubation. IIA ("intermediate") fibres stain more intensely for the mitochondrial enzyme than IIB ("white") fibres, but the difference is less pronounced than in cat or rat muscles (Figs. 78, 80). Note that in the HE-stained section, type I fibres appear slightly darker than type II fibres. Bar, 100 μm. (Preparation courtesy of Dr. Kamieniecka, Copenhagen)

that the histochemical method is not specific, and that mitochondrial ATPase may contribute to the reaction. The localization of the reaction product varies with the physicochemical conditions, and probably also depends on different structural properties of the myofibrils of fast- and slow-twitch muscle fibres. Hence, the myofibrillar localization of the product does not assure that the enzyme demonstrated is localized in the myofibrils. Incubation at 20° C and at 37° C may result in a different intracellular localization (Schmalbruch, unpublished). According to Guth (1973), it is possible to imitate the staining pattern for "myosin" ATPase by incubating a section with inorganic phosphate, instead of ATP. Similar results were obtained at the electron microscopic level; when inorganic phosphate and calcium diffused into the muscle fibres, the precipitates were mainly on the myosin filaments, and they were indistinguishable from those produced by incubation with ATP and Ca^{2+} to show the activity of the myosin ATPase (Tunik 1971). Guth (1973) emphasizes that, although the product of the histochemical reaction for "myosin" ATPase at pH 9.4 undoubtedly is the result of hydrolysis of ATP, it demonstrates the total ATPase activity of the fibre, and not only that of the myosin ATPase, regardless where the precipitate eventually is localized. TAYLOR et al. (1974), however, found a linear correlation between the biochemically determined myosin ATPase in needle biopsies of human leg muscles, and the percentage of fibres reacting intensely for the histochemical "myosin" ATPase. In spite of its shaky theoretical basis, staining for ATPase at pH 9.4 has become a useful marker for fast- and slow-twitch fibres in normal adult mammalian muscles.

Acidic preincubation may invert the staining pattern (BROOKE and KAISER 1970a, b), and at different pH (3.9–4.6) IIA and IIB fibres are distinguished (Fig. 79). In human muscles, after preincubation at pH 4.6, the previously weakly reacting type I fibres show an intense reaction for ATPase at pH 9.4. IIA fibres (intermediate reaction for mitochondrial enzymes) do not react for ATPase at pH 9.4 when they have been incubated at pH 4.6, whereas IIB fibres (weak reaction for mitochondrial enzymes) still show an intermediate reactivity. Preincubation at pH 4.3 completely inhibits the reaction for ATPase at pH 9.4 in all type II fibres whereas the reaction has become most intense in type I fibres. The staining pattern after acidic preincubation is extremely sensitive to small variations of pH, and even in the same laboratory the results may vary when applied to different samples; the acidic pH values that distinguish the subtypes are different in different species (REICHMANN and PETTE 1982). In immature and also in diseased muscles some fibres may stain intensely both after alkaline and acidic preincubation (IIC fibres). These IIC fibres which are regarded immature should not be confused with segmentally contractured fibres which in cross-sections stain intensely with any procedure. Segmental contractures are frequently seen as artefacts; they are identical with "hyalinic" fibres in diseased muscles (SCHMALBRUCH 1973). Some authors distinguish also subtypes of type I fibres (IA and IB; ASKANAS and ENGEL 1975b).

Type IIA fibres tend to stain more intensely for mitochondrial enzymes than do type IIB fibres; thus they correspond to intermediate fibres after staining for mitochondrial or glycolytic enzymes (BROOKE and KAISER 1970b). There is considerable overlap when the fibre types are compared in serial sections (SJØGAARD et al. 1978a, b), and when the biochemically quantitated oxidative capacity of individual fibres is compared with ATPase typing (NEMETH and PETTE 1981; REICHMANN and PETTE 1982). For the sake of simplicity, IIA and IIB fibres are often called "fast-red" and "fast-white" fibres, respectively. IIA and IIB fibres of the abdominal muscles of man stain equally weakly for mitochondrial enzymes (HÄGGMARK and THORSTENSSON 1979).

2. Myosin Heterogeneity and Immunofluorescence

Morphological evidence for molecular differences between myosin from fast and slow muscles was presented by NAKAMURA et al. (1971), who found that uranyl acetate reacted differently with paracrystals of light meromyosin. Electrophoresis of myosin on sodium dodecyl sulphate (SDS) polyacrylamid gel splits the molecule into a heavy chain, with a mol. wt. of about 200,000 daltons, and into different subunits (light chains) with weights in the range of 20,000 daltons. Three different species of light chains are obtained from fast myosin (LC_{1f}, LC_{2f}, LC_{3f}) with mol. wts. of 25,000, 18,000, and 16,000 daltons. Slow myosin contains two species of light chains (LC_{1s}, LC_{2s}) with mol. wts. of 27,000 and 20,000 (19,000) daltons. LC_{1s} is heterogeneous and represents two light chains (LC_{1as}, LC_{1bs}) of 26,500 and 27,500 daltons (LOWEY and RISBY 1971; SARKAR et al. 1971; WEEDS 1976). Each myosin molecule consists of two heavy chains forming the head region and (in an α-helically coiled coil) the tail region, and four light chains non-covalently attached to the head region (Fig. 80). The light chain LC_{2f} is liberated from myosin with 5′5′-dithiobis (2-nitrobenzoic acid) (DTNB), and hence has also been designated DTNB light chain. Each fast myosin molecule contains two DTNB chains; they are not essential for the ATPase activity of myosin. LC_{1f} and LC_{3f}, however, cannot be removed without total loss of enzyme activity: they have been termed alkali light chains 1 and 2 (A1, A2) because they are released under alkaline conditions (LOWEY and RISBY 1971; WEEDS and POPE 1971). The alkali light chains resemble each other, but alkali 1 has an additional terminal peptide sequence, the "difference peptide" which has been used as antigenic determinant for immuncytochemistry (see below). The role of LC_{1as} and LC_{1bs} resembles that of the alkali light chains of fast myosin (FRANK and WEEDS 1974; WEEDS 1976).

Each myosin molecule consists of a fixed number of light and heavy chains (Fig. 80); if two myosins differ in one or several of the components, they represent "isoenzymes" with different antigenic properties. The antigenic properties have been exploited for their differences, to demonstrate fast and slow fibres in histological sections, by staining with antibodies coupled to fluorescent dyes (GRÖSCHEL-STEWART and DONIACH 1969; ARNDT and PEPE 1975) (Figs. 81, 82). GAUTHIER and LOWEY (1977) raised in rabbits antibodies against myosin from the fast chicken pectoralis muscle, and found that in rat diaphragm all fibres except half of the mitochondria-rich (red) fibres were reactive. All reactive fibres were of type II (ATPase at pH 9.4). The authors conclude that there are four types of fibres in rat diaphragm; fast-white, fast-intermediate, fast-red, and slow-red. In the soleus muscle (Fig. 83), only the few type II fibres are stained by the anti-fast antibodies. The type II fibres of rat muscles also react with purified antibodies against proteolytic subfragments and low-molecular subunits of pectoralis myosin. Non-reactive red fibres (slow-red) react strongly with an antibody against myosin isolated from the slow-tonic anterior latissimus dorsi muscle (ALD) of chicken. A minority of fibres of the soleus muscle react with antibodies to pectoralis as well as ALD myosin. Fast-red fibres do not react as strongly as the fast-white and fast-intermediate fibres with antibodies against pectoralis S1 subfragment and pectoralis alkali 1 light chain (GAUTHIER and LOWEY 1979).

By contrast, PIEROBON-BORMIOLI et al. (1980) reported that slow-twitch fibres of mammals only bind antibodies against mammalian soleus muscle, but not antibodies against chicken ALD. Only a minority of fibres of the extraocular muscles and the nuclear-bag fibres of the muscle spindles are stained by anti-ALD myosin (Figs. 93, 96). This suggests that there are two slow myosins,

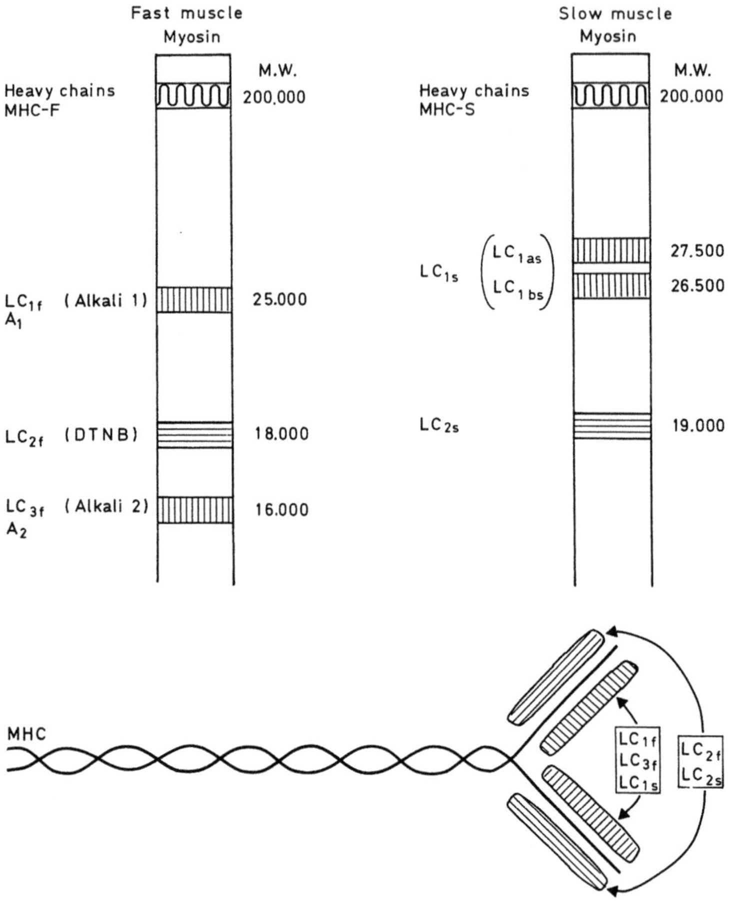

Fig. 80. Scheme illustrating the composition of the myosin molecule and the various elements giving rise to myosin isoenzymes which characterize muscle fibre types. *Top*: The two bars represent one-dimensional polyacrylamid gels of fast- and slow-twitch muscle myosin fragments, respectively. The molecular weights (MW) and the various nomenclature in use are indicated. *Bottom*: Two heavy chains form the tail and together with four light chains the two S1 subunits of the head region. Two of the light chains are "essential" (*cross-hatched*) and two are "non-essential" (*longitudinally hatched*). The structure of the heavy chains and also the light chain composition is different in muscle fibres of different origin. With respect to the essential light chains, A 1 and A3, fast myosin may occur as homodimer (A1-A1, or A2-A2) or as heterodimer (A1-A2). (Adapted from WEEDS 1976)

Fig. 81 A–D. Extensor digitorum longus muscle of rat. Serial cross-sections stained by indirect immunofluorescence with antibodies against the myosins of guinea pig soleus muscle (**A**), of guinea pig masseter muscle (**B**), of guinea pig tensor fasciae latae muscle (**C**), and reacted for succinodehydrogenase (**D**). Only type I fibres (*I*) react with the antibody against soleus myosin; staining with antibodies against masseter (fast-red) and tensor fasciae latae (fast-white) myosins distinguishes IIA (*IIa*) and IIB (*IIb*) fibres rich and poor in mitochondria, respectively. Bar, 100 μm. (Micrographs courtesy of Drs. PIEROBON-BORMIOLI and SCHIAFFINO, Padova)

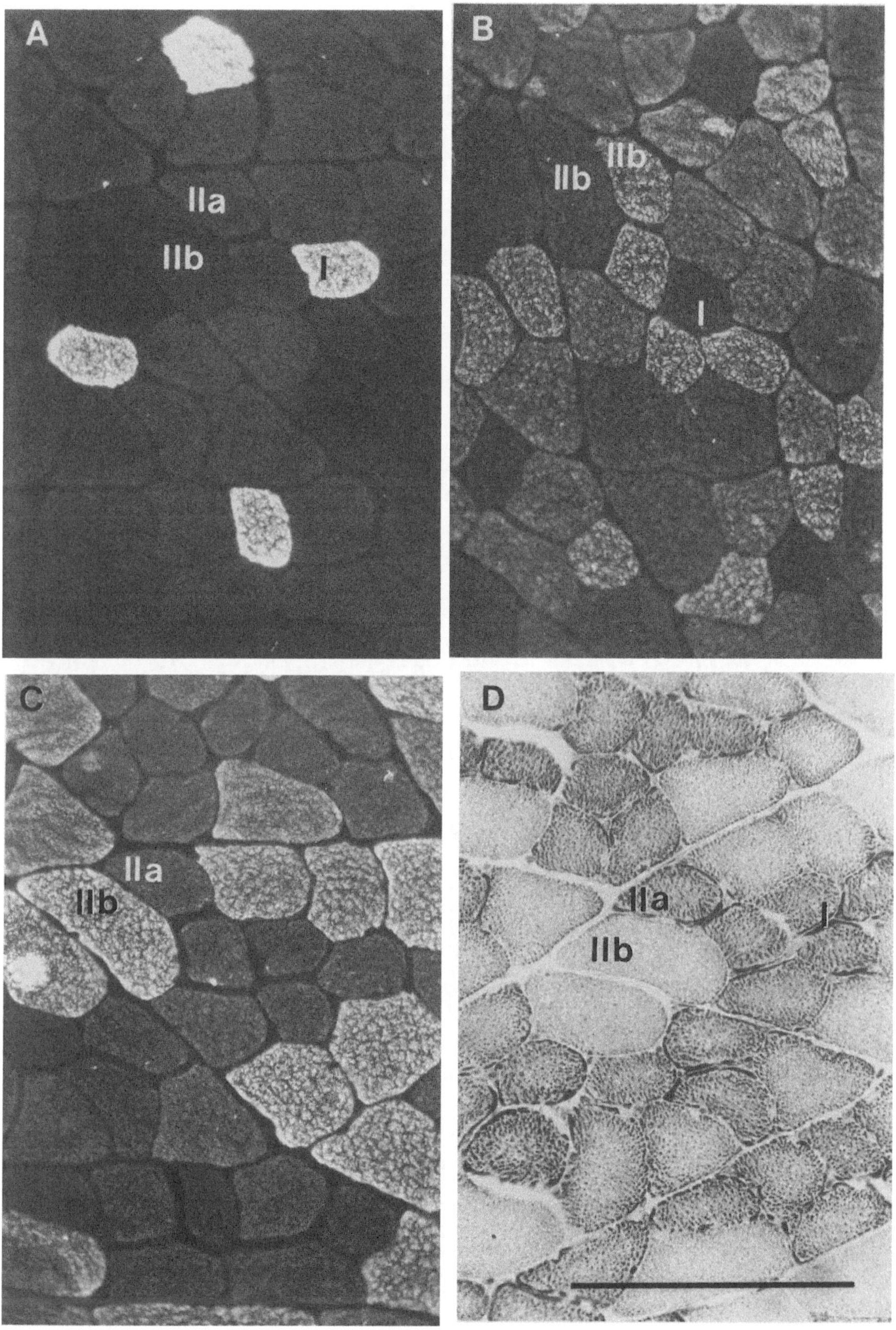
A
IIa
IIb
I
B
IIb
IIb
I
C
IIa
IIb
D
IIa
IIb
I

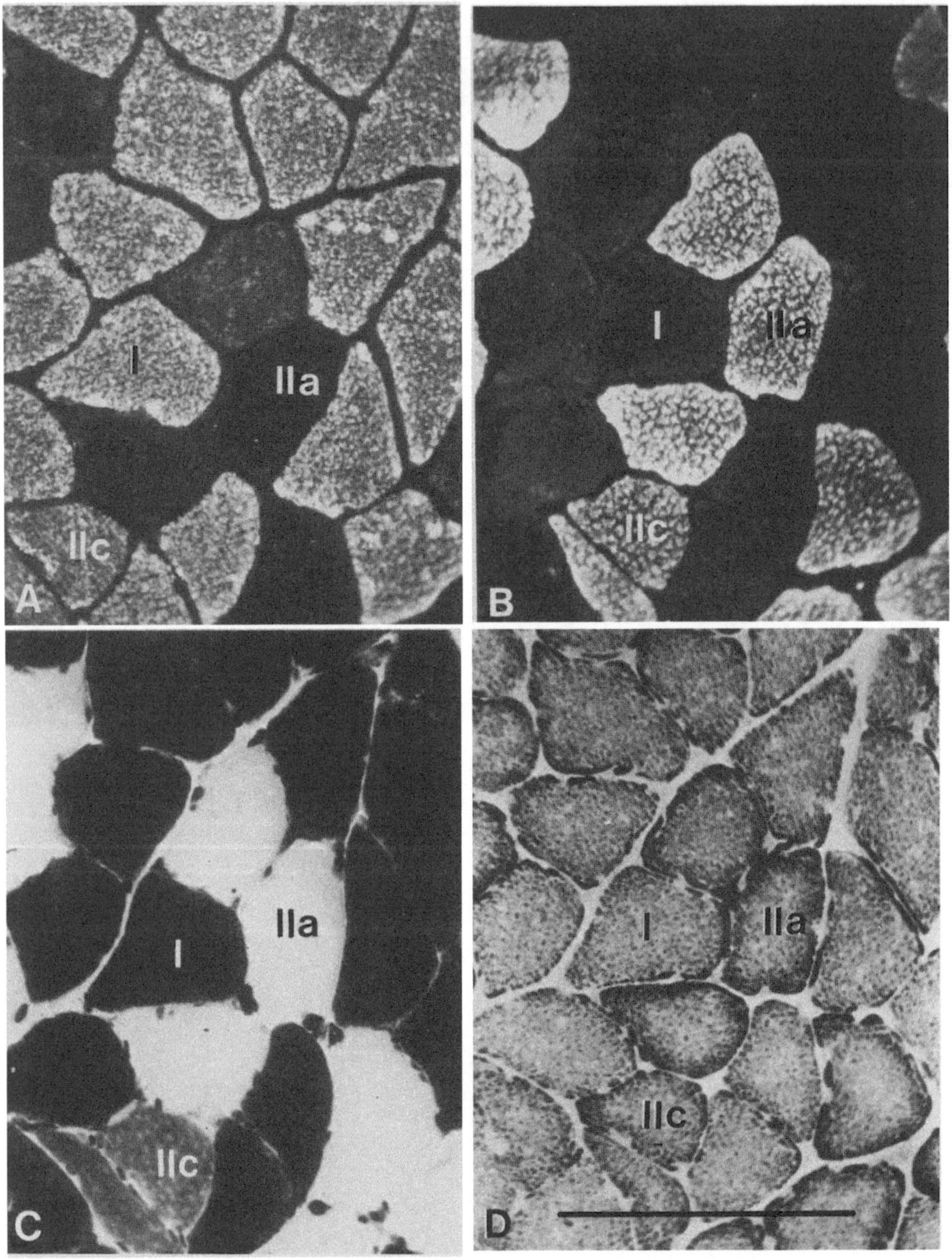

Fig. 82 A–D. Soleus muscle of rat. Serial cross-sections stained by indirect immunofluorescence with antibodies against guinea pig soleus myosin (**A**); guinea pig masseter myosin (**B**), for ATPase at pH 9.4 after acid (pH 4.3) preincubation (**C**), and for succinodehydrogenase (**D**). The area shown contains rather many type II fibres (for comparison, see Fig. 83) which all are rich in mitochondria and react with the antibody against masseter myosin, i.e. are type IIA fibres (*IIa*). Type I fibres (*I*) react only with the antibody against soleus myosin. Two fibres are IIC (*IIC*) and react both with anti-fast and anti-slow myosin; the ATPase reaction in this fibre type is incompletely inhibited by acid preincubation. Bar, 100 μm. (Micrographs courtesy of Drs. PIEROBON-BORMIOLI and SCHIAFFINO, Padova)

slow-twitch and slow-tonic. Nevertheless, the cross-reactivity of slow-tonic and slow-twitch fibres described by GAUTHIER and LOWEY (1979) has been confirmed in a recent paper (GAUTHIER et al. 1983).

The white tensor fasciae latae muscle of guinea pigs consists of fast-twitch fibres poor in mitochondria; according to the ATPase stain at pH 9.4, the fibres are IIB. The red masseter muscle consists mainly of IIA fibres. Both muscles contain the same light chains, but two-dimensional electrophoresis of the heavy chains after tryptic digestion reveals that the peptide maps are different (DALLA LIBERA et al. 1980). This explains the different reactivity of fast-red and fast-white fibres of rat with antibodies against the S1 subfragment of chicken pectoralis myosin (GAUTHIER and LOWEY 1979). PIEROBON-BORMIOLI et al. (1981) raised antibodies against the masseter and tensor fasciae latae myosin of guinea pigs and succeeded in differential staining of IIA and IIB fibres in rat (Figs. 81, 82). These findings indicate that the IIA and IIB myosins differ in heavy chain structure, but are identical with respect to light chains; only the molar ratios differ (SALVIATI et al. 1982; see also below). By contrast, GAUTHIER et al. (1983) report that IIA and IIB fibres of cat react differently with an antibody raised against the light chains of fast myosin from chicken pectoralis muscle.

The diversity of the alkali light chains implies that there may be homo- and heterodimers and still more myosin isoenzymes. D'ALBIS et al. (1979) performed electrophoresis under non-dissociating conditions on pyrophosphate polyacrylamid gel. They separated about a dozen different myosins from various sources. Fast-twitch skeletal muscles contain three alkali light-chain isoenzymes; two are homodimers (A1+A1, A2+A2) and one is a heterodimer (A1+A2). Guinea pig soleus muscle contains only one isoenzyme, but in rat soleus muscles two are found. Electrophoresis of myosin of the chicken ALD produces a major band corresponding to slow myosin, and a minor band corresponding to one of the fast isoenzymes. The fast chicken pectoralis and posterior latisssimus dorsi muscles have three isoenzymes corresponding to the three fast isoenzymes in mammalian fast muscles. FITZSIMONS and HOH (1981) find five isoenzymes in human muscles and associate three of them (designated FM_{1-3}) with type II fibres, and two of them (SM_{1-2}) with type I fibres.

HOH and YEOH (1979) mapped the peptides of the heavy chains of myosins from fast and slow muscles of adult chicken and rabbits, and the myosin of immature rabbit muscles; fast, slow, and fetal myosins have different heavy chains. WHALEN et al. (1981) found in developing fast rat muscles three sorts of heavy chains; a neonatal form occurs after the fetal type and precedes the adult type. ROWLERSON et al. (1981) described still another heavy chain in jaw closing muscles of carnivores, which is not present in the corresponding muscles of man (THORNELL et al. 1984). Thus, we are faced with the existence of six different light chains (three fast and three slow) and an apparently still-increasing number of different heavy chains (slow, fast-white, fast-red, fetal, neonatal, masticatory).

It is unresolved whether different myosins may occur in the same fibre. This has been suggested by GAUTHIER and LOWEY (1979) for type II fibres of rat soleus muscle, but is disputed by DALLA LIBERA et al. (1980). In rabbit and rat (LUTZ et al. 1979; RUBINSTEIN and KELLY 1981), fast IIA but not fast IIB fibres show a weak reaction with antibodies against slow myosin as well. This is not the case in human (BILLETER et al. 1980) and horse (SNOW et al. 1981) muscles. In mice, the fast-twitch extensor digitorum longus muscle contains the known three fast-myosin isoenzymes; the soleus muscle, which like in rat consists of type I and type IIA fibres, contains two isoenzymes termed SM and IM. SM is associated with the type I fibres and IM with the IIA fibres. Soleus IM is a hybrid and contains LC_{1f} and LC_{1s} in approximately equal proportion, together with LC_{2f}, but no detectable LC_{2s} (FITZSIMONS and HOH 1983). Whether the slow and fast light chains are in the same myosin molecule, or whether there are two sorts of myosin molecules in the same fibre, is unknown. It should be noted that many type II fibres of rat and possible also of mouse soleus muscle undergo a prolonged developmental transformation into type I fibres (KUGELBERG 1976), hence the soleus IIA fibres might be different from the IIA fibres in permanently heterogeneous muscles.

SALVIATI et al. (1982) assessed the contractile proteins of adult rabbit muscles by electrophoresis of myofibrillar proteins extracted from 5- to 25-mm-long fragments of individual muscle fibres. Each fibre was histochemically classified by staining a short segment using the ATPase method after acid or alkaline preincubation. Four fibre types were distinguishable. Type I, IIA, and IIB fibres each contained unique myosin heavy chains. Type I and IIB fibres contained sets of "slow" and "fast" light chains, respectively. Type I fibres from the masseter muscle, however, lacked LC_{1as}. Soleus type I fibres had the full complement of slow light chains (LC_{1as}, LC_{1bs}, LC_{2s}). Among these "typical" type I fibres the stoichiometry of LC_{1as} and LC_{1bs} was heterogeneous. Also IIB fibres contained different molar ratios of fast light chains. IIA fibres were characterized by a small amount of LC_{3f}, and 10% of the IIA fibres contained LC_{1bs}. The LC_{1bs} found in some IIA fibres comigrated with LC_{1bs} from type I fibres, but there was evidence that the immunological properties differed. Type I and II fibres had specific isoforms of the troponin subunits as well; in this respect, IIA and IIB fibres were identical. The fourth fibre type, which histochemically stained like IIC fibres, showed coexistence of fast-twitch and slow-twitch fibre forms of myosin light chains, and, occasionally, of troponin subunits (see also Fig. 82). The results reported by SALVIATI and co-workers answer several questions raised by immunocytochemistry of myosin isoenzymes: They clearly indicate that the fibre "types" established by empirically developed and modified histochemical reactions for "myosin" ATPase, useful as they might be to relate fibre morphology and motor unit properties (see below), are gross oversimplifications. On the other hand, if the quantitative aspects of the heterogeneity of myosin were really exploited to classify muscle fibres, one probably would end up with an impractically large number of types. It would be most interesting to know whether a stoichiometric heterogeneity of the light chains exists between the muscle fibres of the same motor unit.

The whole field is in a state of flux, but a consistent picture is arising. In view of the diversity of heavy chains, one may ask whether the myosins are not only tissue- but to a certain extent also species-specific. This could explain some of the controversial results of immunocytochemistry. Nevertheless, several of the early studies were certainly afflicted by impurities of the antibodies.

IV. Fibre Type Classification and Physiological Properties of Motor Units

1. First Attempts and Confusing Nomenclatures

Skeletal muscle fibres have initially been grouped into three types according to their content of lipids or mitochondrial enzymes ("white", intermediate, "red"; A, B, C) (BULLARD 1919; WACHSTEIN and MEISEL 1955; GAUTHIER and PADYKULA 1966). Three types of motor units were found in cat muscles when it became possible to study the contractile properties of individual motor units (HENNEMAN and OLSON 1965). Two of the motor unit types are fast, of which one is easily fatigued and one is fatigue-resistant. All slow-twitch motor units are resistant to fatigue. All three types of motor units are present in the gastrocnemius muscle, but only slow-twitch units are found in the soleus muscle. The concept that skeletal muscles consist of motor units with different functional properties, and different excitability during reflex contractions, has been elaborated in other papers of the same group of workers (HENNEMAN et al. 1965; WUERKER et al. 1965; OLSON and SWETT 1966; MCPHEDRAN et al. 1965; for further references, see BUCHTHAL and SCHMALBRUCH 1980). All fibres of the cat soleus muscle were classified as "intermediate" according to their reaction for mitochondrial ATPase, and HENNEMAN and OLSON (1965) concluded that the intermediately stained muscle fibres of the mixed gastrocnemius muscle belong to slow-twitch motor units.

This was erroneous, and has given rise to considerable confusion. Subsequently, findings in different species were uncritically pooled, and physiological properties were inferred from the staining behaviour. EDGERTON and SIMPSON (1969) typed the muscle fibres of different animals using histochemistry only, and classified fibres that stained equally intensely for mitochondrial enzymes as "intermediate" and "slow", or "red" and "fast", depending on the reaction for ATPase at pH 9.4. Soleus muscle fibres were denoted as prototypes for "intermediate" fibres. ARANGIO and HAGSTROM (1969) described, in rabbits, the slow-twitch soleus muscle fibres as "heavily" stained for succinate dehydrogenase, and recommended the soleus muscle as a source for "red" fibres. YELLIN and GUTH (1970) introduced a classification system based on staining for succinate dehydrogenase (A, B, C) and for ATPase after preincubation at different pH (α, acid labile; β, alkali labile; $\alpha\beta$, hybrid), which was later amplified by adding a gradation of the ATPase-staining intensity (M, moderate; L, light; D, dark) and the mean diameter of the fibre type (YELLIN 1975). One disadvantage of this system is that a fibre classification as "A$\alpha\beta$M90" (YELLIN 1975) simply is not practicable, but the main problem is that different authors classified the same fibres as "intermediate" or "red" or as B or C (e.g. STEIN and PADYKULA 1962; EVERSOLE and STANDISH 1970; CLOSE 1972). Attempts to "synchronize" the nomenclatures (CLOSE 1972; TELERMAN-TOPPET and COËRS 1973; BUCHTHAL et al. 1974) failed because the three-type system of cat does not fit for all species (see below). At the same time, histochemistry of human muscle experienced a boom and became a diagnostic tool in clinical pathology. How desperate the situation was then is illustrated by a quotation from Brooke and Kaiser (1970a): "Classification of striated muscle into different types has always been somewhat confusing but recently has shown an alarming trend towards the incomprehensible ... If this is confusing in human muscles, the situation totally disintegrates when animal muscle is considered". The demands for any classification and nomenclature system set forward by the same authors are still worth emphasizing. "(1) It must be useful when applied to experimental or pathological situations. (2) It should be based on the properties being examined, not derived by inference from other properties of the fiber. (3) The various categories should be easily differentiated, without a gradual transition from one fiber type to another or many indeterminate fibers. If the system of classification is histological, it should not use words like 'slow' or 'fast'".

2. Species Differences

For cat muscles, mainly Burke and co-workers established the relation between physiological and histochemical properties of motor units. These authors made use of the glycogen-depletion method. The fast-twitch motor units of the heterogeneous gastrocnemius muscle (Fig. 78) consist of type II muscle fibres (ATPase pH 9.4 positive), which vary in staining for mitochondrial enzymes. Easily fatigued fast-twitch units consist of muscle fibres that stain weakly for mitochondrial enzymes, and the fibres of fatigue-resistant fast-twitch units stain intermediately for mitochondrial enzymes. The two types of fast motor units have been termed FF (fast, fatiguable) and FR (fast, fatigue resistant) units (Burke et al. 1971, 1973). Reinking et al. (1975) questions the existence of distinct categories of fast-twitch units and attributes the absence of units with intermediate fatiguability to the fact that the cats of Burke et al. (1971, 1973) had been caged. Stephens and Stuart (1974) advocate a "less ambiguous" fast-slow scheme with FF and FR units as extremes of a spectrum. In 1973, Burke et al. found that 2.5% of the motor units in the medial gastrocnemius muscle were fast but did not fit into their FF-FR system because they had intermediate properties with respect to fatigue. They were named F(int) units.

Fig. 83. Rat gastrocnemius (*left*) and soleus muscles (*right*). Serial cross-sections stained for different enzymes. Identical fibres are marked by identical symbols. *Left* (top to bottom): The reactivity for the mitochondrial enzymes, nicotinamide adenine dinucleotide-linked lactic dehydrogenase and succinic dehydrogenase, is similar and roughly inverse to that for menadione-linked alpha-glycerophosphate dehydrogenase (second from *bottom*). Staining for adenosine triphosphatase at pH 9.4 (*bottom*) identifies relatively few type 1 fibres (*arrowheads*, weakly stained). Type II fibres may stain weakly (*bent arrows*) or intensely (*straight arrows*) for the mitochondrial enzymes. *Right* (top to bottom): The soleus muscle consists mainly of fibres with rather intense reaction for a mitochondrial enzyme (nicotinamide adenine dinucleotide-linked lactic dehydrogenase); they react weakly for the glycolytic enzyme menadione-linked alpha-glycerophosphate dehydrogenase (*middle*). These fibres are type I fibres according to the staining for adenosine triphosphatase at pH 9.4 (*bottom*). Few fibres react intensely for all three enzymes and are type II fibres. Note the intense reaction of the blood vessels for ATPase at pH 9.4 (*bottom right*). Bars, 100 µm. (From Schmalbruch and Kamieniecka 1975, with permission of the Histochemical Society)

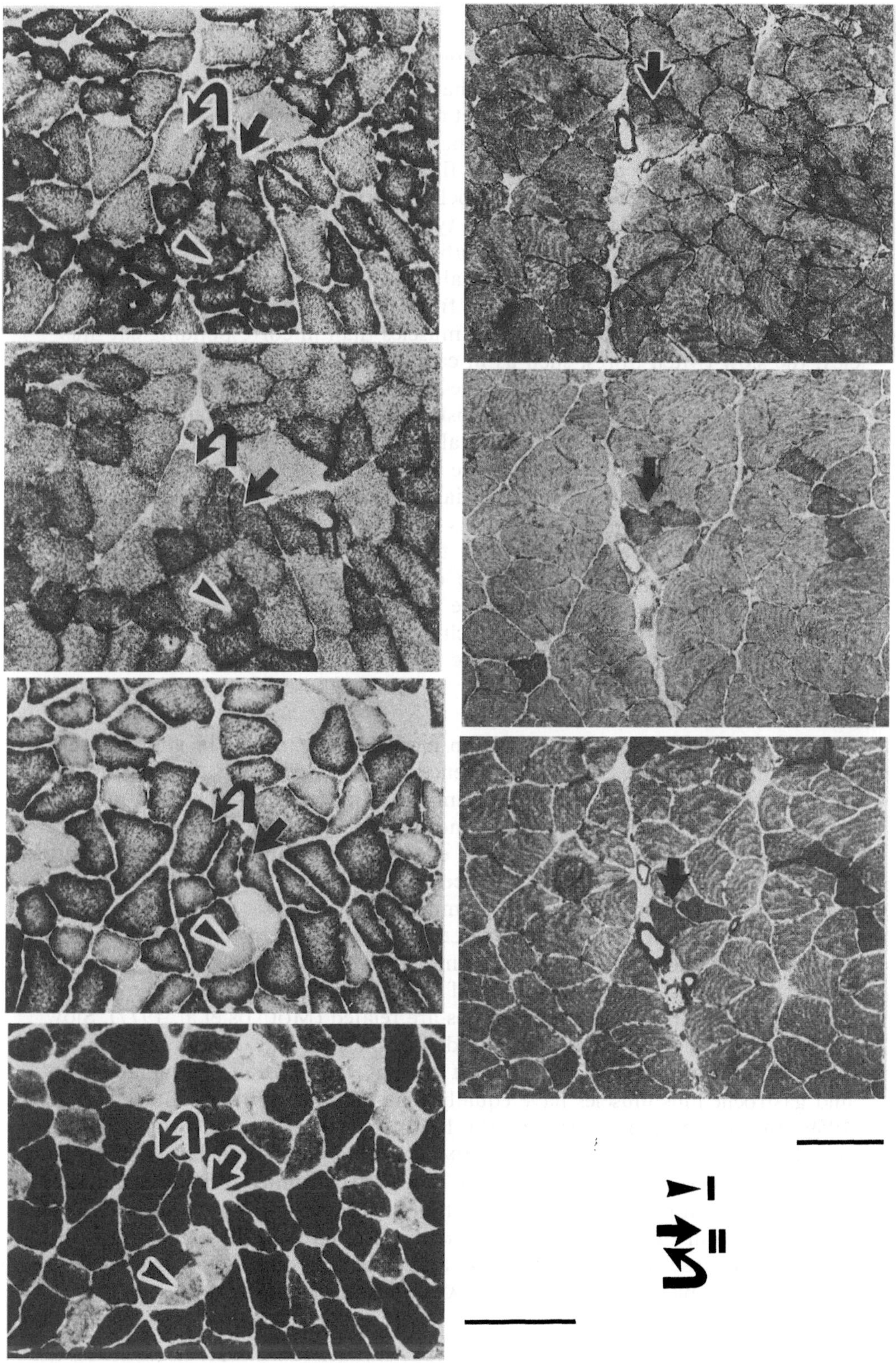

I
II

Later, the same authors reported that 10% of the motor units were F(int) units (Burke et al. 1976, 1977). F(int) units have been observed in other cat muscles as well (Dum and Kennedy 1980; Jami et al. 1982). The fibres of these F(int) units have no specific histochemical staining pattern (Jami et al. 1982).

Slow-twitch motor units consist of fibres that stain weakly for ATPase at pH 9.4 (type I), and intensely for mitochondrial enzymes. All motor units of the soleus muscle are of the slow-twitch type; they have even longer contraction times and are even more resistant to fatigue than the slow-twitch motor units of the gastrocnemius muscle (Burke et al. 1974).

Rat muscles differ in several aspects from cat muscles (Fig. 83). Slow-twitch motor units are less frequent in mixed muscles than in corresponding cat muscles; the slow-twitch fibres, such as in cat, stain weakly for ATPase at pH 9.4 and intensely for mitochondrial enzymes. Nevertheless, staining for mitochondrial enzymes distinguishes, in the almost purely fast extensor digitorum longus muscle, three fibre types (Close 1967); all stain intensely for ATPase at pH 9.4. Seventy-five percent of the "adult" (see below, and also Fig. 85) soleus muscle fibres stain moderately for mitochondrial enzymes and are not stained by antibodies against "fast" myosin, whereas 25% of the fibres stain intensely for mitochondrial enzymes and react with the "fast" antimyosin (Gauthier and Lowey 1977) (Fig. 82). The incidence of fast-twitch fibres and motor units in the soleus muscle decreases with age, and 8 months after birth 90% of the soleus motor units are of the slow-twitch type (Kugelberg 1973a, b, 1976).

No motor unit studies are available for rabbit muscles, but histochemistry suggests that the fibre types essentially resemble those in rat muscles (James 1972a).

The contractile properties of human motor units can only be studied using indirect methods. The essential difference between recording the contraction time of motor units in animals and humans is the fact that in human muscles it is not possible to activate a single motor unit by electrical stimuli, whether they are applied to the nerve or to the endplates. The presence of motor units with different contraction times has been inferred from measurements of contraction times of excised bundles of muscle fibres (Brust and Cosla 1967; Eberstein and Goodgold 1968; Moulds et al. 1977) or by applying threshold stimuli to the endplates (Buchthal and Schmalbruch 1969, 1970) or nerve (Sica and McComas 1971). The rate of tension development and the incidence of type II fibres in excised fibre bundles are related (Moulds et al. 1977). Slow contractions are most frequently recorded in muscles that contain many fibres rich in mitochondria. In contrast to rat and cat, the fibres of the human soleus and gastrocnemius muscles have equally long contraction times (Buller et al. 1959; Buchthal and Schmalbruch 1970). In both muscles, about 90% of the fibres are rich in mitochondria (Schmalbruch 1970). Nevertheless, a considerable portion of the muscle fibres of the gastrocnemius and soleus muscles react intensely for ATPase at pH 9.4 (type II) (Edström and Nyström 1969). The contraction times of these presumably fast fibres in the calf muscles are longer than those of presumably fast- and possibly even of slow-twitch fibres of the brachial biceps muscle (Buchthal and Schmalbruch 1970).

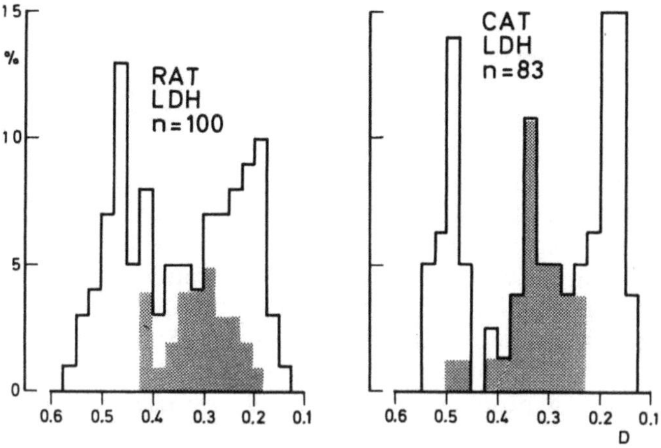

Fig. 84. The optical densities of cross-sections of muscle fibres of the rat and cat gastrocnemius muscles stained for nicotinamide adenine dinucleotide-linked lactic dehydrogenase (LDH), a mitochondrial enzyme, for green light (wave length 548 nm, 8-μm-thick sections). The three peaks in the histogram for the cat muscle coincide with the visually established fibre types. All type I fibres (ATPase pH 9.4) are in the high-density group. In the rat muscle only two peaks are found, and the high-density peak comprises type I and type II fibres. *Shaded areas,* visually classified intermediate fibres; *n*, number of fibres assessed. (From SCHMALBRUCH and KAMIENIECKA 1975, with permission of the Histochemical Society)

Twitches of voluntarily activated motor units have been recorded when the rate of activation was low; the action potential triggered an electronic averager for the output of the strain gauge (BUCHTHAL and SCHMALBRUCH 1970; MILNER-BROWN et al. 1973a; DESMEDT and GODAUX 1977). Also this procedure has demonstrated that there are motor units with short and long contraction times in the same human muscle.

3. How Many Fibre Types Can Be Distinguished?

There is convincing evidence that in normal adult mammals, the reaction for ATPase at pH 9.4 distinguishes fast- and slow-twitch fibres. With this indicator, it is possible to test, using microphotometry, how the slow-twitch fibres in mixed muscles stain for mitochondrial enzymes, and whether staining for mitochondria alone provides a basis for fibre-type categorization. Three peaks of densities have been found in the cat gastrocnemius muscle, corresponding to the three fibre types established by inspection. Those fibres that reacted weakly for ATPase at pH 9.4 (type I) formed the peak with the highest activity for mitochondrial enzymes. Only two peaks of densities have been noted in the rat gastrocnemius muscle. Ten percent of all fibres are of type I (presumably slow-twitch) and together with some of the type II fibres (presumably fast) they constitute the high density peak (SCHMALBRUCH and KAMIENIECKA 1975) (Fig. 84). KUGELBERG and LINDEGREN (1979) find in the rat anterior tibial muscle, a continuous spectrum of densities, rather than two or three peaks. Most motor units are fast; the fatiguability of the individual motor units is inversely almost linearly related to the staining intensity of the muscle fibres. Neither fatiguability nor staining for succinic dehydrogenase provides a basis for categorization. The fibres of the human quadriceps muscle, when stained for mitochondrial enzymes, give two separate peaks of densities (HALKJAER-KRISTENSEN and INGEMANN-HANSEN 1977). Sometimes two peaks and sometimes a spectrum has been found in pig muscles. The type I fibres are always those with the highest oxidative capacity (SWATLAND 1978). One can visually distinguish, with about 90% reproducibility, three muscle fibre types in cat, rat, and human muscles (SCHMALBRUCH and KAMIENIECKA 1974, 1975). This indicates, that not only staining intensities, but also other properties, such as fibre size or the array of the reaction product, are unwittingly assessed when the fibres are visually typed.

In conclusion, staining for mitochondrial enzymes alone does not definitely ascertain whether there are two or three different fibre types, or whether there is, in fact, a spectrum.

The contraction times of the motor units of some muscles fall into two distinct groups, which favours a two-type system. In other muscles, the histogram of contraction times is skewed rather than bimodal (KERNELL et al. 1983a), and additional physiological criteria are necessary to distinguish fast- and slow-twitch units. Training may transform IIB into IIA fibres, but type II and I fibres are stable (ANDERSEN and HENRIKSSON 1977a). This supports a fast-slow system. Other properties of the motor units (size, axon diameter, axonal conduction velocity, electrical properties of the neuron, synaptic connections, post-tetanic potentiation; for references, see BUCHTHAL and SCHMALBRUCH 1980) may justify more elaborate classifications, but such differences are apparently not expressed in the morphology of the muscle fibres.

BURKE et al. (1973) have established a physiological four-type system for cat motor units (FF, FR, F(int), S) (see above). They distinguish fast- (F) and slow-twitch (S) units by the "sag" criterion: the force of fast units rises in response to a standardized stimulus rate to a maximum, whereafter it rapidly declines ("sags") to reach a steady-state; S units do not "sag". The subgroups are then established according to the fatiguability of the motor units. This means that the entire system of classification rests on fatiguability, because "sag" is related to fatigue as well. The relative force of the muscle produced by the S units identified by "sag" is smaller than the relative cross-sectional area of the gastrocnemius muscle occupied by histochemically slow-twitch fibres. This induced the authors to conclude that the specific force of the slow-twitch fibres is smaller than that of fast-twitch fibres. Surprisingly, the specific force of the slow-twitch fibres in the homogeneous soleus muscle does not differ from that of fast-twitch fibres in the gastrocnemius muscle (BURKE and TSAIRIS 1973; BURKE et al. 1973). One might ask whether the absence of "sag" identifies all motor units with histochemically slow-twitch fibres. The force contribution of the S units to the total muscle force, and thereby the specific force of slow compared with fast fibres, would have been underestimated if also some slow-twitch units "sagged". KERNELL et al. (1983a) classified, in the cat peroneus longus muscle, all motor units as S units that had longer contraction times than the slowest fatiguable motor units. The relative force contribution by the S units was then almost three times larger than when the "sag" criterion was used, and the difference in specific force disappeared. This result disputes the value of the "sag" criterion, and also the putative difference in specific force.

Most results, in my opinion, support the conclusion that there are two main fibre types in all mammalian species: slow-twitch type I fibres which stain weakly for ATPase at pH 9.4 and intensely for mitochondrial enzymes, and fast-twitch type II fibres which stain intensely for ATPase at pH 9.4 and weakly, intermediately, and in some species even intensely for mitochondrial enzymes.

V. What Determines the Specialization of Muscle Fibres?

There is no doubt that the contractile properties of a motor unit are predominantly governed by its motoneuron (GUTH 1968). When the central stump of a nerve innervating a muscle with a fast twitch is connected to the distal stump of the nerve supplying a muscle with a slow twitch or vice versa, the cross-innervated fast muscle acquires the properties of a slow-twitch muscle, and the cross-innervated slow-twitch muscle acquires, to a certain extent, those of a fast muscle (cat, BULLER et al. 1960b, 1969; rat, CLOSE and HOH 1969; BARANY and CLOSE 1971; guinea pig, ROBBINS et al. 1969). The cross-over of mechanical and biochemical properties is parallelled by a conversion of the histochemical fibre types (DUBOWITZ 1967; ROMANUL and VAN DER MEULEN 1967; YELLIN 1967) and of the extent of vascularization (ROMANUL and POLLOCK 1969). It appeared as if the transformation of a slow-twitch into a fast-twitch fibre was incomplete, but WEEDS et al. (1974) succeeded in a complete transformation including the myosin isoenzymes. The points of doubt were initially whether the force-velocity relation and the intrinsic speed of shorting (μm/s/sarcomere, CLOSE 1965) of a slow-twitch muscle really could be changed. The inconsistencies were probably due to the fact that the nerve of the fast cat muscle contains few "slow" motor axons. These axons innervate few muscle fibres but expand their field in the cross-innervated originally slow-twitch muscle, and thus blur the transformation into a fast-twitch muscle (CLOSE and HOH 1969; LUFF 1975). LEWIS et al. (1982b) find in the cross-innervated soleus muscle of cat, after 3 years, large slow-twitch and small fast-twitch motor units. Slow-twitch units of normal mixed muscles are smaller than fast-twitch units. Nerves innervating fast rat muscles contain fewer "slow" axons than nerves innervating fast cat muscles; hence, the transformation of the slow-twitch soleus muscle by cross-innervation is more complete in rat than in cat. It is questionable whether this explanation is sufficient. CHAN et al. (1982) report that the motor units of the cat soleus muscle, 1 year after cross-innervation by the nerve to the fast flexor hallucis longus muscles, have short contraction times, but that with respect to fatiguability and size the motor units resemble normal soleus motor units. All muscle fibres stain intensely for mitochondrial enzymes, but only

14% have become type II fibres. This indicates that fatiguability and unit size may be subject to non-neural influences, but also that the "myofibrillar" ATPase stain is no reliable marker for fast fibres in "abnormal" muscles.

GAUTHIER et al. (1983) achieved a complete transformation of the fast cat flexor digitorum longus muscle innervated by the soleus nerve with respect to contractile properties, ATPase staining, and myosin isoenzymes. The soleus muscle innervated by the flexor digitorum nerve, however, did not change with respect to ATPase staining although the motor units became fast, and the fibres continued to react with antibodies against "slow-tonic" myosin; in contrast to normal soleus muscles, they now showed a positive response to an antibody specific for one of the light chains of "fast" myosin.

YELLIN (1975) transplanted the hypoglossal nerve of rat from the tongue to the sternomastoid muscle and found incomplete conversion of the histochemical properties of the fibre types. This may reflect unusual properties of the tongue motoneurons.

Injured motoneurons regenerate and their sprouting axons re-innervate the denervated muscle fibres. The regenerated motoneuron induces the type differentiation of the re-innervated muscle fibre. This causes a change in the fibre type array because re-innervation is non-selective (MILEDI and STEFANI 1969). A re-innervated muscle does not display the usual checkerboard pattern of fibres of different types (see also Chap. H); fibres of the same motor units which are histochemically identical are arranged in fields ("type grouping").

The motoneurons of fast-twitch motor units activated during a stretch reflex, fire in high-frequency bursts, and motoneurons of slow-twitch motor units have low-frequency discharges (ECCLES et al. 1958). The neuronal impulse activity apparently determines the specialization of the muscle fibres (BULLER et al. 1960a; ECCLES et al. 1962). A fast muscle is changed into a slow-twitch muscle when the nerve is chronically stimulated at 10 Hz (SALMONS and VRBOVA 1969; AL-AMOOD et al. 1973; RILEY and ALLIN 1973; SRETER et al. 1973; SALMONS and SRETER 1976; BROWN et al. 1976; PETTE et al. 1972, 1973, 1976; RAMIREZ and PETTE 1974; HUDLICKA and TYLER 1980). Transformation of a slow-twitch muscle into a fast-twitch muscle using high frequency stimulation is less successful, possibly because of the underlying low frequency discharges of the intact neuron. LØMO et al. (1974) denervated the slow-twitch soleus muscle of rat and stimulated the muscle directly with brief trains of stimuli. They succeeded in inducing many of the properties of a fast muscle. Nevertheless, HUDLICKA et al. (1982), comparing the effect of different stimulation patterns on fast-twitch rabbit muscles with intact innervation, found that after 2 months the myosin light chains characteristic for slow-twitch fibres occurred, independently of whether the muscle had been stimulated continuously at 10 Hz or intermittently at 40 Hz. The muscle became slow in both experiments, but the change was most pronounced after 10 Hz stimulation. Also the activity of succinate dehydrogenase increased after both types of stimulation (PETTE and TYLER 1983). SRETER et al. (1982) stimulated the nerve to the fast anterior tibial muscle of rabbit for 5 weeks intermittently at 60 Hz; histochemically, the muscle became a slow-twitch muscle and both "slow" and "fast" myosin light chains were present. The authors concluded that not the pattern of stimulation, but the total activity, determines the transformation of fast- into slow-twitch muscle.

In some types of experiments, inactivity converts a slow-twitch muscle into a fast-twitch muscle (see below).

The results of the stimulation experiments suggest that the nerve exerts its influence by electrical activity rather than by a trophic substance other than acetylcholine. INESTROSA and FERNANDEZ (1976) succeeded in reducing the activity of mitochondrial enzymes by blocking the axoplasmic flow by topically applied colchicine, which does not interfere with the potential conduction of the axon. Nevertheless, it is difficult to avoid systemic colchicine effects. A discussion of the neurotrophic effect of the motoneuron is beyond the scope of this review; the parameters studied usually are the electrical properties of the membrane, rather than fibre-type differentiation.

The changes found after inactivation are inconsistent. Denervation, transection of the spinal cord, or prolonged general anaesthesia prolongs the contraction time, predominantly in fast muscles (LEWIS 1972; KEAN et al. 1974; DAVIS and MONTGOMERY 1977). Nevertheless, the intrinsic speed of shortening of the denervated rat extensor digitorum longus muscle decreased by only 13% (from 49 µm/s to 43 µm/s), whereas that of the denervated soleus muscle decreased by 46% (from 28 µm/s to 15 µm/s). The increase in isometric twitch contraction time was 76% and 46% in the extensor digitorum longus and soleus muscles, respectively (BAKER and LEWIS 1983). The results indicate that the slowing of the isometric contraction of the fast muscle is not due to alterations of the contractile proteins, and that the contractile properties of the slow soleus muscle, more than those of the fast muscle, depend on neural influences. Spinal isolation in kitten causes shortening of the contraction times, predominantly in slow-twitch muscles (BULLER et al. 1960a). Shortened contraction times are also found after tenotomy (BULLER and LEWIS 1965a), denervation of the antagonist (GUTH and WELLS 1972), or after fixation of the joint (FISCHBACH and ROBBINS 1969; BOOTH and KELSO 1973). Chordotomy in young rats, guinea pigs, and cats increases the percentage of histochemically fast soleus muscle fibres (KARPATI and ENGEL 1968; RUBINSTEIN and KELLY 1978; JOHNSON et al. 1982). The amount of "fast" myosin light chains in the rat soleus muscle investigated 2–4 weeks after the operation was much larger than normal (RUBINSTEIN and KELLY 1978).

Sustained overactivity has been produced in rat muscles by systemic administering of reserpine. This compound induces excessive discharge of motoneurons by suprasegmental disinhibition. Both slow- and fast-twitch muscles became slower; surprisingly, the percentage of type I fibres in the soleus muscle decreased at the same time (EISEN et al. 1975).

VI. The Development of Muscle Fibre Types

1. Histochemistry

The age at which different fibre types can be distinguished depends on the species and on the muscles studied. Respiratory muscles are generally the first to mature. The fibre types of kitten hindlimb muscles can be identified, using

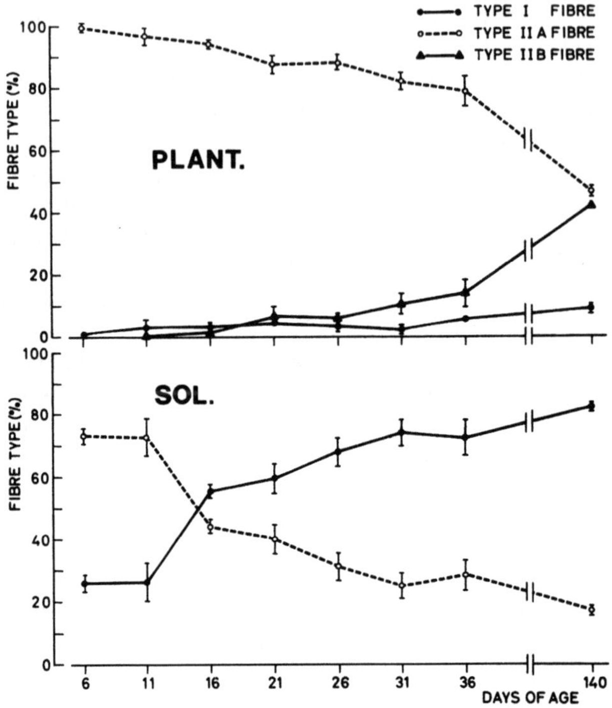

Fig. 85. The age-related change of the proportion of fibre types in the plantaris (*PLANT*) and soleus (*SOL*) muscles of rat. The plantaris muscle is a predominantly fast-twitch muscle with only few slow-twitch type I fibres; the incidence of IIA fibres ("fast-red") decreases, and that of IIB fibres ("fast-white") increases over several months. The soleus muscle of very young rats contains almost 80% IIA ("fast-red") fibres, which increasingly are transformed; eventually the muscle contains > 80% slow-twitch type I fibres (see also Fig. 83, *right*). These diagrams may explain controversial results, in the literature, with respect to the incidence of fast- and slow-twitch fibres in "adult" rat soleus muscles. Each *point* and *bar* represents the mean ± SD of six rats, unshown SD were < 1%. (All data from Ho et al. 1983)

histochemistry, 6 weeks after birth, but the physiological maturation of motor units is not complete until 10 weeks after birth. The histochemical differences in respiratory muscles are apparent at birth (Nyström 1968b; Hammarberg 1974, 1975). The fibres of the hindlimb muscles of newborn rats and guinea pigs are undifferentiated, whereas in man (Dubowitz 1965) and rhesus monkey (Beatty et al. 1967), the histochemical differentiation takes place before birth. A checkerboard pattern of differently stained fibres occurs in human muscle at 26–30 weeks' gestation (Fenichel 1966; Dubowitz 1968). Chicken muscles are differentiated at the time of hatching. The glycolytic enzymes in "red" and "white" muscles of pigs reach their final activities 2 weeks after birth (Davies 1972; Dalrymple et al. 1974). The soleus muscle of rat has a long postnatal period of maturing, and the percentage of fast- and slow-twitch fibres changes over a period of several months (Kugelberg 1976). In the plantaris muscle, the percentage of IIB fibres increases at the expense of IIA fibres, but the number of type I fibres remains small (Ho et al. 1983) (Fig. 85).

Histochemistry of immature muscles is hampered by the facts that the fibre diameters are small compared with the thickness of the section, and often "fibres" seen by light microscopy in reality are clusters of myofibres, myotubes, and myoblasts within a common basal lamina. The total diameter of such a cluster may be 10–20 μm. Individual cells can hardly be identified, and the relative coarseness of the reaction product usually provokes the impression that the preparation is technically poor; there is a strong degree of arbitrariness in classifying the muscle cells. The matter is further confused by the fact that all immature muscle fibres stain intensely for ATPase at pH 9.4 (GUTH and SAMAHA 1972). This is surprising because the biochemically determined activity of myofibrillar ATPase is low, regardless of whether the muscle is determined to become a fast- or slow-twitch muscle (PELLONI-MUELLER et al. 1976), and because immature muscles have slow contractions (BULLER et al. 1960a; ÄNGGARD and OTTOSON 1963; BULLER and LEWIS 1965b; HAMMARBERG 1975).

2. Myosin Isoenzymes

There is general agreement that fast- and slow-twitch muscle fibres of adult mammals contain different myosin isoenzymes (see Sect. D.III.2). The myosins of developing muscles have been a matter of controversy.

According to GAUTHIER et al. (1978), all fibres of the immature rat diaphragm react with antibodies specific for both slow and fast myosin. They interpret this to mean that myosins with slow as well as fast characteristics co-exist in the same fibre. At later stages, myosins are segregated into different fibre types. The high (histochemically assessed) ATPase activity after acid *and* alkaline preincubation in immature fibres is regarded as being consistent with the co-existence of slow and fast myosins. The results have been elaborated in a subsequent study of rat and chick embryonic muscles (GAUTHIER et al. 1982). Antibodies were prepared against the "difference peptide" of the alkali light chain 1 (LC_{1f}), the slow ALD myosin, and the ALD heavy chain. One day after birth single fibres of rat soleus muscle cease to react with anti-alkali 1. In chicken most fibres of the future fast posterior latissimus dorsi muscle lose their reactibility with anti-ALD at day 18 of gestation, the response of the future fast pectoralis fibres to anti-ALD decreases 1 day posthatch. RUBINSTEIN and KELLY (1981) used antibodies against the heavy chains of rabbit slow and fast myosin, and find that in fetal rats of 15 days gestation all fibres react with the fast but not with the slow anti-myosin. At day 17, a gradual increase in the binding of the antibody to slow myosin occurs in the primary myotubes. With maturation, the primary myotubes lose their binding to anti-fast myosin, and become type I fibres. These fibres differentiate into slow fibres, and since they are the first to be innervated, they dominate the contractile response of the muscle during this stage of development (KELLY and RUBINSTEIN 1980). The secondary myotubes of future fast muscles become type II fibres; whereas, in the future slow soleus muscles, most of the second generation cells eventually change to type I fibres over a prolonged postnatal period (KUGELBERG 1976). Antibodies against cat soleus myosin (slow), 3 days before birth, in the rat gastrocnemius muscle stain only few fibres. At birth, 20% of the soleus muscle fibres react with slow anti-myosin; there is a clear distinction between fibres which contain slow myosin and those which do not (ROWLERSON 1980).

Cultured chick muscle cells, whether derived from fast or slow muscles, according to RUBINSTEIN and HOLTZER (1979), react with antibodies to fast heavy chains only, whereas MASAKI and YOSHIZAKI (1974) and CANTINI et al. (1980) report that the myotubes are stained both by fast and slow anti-myosins.

Almost all workers report that the principal light chains in immature muscle cells are LC_{1f} (alkali 1) and LC_{2f}, and that LC_{3f} is scarce and occurs in substantial amounts first at late stages of development (CHI et al. 1975a, b; PELLONI-MUELLER et al. 1976; HOH and YEOH 1979; RUBINSTEIN et al. 1977; RUBINSTEIN and KELLY 1981; KELLER and EMERSON 1980, 1982; GAUTHIER et al. 1982; SHELTON and BANDMAN 1983). There is, however, disagreement whether these are the only light chains. CHI et al. (1975a, b) attribute the presence of LC_{3f} in standard cultures to contaminating fibroblasts and emphasize that the myofibrils are able to assemble and to contract in the absence of this myosin component. LC_{1s} occurs in rats in late fetal life and in future slow muscles successively replaces LC_{1f} (RUBINSTEIN and KELLY 1981; GAUTHIER et al. 1982). KELLER and EMERSON (1980, 1982), however, observed that synthesis of all fast and slow light chains is initiated as soon as the myoblasts fuse. This harmonizes with the results of OBINATA et al. (1980) and STOKDALE et al. (1981): future fast and slow embryonic chicken muscles synthesize an overweight of LC_{1f} and LC_{2f} but considerable amounts of LC_{1s} and LC_{2s} as well. The synthesis of the slow light chains ceases in the embryonic pectoralis muscle (future fast) at 14 days incubation.

KELLER and EMERSON (1980, 1982), OBINATA et al. (1980), and STOCKDALE et al. (1981) do not find the specific embryonic light chain LC_{emb} observed by WHALEN et al. (1978), RUBINSTEIN and KELLY (1981), GAUTHIER et al. (1982), and STROHMAN et al. (1983) both in vitro and in vivo. This embryonic myosin light chain seems to be restricted to mammalian muscles (rat, man), since those studies that failed to find it were performed on avian muscles.

The controversies concerning development of the light chain pattern seem less grave than those with respect to immunofluorescence; one may suspect that the source of these controversies rests with the heterogeneity of the heavy chains. WHALEN et al. (1979, 1981) and HOH and YEOH (1979) described distinct fetal and neonatal heavy chain types. The fetal heavy chain has peptide sequences in common both with the adult fast and adult slow isoforms (DALLA LIBERA 1981), and RUBINSTEIN and KELLY (1981) suggest that GAUTHIER's et al. (1978) labelled antibodies against fast and slow adult myosin might have cross-reacted with the fetal heavy chain and, hence, both have stained immature muscle fibres.

The conversion of the myosin types during development is apparently nerve-induced. Muscle cells cultured in the absence of nerves synthesize fetal myosin heavy chains only (BADER et al. 1982). It has been proposed that the neonatal heavy chain is actively induced as a consequence of polyneural innervation, which is a transient stage of the development of the motor unit, and that the accumulation of adult myosin isoenzymes occurs only in response to the adult motor innervation pattern (WHALEN et al. 1981). Nevertheless, the change from neonatal to fast adult myosin heavy chains in the rat gastrocnemius muscle takes place even when the muscle is denervated at age 1 week (BUTLER-BROWNE et al. 1982). CARRARO et al. (1983) investigated regenerating non-innervated soleus muscles of rat in order to test whether other factors not operative in culture might influence the development of the specific myosin isoenzymes. LC_{emb} was found already after 4 days regeneration; LC_{1f} and LC_{1s} and LC_{3f} occurred during the next 2 weeks, whereas LC_{emb} disappeared. The heavy chains of 2-week regenerates differed from fetal and adult slow heavy chains, and were possibly identical with the neonatal form. The authors conclude that the different genes are sequentially activated also without nerve influence, and that only the adult type differentiation is under the influence of the motor nerve activity.

Not only myosin but also other myofibrillar proteins occur in different forms in fast and slow muscles (for references, see JOLESZ and SRETER (1981). DHOOT and PERRY (1980) and DHOOT (1983) report that antibodies against the individual components of troponin from slow muscles allowed distinguishing of fibre types in rat muscles at day 15–18 of gestation, i.e. well before an appropriate motor innervation is established.

Rat soleus and anterior tibial muscles denervated at birth consist of ATPase-positive type II fibres only; the myosin light chain pattern is that of fast muscles (RUBINSTEIN and KELLY 1978). Chordotomy in newborn rats affects the fibre type distribution and the light chain pattern of the developing muscles in the same way as denervation, but cytoplasmic growth of the fibres is not blocked (MARGRETH et al. 1972, 1980). Muscle fibres of the diaphragm denervated in adult rats are atrophic and contain predominantly "fast" myosin heavy chains (CARRARO et al. 1982).

An unexpected finding of the myosin isoenzyme studies is that differentiation of slow-twitch fibres apparently depends more on innervation and activity than differentiation of fast-twitch fibres does. This is at variance with the long-held view that slow-twitch fibres represent a more "primitive" form of muscle cells (BULLER et al. 1960a; BULLER and LEWIS 1965b). Nevertheless, it harmonizes with the observation that denervation of adult rat muscles reduces the intrinsic speed of shortening more in the soleus than in the extensor digitorum longus muscle (BAKER and LEWIS 1983) (see Sect. D.V). KELLY and RUBINSTEIN (1980) explain the paradox that fetal muscles contract slowly although they contain predominantly "fast" myosin, by the fact that the large primary myotubes along which the secondary myotubes develop are the first to become innervated and to synthesize "slow" myosin. They suggest that these primary myotubes constitute the fundamental motor units of the developing muscle, and dominate the contractile response.

VII. Non-Neural Influences on Fibre Types

For a variety of species it has been shown that training of a muscle increases its capillarization and the amount of oxidative enzymes in the muscle fibres

(Holloszy 1967; Edgerton et al. 1969; Faulkner et al. 1972; Fitts et al. 1973; Henriksson 1976; Gollnick and Saltin 1982; see also Sect. C.IV). This is consistent with the increased resistance to fatigue after training. In endurance-trained muscles, all fibres stain more or less intensely for oxidative enzymes and the histochemical differences between the fibre types are less distinct than in untrained muscles. A similar effect is seen in detrained human muscles in which all fibres become more or less poor in oxidative enzymes. Endurance training increases the oxidative capacity of slow-twitch fibres, whereas strength training affects predominantly fast-twitch fibres (Henriksson and Reitman 1976).

In the developing rat soleus muscle, in which fast fibres are transformed into slow-twitch fibres, IIC fibres that stain intensely both after acidic and alkaline preincubation are common. IIC fibres contain slow and fast myosin isoenzymes and are suspected to represent a transitional form (Kugelberg 1976; Pette and Schnez 1977; Rubinstein et al. 1977; Snow et al. 1981) (Fig. 82).

There is no physiological evidence for a change of the contraction time of trained muscles. Training-induced type II to type I conversions were reported in a needle-biopsy study in horses (Guy and Snow 1977) and in the soleus muscle of weight-lifting rats in which the incidence of type I fibres increased from 84% to 92% (Jaweed et al. 1977). Needle biopsies are subject to error (see below), and a type II to type I conversion is a normal developmental process in the soleus muscle of adult rats (Kugelberg 1976). Training does convert the IIA and IIB subtypes of fast-twitch fibres (Henriksson 1976; Andersen and Henriksson 1977a). Green et al. (1979) observed, in elite ice hockey players, fewer IIB fibres (glycolytic, fast-twitch) than in control subjects (3.9% vs 12.2%), but no change in the type I-type II proportion. An increased incidence of IIC fibres, a possible sign of commencing transformation (Brooke et al. 1971) from 0.1% to 5.5% ($P<0.05$) was seen in a needle-biopsy study of six young men who had been subjected to extreme endurance training (Henriksson et al. 1980); the low probability (95%) for a 55-fold increase illustrates the inconsistency of the results. Müntener (1982) found in the rat sternohyoid muscle, 6–9 days after various unspecific manipulations (excision of the contralateral muscle, incision of skin or fascia, anaesthesia), a histochemical and immunological transformation of fast- into slow-twitch fibres and vice versa. These results need confirmation.

The physiological "stability" of fibre and motor unit types was demonstrated by Walsh et al. (1978) in cat. The medial gastrocnemius muscle was functionally isolated (i.e. overloaded) by denervating or removing its synergists. After 14–32 weeks, individual motor units had the same contraction time and fatiguability as in normal muscles; only the tetanic tension of the units had increased, reflecting the compensatory hypertrophy of the isolated muscle. The histochemistry was unchanged. Fast-twitch rat muscles which were functionally isolated at birth, however, contained an increased number of type I (slow-twitch) fibres and the type II fibres were rich in mitochondrial enzymes (Schiaffino and Pierobon-Bormioli 1973). Undernutrition of hamsters during development produced selective and reversible atrophy of the glycolytic type II fibres; weight lifting induced hypertrophy of the same fibres without any evidence of fibre type conversion (GOLDSPINK and WARD 1979).

PETTE (1984) reviewed the biochemical changes during experimental transformation of the fast rabbit anterior tibial muscle into a slow-twitch muscle by means of chronic stimulation (see Sect. D.V). The metabolic "white to red" transformation always precedes the change in the Ca^{2+}-accumulating capacity of the SR; the change in myosin type is a late event. In a study from PETTE's laboratory (GREEN et al. 1983, 1984), high-intensity endurance training of rats led, although to a much lower extent, to similar changes as chronic stimulation. The shift in enzymes of the aerobic and aerobic pathways of metabolism was pronounced, whereas there was only a small but significant increase in the amount of slow myosin light chains. PETTE (1984) concludes that the histochemical fibre types represent "metastable entities" within a dynamic equilibrium, and that increased activity, in an ordered sequence of events, induces fast to slow conversion. Decreased activity might favour a slow to fast transition. This concept, if generally applicable, appears attractive because it renders the distinction of neural and non-neural influences irrelevant. Nevertheless, the rat experiments of GREEN et al. (1983, 1984) do not support the notion that human muscle fibre types may be converted by training. The concept described could explain how motor units differentiate under the influence of their motoneurons, but there is no evidence that the change in activity level during voluntary activity suffices to bring about a change in fibre type. In rabbit, complete transformation of a fast muscle required 160-day stimulation at 10 Hz, 24 h each day (PETTE 1984).

Several rat, cat, and pig muscles depend on androgens (levator ani muscle, pubocaudalis muscle, bulbucavernosus muscle) and are atrophic in females, and in castrated or oestrogen-treated males (VENABLE 1966a, b; GORI et al. 1967, 1969; GALAVAZI and SZIRMAI 1971a, b; GUTMANN and HANZLIKOVA 1973; CARLSON et al. 1979c; BLEISCH et al. 1982; HUGHES et al. 1983a, b). No consistent effect on fibre types has been reported. The only hormone known to affect the fibre-type differentiation is thyroid hormone. The biochemically determined myosin ATPase activity and the number of type II fibres in soleus muscles of thyroidectomized rats increases after administration of thyroid hormone (IANUZZO et al. 1977). It is unknown whether thyroid hormone affects the muscle fibres directly, or exerts its influence through the motoneuron. After thyroidectomy the rat soleus muscle has slower contractions than normal, and the type II fibres disappear. No fibre type conversion occurs when the muscle is denervated before thyroidectomy. The fast-to-slow transformation of the fibres is reflected in a change of the myosin light chain pattern; fast myosin is almost completely lost (JOHNSON et al. 1980). Experimental thyroidism in rats increases the oxidative capacity of the fast-twitch fibres of the extensor digitorum longus muscle, but does not change its contractility; the contraction time of the soleus muscle shortens (NICOL and BRUCE 1981).

MARTIN (1979) reports that, in the rectus abdominis muscle of pregnant rats, type I fibres undergo selective hypertrophy, but the proportion of type I and type II fibres is unchanged. The author attributes this increase in fibre diameter to stretch rather than to hormonal influences.

VIII. The Fibre Type Composition of Different Muscles in Different Species

Most workers agree now that at least two histochemical staining procedures are necessary in order to categorize a given muscle fibre – one for a mitochondrial enzyme to grade its presumed resistance to fatigue, and the other, the ATPase reaction at pH 9.4, to mark fast- and slow-twitch fibres. By these two methods, possibly aided by staining for ATPase after acidic preincubation, in most animals three fibre types can be distinguished. A fourth fibre type is identified in rat and mice, but not in cat or man; it reacts intensely both for ATPase pH 9.4 (type II, presumably fast) and for mitochondrial enzymes (fatigue-resistant) (KUGELBERG 1973a) (Fig. 83). Also the muscles of the three shrew (*Tupaia glis*) contain few of these fibres (about 2%–3% of all), but they are absent in other prosimian primates (SICKLES and PINKSTAFF 1981a). The problem of fibre categorization according to the activity of mitochondrial enzymes (classes vs continuous spectrum) has been discussed (see above).

Slow-twitch fibres and glycolytic fast-twitch fibres tend to be more frequent in large animals than in small animals (Table 5); the percentage of oxidative fast fibres is higher in small animals than in large animals (GAUTHIER and PADYKULA 1966; GUNN and DAVIES 1971). The high incidence of fast fibres in small animals relates to the fact that the load is smaller and that the muscles are faster in small than in large animals. The strength of the non-muscle elements (tendons and tendon insertions) and the inertia of the limb limits the acceleration. The inverse relation between allowable acceleration and body size explains why small animals move with about the same speed as large animals (HILL 1950b). In addition, the specific locomotor pattern of an animal is reflected by the fibre type composition of its muscles. Fast-glycolytic fibres that are able to develop force during rapid movements even when the blood supply is shut off, are lacking in limb muscles of the slowly moving skunk (*Mephitis mephitis*) (VAN DE GRAAF et al. 1977) and the slow loris (*Nycticebus couang*) (SICKLES and PINKSTAFF 1981b). The difference between the soleus muscles composed almost exclusively of slow-twitch fibres and the fast gastrocnemius muscle is less conspicuous in these species than in cat or rat. In Tables 5–9 information with respect to the fibre type composition of different muscles in different mammals and in man are compiled. The muscles listed were selected with respect to physiological, experimental, or clinical importance.

The findings in large human muscles have to be viewed with caution because regional differences within a muscle may give rise to varying results. This applies in particular to needle-biopsy studies. The incidence of type I fibres within a given muscle may range from 20% to 80% (NYGAARD and SANCHEZ 1982). The deep part of most mixed muscles contains more slow-twitch fibres than the superficial part. This is particularly striking in anterior tibial muscles of rodents and also cats (Fig. 86), but is true to a certain extent for human muscles as well. ENGLISH and LETBETTER (1982) found that the cat lateral gastrocnemius muscle consists of four discrete compartments supplied by separate nerve

Table 5. The fibre type compostion, and the percentage of motor units of different types in various mammalian muscles. Data for mean fibre diameter, motor unit properties, number of motor units, and number of muscle fibres in the muscle are listed as well. Note different classification systems; "Red", intermediate, "white" means that the fibres were typified according to their activity for mitochondrial enzymes; *I, II* means that the staining for ATPase at pH 9.4 served as the basis for classification

Muscle	Age (weight) of animal	% of fibre type (mean diameter in μm) "red"	intermediate	I	II	"white"	% of motor units (mean contraction time in ms) slow-twitch S	fast-twitch FR	FF	Total number of motor units/muscle fibres	References
Mouse											
Extensor digit. long.	80 days				100			100 (8.7)		20/?	Bateson and Parry (1983), Fitzsimmons and Hoh (1983)
Soleus	21 weeks			41 (31)	59 (52)		46 (22)	54 (13)		21/850	Lewis et al. (1982a, b) Timson (1982)
Rat											
Med. gastrocn.	16–50 weeks	38	32	13	87	30					Schmalbruch and Kamieniecka (1975)
Soleus	5 weeks			66 (44)	34 (36)		67 (27–40)	33 (15–26)		30/2,900	Kugelberg (1973a, b), Kugelberg (1976)
	12 weeks						90 (38)	10 (18)			Close (1967)
	34 weeks			93 (71)	7 (75)		90 (27–40)	10 (15–26)			Kugelberg (1973a, b), Kugelberg (1976)
	21 weeks			70 (67)	30 (51)						Kelly (1978a)
Anterior tibial	?			5	95		6 (20)	94 (11)		73/9,000	Edström and Kugelberg (1968), Kugelberg and Lindegren (1979)
Anterior tibial. superf.	350–400 g	22 (36)	29 (48)			49 (59)					Schmalbruch (1971)

Table 5 (continued)

Muscle	Age (weight) of animal	% of fibre type (mean diameter in μm) "red"	intermediate	I	II	"white"	% of motor units (mean contraction time in ms) slow-twitch S	fast-twitch FR	FF	Total number of motor units/muscle fibres	References
Anterior tibial. deep		58 (29)	26 (38)			16 (49)					
Extensor digit. long.	21 weeks			7 (46)	93 (61)						KELLY (1978a)
	12 weeks						2 (24)	98 (11)		40/3,200	CLOSE (1967), LOWRIE et al. (1982)
Semitendinosus	12–15 weeks			7	93						ARIANO et al. (1973)
Red part	?	52 (45)	40 (52)			8 (59)					GAUTHIER (1974)
White part	?	4 (47)	14 (47)			82 (69)					
Diaphragm	12–15 weeks	60 (27)	20 (34)			20 (44)					GAUTHIER (1969), GAUTHIER (1974)
Sternomastoid	150–170 g										
Red part		66 (40)	10 (62			24 (68)					GOTTSCHALL et al. (1980a)
				11	89						
White part		0	9 (60)			91 (79)					
				0	100						
Plantaris	?			6	94						ARIANO et al. (1973)
Vast. later./med.				1	99						
Vast. intermed.				36	64						
Adductor long.				88	12						
Gracilis posticus				36	64						
Gracilis anticus				9	81						
Post. cricoarytenoid	?			3	97						HINRICHSEN and DULHUNTY (1982)
Cricothyroid				16	84						

Cat											
Med. gastrocn.	3–4 kg	39 (40)		25 (48)		36 (55)					SCHMALBRUCH (1971)
	adult		25		75		28 (73)	28 (43)	44 (34)	280/170,000	BOYD and DAVEY (1968), BURKE et al. (1971), BURKE et al. (1974), BURKE and TSAIRIS (1973), CLARK and LUSCHEI (1981)
Soleus	2–3.3 kg		100 (74)		0		100 (97)			155/30,000	BOYD and DAVEY (1968), BURKE et al. (1974), CLARK (1931)
Anterior tibial	2.3–3.7 kg		12		88		11 (42)	50 (31)	39 (27)	200/55,000	DOYLE and MAYER (1969), DUM and KENNEDY (1980)
Extensor digit. long.	2.3–3.7 kg						6 (37)	57 (29)	(29)	130/40,000	BOYD and DAVEY (1968), CLARK (1931), DOYLE and MAYER (1969) OLSON and SWETT (1966)
Flexor digit. long.	2.2–4.2 kg	14 (49)		36 (55)		50 (52)					
Flexor halluc. long.		28 (38)		31 (50)		41 (59)					
Flexor carpi rad.	3.7 kg		36		64						GONYEA (1979), GONYEA et al. (1981)
Palmaris long.			19		81						
Flexor carpi uln.											
Humeral head			49		51						
Ulnar head			35		65						
Flexor digit. prof.											
1.+5. head			1		99						
3.+4. head			10		90						
2. head			18		82						
Tensor fasciae latae			1		99						
Rectus femoris			22		78						
Vastus lat.			27		73						
Vastus intermed.			98		2						
Vastus med.			13		87						
Cruralis			83		17						
Popliteus			4		96						

Table 5 (continued)

Muscle	Age (weight) of animal	% of fibre type (mean diameter in µm)				% of motor units (mean contraction time in ms)			Total number of motor units/muscle fibres	References
		"red"	intermediate		"white"	slow-twitch	fast-twitch			
			I	II		S	FR	FF		
Flexor digit. long.			7	93						
Biventer cervicis			50	50						RICHMONDS and ABRAHAMS (1975)
Occipitoscapularis			55	45						
Splenius			25	75						
Rectus capitis major			24	76						
Complexus cervicis			40	60						
Dog										
Med. gastrocn.	?		51	49						ARMSTRONG et al. (1982)
Extensor digit. long.			29	71						
Flexor digit. superf.			89	11						
Extensor carpi rad.			14	86						
Vastus lat.			43	57						
Vastus intermed.			88	12						
Vastus med.			39	61						
Quadratus femoris			100	0						
Biceps brachii			48	52						
Anconeus			100	0						
Longissimus thorac.			39	61						
Rectus abdominis			48	52						
Diaphragm			51	49						
Monkeys										
Tree shrew (*Tupaia glis*)										
Psoas major	?		10	90						SICKLES and PINKSTAFF (1981 a, b)
Med. gluteus			2	98						
Rectus femoris										
Superficial			10	90						
Deep			21	79						

Vastus med.		1	99	
Vastus intermed.		92	8	
Anterior tibial		6	94	
Extensor digit. long.		6	94	
Med. gastrocn.		10	90	
Soleus		97	3	
Lesser bushbaby (*Galago senegalensis*)				
Psoas major		20	80	
Med. gluteus		12	88	
Vastus med.		1	99	
Vastus intermed.		49	51	
Anterior tibial		5	95	
Extensor digit. long.		7	93	
Med. gastrocn.		20	80	
Soleus		94	6	
Slow loris (*Nycticebus couang*)				
Psoas major		73	27	
Med. gluteus		69	31	
Vastus med.		65	35	
Vastus intermed.		86	14	
Anterior tibial		40	60	
Extensor digit. long.		29	71	
Med. gastrocn.		45	55	
Soleus		78	22	
Macaca fascicularis				
Temporalis posterior	?	16	84	CLARK and LUSCHEI (1981)
Temporalis anterior		65	35	
Masseter		45	55	
Pterygoideus med.		100	0	
Gastrocnemius		15	85	
Horse				
Triceps brachii	1–28 years	23	77	ESSÉN et al. (1980)
Vastus lat.		27	73	
Gluteus		21	79	
Semitendinosus		16	84	

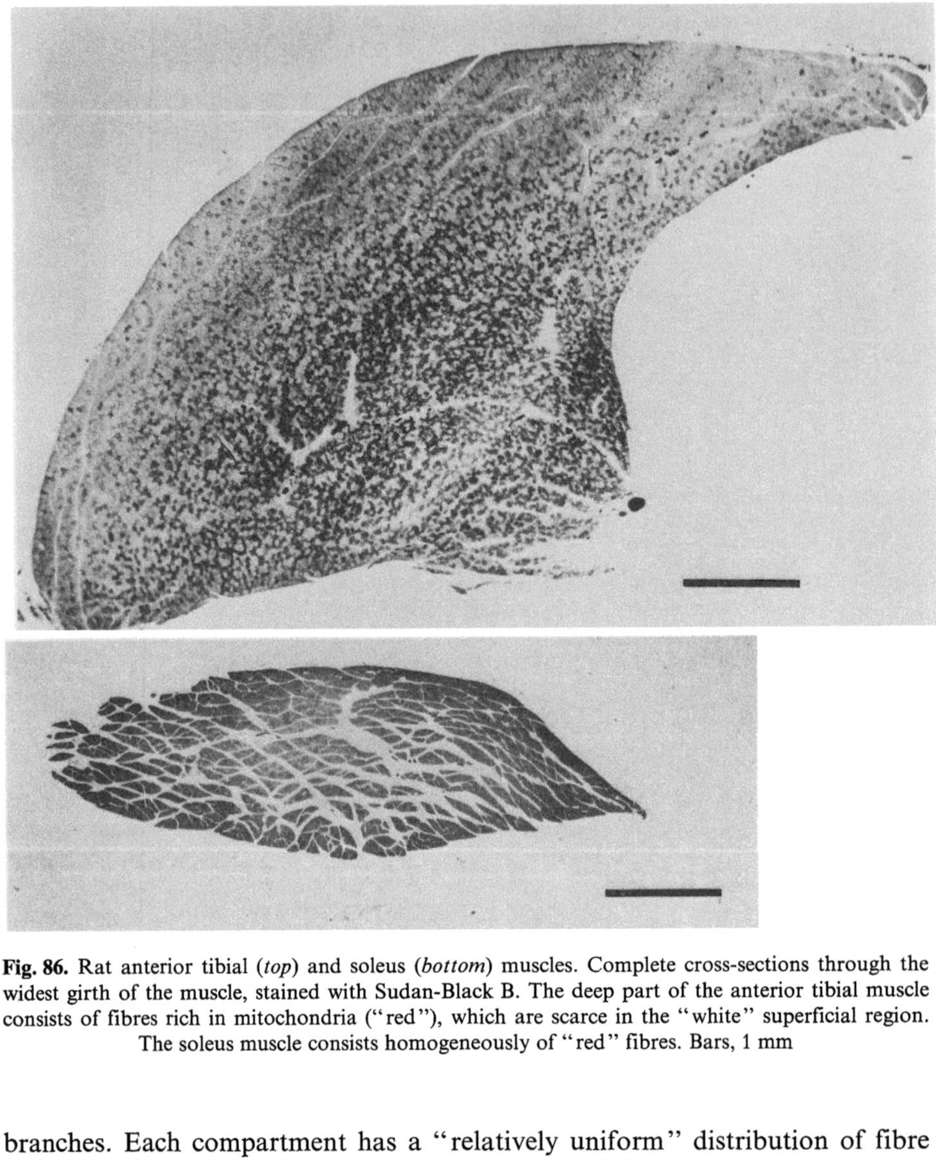

Fig. 86. Rat anterior tibial (*top*) and soleus (*bottom*) muscles. Complete cross-sections through the widest girth of the muscle, stained with Sudan-Black B. The deep part of the anterior tibial muscle consists of fibres rich in mitochondria ("red"), which are scarce in the "white" superficial region. The soleus muscle consists homogeneously of "red" fibres. Bars, 1 mm

branches. Each compartment has a "relatively uniform" distribution of fibre types, but the variation between the compartments is large.

Interindividual differences of the fibre-type composition have been reported in athletes, and the predominance of either type I or type II fibres has been related to specific performances. Elite marathon runners have a high proportion of type I fibres with high oxidative capacity (FINK et al. 1977; SALTIN 1977). It is now widely recognized that training does not change the fast- and slow-twitch fibre type composition. Hence, the differences have been attributed to selection: only a person with a genetically determined high incidence of slow-twitch, fatigue-resistant type I fibres is likely to become a successful marathon runner. It is, however, noteworthy, that the mean fibre type percentage differed

Table 6. Diameter of muscle fibres and incidence of type II fibres in different human muscles. Frozen sections were stained for ATPase at pH 9.4, and 200 to 400 fibres were assessed. S.D.: standard deviation of the mean values of the muscles giving the interindividual variation

Subjects (age in years)	Muscle	Type I/S.D. (μm/μm) diameter	Type II/S.D. (μm/μm) diameter	% Type II mean/range	Reference
16 males (19–58)	Brachial biceps	64/6	73/5	63/38-83	Brooke and Engel (1969)[a]
25 females (19–57)		57/5	47/5	57/50–77	
14 males (14–58)	Lateral vastus	60/6	65/8	63/48-77	
14 females (19–61)		59/6	50/6	67/45–77	
5 males (25–55)	Deltoid	51/6	47/6		
4 females (53–68)	Anterior tibial	51/7	56/7	24/20–35	Schmalbruch (unpublished)[b]
9 females	Rectus abdom.	50	52	45/32–51	Häggmark and
and	Obliquus ext.	50	53	42/23–60	Thorstensson (1979)[a]
4 males	Obliquus int.	50	52	44/26–64	
(24–55)	Transvers. abdom.	49	50	44/36–49	

[a] Surgical biopsies [b] Autopsies

considerably less between individuals when the percentages of fibre types, at autopsy, are determined at numerous sites, and pooled. The variability decreases with increasing sample size and the interindividual variation expressed as standard deviation is smaller than between usual needle biopsies (see also Tables 6, 7, 9) (Nygaard and Sanchez 1982: five lateral vastus muscles, 45% type I fibres, SD 7.4%; five brachial biceps muscles, 52% type I fibres, SD 7.7%; Lexell et al. 1983a: five lateral vastus muscles, 52% type I fibres, SD 7%). Therefore, the relation between fibre type percentages in needle biopsies and motor performance needs confirmation.

IX. Fibre Types and Electron Microscopy

After it was ascertained by electron microscopy that several of the distinguishing light microscopic features (see above) were real, numerous extremely specialized muscles were investigated. The aim was to relate the physiological properties to the abundance or scarcity of certain organelles, and to gain insight into the function of these organelles. High frequency muscles from fish and bat, extremely fast crab muscles, flight muscles of hummingbirds, and a multi-

Table 7. The 95% confidence ranges for mean fibre diameters and fibre type percentages in normal human muscles. All data from JOHNSON et al. (1973) and POLGAR et al. (1973). The material stems from autopsies of 6 males aged 17 to 30 years who died without signs of affection of the neuromuscular system

Muscle	Diameter in μm		Percentage	
	Type I	Type II	Type I	Type II
Abductor digiti minimi	26–52	28–60	35–72	31–65
Abductor pollicis brevis	14–70	27–79	52–74	26–48
Adductor magnus (surface)	50–80	48–83	42–65	35–58
Adductor magnus (deep)	45–78	34–90	50–76	24–50
Adductor pollicis	31–65	44–80	71–90	11–29
Biceps brachii (surface)	31–70	31–72	34–51	49–66
Biceps brachii (deep)	45–77	52–82	41–61	39–60
Biceps femoris	29–74	27–83	56–78	22–44
Brachioradialis	35–69	26–98	30–53	47–73
Deltoid (surface)	33–70	27–90	43–63	37–57
Deltoid (deep)	29–68	49–57	46–76	24–54
First dorsal interosseus	35–66	47–86	51–63	37–49
Erector spinae (surface)	41–78	32–83	33–84	17–67
Erector spinae (deep)	49–73	34–74	32–78	22–68
Extensor digitorum	36–58	34–68	42–53	47–58
Extensor digitorum brevis	28–53	40–67	29–61	39–71
Flexor digitorum brevis	42–78	45–84	34–55	45–66
Flexor digitorum profundus	28–76	32–85	27–68	32–73
Frontalis	14–28	19–33	31–97	3–69
Gastrocnemius lat. (surface)	43–54	40–61	37–50	50–63
Gastrocnemius lat. (deep)	37–60	42–60	43–57	43–57
Gastrocnemius med.	36–92	43–90	46–56	44–54
Gluteus maximus	31–84	32–87	36–67	33–62
Iliopsoas	39–69	34–70	40–59	41–61
Infraspinatus	30–60	24–70	37–54	46–63
Latissimus dorsi	31–74	30–92	38–63	37–62
Orbicularis oculi	11–31	14–35	4–27	73–96
Pectoralis major (clavicul. head)	35–75	25–94	32–52	48–68
Pectoralis major (sternal head)	29–86	25–96	29–58	42–72
Peroneus longus	40–67	42–71	53–73	27–48
Rectus abdominis	32–55	35–75	35–57	43–65
Rectus femoris lat. (surface)	56–76	50–96	22–37	63–78
Rectus femoris lat. (deep)	51–85	56–93	36–49	52–64
Rectus femoris med.	48–80	55–94	34–52	49–66
Rhomboid	39–75	37–73	34–55	45–66
Sartorius	39–57	34–63	40–60	40–60
Soleus (surface)	47–87	45–95	75–98	2–26
Soleus (deep)	56–82	44–97	80–98	2–20
Sternomastoid	36–64	29–78	28–43	57–73
Supraspinatus	31–76	28–76	41–78	23–59
Temporalis	25–51	10–50	33–60	40–67
Tibialis anterior (surface)	38–71	46–85	63–84	16–37
Tibialis anterior (deep)	36–76	48–80	67–78	22–33
Trapezius	47–71	48–86	33–75	25–67
Triceps (surface)	42–77	29–95	17–49	51–84
Triceps (deep)	37–84	25–97	8–58[a]	42–92[a]
Vastus lateralis (surface)	40–87	48–78	18–57[a]	43–82[a]
Vastus lateralis (deep)	44–85	58–70	38–56	44–63
Vastus medialis (surface)	50–73	55–74	36–51	49–64
Vastus medialis (deep)	47–77	50–81	52–72	29–49

[a] Recalculated from individual data in JOHNSON et al. (1973)

Table 8. Diameter of muscle fibres and incidences of type II fibres in the deltoid and lateral vastus muscle of children (autopsy samples, frozen sections). Of each sample 200 fibres were assessed. *S.D.*, standard deviation of the mean in the different subjects giving the interindividual variation. (Modified from OERTEL 1983)

Age group (years)	Number of subjects	Deltoid muscle mean diameter/S.D. (μm)		% Type II fibres	Lateral vastus muscle mean diameter/S.D. (μm)		% Type II fibres
		Type I	Type II		Type I	Type II	
0–0.1	2	11	10	51	13	11	60
0.2–0.5	13	15/3	13/3	–	13/2	12/2	–
0.5–1.5	12	20/2	17/2	46	17/3	17/3	47
1.5–3	4	24/4	17/3	38	26/8	22/4	45
3–6	6	26/3	18/3	41	34/7	27/4	40
6–10	4	33/2	26/2	40	41/3	33/4	45
10–15	6	42/7	40/10	47	51/6	48/6	55
15–20 (male)	13	54/7	58/10	–	62/9	62/11	–
15–20 (female)	15	49/10	44/10	–	60/37	51/7	–

Table 9. Percentages of type I, IIA, and IIB fibres in lateral vastus muscles of men at different ages. (Needle biopsies, frozen sections). *S.D.* , standard deviation in the different groups giving the interindividual variation

Number of subjects	Mean age (years)	% Type I fibres/S.D.	% Type IIA fibres/S.D.	% Type IIB fibres/S.D.	References
Males					
6	6	62/14	17/10	21/10	BELL et al. (1980)
69	16	54/12	32/9	13/7	HEDBERG and JANSSON (1976)
26	25	54/11	32/11	14/10	ANIANSSON et al. (1980)
14	61	53/14	29/11	18/9	
17	70	48/11	36/8	16/8	HEDBERG and JANSSON (1976)
18	78	60/15	23/13	17/16	ANIANSSON et al. (1980)
Females					
7	6	56/8	22/9	22/8	BELL et al. (1980)
47	16	52/9	33/9	15/7	HEDBERG and JANSSON (1976)
25	24	51/10	33/9	15/8	ANIANSSON et al. (1980)
8	70	50/13	44/15	5/9	HEDBERG and JANSSON (1976)

tude of insect muscles were the subjects of these studies. The differences were remarkable (see Sect. C.III.5), but the results eventually became merely confirmatory and contributed little to the understanding of muscle fibre types in mammals.

In mammalian muscles, the "white", intermediate, and "red" fibres, defined histochemically, were shown to vary in mitochondrial content (PELLEGRINO and FRANZINI 1963; GAUTHIER and PADYKULA 1966; BUBENZER 1964, 1966; SHAFIQ et al. 1966, 1969; SCHIAFFINO et al. 1970; SCHMALBRUCH 1967b, 1970, 1971; GAUTHIER 1969, 1974) (Fig. 87). The overlap between the fibre types is large,

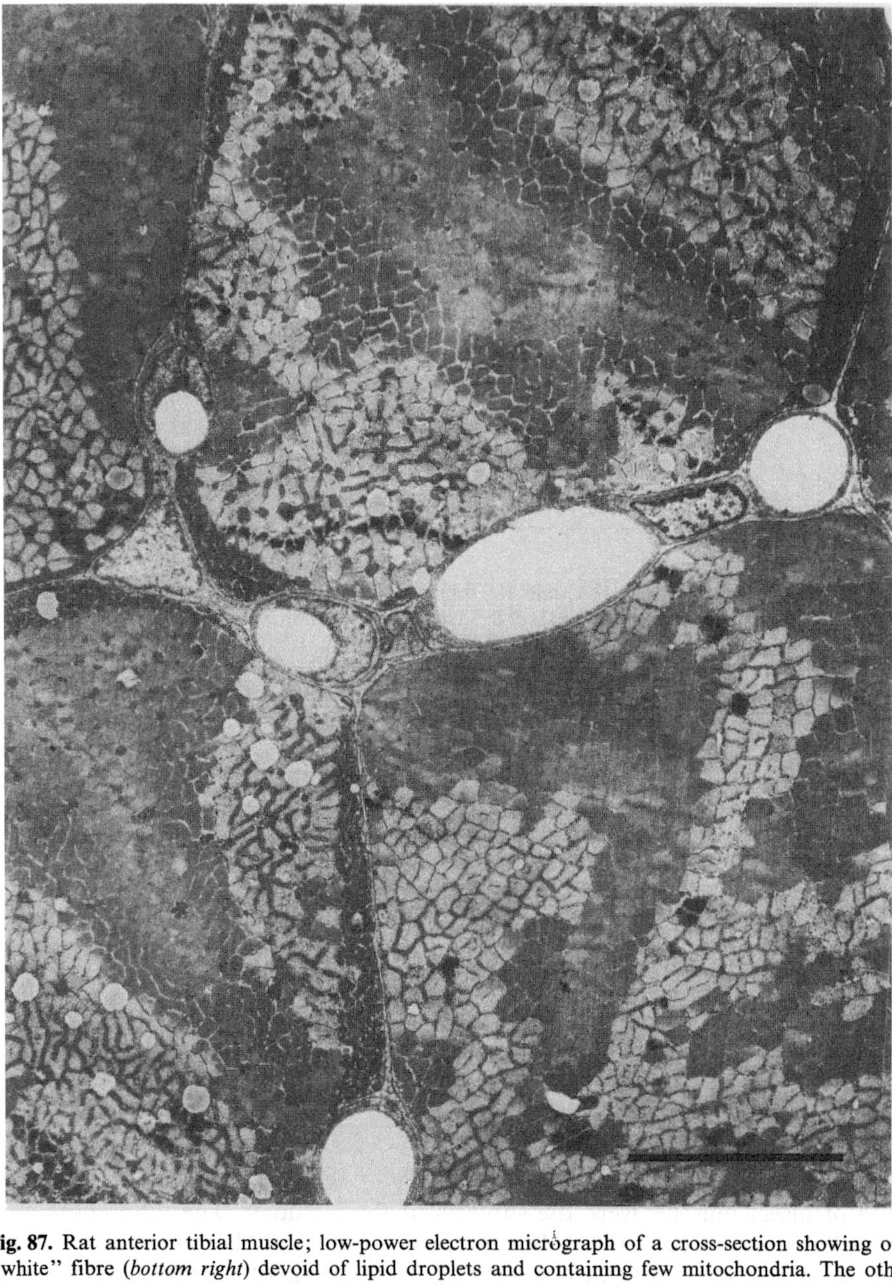

Fig. 87. Rat anterior tibial muscle; low-power electron micrograph of a cross-section showing one "white" fibre (*bottom right*) devoid of lipid droplets and containing few mitochondria. The other fibres contain lipid droplets (unstained) and prominent subsarcolemmal aggregations of mitochondria. The "dark" areas are cross-sections through the A-band and Z-disc levels, the "light" areas are sections passing through the I-band level. The muscle was fixed by vascular perfusion and the capillaries are extremely dilated. Bar, 10 μm

and the mitochondrial content provides a poor basis for classification; only few qualitative differences in the distribution of mitochondria and lipid droplets have been described.

The terminal cisterns of the sarcoplasmic reticulum and the T system are involved in excitation-contraction coupling and more of these membranes are found in fast- than in slow-twitch muscles (SCHIAFFINO et al. 1970; LUFF and ATWOOD 1971; SCHMALBRUCH 1971; EISENBERG et al. 1974; EISENBERG and KUDA 1975, 1976). The surface and volume densities of the membranes vary continuously and do not allow classification of the muscle fibres (for review, see EISENBERG 1983). KUGELBERG and THORNELL (1983) reported that the volume density of the terminal cisterns in fibres of physiologically identified rat motor units is inversely related to the isometric twitch contraction time. This relation is valid over the whole range of motor units, irrespective of the type of muscle.

Slow-tonic (non-twitch) muscle fibres of amphibia lack an M line and their Z line is wider and less regular than the Z line of twitch fibres (see Chap. E). FORSSMANN and MATTER (1966) found no M line in some fibres of rat diaphragm. The physiological properties of these fibres, which have also been seen in human laryngeal muscles (SCHMALBRUCH 1970), are, however, unknown. It is unlikely that they are slow (non-twitch) fibres. Most fibres of the diaphragm, whether rich or poor in mitochondria, and the slow-twitch fibres of the soleus muscle, have distinct M lines. PAYNE et al. (1975) found wider M lines in human muscle type I fibres than in type II fibres. This confirmed a previous observation in rat (SCHIAFFINO et al. 1970). The M line as distinguishing criterion for mammalian fibre types has been revived by SJÖSTRÖM et al. (1982a,b,c). They observed that in human muscles studied in ultrathin frozen sections, only the type I fibres display five strong lines corresponding to the five M-line bridges. IIA fibres have three strong central and two weak outer lines, and IIB fibres have only the three central lines. The same differences could be seen in plastic-embedded sections (Fig. 38).

The Z lines of the fibres of a mammalian slow-twitch muscle are wider than those of the fibres of a fast-twitch muscle (DUCHEN 1971; SCHMALBRUCH 1971) (Figs. 88, 89). However, in a homogeneous fast rat muscle, the Z-line width varies from 60 to 120 nm, perhaps implying that the Z-line width varies independent of the speed of contraction. Fibres rich in mitochondria have the widest Z lines (SCHIAFFINO et al. 1970). Fast-red and slow-red fibres in rats, distinguished by an antibody against fast myosin, did not differ in electron micrographs (GAUTHIER 1979). The Z-line width in human muscles varies continuously (PAYNE et al. 1975; CULLEN and WEIGHTMAN 1975) and does not allow separation of fibre types.

GALVAS et al. (1982) classified individual fibres of the cat flexor carpi radialis muscle using histochemistry and assessed the same fibres using electron microscopy. This muscle consists of a red and a white region, both containing type I and type II fibres. There was wide overlap betwen the fibre types with respect to mitochondrial volume, volume of lipid droplets, amount of reticulum and T-system membranes, and Z-line width. The mean membrane volume (SR + T system) was smaller in type I fibres than in type II fibres, and the mean Z-line

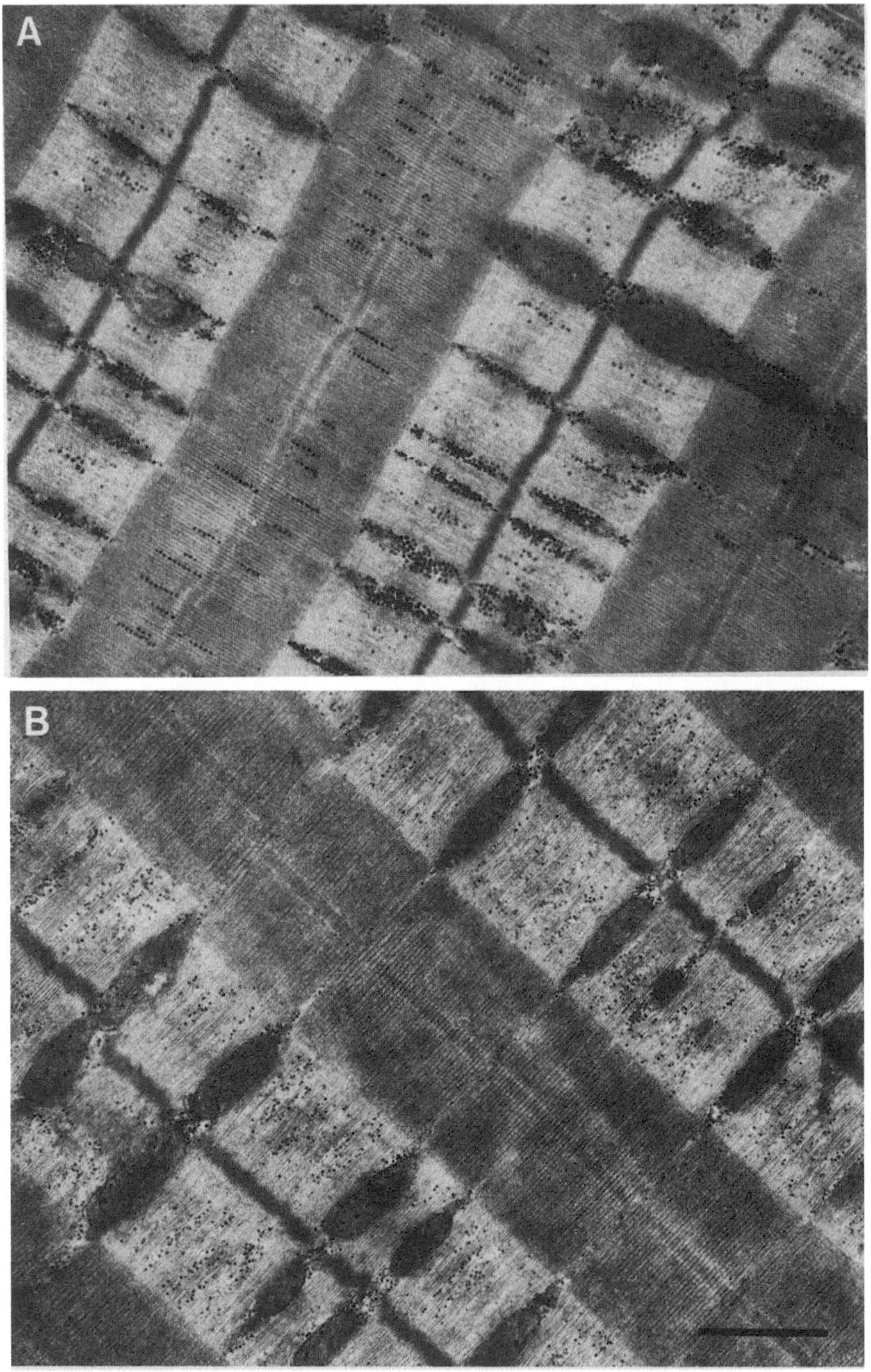

Fig. 88 A, B. Cat gastrocnemius muscle. The micrographs show presumably fast-twitch (**A**) and slow-twitch muscle fibres (**B**). Note the different glycogen distribution; the Z lines are wider in the slow-twitch soleus fibre. Bar 1 μm. (From SCHMALBRUCH 1971)

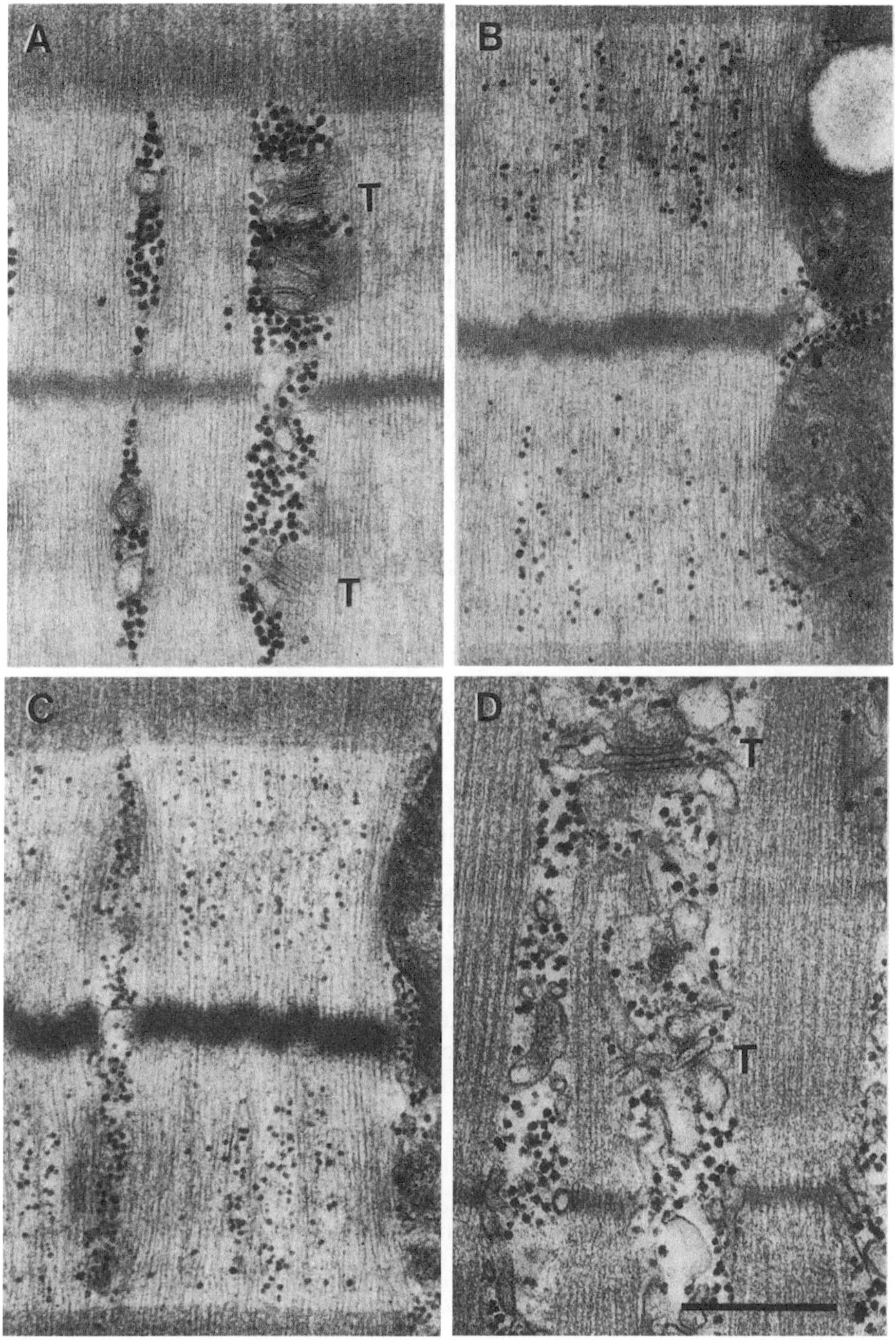

Fig. 89 A–D. Z-line width and glycogen distribution in different muscle fibre types. **A** Cat gastrocnemius muscle, presumably fast-twitch fibre. Note T tubules (*T*). Narrow Z line. **B** Cat gastrocnemius muscle, presumably slow-twitch fibre. Wide Z line. **C** Cat soleus muscle, slow-twitch fibre. Wide Z line. **D** Rabbit vocalis muscle, fast-twitch fibre. Note elaborate sarcoplasmic reticulum and T tubules (*T*). Narrow Z line. Bar, 0.5 µm. (From SCHMALBRUCH 1971)

width was smaller in IIB fibres than in IIA and I fibres. The other parameters studied did not significantly differ for the various fibre types. Hence, the morphometric data do not allow distinguishing of fibre types, not even the mean values of most parameters are different. The authors found the highest mitochondrial content in the type I fibres of the red muscle region, but nonetheless denoted this fibre type as "intermediate" (see Sect. D.IV.1).

Some workers combined two criteria to test whether this allowed fibre type discrimination (Fig. 90). The Z-line width in human muscles increases continuously with mitochondrial content (JERUSALEM et al. 1975; ENGEL et al. 1979; PRINCE et al. 1981; SJÖSTRÖM et al. 1982a). Training changes the slope of this relation because it increases only the mitochondrial volume and not the Z-line width (PRINCE et al. 1981). Both, however, increase in rabbit anterior tibial muscles chronically stimulated to induce a fast-slow transformation (EISENBERG and SALMONS 1981) (see Sect. D.V). The relation between Z-line width and mitochondrial content in guinea pig and rabbit applies only to fast fibres; slow-red fibres with the same mitochondrial volume as fast-red fibres have wider Z lines (EISENBERG and KUDA 1976, 1977; EISENBERG and SALMONS 1981; EISENBERG 1983). Mitochondrial content and sarcoplasmic reticulum are not related, probably because fast-twitch fibres may be oxidative or glycolytic. The Z-line width is inversely related to the number of T tubules (Fig. 90). According to EISENBERG (1983), the fibres of different guinea pig muscles can be identified with 70%–80% certainty when statistical methods are applied using several morphometric parameters.

Another distinguishing feature is the array of glycogen granules (Figs. 88, 89). In fast-twitch fibres of rat, cat, and rabbit, strands of granules surround the myofibrils at the I-band level; in addition single but rather prominent granules are arranged in longitudinal rows between the filaments of the A band. These rows are lacking in slow-twitch fibres, and the glycogen granules at the I-band level lie mainly within the myofibrils between the individual I filaments (SCHMALBRUCH 1971, 1979a). This difference is best seen in muscles fixed by vascular perfusion; it is less distinct in human muscle fibres (SCHMALBRUCH and KAMIENIECKA 1974), probably because glycogen is consumed during the anoxic phase between excision and fixation.

The present contribution of electron microscopy to the morphological identification of physiological fibre types in heterogeneous mammalian muscles con-

Fig. 90. Morphometry of human muscle fibres (quadriceps femoris muscle). *Top left*: The histogram of Z line widths displays two peaks. It is assumed that the fibres with wide Z lines (*shaded area*) are slow-twitch and those with narrow Z lines are fast-twitch fibres (limit arbitrarily set at 100 nm). *Top right, middle:* The histograms of the surface areas of the T system and the terminal cisterns also reveal two peaks; fibres with wide Z lines (*shaded area*) are those with relatively few membranes of these systems. The surface area of the longitudinal sarcoplasmic reticulum without the terminal cisterns (*middle right*) is about the same in fibres with wide and narrow Z lines. *Bottom left:* The Z line width increases continuously with the relative amount of mitochondria. Distinct fibre types cannot be identified. *Bottom right:* Z line width and amount of T tubules are inversely related. This is in agreement with the assumption that fast-twitch fibres have many T tubules and narrow Z lines. [Unpublished material of Dr. B.R. EISENBERG, Chicago; mean values reported in EISENBERG (1983); diagrams courtesy of Dr. EISENBERG]

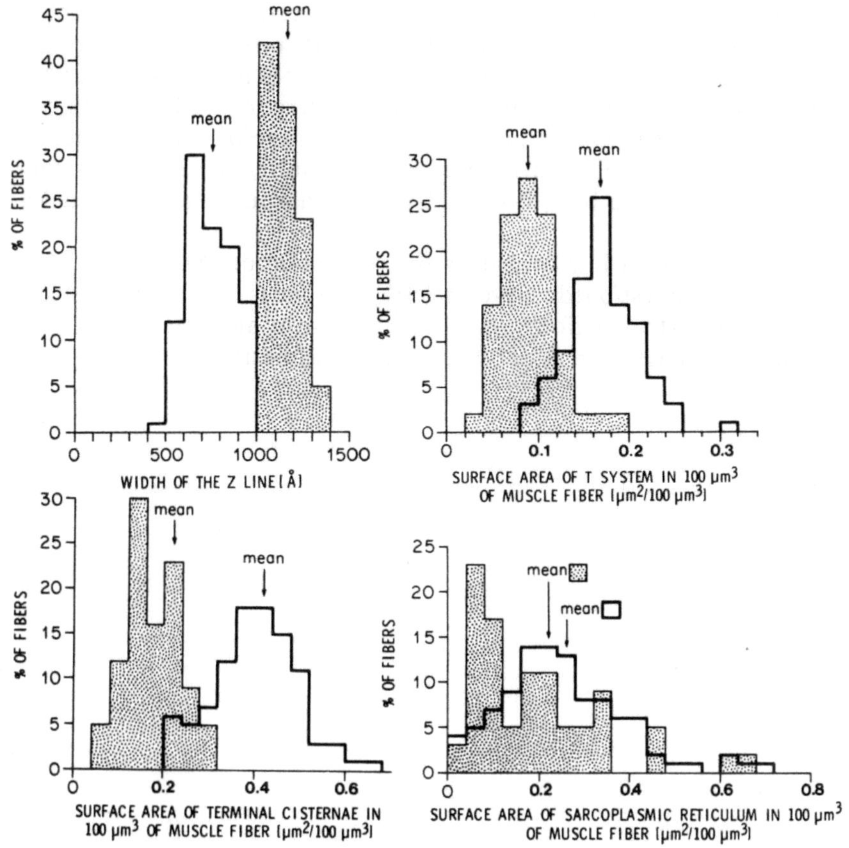
mean
mean
% OF FIBERS
WIDTH OF THE Z LINE [Å]
mean
mean
% OF FIBERS
SURFACE AREA OF T SYSTEM IN 100 μm3 OF MUSCLE FIBER [μm2/100 μm3]
mean
mean
% OF FIBERS
SURFACE AREA OF TERMINAL CISTERNAE IN 100 μm3 OF MUSCLE FIBER [μm2/100 μm3]
mean
mean
% OF FIBERS
SURFACE AREA OF SARCOPLASMIC RETICULUM IN 100 μm3 OF MUSCLE FIBER [μm2/100 μm3]

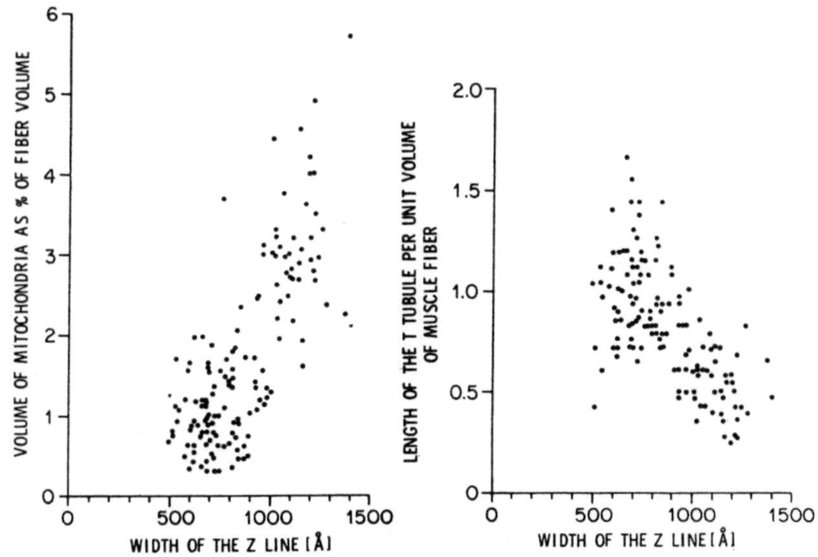
VOLUME OF MITOCHONDRIA AS % OF FIBER VOLUME
WIDTH OF THE Z LINE [Å]
LENGTH OF THE T TUBULE PER UNIT VOLUME OF MUSCLE FIBER
WIDTH OF THE Z LINE [Å]

trasts with the initial optimism. Several ultrastructural features characterizing fast- and slow-twitch, and fatiguable and fatigue-resistant fibres are known; usually it is possible to identify examples of fibre types in electron micrographs, but not all fibres can be typed with reasonable certainty. The fact that fibres of known origin or with known histochemical or physiological properties cluster in a scatter-gram of morphometric data, or occupy different regions in a continuous spectrum, is of little help, because the limits are uncertain in an unknown muscle.

The array of the M-line bridges is the only criterion the validity of which has not yet been questioned. Nevertheless, whether it can be used as marker for fast- and slow-twitch fibres in all muscles from all species, like the light microscopical ATPase stain at pH 9.4, needs confirmation.

E. Slow Muscle Fibres

I. Amphibia

1. Felderstruktur Fibres

The discussion how mammalian muscle fibres could be subdivided into fibre types was provoked not only by observations on red and white mammalian muscles (see Chap. D) but also by the fact that amphibia have some muscles with outstanding functional properties. These muscles contract slowly, and respond with lasting contractions to acetylcholine and not, like twitch muscles, with short and rapid contractions, or with series of contractions (SOMMERKAMP 1928). The occurrence of slow-tonic muscles is related to the specific motor behavior of the animals (WACHOLDER and VON LEDEBUR 1930). KRÜGER (1929) found that the "tonic" muscles contain a minority of fibres which in cross-sections display ribbon-shaped myofibrils; he postulated that these fibres were the specific tonic (slow) fibres and named them "Felderstruktur" fibres in contrast to phasic (twitch) "Fibrillenstruktur" fibres. Fibrillenstruktur refers to the dotted appearance of the fibres in cross-sections. The Felderstruktur fibres carry several nerve endings each resembling a bunch of grapes (*terminaisons en grappe,* TSCHIRIEW 1879); the Fibrillenstruktur fibres have one or a few solid endplates (*terminaisons en plaque,* TSCHIRIEW 1879) (KRÜGER 1952). The different innervation pattern is also described as "multiple" and "focal", respectively.

Fibrillenstruktur fibres resemble twitch muscle fibres in mammalian muscles. Amphibian twitch muscles once commanded great interest because they are durable experimental models in muscle physiology; nevertheless, modern physiological techniques allow to work almost equally well with mammalian muscle and only for single fibre studies are amphibian muscles still indispensible. Therefore, in this review, non-mammalian twitch fibres will only be described when the results obtained appear important for the understanding of mammalian muscle. On the other hand, slow (tonic, non-twitch) fibres are rare in mammals and occur in specialized muscles only. There even exist intermediate forms which partly resemble twitch fibres and partly slow fibres. The role slow fibres play in mammalian muscles is not clear, in some cases their very existence is disputed (see below). Slow fibres have been described in invertebrates and also in fish muscles, but only the results obtained in amphibia and birds seem to be relevant for the understanding of mammalian slow fibres.

Following the observations by TASAKI and MITZUTANI (1944), that some toad muscle fibres are innervated by small-diameter nerve fibres that elicit only slow contractions, KUFFLER and VAUGHAN WILLIAMS (1953) demonstrated that certain frog muscle fibres are multiply innervated by γ-axons; these muscle fibres do not propagate action potentials and respond to repetitive stimulation with slow and graded contractions. By contrast, all singly innervated fibres react with all-or-nothing responses. The identification of the tonic fibres with KRÜGER'S (1929) Felderstruktur fibres was positively established by PEACHEY and HUXLEY (1962).

For many years, KRÜGER'S (1929) observations were disputed or simply neglected (for review, see HESS 1970). The main obstacle to the acceptance of Felderstruktur fibres as slow (tonic) fibres was probably KRÜGER'S (1952) claim that also mammalian skeletal muscles contained Felderstruktur fibres and that these were equivalent to slow fibres in frogs (HUXLEY 1964). Mammalian skeletal muscle fibres may contract with different speeds, but they are all unquestionably twitch fibres with propagated action potentials.

2. Histochemistry

LÄNNERGREN and SMITH (1966), KIESSLING and WOHLRAB (1969), ASMUSSEN and KIESSLING (1970, 1974), and SMITH and OVALLE (1973) distinguished histochemically, in frog and toad muscles, three types of Fibrillenstruktur fibres and one or two types of Felderstruktur fibres. All Fibrillenstruktur fibres react intensely for ATPase at pH 9.4 (type II fibres), but differ with respect to oxidative enzymes and fat. The fibre diameter ranges from 60 to 170 μm; the mitochondrial and lipid contents decrease with increasing diameter. The frog sartorius muscle is a pure twitch muscle containing large easily fatiguable motor units, and small ones with high resistance to fatigue. The isometric twitch contraction times range from 21 to 36 ms (22° C), which is short compared with mammalian muscle fibres – considering the low temperature (see Table 5). The fibres receive synaptic inputs from several motoneurons (LUFF and PROSKE 1976). This polyneural innervation pattern (which should not be confused with the multiple innervation of the slow fibres) has been explained by the fact that the fibres are long and run all the way through the muscle. There are four distinct endplate zones in the sartorius muscle, but only one in the shorter ileofibularis muscle. Felderstruktur fibres occur only in short muscles. These fibres are thin and react weakly both for ATPase at pH 9.4 and for mitochondrial enzymes; they contain almost no fat. Felderstruktur fibres are regionally concentrated (tonus bundle) and are always associated with thin twitch fibres rich in mitochondria and fat (ASMUSSEN and KIESSLING 1970).

3. Twitching and Non-Twitching Slow Fibres

LÄNNERGREN and SMITH (1966) and SMITH and OVALLE (1973) distinguish "small pale"and "small clear" Felderstruktur fibres. "Small pale" fibres contain slightly more fat and more mitochondria than "small clear" fibres, and the reaction for ATPase at pH 9.4 is weak rather than absent.

The question of whether there are really two types of slow (non-twitch) fibres or transitional forms between slow and twitch fibres has been discussed for several years. PEACHEY and HUXLEY (1962) occasionally encountered fibres that twitched although they appeared to be Felderstruktur fibres. SHAMARINA (1962, 1966) found in the frog ileofibularis muscle, non-twitch fibres that only propagated action potentials when they were repetitively stimulated; the fibres were focally innervated (NASLEDOW 1965). The classification of these focally innervated fibres as slow fibres has been debated (ORKAND 1963; HUXLEY 1964; PEACHEY 1968). ORKAND (1963) suspected that they were twitch fibres with a low safety factor for synaptic transmission, and PEACHEY (1968) raised the question of whether the recordings were from the tonus bundle of the muscle.

LÄNNERGREN (1979) classified living slow fibres of toad muscles in dark-field illumination as "small pale" and "small clear" fibres; "small pale" fibres but not "small clear" fibres contain lipid droplets. Only the "small clear" fibres are typical slow (non-twitch) fibres; "small pale" fibres twitch, but, unlike twitch-fibres, react with lasting reversible contractures to KCl or ACh. The twitch contraction time is longer (117 ms) than that of true twitch fibres (30–40 ms). The force-velocity relation (see Fig. 1) of the "twitching slow" ("small pale") fibres is intermediate to those of typical twitch and of typical slow fibres. Unlike the fibres described by SHAMARINA (1962, 1966) and NASLEDOW (1965), "twitching slow" fibres have 5–6 endplates distributed along their length. The axon terminals are beaded with incomplete separation of the beads. Non-twitch slow fibres ("small clear") have 7–10 endplates consisting of aggregates of small separated synaptic areas. The presence of five different fibre types in the iliofibularis muscle of *Xenopus laevis* was confirmed by determining the native myosin isoenzyme composition of isolated muscle fibres with known isotonic contraction properties (LÄNNERGREN and HOH 1984).

Still other multiply innervated twitching fibres have been found in the submaxillaris (submandibularis) muscle of frog. These fibres are 5–50 μm thick and have 5–10 synaptic sites (KORDYLEWSKI 1979a). The properties of their postsynaptic ACh-activated channels are different from those in slow non-twitch or in focally innervated twitch fibres (MILEDI and UCHITEL 1981). A thorough anatomical and physiological study confirmed that the submaxillaris muscle of frog is composed of a distinctive population of fibres with electrical properties like those of twitch muscle fibres, and multiple innervation like that of slow muscle fibres (MILEDI and UCHITEL 1984). In other muscles, these "intermediate" fibres are rare; ELIZALDE et al. (1983) failed to find them in the frog piriformis muscle which contains many slow fibres.

4. Ultrastructure

Initially no T system and no triadic junctions were detected in Felderstruktur (slow) fibres (PEACHEY and HUXLEY 1962; HESS 1965) but both have later been shown (PEACHEY 1965; PAGE 1965; HOYLE et al. 1966; FRANZINI-ARMSTRONG 1973a; LÄNNERGREN 1975; DAUBER 1979). There is one T system per sarcomere, mainly at the Z-line level. T tubules are scarce, and triadic and diadic junctions

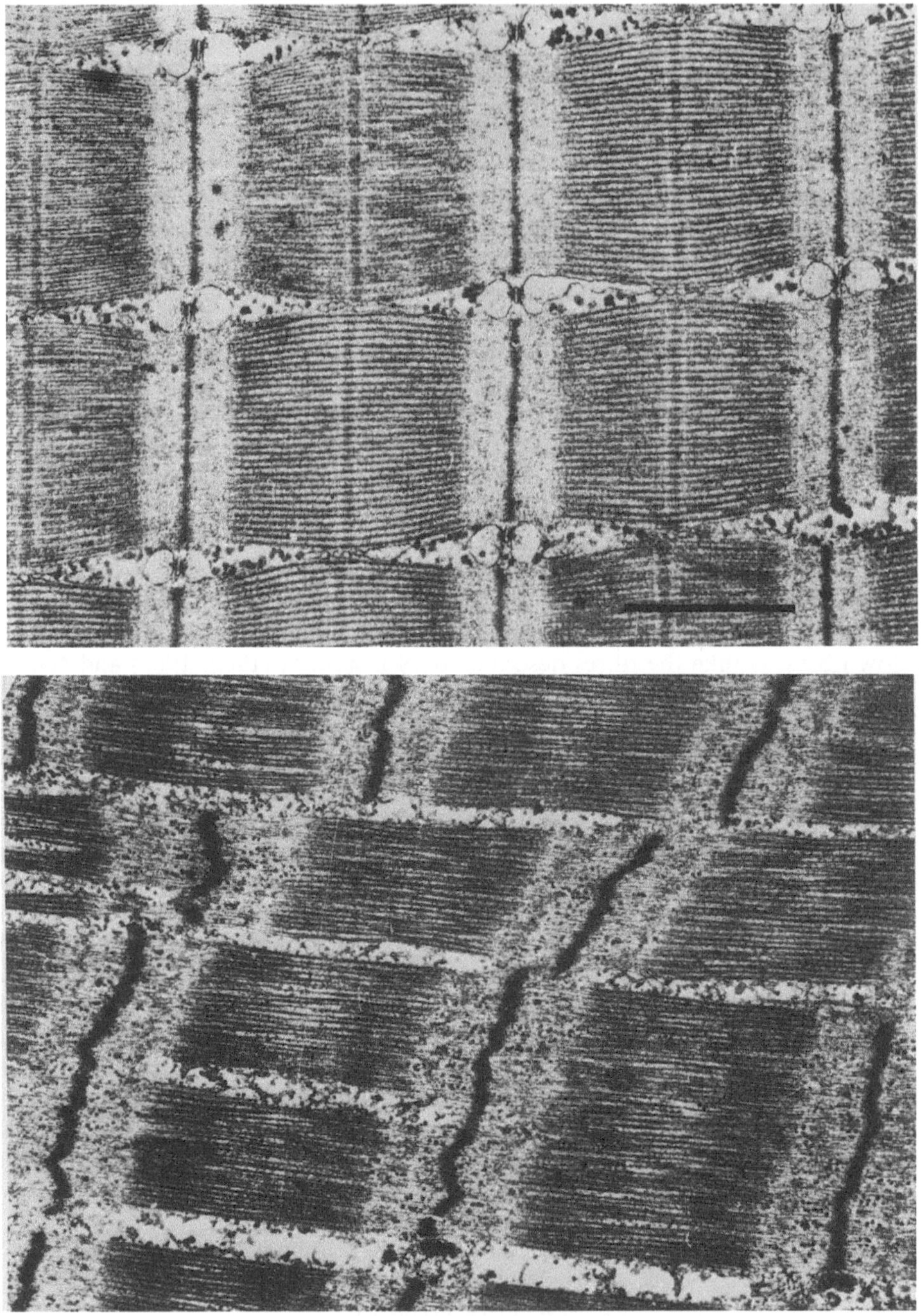

Fig. 91. Twitch (*top*) and slow (*bottom*) fibres from the pelvic segment of the rectus abdominis muscle of frog. The two micrographs show the essential features distinguishing the sarcomere structure of slow and twitch fibres; slow fibres in other species differ only in details from the slow fibre shown here. M lines are absent in the slow fibre, and the Z lines are wide and have an irregular course; triadic junctions are regularly present at the Z line level in the twitch, but not in the slow fibre. Glycogen granules are situated between the myofibrils in the twitch fibre; in the slow fibre they appear between the thin filaments. This difference is found in fast- and slow-twitch fibres of mammals as well (Figs. 88, 89). Bar, 1 μm. (From Uhrik and Schmidt, 1973, with permission of the authors)

are often longitudinally oriented. The paucity of T-system membranes in slow fibres is reflected by the high specific resistance and the low specific capacity of the sarcolemma, which, unlike in twitch fibres, is not enlarged by the T tubules (ADRIAN and PEACHEY 1965; PROSKE and VAUGHAN 1968) (see Sect. C.III.3b). The SR consists of irregularly arranged tubules without prominent terminal cisterns. The gap of the triadic junctions is slightly wider than in twitch fibres but otherwise the structure is the same (FRANZINI-ARMSTRONG 1973a). The Ca^{2+}-storage capacity of the SR is insufficient. Ca^{2+} necessary for contraction is transported across the sarcolemma, rather than exclusively liberated from internal stores; in contrast to twitch fibres, slow fibres lose their contractility in Ca^{2+}-free Ringer's solution (LÜTTGAU 1963). The Ringer-glycerol-Ringer treatment, which in twitch fibres destroys the T tubules and uncouples excitation and contraction (see Sect. C.III.3b), has no effect on the contractility of slow fibres (STEFANI and STEINBACH 1968).

The fine structure of the myofibrils of slow and twitch fibres is different (Fig. 91). The lengths of the myofilaments are the same (PAGE 1965), but the Z lines of slow fibres are much wider than in twitch fibres and usually jagged (PEACHEY and HUXLEY 1962; HESS 1963, 1967; PEACHEY 1965; PAGE 1965; HOYLE et al. 1966; UHRIK and SCHMIDT 1973; GILLY 1975). The M line is lacking or indistinct. SMITH and OVALLE (1973) and LÄNNERGREN (1979) find rudimentary M lines in "small pale" fibres (twitching slow), but never in "small clear" fibres (non-twitch slow). In twitch fibres, the thin filaments when approaching the Z disc are in tetragonal array (Sect. C.I.15), in slow fibres they remain irregularly packed right up to their insertion into the Z disc (PAGE 1965). The twitching slow fibres ("small pale"), however, show a square lattice (LÄNNERGREN 1979). ASSMUSSEN and KIESSLING (1970) report that presumed slow fibres of frog muscles mostly have subsarcolemmal myonuclei, whereas presumed twitch fibres mostly have internal nuclei. The mitochondrial content of slow fibres is small, which is in harmony with the weak histochemical reaction for oxidative enzymes, and also with the lack or scarcity of lipid droplets. In twitch fibres of *Xenopus laevis,* the mitochondria occupy 3%–17% of the fibre volume; in slow "small pale" fibres 4%, and in slow "small clear" fibres 1%, of the fibre volume is occupied by mitochondria (SMITH and OVALLE 1973).

II. Birds

In contrast to amphibian muscles, some avian muscles consist almost entirely of multiply innervated Felderstruktur fibres (KRÜGER 1950, 1952). The best investigated example is the anterior latissimus dorsi muscle (ALD) of chicken; by contrast, the posterior latissimus dorsi muscle (PLD) consists to 90% of focally innervated Fibrillenstruktur fibres. The slow fibres of ALD respond to depolarization with prolonged contracture, but unlike typical amphibian slow fibres they propagate action potentials. Tetanic tension and lasting ACh contracture have the same amplitude, indicating that no twitch fibres are present.

The PLD develops a contracture of at most 10% of the maximal tetanic tension, because there are only few slow fibres in this muscle (GINSBORG 1960). The structure of avian slow fibres is similar to that of amphibian slow fibres, but there are two T systems per sarcomere and all fibres have an M line. Thin motor axons form en grappe endings (HESS 1961a, 1967; GINSBORG and MACKAY 1961; PAGE 1969). ZENKER and GRUBER (1967) count up to 90 endplates per fibre in the chicken ALD. The endplate structure varies considerably, and the authors emphasize that the en plaque vs en grappe classification is an oversimplification.

According to PIEROBON-BORMIOLI et al. (1980), antibodies against myosin from the slow chicken ALD react with slow fibres of frog, toad, lizard, and tortoise, but not with any of the fibre types in human or rat muscles, because the myosin isoenzyme in avian and amphibian slow muscle fibres is structurally different from that of mammalian slow-twitch fibres. By contrast, GAUTHIER and LOWEY (1979) and GAUTHIER et al. (1983) report that mammalian slow-twitch fibres bind antibodies against the myosin of chicken ALD. Antibodies against mammalian slow-twitch myosin identify two fibre types in the chicken ALD; both react weakly for ATPase at pH 9.4, but only some show reversal after acid preincubation, and bind antibodies against guinea pig soleus myosin (PIEROBON-BORMIOLI et al. 1980). Physiological evidence for two sorts of ALD fibres is not available. SHAFIQ et al. (1974) observed differences in the Z-band width of chicken ALD fibres and distinguished tonic I (135-nm-wide Z line) and tonic II (67-nm-wide Z line) fibres.

III. Mammals

1. Extraocular Muscles

a) Two Sorts of Slow Fibres

Mammalian extraocular muscles contain some fibres with Felderstruktur (SIEBECK and KRÜGER 1955). KUPFER (1960) found multiple nerve endings of γ-axons, and HESS (1961b, 1967) and PILAR and HESS (1966) established the connection between Felderstruktur, en grappe nerve endings, jagged Z lines, and scarce SR.

It had long been known that ACh causes contracture of extraocular muscles; HESS and PILAR (1963) and MATYUSHKIN (1964) demonstrated in vitro that this is a property of multiply innervated fibres which do not conduct action potentials, and which are present in all extraocular muscles, but not in the levator palpebrae and retractor bulbi muscles. BACH-Y-RITA and ITO (1966) studied the muscle in situ and failed to find non-propagating fibres. All fibres twitched, but some were focally and others multiply innervated. There was no evidence for polyneural innervation (BACH-Y-RITA and LENNERSTRAND 1975).

PEACHEY (1968) proposes that these conflicting results are due to the presence of two sorts of slow fibres in extraocular muscles comparable to those described in amphibia (non-propagating) and birds (propagating). This has been confirmed by LENNERSTRAND (1974a), who recorded intracellularly from muscle fibres of individual motor units in cat, labelled the same muscle fibres, and anatomically determined the pattern of innervation. All focally innervated fibres belong to twitch motor units; some multiply innervated fibres conduct action potentials and twitch, and some do not propagate potentials. The singly innervated fibres conduct faster than the multiply innervated fibres (3 and 2 m/s, respectively) because they are thicker (HÅKANSSON 1956); the isometric twitch contraction time is <10 ms in singly and 5–15 ms in twitching multiply innervated units. The multiply innervated non-propagating fibres, during repetitive stimulation, contract slowly in a graded fashion. The multiply innervated twitching units differ from singly innervated units in that the tetanic force increases with the stimulus frequency, even beyond the fusion frequency. Singly innervated fibres reach their maximum force at fusion frequency; a further increase in stimulus rate initially increases the rate of tension development, but eventually the force decreases because endplates or muscle fibres are unable to follow electrically. The increase in tension beyond the tetanic fusion frequency has been used as criterion for the presence of slow fibres in other mammalian muscles (see below) (DALE and MUID 1978).

b) Histochemistry and Ultrastructure

The histochemical picture of extraocular muscles appears confusing and a large and variable number of fibre types have been described (see Sect. F.I). The distribution of different fibre types is neither random nor uniform. There is a tendency for certain fibre types to be concentrated in particular regions of the muscles or to be absent from other regions of the muscle. One distinguishes the orbital layer of the muscle cross-section which is near the orbital surface of the muscle, and the global layer which is adjacent to the globe. The orbital layer comprises one tenth to one fifth of the cross-sectional area and contains predominantly thin fibres (KERN 1965). ASMUSSEN (1974) and ASMUSSEN and WOHLRAB (1974) have reviewed the findings in cat, rabbit (ASMUSSEN et al. 1971), sheep (HARKER 1972a; BARKER and HARKER 1972), and rat (MAYR 1971). All singly innervated twitch fibres react intensely for ATPase at pH 9.4; about 20% of the fibres of the orbital layer, and 5%–10% of the fibres of the global layer react weakly or not at all for ATPase at pH 9.4, and are poor in mitochondrial enzymes. These fibres are multiply innervated. In the global layer, they have roughly the same diameter as the surrounding twitch fibres (15–25 μm in rat; 33 μm in cat, 30 μm in sheep); in the orbital layer they are thinner (5–10 μm in rat, 22 μm in cat, 10–25 μm in sheep) than the surrounding fibres. The Z lines are wide and irregular, the M lines are usually lacking, and the SR and T system are sparse. The thin ATPase-negative fibres of the orbital layer are probably the multiply innervated twitching fibres assessed in situ by BACH-Y-RITA and ITO (1966) and LENNERSTRAND (1974a); the presumed slow fibres of the global layer are probably the non-propagating slow fibres found in vitro by HESS and PILAR (1963).

The histochemical and electron microscopic results described are basically in harmony with those of other studies (bank vole, KACZMARSKI 1970b; rabbit, CHENG-MINODA et al. 1968; HOPKINS 1975; DAVIDOWITZ et al. 1980a; rat, YELLIN 1969a; HANSON and LENNERSTRAND 1977; NAG and CHENG 1982; cat, PEACHEY et al. 1974; HANSON et al. 1980; LENNERSTRAND 1980; rhesus monkey, CHENG and BREININ 1965, 1966; MAYR et al. 1966; MILLER 1967, 1971). There is, however, disagreement with respect to the presence or absence of M lines in slow fibres of mammalian extraocular muscles. CHENG-MINODA et al. (1968) find no M line in the rabbit; whereas according to HESS (1967), slow fibres in cats (like avian slow fibres) have M lines. In rats, not only the slow fibres but also some of the twitch fibres lack M lines (MAYR 1971). In sheep, presumed slow fibres have no M lines, whereas all presumed twitch fibres have M lines (HARKER 1972a); BARKER and HARKER (1972), however, report that the multiply innervated fibres of the global layer differ from those of the orbital layer by having an M line. This is unexpected because the twitching slow fibres which are the more likely candidates for M lines are in the orbital layer.

ZENKER and GRUBER (1967) find 15–18 synapses on each multiply innervated fibre of the rhesus monkey, and NAMBA et al. (1968) find in rat superior rectus muscle 10–16 synapses per fibre. The basal lamina separating pre- and postsynaptic membranes is occasionally interrupted in multiply innervated fibres such that axon and muscle fibre are in direct contact; gap junctions like in electrotonic or mixed synapses have not been found. Subsynaptic folds are scarce or lacking (CHENG and BREININ 1965; TERÄVÄINEN 1968, 1969b; MAYR and ZENKER 1969).

The slow fibres of the orbital layer vary in diameter and mitochondrial content along their length; close to both tendon ends the fibres are thicker than towards the middle (BARKER and HARKER 1972; DAVIDOWITZ et al. 1980a). In sheep they branch and reunite, sometimes branches pass through peripherally distributed muscle spindles and become nuclear bag fibres with equatorial nucleation and sensory innervation (BARKER and HARKER 1972; BARKER et al. 1972).

All multiply innervated fibres in extraocular muscles of man and guinea pig react strongly with antibodies against myosin from the slow-tonic avian ALD, but only weakly with antibodies against the slow-twitch guinea pig soleus muscle (PIEROBON-BORMIOLI et al. 1979, 1980) (Figs. 92, 93).

DAVIDOWITZ et al. (1980b, 1983) described in multiply innervated fibres of rabbit extraocular muscles peculiar subsarcolemmal membrane-glycogen complexes. They are up to 40 μm long and consist of concentric cisterns connecting to Golgi cisterns, and rows of large β-glycogen granules. The significance of these structures, and whether they occur only in rabbits, is unknown. They are identical to the "glycogen bodies" found in sensory nerve fibres of rabbit muscle spindles (CORVAJA et al. 1971) and in neonatally denervated extrafusal rat muscle fibres (SCHIAFFINO and HANZLIKOVA 1972a).

c) The Force Contribution of Slow Fibres

The tension developed during a contracture induced by ACh or succinylcholine is 30%–40% of the maximum tetanic force of the entire extraocular muscle

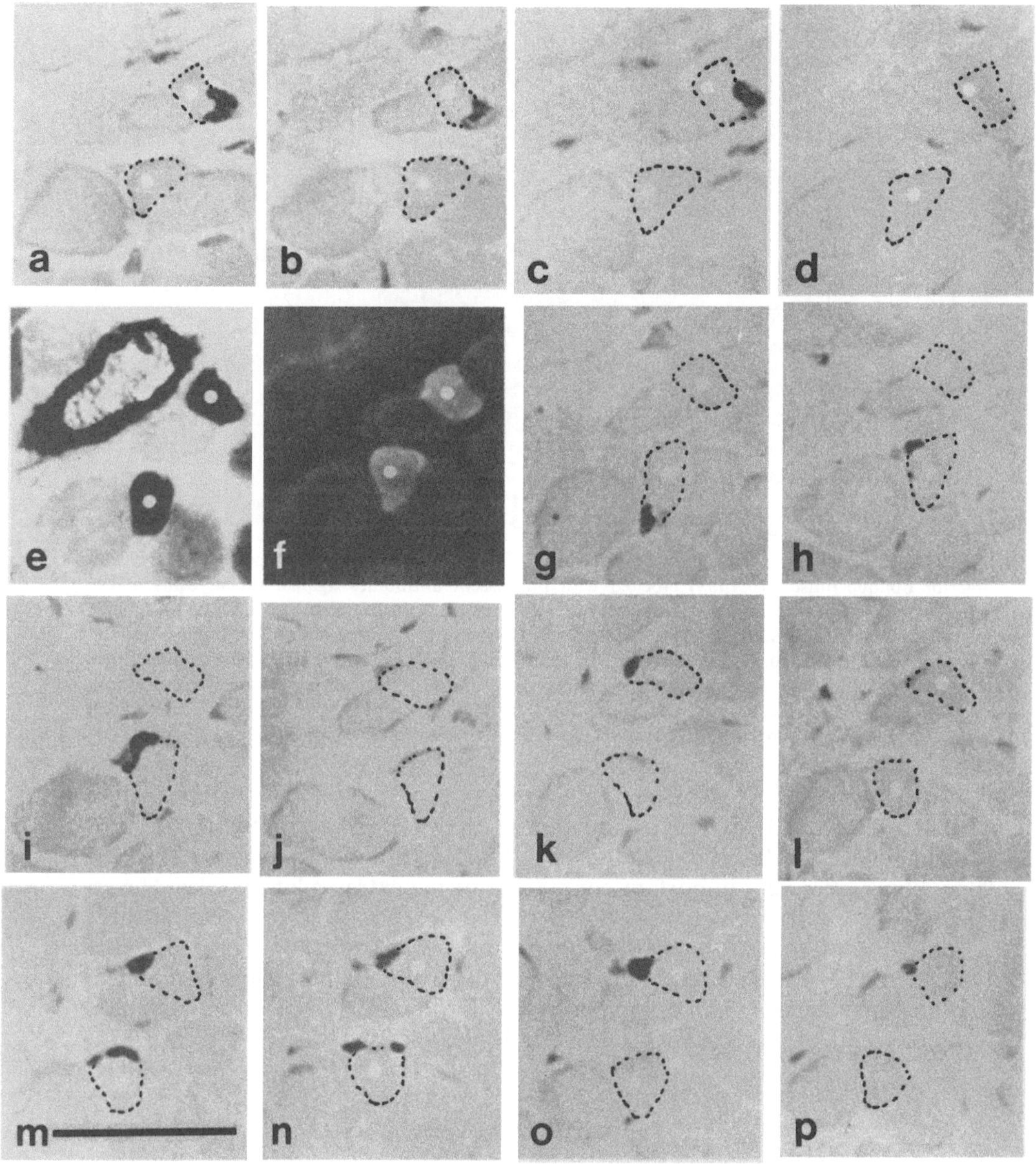

Fig. 92 a–p. Inferior oblique muscle of a child operated on for retinoblastoma. Spaced serial cross-sections covering 800 μm fibre length. All sections are stained for acetylcholinesterase except for **e** and **f**. Two presumed slow-tonic fibres are outlined and marked by *white dots*. Both fibres carry several endplates and the series of sections passes through the discontinuous reactive areas. Section **e** has been stained for ATPase at pH 9.4 after preincubation at pH 4.3, and shows that only the two *marked* fibres react. The same fibres (section **f**) are labelled by antibodies against the myosin of the slow-tonic anterior latissimus dorsi (ALD) muscle of chicken. Bar, 50 μm. (From PIEROBON-BORMIOLI et al. 1979, with permission of the authors and J.B. Lippincott; original plate kindly provided by Dr. PIEROBON-BORMIOLI, Padova)

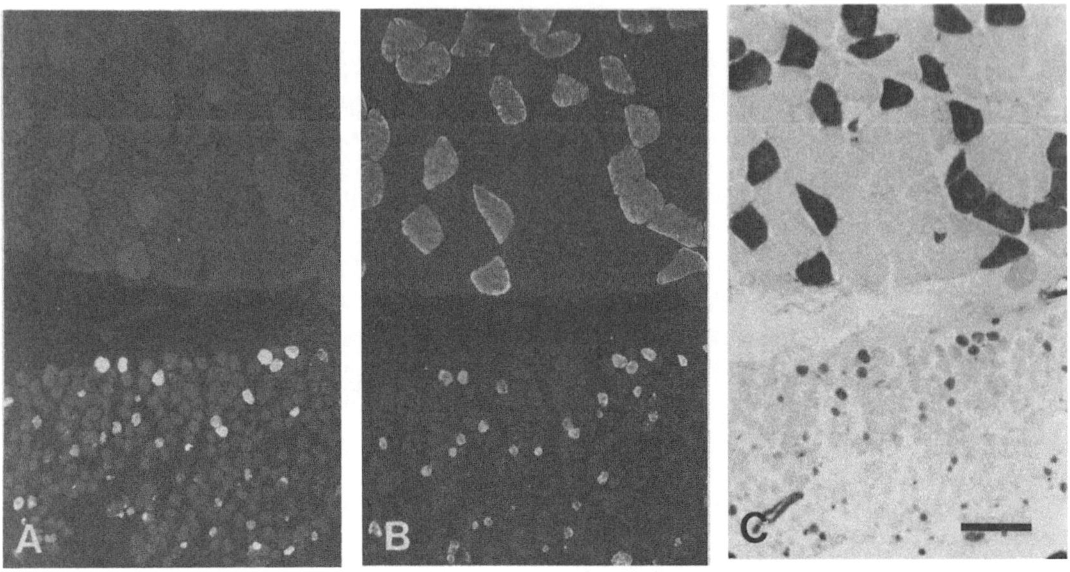

Fig. 93 A–C. Serial cross-sections through a "sandwich" composed of human pectoralis (*top*) and inferior oblique muscle (*bottom*) labelled with fluorescent antibodies against the myosins of the slow-tonic chicken anterior latissimus dorsi muscle (anti-ALD) (**A**) and of the slow-twitch guinea pig soleus muscle (anti-SOL) (**B**), and stained for ATPase at pH 9.4 after acid (pH 4.3) preincubation (**C**). Anti-ALD stains only few fibres of the extraocular muscles, anti-SOL binds to these fibres as well. Only anti-SOL reacts with the type I fibres of the pectoralis muscle; these fibres are dark in **C**. The ATPase reaction of the slow fibres of the extraocular muscle seems inconsistent. Bar, 100 μm. (Micrographs courtesy of Drs. Pierobon-Bormioli and Schiaffino, Padova)

(Bach-Y-Rita and Ito 1966; Lennerstrand and Hanson 1978a, b). This should reflect the force contributed by the slow fibres and suggests that the slow fibres are a very important component of the extraocular muscles. Nevertheless, the contribution of presumed slow fibres to the cross-sectional area is only 5% (Kaczmarski 1970a, b). Fuchs and Luschei (1971) and Close and Luff (1974) failed to find a relevant contribution of a slow component to the force during tetanic stimulation of the lateral rectus of monkey and inferior rectus of rat. The reason for this discrepancy is obscure.

2. Inner Ear Muscles

It has been claimed that inner ear muscles contain slow fibres, but the morphological data for the stapedius and tensor tympani muscles are conflicting. Erulkar et al. (1964), Fernand and Hess (1969), and Seiden (1971) find in both muscles in cats, Felderstruktur fibres with corresponding electron microscopic properties. Surprisingly, the supposed slow fibres have an M line, whereas the presumed twitch fibres lack an M line (Fernand and Hess 1969).

Fernand and Hess (1969) also observed multiply innervated fibres, but according to Blevins (1963, 1964), Erulkar et al. (1964) (cat), and Malmfors and Wersäll (1960a, b) (rabbit), all endplates are concentrated in a narrow endplate zone. BURGENER and MAYER (1980) find no multiply innervated fibres in the stapedius and tensor tympani muscles of man, cat, rabbit, guinea pig, and rat. All fibres of guinea pig stapedius have distinct M lines; staining for ATPase at pH 9.4 identifies in both muscles only type I and type II fibres with histochemical "twitch" properties (rabbit, ASMUSSEN and WOHLRAB 1971; VITA et al. 1983; cat, TEIG and DAHL 1972; guinea pig, BURGENER and MAYR 1980). Also the electrophysiological evidence for a slow fibre system (ERULKAR et al. 1964) was not confirmed (TEIG 1972a, b). By contrast, SHAW and BAKER (1983) maintain that the studies of TEIG (1972b) and TEIG and DAHL (1972) confirm both physiologically and histologically that the cat stapedius muscle contains slow fibres as claimed by ERULKAR et al. (1964) and FERNAND and HESS (1969).

3. Cremaster

DALE and MUID (1978) report that the guinea pig cremaster muscle responds with a lasting dose-related contracture to ACh or KCl. The muscle twitches after single stimuli, and develops tetanus at 20 Hz, but, like in twitching slow motor units in eye muscles, the force continues to increase at higher stimulus frequency. The morphological evidence for a slow fibre system is weak: "virtually all" fibres have en plaque endplates, "in only a very few instances" are several small endplates seen on a single fibre (DALE and MUID 1978). On the other hand, NINOMIYA et al. (1981) describe both singly and multiply innervated fibres with many or no subsynaptic folds, respectively, and Fibrillenstruktur and Felderstruktur fibres. The accompanying micrographs do not show these differences.

4. Oesophagus

Another muscle which has been suspected to contain slow fibres is that of the oesophagus (PEACHEY 1968). Nevertheless, the lasting ACh contracture previously reported is caused by the smooth muscle cells of the muscularis mucosae (ASMUSSEN and GAUNITZ 1978); the suspected multiple innervation sites shown by AChE staining (FLOYD 1973) represent AChE-positive myotendinous and myomyal junctions (GRUBER 1978). The contractile properties (FLOYD and MORRISON 1975; ASMUSSEN and GAUNITZ 1978), the histochemistry, the ultrastructure, the response to nerve stimulation, and the endplate configuration (GRUBER 1968, 1978) indicate that all cross-striated oesophagus fibres are twitch fibres.

IV. Comments

Initially the distinction between propagating twitch and non-propagating slow fibres seemed unambiguous, but the definition of slow fibres has become blurred since multiply innervated propagating fibres have been found in amphibia, birds, and mammals.

All multiply innervated muscle fibres develop a lasting but reversible contracture after application of a depolarizing agent (ACh, succinylcholine, KCl); non-propagating fibres respond with a graded contraction to repetitive stimulation. Propagating multiply innervated fibres (twitching slow fibres) twitch after a single stimulus, but the force increases further when the stimulus frequency is raised above the fusion frequency. The Z line is wide and often jagged in both propagating and non-propagating "slow" fibres; SR and T system are scarce, and the M line is either lacking or less distinct than in typical twitch fibres. The reactions for ATPase at pH 9.4 and for mitochondrial enzymes are both weak; this distinguishes tonic slow fibres from slow-twitch fibres in which a weak reaction for ATPase at pH 9.4 usually is combined with a strong reaction for mitochondrial enzymes. Tonic "slow" fibres are intensely stained with labelled antibodies against chicken ALD myosin, but react less intensely with antibodies against mammalian soleus myosin.

Twitching slow fibres of amphibia and birds contract much more slowly than do true twitch fibres, but in cat extraocular muscles the motor units of multiply innervated propagating fibres have contraction times only slightly longer than those of focally innervated twitch units.

The twitching slow muscle fibres present a classification problem. If the line were drawn between twitch and non-twitch fibres, multiply and singly innervated fibres, and Felderstruktur and Fibrillenstruktur fibres would end up in the same class. This is not sensible. Hence, until a better nomenclature comes into common usage, one has to distinguish non-twitching *slow* fibres, twitching *slow* fibres, slow-*twitch* fibres, and fast-*twitch* fibres.

The presence of slow fibres (non-twitching and twitching) in skeletal muscles of amphibia and birds, and in mammalian extraocular muscles and muscle spindles (see Chap. F) is unquestioned. Whether slow fibres are present in inner ear muscles and in the cremaster needs confirmation. In fish and reptiles the existence of slow fibres is proven (for review, see HESS 1970), but the relationships between morphology and physiology have not been studied as thoroughly as in amphibia. The information available indicates that slow and twitch fibres in fish, with respect to mitochondrial content and innervation, differ from those in mammals, birds, and amphibia (HESS 1970).

F. Non-Skeletal Muscles

Although the delineation between "skeletal" and "non-skeletal" muscles is debatable, it is obvious that some specialized cross-striated muscles or muscle fibres differ morphologically and physiologically from typical skeletal muscles moving the body or maintaining posture. The findings on these muscles are reviewed here.

I. Extraocular Muscles

The fibres of extraocular muscles are rich in mitochondria and richly supplied by capillaries. Histochemistry never shows a checkerboard pattern of fibre types as in skeletal muscles; the muscles are layered with relatively thick fibres in the global and thin fibres in the orbital part of the muscle. Mammalian and frog extrocular muscles are extremely fast and have a high resistance to fatigue. Physiological studies in other species have not been performed.

1. Fish

"Thin red" and "thick white" fibres have been distinguished (*Perca fluviatilis* (perch), Kilarski 1965; *Esox lucius* (pike), Kilarski 1966; *Tinca tinca* (tench), Kilarski 1967; *Carassius carassius* and *auratus* (carp and goldfish), Kilarski and Bigaj 1969; Davey et al. 1975). "Thin red" fibres have polygonal well- delineated myofibrils, are rich in mitochondria and SR, and have numerous triadic junctions at the A-I border; the "large white" fibres are poor in mitochondria, have ribbon-shaped myofibrils arranged in a radial fashion, are rich in SR, and have triadic junctions only at the Z-line level. Hence "red" fibres have two and "white" fibres one T system per sarcomere. Motor axons spiral around "white" fibres and make numerous synaptic contacts; "red" fibres are focally innervated by one motor axon running transversely across the muscle. Both types have distinct M lines. No physiological data are available. The "red" fibres are claimed to be slow fibres without propagated action potential, but both fibre types have morphological properties of twitch (phasic) fibres (Asmussen and Wohlrab 1974).

2. Reptiles

All extraocular muscle fibres of lizards have T tubules at the A-I border. Three fibre types have been described. Type A fibres are thin, are innervated by *en grappe* endings, and are in the orbital layer. The Z line is wide, the M line is lacking, SR is scarce, and the postsynaptic membrane is smooth. Type B fibres resemble A fibres, but they have a central accumulation of mitochondria. B fibres are not found in the retractor bulbi muscle. C fibres are large, are in the global layer, and have narrow polygonal myofibrils with distinct M line. Both A and B fibres are supposed to be slow fibres, and the C fibres are supposed to be twitch fibres (KACZMARSKI 1969).

3. Amphibia

Only the oculorotatory muscles, but not the retractor bulbi muscle, contain physiologically (ASMUSSEN 1978b) and morphologically (KILARSKI and BIGAJ 1969) identifiable slow fibres which are in the orbital layer. The slow fibres are rich in mitochondria ("red"); their diameter is 10–15 μm. Twitch ("white") fibres range in diameter from 30 to 60 μm. "White" but not "red" fibres have an M line; in all fibre types the T system is at the Z-line level. The retractor bulbi muscle consists of 10- to 150-μm-thick fibres with variable mitochondrial content, forming a mosaic pattern as in skeletal muscles. The twitch contraction time of extraocular muscles of frog is 15 ms (20° C), compared with 30 ms for the ileofibularis muscle. The resistance to fatigue is high, and the force per unit cross-sectional area is about half that of a skeletal muscle (1.5 vs 3.0 kg per cm^2) (ASMUSSEN 1978a, b, c). The twitch contraction time of the retractor bulbi muscle is comparable to that of the sartorius muscle (ASMUSSEN 1978c).

4. Birds

Three types of fibres have been identified (sparrow: "white", "red", "pink", KACZMARSKI 1970a; pigeon: "first", "second", "third", ALVARADO-MALLART 1972). Two T systems per sarcomere are regularly present. "White" fibres are in the orbital layer, are thin (6-μm diameter), and constitute 12% of the fibre population but occupy only 1% of the cross-sectional area. Morphologically, these fibres resemble slow fibres; they correspond to the "second" type in pigeons. "Red" fibres are rich in large mitochondria and in SR, have a mean diameter of 16 μm, and occupy 80% of the cross-section. The "red" fibres in sparrows probably correspond to the "third" type in pigeons; the M line is absent in sparrow muscles but in pigeons an M line is present. "Pink" fibres, about 8 μm thick ("first" type in pigeon), have small mitochondria, narrow and straight Z line, distinct M lines, and abundant SR. "Pink" fibres are typical twitch fibres, whereas "white" ("second") and "red" ("third") fibres resemble slow fibres in amphibian and avian skeletal muscles, respectively (see Chap. E). If this morphological identification is correct, slow fibres are far more numerous in bird extraocular muscles than in those of mammals.

5. Mammals

Most authors distinguish 5–6 different fibre types in oculorotatory muscles. Two fibre types are multiply innervated and are slow fibres, they occupy

5%–10% of the cross-sectional area (see Chap. E). The twitch fibre types vary in diameter and mitochondrial content, all are type II fibres (rat, Mayr 1971; Nag and Cheng 1982; sheep, Harker 1972a; rhesus monkey, Miller 1967, 1971; cat, Asmussen et al. 1971; Hanson and Lennerstrand 1977; Hanson et al. 1980; rabbit, Asmussen et al. 1971). The global layer of the muscles consists of rather thick and the orbital layer of rather thin fibres; there is no checkerboard pattern.

Like all cross-striated muscle fibres in mammals, the fibres have two T systems per sarcomere. The twitch fibres have distinct M lines; only in rats, twitch fibres rich in mitochondria are, like slow fibres, devoid of an M line (Mayr 1971). The mitochondrial content of all twitch fibres is high; the relative mitochondrial volume per unit fibre volume in rhesus monkey is up to 6 times larger than in mitochondria-rich skeletal muscle fibres (Miller 1971). The mitochondria are ovoids or longitudinally arranged bars or chains and never form mitochondrial grids or networks like in skeletal muscle fibres. Sometime clusters of mitochondria occupy the centre of a fibre.

Hoogenraad et al. (1979) find the same layer organization known in other mammals in the human inferior oblique muscles; there are 5- to 40-μm-thick fibres in the global part and 5- to 20-μm-thick fibres in the orbital part. Most fibres stain intensely for ATPase at pH 9.4. By employing different staining procedures, the authors arrive at 18 different fibre types which makes it impossible to relate the findings to those in other species.

The morphological inhomogeneity of the twitch fibres is not reflected by different motor unit properties. All muscles and also all motor units are invariably fast, faster than the fastest skeletal muscle fibres (Table 10). Fusion frequencies of up to 400 Hz have been found; unlike in twitching slow fibres the force does not increase with higher frequencies. Only the initial rate of tension development is frequency-dependent. The fibres are able electrically to follow stimulus frequencies of up to 1,000 Hz; the motoneurons may discharge spontaneously at frequencies of up to 400 Hz (Bach-y-Rita and Ito 1966; Hanson and Lennerstrand 1977; Lennerstrand and Hanson 1978a, b; Asmussen and Gaunitz 1981). In the lateral rectus muscle of monkeys, stimulation with 30 Hz which should activate a slow fibre system does not produce any tension (Fuchs and Luschei 1971).

The retractor bulbi muscle of cat is devoid of slow fibres and shows a checkerboard pattern of histochemically different fibre types. The contraction times of its motor units are somewhat longer than in oculorotatory muscles, but compared with skeletal muscles the retractor bulbi is still a very fast muscle (Lennerstrand 1974b). Surprisingly, 7 out of 40 abducens motoneurons sent collaterals to muscle fibres of the retractor bulbi and lateral rectus muscles (Crandall et al.1981). This would imply that in extraocular muscles one motoneuron innervates muscle fibres with different functional properties.

In some species, the oculorotatory muscles contain spindles (man and monkey, Cooper and Daniel 1949; Greene and Jampel 1966; mouse, Mahran and Sakla 1965; sheep, Harker 1972a), in others none have been found (cat, dog, rabbit, and rat, Cilimbaris 1910; birds, Maier et al. 1971).

Human and monkey extraocular muscles contain myelinated nerve fibres spiralling around individual muscle fibres. It has been assumed that these nerve

Table 10. Isometric twitch contraction time and isotonic speed of shortening of extraocular muscles in comparison to corresponding data in skeletal muscles. (All recordings at about 37° C)

Species and muscle	Isometric twitch contraction time (ms)	Fusion frequency (Hz)	Intrinsic speed of shortening (μm/s/ sarcomere)	References
Cat				
Inferior oblique				
Singly innvervated motor units	3.5–9	150–350		LENNERSTRAND (1974a)
All motor units	3.5–20	80–350		
Whole muscle	5.7			LENNERSTRAND and HANSON (1978a)
Lateral rectus				
Motor units	3–14			GOLDBERG et al. (1976)
Retractor bulbi				
Motor units	7–13	200		LENNERSTRAND (1974b)
Peroneus longus muscle				
Motor units	15–42	80–100		KERNELL et al. (1983a, b)
Rat				
Inferior rectus				
Motor units	5–6			CLOSE and LUFF (1974)
Whole muscle	4.38		62	
Ext. digit. longus	11		49	
Mouse				
Inferior rectus	3.7		49.5	LUFF (1981)
Ext. digit. longus	6.9		51.6	
Soleus	20.9		23.1	
Diaphragm	13.2		39.7	
Macaca mulatta				
Lateral rectus	7.0	400		FUCHS and LUSCHEI (1971)

fibres represent a receptor system different from muscle spindles (DANIEL 1946). Similar spiralling nerve fibres are present in fish extraocular muscles where they have been shown to be motor axons (see above). RUSKELL (1984) unambiguously established that also in man and monkey these spiral nerve endings terminate in motor endplates.

6. Conclusions

The data reviewed indicate that mammalian extraocular muscles consist largely of fast-twitch motor units; the few slow motor units are probably of two different types. It is not clear how much the slow fibre units contribute to the function of the extraocular muscles. The twitch contraction time of the cat inferior oblique muscle is dominated by the fastest units (Table 10). Despite the scientific interest slow fibres in mammalian extraocular muscles have commanded, it may very well be that these fibres are mere developmental relics.

The most striking property of the twitch fibres is that they are extremely fast and at the same time high-oxidative fatigue-resistant fibres. The uniformity of the contraction speed contrasts with the multitude of morphological fibre types that have been described. The different fast-twitch units vary slightly with respect to fatiguability (LENNERSTRAND 1974b).

The extraocular muscle fibres in mammals, birds, and amphibia are comparable; this is not so for fish. In all species, the retractor bulbi muscle differs from the oculorotatory muscles in that it resembles a skeletal muscle.

II. Intrafusal Muscle Fibres

Intrafusal fibres elicit responses from sensory nerve endings attached to their equatorial midportion; they do so either by active contraction, or by passively transmitting tension from the connective tissue of the muscle. Intrafusal fibres are innervated at one or both poles, usually by several fusimotor axons. The spindle-shaped capsule ensheathes the sensory endings in the equatorial region and also the efferent innervated polar regions of the fibres; the ends of the intrafusal fibres may extend into the extracapsular space.

A spindle responds to the amplitude of stretch (static), or to the rate of stretch (dynamic) with rapid adaptation after a certain degree of stretch has been reached. It is reasonable to assume – but has not been proven – that the stimulus to the fibre is modulated in a specific fashion by the viscoelastic properties of the intrafusal fibres such that the structure of the fibre determines whether the response is static or dynamic. The viscoelastic properties of resting and contracted fibres are different; this is probably the mechanism by which the efferent innervation of the intrafusal muscle fibres influences gain and response characteristic of the receptor system.

The structure of the muscle spindles and the number of intrafusal fibres they contain differ in different species. Comparison of the simple spindles in reptiles and amphibia may contribute to the understanding of more complex systems in mammals. To my knowledge, studies of fish muscle spindles do not exist and one might ask whether fish muscles contain muscle spindles. Muscle receptor systems in invertebrates are basically different from muscle spindles and shall not be discussed in this review.

1. Reptiles

Snake muscle spindles are very simply built and contain only one intrafusal fibre which is long and runs the entire length of the muscle. The sarcomere length exceeds that in extrafusal fibres, indicating that the spindles are prestretched. The intrafusal fibres are innervated by collaterals of axons to extrafusal muscle fibres (skeleto-fusimotor innervation); the neuromuscular junctions have no or only few postsynaptic folds. The M line is lacking and the Z lines are wide; triadic junctions are inconspicuous. Some intrafusal fibres are of equal diameter along their length, others show a central enlargement with few clustered myonuclei and a reduction of the mass of contractile filaments. The spindles with a centrally enlarged muscle fibres respond to rate, rather than amplitude, of stretch (dynamic or phasic-type spindle), those without enlargement respond to amplitude of stretch (static or tonic-type spindle) (FUKAMI and HUNT 1970; FUKAMI 1970; PALLOT and RIDGE 1972). FUKAMI (1970) observed twitch-like contractions and propagated action potentials in both types of spindles, although in both the fibres morphologically resemble extrafusal slow fibres. The contractile responses of the intrafusal fibres do not depend on whether the efferent fibres to the spindle are collaterals of axons to extrafusal slow or twitch fibres (HUNT and RIDGE 1974). By contrast, PALLOT (1974) distinguished ATPase-positive intrafusal fibres with M line and ATPase-negative intrafusal fibres without M line. According to PALLOT, the ATPase-positive intrafusal fibres are indiscriminately innervated by collaterals of axons to extrafusal twitch and slow fibres, whereas ATPase-negative intrafusal fibres are only innervated by collaterals of axons to extrafusal slow fibres.

Also lizard muscle spindles contain only one intrafusal fibre. There are "short capsule" and "long capsule" types with and without central thickening of the intrafusal fibre. The intrafusal muscle fibres are likewise innervated by collaterals of axons to extrafusal fibres (PROSKE 1969).

2. Amphibia

Frog spindles contain 3–5 intrafusal fibres with a chain of central myonuclei. Along their length, the fibres vary considerably in diameter. Most myofibrils terminate in the equatorial sensory region in finger-like projections of the sarcolemma ("reticular region"), and for a distance of 100–200 μm the fibres are practically devoid of myofilaments. This probably renders the reticular region, to which the sensory nerve fibres adhere, less stiff than the polar regions (references KARLSSON et al. 1966; HOUK et al. 1966). KARLSSON et al. (1966) and KARLSSON and ANDERSSON-CEDERGREN (1966) found M lines in all fibres. Thin fusimotor axons form endplates with shallow but distinct subsynaptic folds; occasionally the basal lamina is lacking and pre- and postsynaptic membranes are in direct contact. All intrafusal fibres morphologically resemble twitch fibres.

STERLING (1974), however, distinguishes two different fibre types: one reacted intensely for phosphorylase like extrafusal slow-twitch fibres, and one reacted weakly for phosphorylase like extrafusal slow (non-twitch) fibres. Slow intrafusal fibres receive collaterals from nerves to extrafusal slow fibres only, and hence spindles with slow intrafusal fibres are present only in muscle regions with slow fibres. OVALLE and SMITH (1974) do not find histochemical differences, and report that all intrafusal fibres of anuran spindles stain for ATPase the same as fast-twitch extrafusal fibres do. The fibres differ, however, in diameter: "thick" fibres have M lines like twitch fibres, whereas "thin" fibres have no M lines and resemble slow fibres. In view of the caliber variations along the length of the fibres documented by KARLSSON et al. (1966), classification of amphibian intrafusal fibres by their diameter would appear debatable.

3. Birds

Avian muscle spindles contain 3–6 fibres which are 8–9 μm thick and vary little in diameter along their length. The postsynaptic membrane of the neuromuscular junctions is smooth; there are fibres with and fibres without M line. Occasionally, adjacent fibres are closely apposed, which by light microscopy might be mistaken as branching (JAMES and MEEK 1973; ADAL 1973). The intrafusal fibres in chicken anterior latissimus dorsi (ALD) stain for ATPase at pH 9.4 like typical extrafusal type I or II fibres; acid preincubation reverses the staining pattern (OVALLE 1978).

4. Mammals

a) Fibre Types

It has long been recognized that there are at least two diameter classes of mammalian intrafusal fibres. Thick fibres show an equatorial accumulation of myonuclei and have been named nuclear bag fibres, in contrast to thin fibres which have a central chain of myonuclei and hence have been named nuclear chain fibres (COOPER and DANIEL 1956; BOYD 1956) (Fig. 94). BARKER and GIDUMAL (1961) also distinguish fibres with intermediate diameter which belong to the nuclear bag fibres. By electron microscopy (Fig. 95), nuclear chain fibres resemble extrafusal twitch fibres, and nuclear bag fibres resemble extrafusal slow (tonic) fibres. The bag fibres have no distinct M lines; the equatorial "bag" is stuffed with myonuclei, such that myofilaments are reduced to a narrow peripheral rim. These differences have been confirmed for all species studied: (rat, LANDON 1966; HENNIG 1969; VON DÜRING and ANDRES 1969; RUMPELT and SCHMALBRUCH 1969; OVALLE 1971, 1972; cat, CORAVAJA et al. 1969; guinea pig, BANKS and JAMES 1973; tupaia, VON DÜRING and ANDRES 1969; sheep, HARKER 1972b; mouse, EDWARDS 1975; man, RUMPELT and SCHMALBRUCH 1969).

The number of intrafusal fibres per spindle varies: rat, 1–4, mostly 2, nuclear bag and 2–4, mostly 2, nuclear chain fibres (PORAYKO and SMITH 1968; YELLIN 1969b; RUMPELT and SCHMALBRUCH 1969; BANKS et al. 1977) (Fig. 94); cat, 1–3, mostly 2, nuclear bag and 1–10 nuclear chain fibres (VON DÜRING and ANDRES 1969; BANKS et al. 1977; BAKKER and RICHMOND 1981); opossum, 2 or more nuclear bag and 4–12 nuclear chain fibres (JONES 1966); man, 1–3 nuclear bag and 2–7 nuclear chain fibres (RUMPELT and SCHMALBRUCH 1969), total 9–11 (GRUNER 1961) or 2–14 fibres (SWASH and FOX 1972); mouse, total

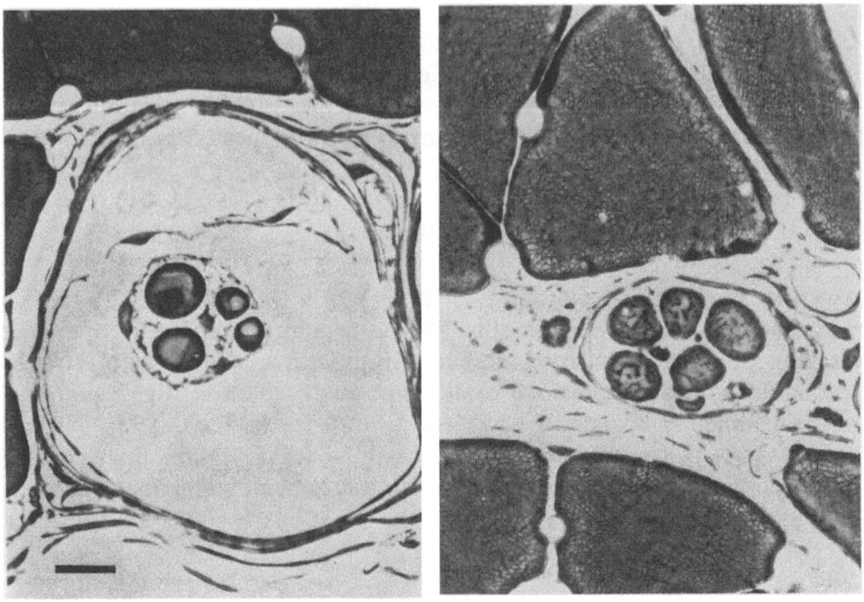

Fig. 94. Muscle spindles. Rat soleus muscle, fixed by vascular perfusion; 3-μm plastic sections stained with p-phenylenediamine and photographed with phase optics. *Left:* Cross-section through the equatorial region showing two nuclear bag and two nuclear chain fibres. The inner spindle capsule and the wide outer spindle capsule are distinct. Capillaries are restricted to the outer spindle capsule. The wide capsule space usually collapses in specimens fixed by immersion. Note large distance between capillaries and intrafusal muscle fibres. *Right:* Cross-section through the polar region showing five intrafusal fibres enclosed by a narrow capsule. The fibres are of equal diameter; the two fibres to the right seem to contain fewer mitochondria than do the three fibres to the left, and probably are nuclear bag fibres. The fact that in the polar region the difference in diameter is lost may have contributed to the erroneous statement that in some species the spindles lack nuclear chain fibres (see text). Bar, 10 μm

2–4 fibres (EDWARDS 1975); guinea pig, total 4.8 fibres (mean) (BANKS and JAMES 1973); dog, total 8 fibres (BANKER and GIRVIN 1971).

In 1964, BARKER and HUNT reported that rabbit muscle spindles are devoid of nuclear chain fibres, but electron microscopy shows different types of intrafusal fibres also in rabbits (VON DÜRING and ANDRES 1969; CORVAJA and POMPEIANO 1970; BANKS and JAMES 1974). In 1972, BARKER et al. report that rabbit spindles contain 1–2 typical nuclear bag, 1–2 intermediate (nuclear bag) fibres, and 2–3 nuclear chain fibres.

BARKER et al. (1972) and BANKS et al. (1977) describe in cat, rabbit, and rat, differences in the M-line structure of nuclear chain, "typical" nuclear bag, and "intermediate" nuclear bag fibres. The M line of nuclear chain fibres consists of five parallel arrays of bridges; in intermediate fibres only the two outer lines are present, whereas in typical bag fibres no bridges of the M line are found. The matter is complicated by the fact that in bag fibres the M-line appearance varies along the length of the fibre; in chain fibres a distinct M line is seen throughout the fibre. BARKER et al. (1976a) also distinguish two types

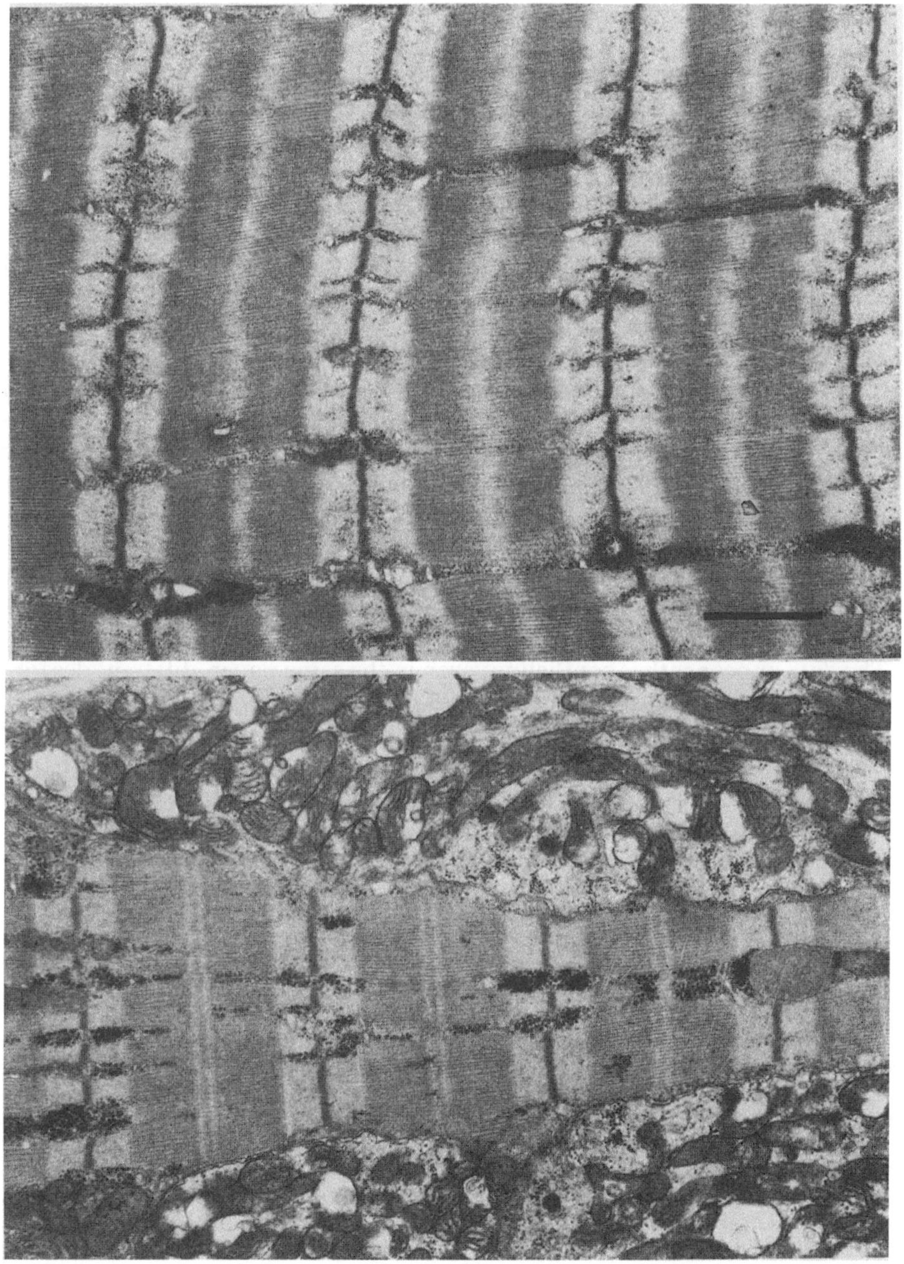

Fig. 95. Intrafusal fibres, rat anterior tibial muscle. The nuclear bag fibre (*top*) lacks M lines and resembles the slow frog fibre shown in Fig. 91. The nuclear chain fibre (*bottom*) resembles a twitch fibre and has distinct M lines. Also shown is a sensory ending wound around the fibre. The ending contains many mitochondria. In contrast to those in adjacent extrafusal fibres, these mitochondria are artefactually swollen, although the muscle was fixed by vascular perfusion. This is probably due to the fact that blood vessels are distant from the intrafusal fibres (see Fig. 94). Bar, 1 μm

of nuclear chain fibres: short ones that terminate together with the capsule and long ones which together with the nuclear bag fibres extend for more than 1 mm into the connective tissue of the muscle.

Three types of intrafusal fibres are distinguished using histochemistry. Nuclear chain fibres stain intensely for mitochondrial enzymes and moderately for glycolytic enzymes; "intermediate" nuclear bag fibres stain moderately for all metabolic enzymes and typical nuclear bag fibres stain weakly for all metabolic enzymes (YELLIN 1969b). Staining for ATPase at pH 9.4 identifies the nuclear chain fibres as "fast-twitch" type II fibres (strong reaction after alkaline and weak reaction after acid preincubation); intermediate nuclear bag fibres (bag_2 fibres) stain like IIC fibres (strong reaction both after alkaline and acid preincubation), and typical nuclear bag fibres (bag_1 fibres) which stain like slow-tonic fibres (weak or moderate reaction after alkaline, and weak reaction after acid preincubation) (JAMES 1971a, b; HARKER 1972b; BANKS et al. 1975, 1977; KUCERA 1977, 1980; KUCERA and DOROVINI-ZIS 1979; BAKKER and RICHMOND 1981). ARENDT and ASMUSSEN (1974b) combined different staining procedures and arrived at six intrafusal fibre types. According to KUCERA (1980), long chain fibres stain less intensely for mitochondrial enzymes than do short chain fibres, but the staining intensity may differ in the two spindle poles. Also nuclear bag fibres may differ in staining along their length such that a bag_1 fibre at one pole might appear as a bag_2 fibre at the other pole (KUCERA 1981).

The data appear rather conflicting; even in the same paper, different descriptions of the staining pattern of bag_1 fibres may be found (BARKER et al. 1972). Assessing the reaction intensity of metabolic enzymes in the thin fibres is subject to error, and the differences between intrafusal fibre types are less clear than those between extrafusal fibre types. Also the ATPase reactions are less constant than stated in most papers (Fig. 96). Nevertheless, it appears safe to conclude that there are nuclear bag and nuclear chain fibres, and that nuclear bag fibres, by staining for ATPase at pH 9.4 (best after acid preincubation), can be subdi-

Fig. 96 A–D. Muscle spindle, human medial vastus muscle. Serial cross-sections stained with haematoxylin eosin (**A**), for nicotinamide adenine dinucleotide-linked lactic dehydrogenase (mitochondrial) (**B**), and for ATPase at pH 9.4 after preincubation at pH 10.3 (**C**) and at pH 4.3 (**D**). The fibres have been numbered. The HE-stained section shows seven intrafusal fibres: *1* and *2* are thicker than *3–7* and are probably nuclear bag fibres. The structure between fibres *5* and *1* is probably a group of axons. A circular structure seen in **D** (*open arrow*) occurs in connection with a blood vessel (see also **C**), and is not contained in the other sections. It may be the end of an intrafusal fibre from another spindle arranged in a tandem-like fashion. Staining for a mitochondrial enzyme gives an intense reaction in all fibres and does not allow distinguishing of types. The difference between the adjacent extrafusal fibres is pronounced. The ATPase reaction after alkaline preincubation (**C**) is strong in all suspected nuclear chain fibres and negative in both suspected nuclear bag fibres (*1* and *2*). Acid preincubation (**D**) leaves the reaction more or less negative in fibre *1*, which thus behaves like a slow tonic bag_1 fibre. Fibre *2*, however, shows reversal of the reaction and thus stains like the extrafusal type I fibre (*I*). The chain fibres *3*, *4*, *6*, *7* lose their reaction after acid preincubation and behave like extrafusal type II fibres. Fibre *5* reacts both after alkaline and acid preincubation, i.e. it is a IIC fibre and conforms to the histochemical profile described for bag_2 fibres, although its diameter suggests that it is a chain fibre. This plate illustrates the complexity of intrafusal fibre types in man, and also shows that the smallness of the fibres and the reactibility

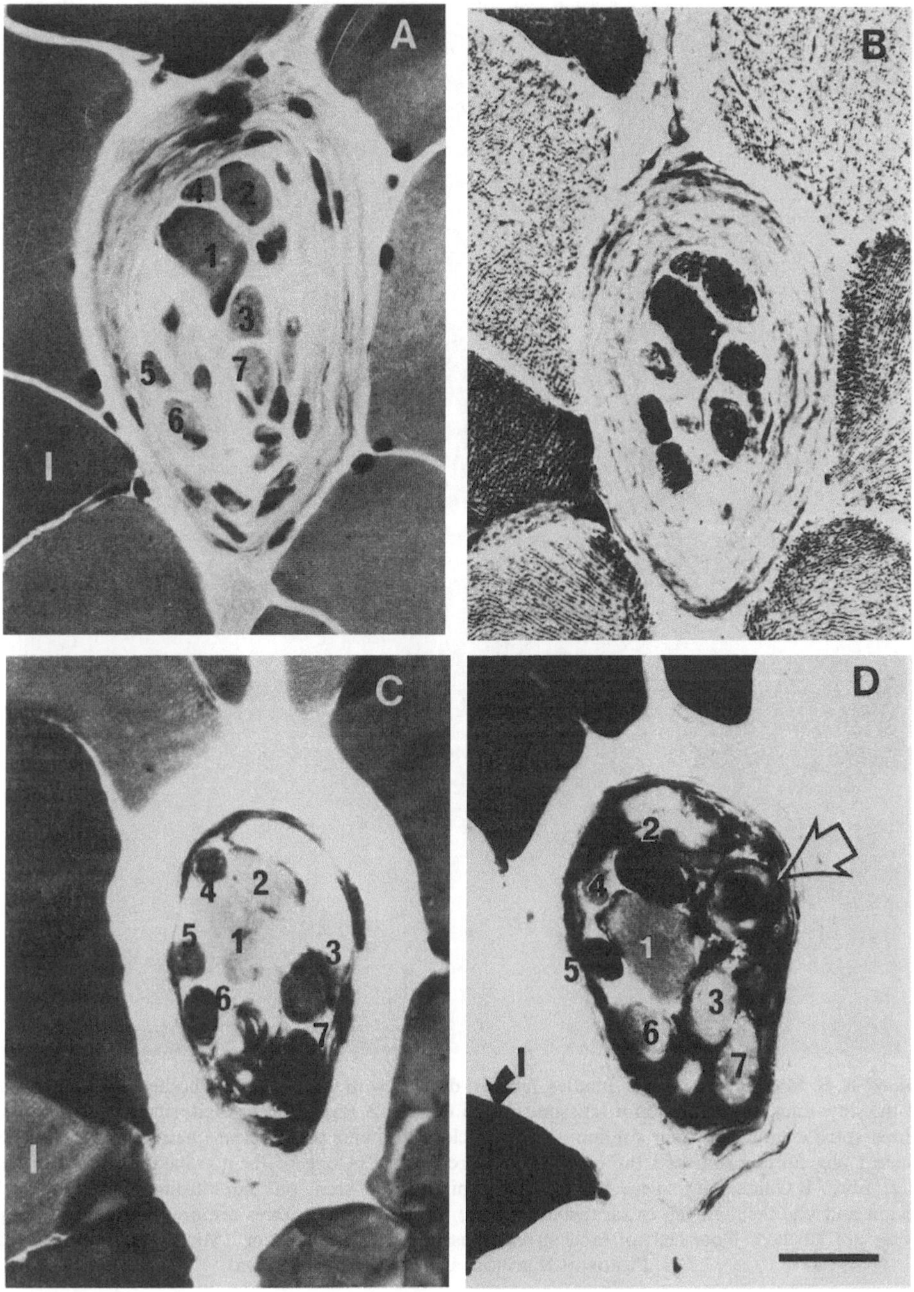

of other spindle structures makes the interpretation of the results rather uncertain. Of the seven fibres shown, one is a typical bag_1 fibre ("slow-tonic") and four are typical chain fibres (type II, "fast-twitch"), one reacts like a bag_2 fibre but resembles with respect to size a chain fibre, and one looks like a bag fibre but stains like an extrafusal type I fibre and differs from all intrafusal fibre types described. Bar, 25 μm. (Preparation courtesy of Dr. Kamieniecka, Copenhagen)

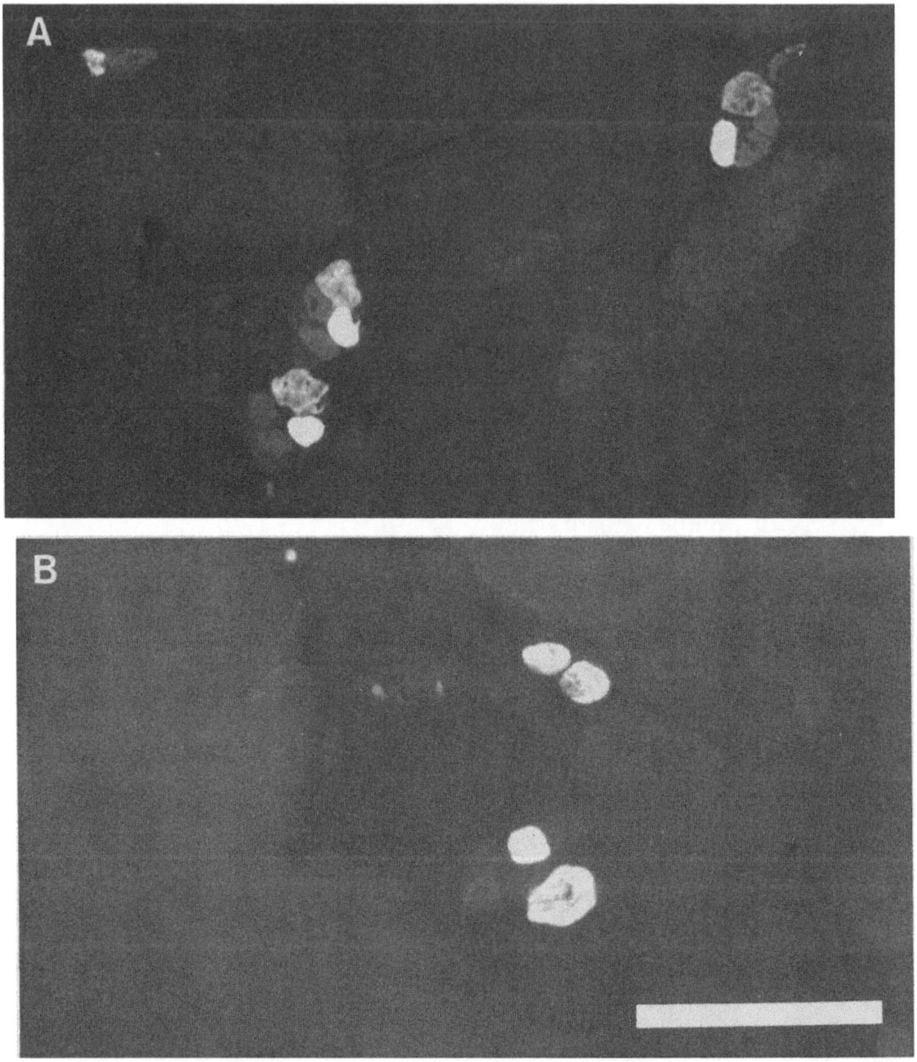

Fig. 97 A, B. Mammalian muscle spindles, labelled directly with fluorescent antibodies against myosin of the slow-tonic chicken anterior latissimus dorsi muscle. **A** Rat extensor digitorum longus muscle. Three spindles are seen each containing two nuclear bag and two nuclear chain fibres. Only the nuclear bag fibres react, and the reaction is more intense in one of them, which probably is the bag_1 fibre. **B** Guinea pig soleus muscle. Two spindles are seen, one cut through the equatorial region and one through the extracapsular region, where only bag fibres are present. Only the bag fibres are labelled. Note lack of label in extrafusal fibres. Bar, 100 µm. (Micrograph courtesy of Drs. PIEROBON-BORMIOLI and SCHIAFFINO, Padova)

vided into bag_1 (typical bag) and bag_2 (intermediate bag) fibres. Only the bag_1 fibres stain for ATPase like slow fibres, and only bag_1 fibres react intensely with antibodies against myosin of the (non-twitch) ALD of chicken (PIEROBON-BORMIOLI et al. 1980) (Fig. 97).

b) Efferent Innervation

Mammalian intrafusal fibres are innervated by γ-fusimotor axons and – to a lesser extent – by β-skeleto-fusimotor axons which are collaterals of motor axons to extrafusal fibres. The motor axons to intrafusal fibres are, according to physiological criteria, classified as "static" or "dynamic". Stimulation of "static" fusimotor axons elicits a static response from the spindle afferents, and diminishes a dynamic response elicited by stretching the spindle. Stimulation of "dynamic" fusimotor or skeleto-fusimotor axons enhances the dynamic response to stretch.

Since there are fusimotor axons influencing either the dynamic or the static receptor response, and since there are two main types of intrafusal muscle fibres, it was postulated that each type of intrafusal muscle fibre with its innervation accounted for dynamic or static responses such that there are two anatomically and physiologically distinct receptor systems (BOYD 1962a, b). This view seems to harmonize with the fact that snake muscle spindles with different morphology are static or dynamic receptors (see above).

The neuromuscular junctions on nuclear chain fibres have a more complicated subneural apparatus and more distinct subsynaptic folds than do those on nuclear bag fibres; most synapses are found on the intracapsular polar region (rat and man, RUMPELT and SCHMALBRUCH 1969; cat, BARKER et al. 1972; mouse, EDWARDS 1975; sheep, HARKER 1972b).

Light microscopy of silver-impregnated specimens shows both en plaque (plate) and en grappe (trail, spray-like, grape) endings with multiple terminals; the proportion of these endplate types to the intrafusal fibre types is a matter of dispute. BOYD (1962a, b) thought that trail endings are present only on nuclear chain fibres, and plate endings only on nuclear bag fibres; BARKER (1962), however, maintained that both fibre types received plate endings. JONES (1966) found, in the opossum, plate and trail endings on both fibre types; sometimes branching axons supplied nuclear bag and nuclear chain fibres simultaneously. BARKER (1966) subdivided the plate endings into p_1- and p_2-plate endings, supplied by skeleto-fusimotor β-fibres and fusimotor γ-fibres, respectively. The p_1-plate endings were always extracapsular. MAYR (1969) found in rat spindles that all p_1-plate endings, one per spindle pole, were on nuclear chain fibres; each bag fibre had up to 2×3 p_2-plate endings. Trail endings were close to the equatorial region and were predominantly on nuclear bag fibres. According to BARKER et al. (1972), two-thirds of the motor axons formed trail endings, the rest formed p_1- and p_2-plate endings. These authors believed that some trail endings belonged to non-myelinated fibres, and that not all intrafusal muscle fibres received trail endings. In human muscle spindles, according to SWASH and FOX (1972), trail endings were preferentially on nuclear chain fibres. The plate endings were overwhelmingly on nuclear bag fibres which may have up to 3 p_1-plate endings per pole. The extracapsular p_1-plate endings were large and difficult to distinguish from motor endings on extrafusal fibres.

It is impossible to derive a consistent picture from these contradictory results. The only conclusion that can be drawn is that plate and trail endings are not exclusively on nuclear chain or nuclear bag fibres. To establish the relations between (a) the fibre type (bag_1, bag_2, chain) (b) the morphological type of motor nerve endings (p_1, p_2, trail), and (c) the physiologically defined type of motor axon (static, dynamic), the following experiments have been performed.

1. All nerve fibres to a muscle, with the exception of one γ-axon, are severed, and, after the nerve has degenerated, the one surviving axon and its ending are morphologically and physiologically characterized (BARKER et al. 1973).

2. A physiologically identified axon is repetitively stimulated to mark the intrafusal fibres it innervates using the glycogen depletion method (BARKER et al. 1976b, 1977; HARKER et al. 1977; EMONET-DENAND et al. 1980).

3. The response of intrafusal fibres to stimulation of a physiologically identified fusimotor axon is recorded intracellularly, and the fibre is then labelled by intracellularly injecting Lucifer yellow, a fluorescent dye (BARKER et al. 1978).

4. The contractile response elicited by stimulation of a physiologically classified γ-axon is filmed (BOYD and WARD 1975; BOYD et al. 1977).

In all cases the cat's tenuissimus and peroneus brevis muscles are used.

The results are as follows: Each static γ-axon supplies 3–8 spindles and distributes only trail endings to both nuclear bag and nuclear chain fibres, usually to one spindle pole only. Each trail ending consists of a mean of 11 terminals on nuclear bag fibres, and of 7 terminals on nuclear chain fibres (BARKER et al. 1973). Repetitive stimulation of static axons depletes 92% of the chain, 67% of the bag_2, and 36% of the bag_1 fibres (peroneus brevis muscle, EMONET-DENAND et al. 1980). In the tenuissimus muscle, bag_1 and bag_2 fibres are depleted equally; the same static γ-axon usually activates nuclear bag and nuclear chain fibres within a spindle, but sometimes it activates nuclear bag or nuclear chain fibres only. Hence there is no evidence for a selective static innervation of any of the fibre types, but trail endings invariably belong to static axons.

Dynamic γ-axons deplete almost exclusively bag_1 fibres (BARKER et al. 1976b). Also dynamic β-axons, which are collaterals of nerve fibres to extrafusal slow-twitch fibres, innervate almost exclusively bag_1 fibres (BARKER et al. 1977).

Static γ-axons elicit propagated action potentials in nuclear chain fibres and non-propagated junctional potentials in bag_2 fibres. Dynamic γ-axons always elicit non-propagated junctional potentials. Electron micrographs of dye-labelled fibres have shown that trail endings of static γ-axons have subsynaptic folds when they are on chain fibres, but not when they are on bag_2 fibres. Dynamic β-axons, with rather slow conduction velocity, terminate on bag_1 fibres at p_1-plate endings without postsynaptic folds (BARKER et al. 1978). Fast-conducting skeleto-fusimotor β-axons are static and terminate on long nuclear chain fibres (HARKER et al. 1977).

Repetitive stimulation of γ-axons produces slow contractions in bag_1, moderately fast contractions in bag_2, and fast contractions in chain fibres. Usually only one of the spindle poles responds. Most spindles have two bag fibres and receive one dynamic γ- or β-axon, activating only one of the bag fibres (the "dynamic bag fibre", probably identical with the bag_1 fibre). The other bag fibre (probably bag_2) and the chain fibres are activated by static γ-fibres. Most spindles receive three static nerve fibres, two of which selectively activate the "static bag fibre" (probably bag_2) and the chain fibres, respectively. The third static γ-fibre is non-selective and sends branches both to chain and bag_2 fibres (BOYD and WARD 1975; BOYD et al. 1977).

The results presented by BOYD and co-workers (1977) indicate that the nuclear bag_1 fibres are selectively innervated by dynamic axons, whereas nuclear bag_2 and chain fibres all receive static axons. This does not agree with the anatomical observation that static trail endings occur indiscriminately on all fibres, and that stimulation of static γ-axons depletes bag_1 and bag_2 fibres equally of glycogen (BARKER et al. 1973, 1976b). BARKER et al. (1976b) argue that BOYD and co-workers (1977) may have overlooked slow contractions of bag_1 fibres when adjacent chain fibres contracted vigorously. This may be the case. Nevertheless, the location of a silver-stained synapse on very thin muscle fibres studied in a teased specimen is not always unequivocal; the glycogen depletion method is subject to error as well, in particular for fatigue-resistant fibres which may switch to oxidative metabolism. Bag_1 fibres are inherently poor in glycogen; stimulation of dynamic β-axons depletes only segments of the bag_1 fibres which might be explained by local activation (BARKER et al. 1977). The explanation, however does not apply to the co-contracting extrafusal twitch fibres which were segmentally depleted as well.

Using a modified stimulation technique, DECORTE et al. (1984) observed that the glycogen depletion of bag_1 fibres by stimulation of static γ-axons is not constant. They attribute the loss of glycogen to unknown non-neural mechanisms, and thus confirm the results of Boyd and co-workers.

KUCERA (1985) investigated the motor innervation of the bag_1 fibres of the cat tenuissimus muscle in serial transverse sections for light and electron microscopy. Of 71 axons, 66 terminated on bag_1 fibres only, whereas 5 co-innervated a chain fibre. The structure of the terminals of γ- and β-axons as seen in cross-sections did not differ. By contrast, in a study using 1-µm serial sections for light microscopy, non-selective motor innervation was found in spindles of the rat soleus muscle; bag_1, bag_2, and chain fibres were co-innervated (WALRO and KUCERA 1984).

The original model of distinct transmitter systems for static and dynamic signals with different viscoelastic properties was probably too simple (BOYD et al. 1977). At least for cat, it seems certain that bag_1 fibres preferentially receive dynamic, and bag_2 and chain fibres preferentially receive static axons. BAKKER and RICHMOND (1981) report that in neck muscles of cat, one-third of the spindles are devoid of bag_1 fibres. Whether this implies that these spindles have a purely static function is unknown.

The comparison of intrafusal and extrafusal muscle fibres uncovers several differences and even paradoxes. Intrafusal fibres are at the same time multiply (trail endings) and polyneurally innervated. The slow bag_1 fibres receive dynamic β- and γ-axons, which have larger diameters and conduct faster than the static axons to fast nuclear chain fibres. The fast conducting dynamic axons terminate on plate endings, which lack postsynaptic folds, whereas the multiple (trail) endings of thin and slowly conducting axons on fast chain fibres have postsynaptic folds.

c) Branching Intrafusal Fibres?

The slow extrafusal fibres of the orbital layer of sheep extraocular muscles may branch, and a branch while passing through the capsule of a muscle spindle may become a nuclear bag fibre with sensory innervation (HARKER 1972a, b; BARKER et al. 1972). Branching nuclear chain fibres have been observed in other muscles as well, at least using light microscopy (BARKER and GIDUMAL 1960, 1961; BOYD 1960). Electron microscopy shows that two adjacent chain fibres often become closely associated; they may share the basal lamina, or a sensory terminal, or a satellite cell. In the light microscope, this simulates branching, but true cytoplasmic connections between fibres have never been found (HENNIG 1969; ADAL 1969; CORVAJA et al. 1969; SCALZI and PRICE 1972; BANKER and GIRVIN 1972). CORVAJA et al. (1967, 1969) saw desmosomes between the attached fibres, but no gap junctions suggesting electrotonic coupling.

III. Laryngeal Muscles

Laryngeal muscles of dog, cat, monkey, rabbit, and sheep, have been shown to contract faster than skeletal muscles. The vocalis muscle is the fastest one of the laryngeal muscles; the isotonic twitch contraction time of the thyroarytenoid is only half as long as that of the cricothyroid or posterior cricoarytenoid muscles (MÅRTENSSON and SKOGLUND 1964).

The cricothyroid muscle of rabbit consists of different fibre types with different activities of the mitochondrial enzymes; the isometric twitch contraction time is comparable to that of a fast limb muscle (HALL-CRAGGS 1968). In rat, 85% of the fibres are type II; the isometric twitch contraction time (7 ms) is longer than that of the posterior cricoarytenoid (3.4 ms), which contains only type II fibres. Both muscles are resistant to fatigue. Mitochondria occupy 10%–15% of the cell volume. The SR is more abundant in the faster posterior cricoarytenoid than in the somewhat slower cricothyroid muscle. The fibre diameter is 30–40 µm in both muscles, and according to Z-line and M-line structure and endplate morphology all fibres are twitch fibres (HINRICHSEN and DULHUNTY 1982).

HALL-CRAGGS (1968) reports that the thyroarytenoid of rabbit (comprising also the vocalis) consists of uniformly large fibres (50- to 80-µm diameter) with high mitochondrial content and rich vascularization. The isometric twitch contraction time is 7 ms. SCHMALBRUCH (1971) finds in the same species that the fibres of the vocalis muscle are only 16 µm thick, compared with 31 µm in the thyroarytenoid muscle. The vocalis muscle has 3,400 capillaries/mm^2 cross-sectional area, compared with 860/mm^2 in the thyroarytenoid (see Sect. B. V). The thyroarytenoid and vocalis muscles of rat consist entirely of type II fibres which stain intensely for oxidative enzymes. The thyroarytenoid, but not the vocalis muscle, reacts intensely for glycolytic enzymes as well (KAMIENIECKA, unpublished) (Figs. 98, 99).

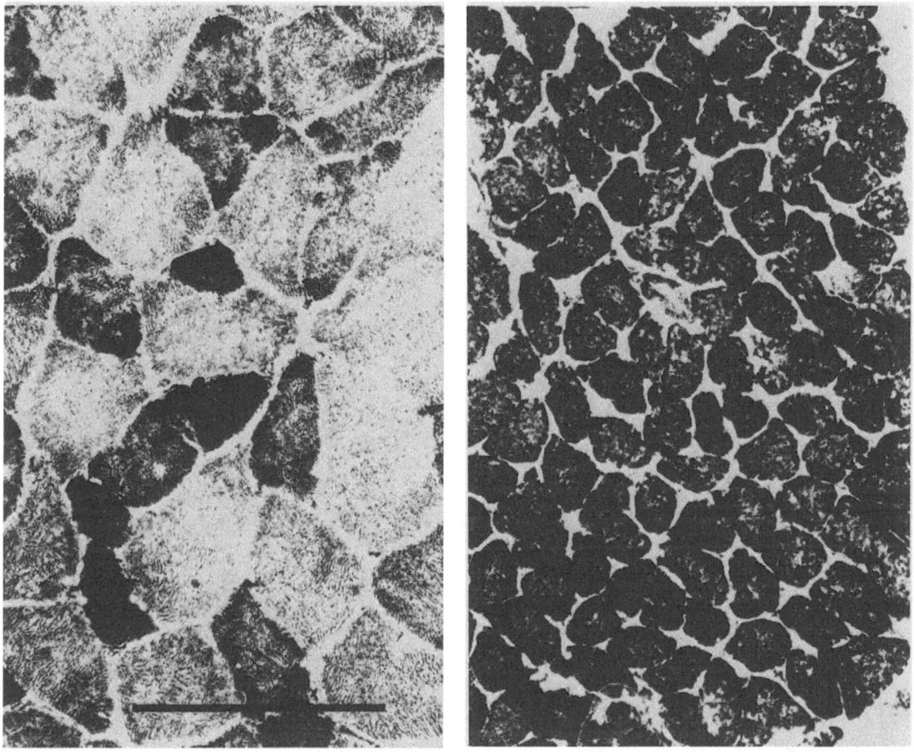

Fig. 98. Rat cricothyroid (*left*) and vocalis (*right*) muscles, stained for nicotinamid adenine dinucleotide-linked lactic dehydrogenase, a mitochondrial enzyme. Note difference in fibre type composition. The vocalis muscle is a fast-"red" muscle, the physiological properties of the cricothyroid are comparable with those of fast skeletal muscles. Bar, 100 μm. (Preparation courtesy of Dr. KAMIENIECKA, Copenhagen)

Electron micrographs of the rabbit thyroarytenoid muscle disclose a richly developed SR and many triadic junctions; the mitochondria are large and are longitudinally arranged between the myofibrils. There is no evidence for transverse mitochondrial systems (SCHMALBRUCH 1971) (Fig. 100). The array of mitochondria resembles that in heart muscle cells. Micrographs of the rat cricothyroid muscle, however, show pairs of mitochondrial profiles at the Z-line level (HINRICHSEN and DULHUNTY 1982), suggesting that they belong to transverse mitochondrial networks.

The large cricothyroid muscle of bat is probably the fastest mammalian muscle. It contains abundant SR and triadic junctions. This muscle is involved in the modulation of the ultrasonic cries used in echolocation. From the frequency of sound modulation it has been calculated that the total twitch duration must be less than 4 ms; the muscle produces up to 200 unfused twitches/s. The mitochondria are large and long isolated bodies (REVEL 1962). Endplates have elaborate subsynaptic folds. CHO et al. (1972) find many "tubular aggre-

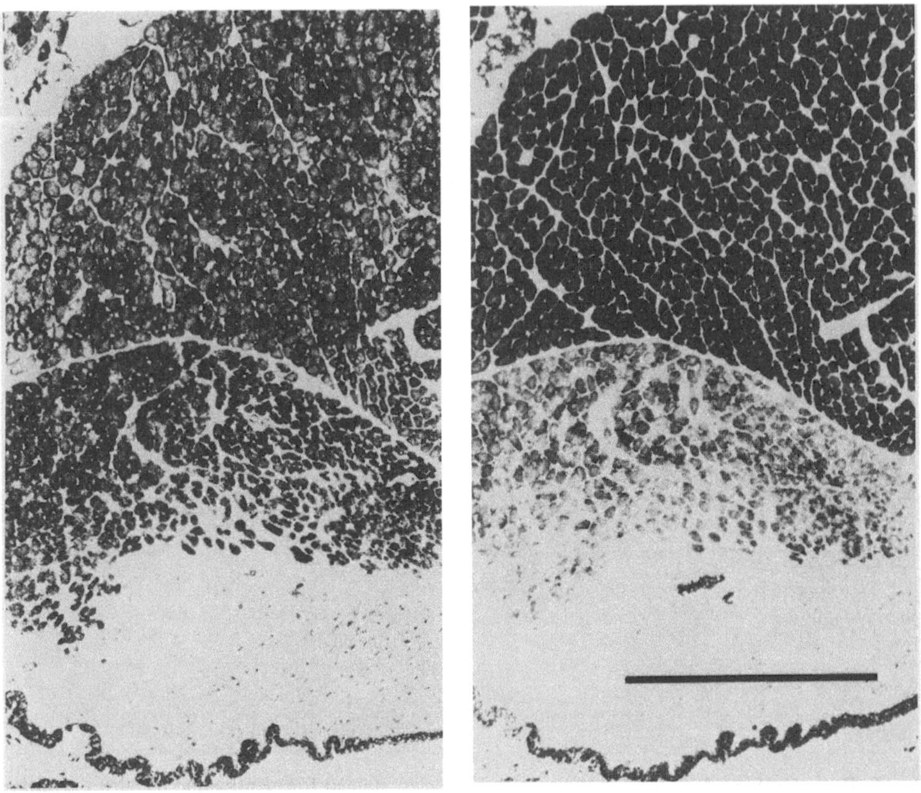

Fig. 99. Rat vocal cord, showing epithelium (*bottom*). The vocalis muscle proper (*bottom*) and the thyroarytenoid muscles are distinct. Serial sections stained for nicotinamide adenine dinucleotide-linked lactic dehydrogenase (*left*) and for menadione-linked alpha-glycerophosphate dehydrogenase (*right*). The fibres of the vocalis muscle are thinner, but no difference compared with the thyroarytenoid muscle is seen with respect to staining for the mitochondrial enzyme. Staining for the glycolytic enzyme reveals that the vocalis muscle is purely oxidative, whereas the thyroarytenoid has a high glycolytic activity as well. Note that the vocalis muscle has no sharp boundary towards the epithelium. Bar, 0.5 mm. (Preparation courtesy of Dr. KAMIENIECKA, Copenhagen)

gates" in this muscle; these structures are well known in muscle pathology. They originate from excess SR.

ANZENBACHER and ZENKER (1962) find in human thyroarytenoid muscles stained for AChE that the endplates are not concentrated in an endplate zone. They suspect that many small additional reactive sites distributed along the fibres are terminals of γ-axons. ZENKER (1966), in a subsequent study of laryngeal muscles obtained during surgery instead of at autopsy, identifies the reactive sites outside the endplate zone as myotendinous junctions. There is no other histological and no physiological evidence for the presence of a slow fibre system in laryngeal muscles.

In tree frogs, the larynx muscles are faster than skeletal muscles and there are many large endplates. The ultrastructure of all fibres is that of twitch fibres;

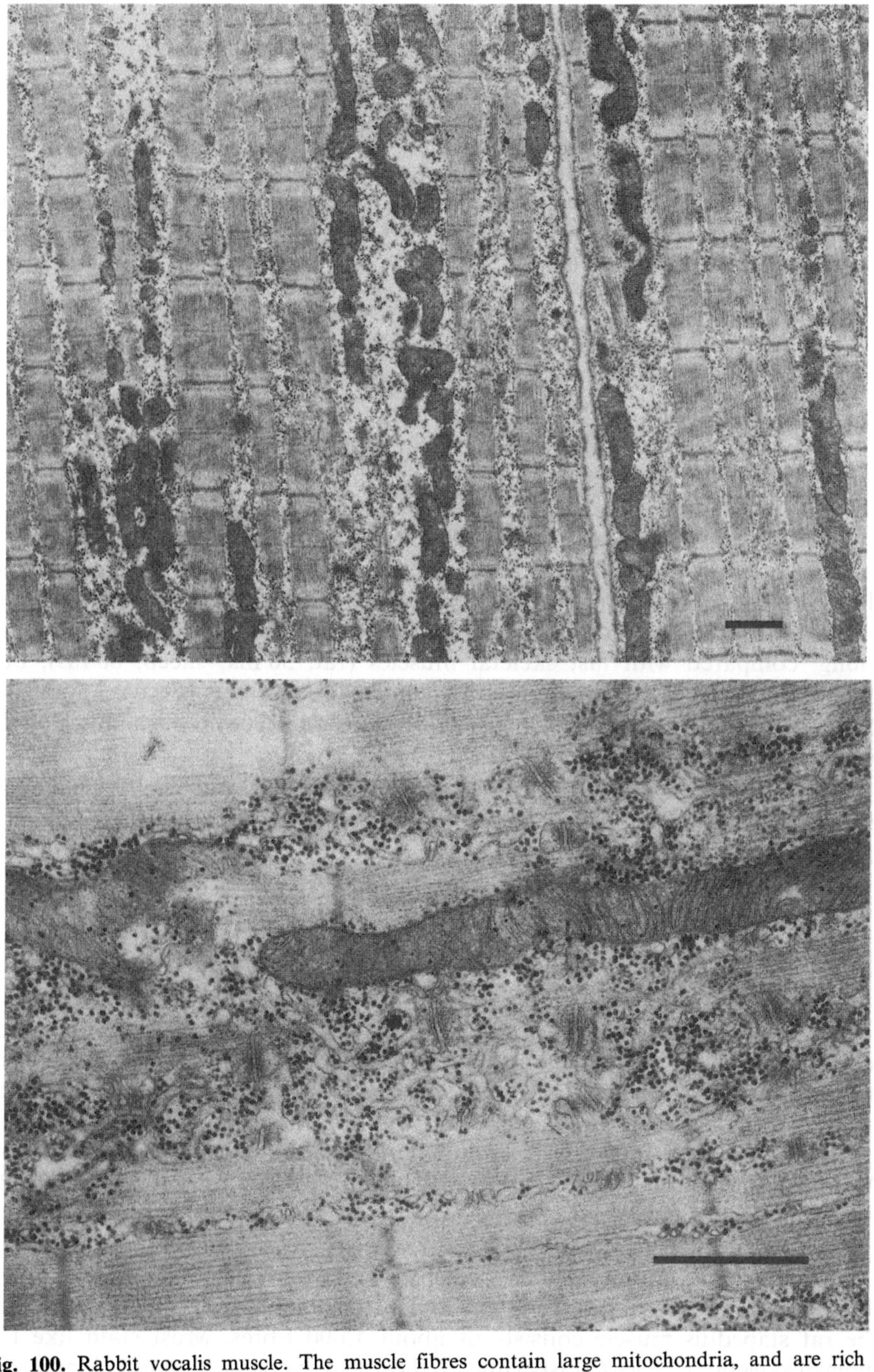

Fig. 100. Rabbit vocalis muscle. The muscle fibres contain large mitochondria, and are rich in sarcoplasmic reticulum and T tubules. Note that the mitochondria do not occur in doublet profiles on both sides of the Z line; they are in the form of longitudinally oriented bodies. A similar array is found in cardiac muscle cells. Bars, 1 μm. (From SCHMALBRUCH 1971)

no difference between the different laryngeal muscles has been found. In male, but not in female frogs, the fibres contain large mitochondria and fat droplets; also the synaptic area of the motor endplate is larger in males than in females. This difference is believed to reflect the high frequency of mating calls produced by the male frogs (EICHELBERG and SCHNEIDER 1973, 1974).

There is uncertainty with respect to muscle spindles in laryngeal muscles. Most authors do not mention them, and I have never seen any in laryngeal muscles of rat, rabbit, and man. KEENE (1961) reports that there are spindles in all muscles of the human larynx.

IV. The Oesophageal Muscle

All cross-striated fibres of the oesophagus stain intensely for ATPase at pH 9.4 (type II) and moderately for mitochondrial enzymes. Prolonged stimulation of the vagus nerves readily depletes the fibres of glycogen (GRUBER 1978). The fibres are focally innervated and have endplates with extensive subsynaptic folds. Also the ultrastructure of the fibres suggests that all are fast-twitch fibres (rat, GRUBER 1968; sheep, FLOYD 1973). The isometric twitch contraction time is long, compared with fast skeletal muscles (rat, 30 ms; sheep, 80 ms), but the fibres show post-tetanic potentiation similar to fast fibres (FLOYD and MORRISON 1975; ASMUSSEN and GAUNITZ 1978). GRUBER (1968) reports that muscle spindles are lacking.

In contrast to oesophageal muscle fibres, pharyngeal muscle fibres are unusually rich in mitochondria (LEESON and LEESON 1969).

V. Inner Ear Muscles

The tensor tympani muscle of rabbit consists of a central part with rather thick fibres (15- to 35-μm diameter) and a peripheral rim of thin fibres (10- to 12-μm diameter). Almost all fibres are type II; the activity of mitochondrial enzymes is high and inversely related to the fibre diameter. The few ATPase-negative fibres stain intensely for mitochondrial enzymes and are type I fibres (ASMUSSEN and WOHLRAB 1971). The cat tensor tympani muscle contains 60% type II fibres and 40% type I fibres, ranging in diameter from 5 to 25 μm and from 5 to 15 μm, respectively (TEIG and DAHL 1972).

The stapedius muscle of cat contains 80% type II fibres and 20% type I fibres, the diameters are 5 to 40 μm and 5 to 20 μm (TEIG and DAHL 1972). The rat stapedius muscle consists of about 1,000 fibres. Most stain like IIB fibres of skeletal muscles ("fast-white"), the mean diameter is 14 μm. A minority of fibres are type I fibres, the mean diameter being 10 μm (VITA et al. 1983). The fibres of the guinea pig stapedius muscle have a mean diameter of only

6 μm. Two-thirds of the fibres resemble myotubes, have central nuclei, and stain for ATPase at pH 9.4 like type IIC fibres (no inhibition at pH 4.35), one-third of the fibres appears more mature and stains like type I fibres. All fibres have rather wide Z lines, distinct M lines, and are focally innervated (BURGENER and MAYR 1980).

No fibres, histochemically resembling slow fibres have been observed in the stapedius or tensor tympani muscles of any species studied. This contrasts with the previously presented evidence for slow fibre systems in both muscles of cats (see Chap. E).

The cat stapedius muscle is a purely fast muscle and consists of 87 fast-twitch motor units with isometric twitch contraction times of 14–39 ms. The tensor tympani muscle contains 92 fast-twitch motor units with isometric twitch contraction times of 23–40 ms, and 40 slow-twitch motor units with contraction times of 58–92 ms (TEIG 1972b). The absence of slow-twitch units in the stapedius muscle with 20% type I fibres may be due to the physiological sampling technique (TEIG and DAHL 1972). Tensor tympani and stapedius muscles in rabbit physiologically resemble the corresponding muscles in cat (TEIG 1972a).

In rabbit (MALMFORS and WERSÄLL 1960a, b) and cat (BLEVINS 1963, 1964), the number and size of the motor units were determined by counting the nerve and muscle fibres. After a correction for sensory fibres, the results indicated that each muscle contains many hundred motor units with less than 10 muscle fibres each. This is clearly not compatible with the single motor unit study by TEIG (1972b). The reason for the discrepancy is obscure (see also Chap. H). Several workers labelled and counted stapedius muscle motoneurons using horseradish peroxidase. The findings in cat range from 200 (LYON 1978) to 670 (SHAW and BAKER 1983) and 1,150 (JOSEPH et al. 1985). These figures would include γ-motoneurons if there are muscle spindles in the stapedius muscle (see below). SHAW and BAKER (1983) and JOSEPH et al. (1985) relate their results to nerve fibre counts rather than to the single motor unit study of TEIG (1972b). In monkeys, THOMPSON et al. (1985) never find more than 100 stapedius motoneurons which always form one discrete group, whereas, in the studies in cat, labelled cells occurred in a variety of places. THOMPSON et al. (1985) state that the discrepancies should be attributed to variations in the application and containment of horseradish peroxidase rather than to species differences.

MALMFORS and WERSÄLL (1960b) find muscle spindles in the rabbit stapedius muscle, which is not confirmed by VITA et al. (1983). DAVID et al. (1966) explicitly state that the guinea pig stapedius muscle is devoid of spindles.

VI. Mandibular Muscles

These muscles appear macroscopically red; the mitochondrial content is high and most fibres react intensely for ATPase at pH 9.4 (IIA fibres, "fast-red") (monkey, CLARK and LUSCHEI 1981; MAXWELL et al. 1981; cat, BOSLEY et al. 1972; TAYLOR et al. 1973; mouse, guinea pig, and rabbit, SCHIAFFINO 1974). Nevertheless, species which masticate regularly and slowly (e.g. sheep) may have a predominance of type I fibres (ROWLERSON et al. 1981), presumed to be slow-twitch fibres.

Staining for ATPase at pH 9.4 discloses that there are not only type I and II fibres, but also fibres with intermediate reaction (cat, TAYLOR et al. 1973; man, RINGQVIST 1973; RINGQVIST et al. 1982). The fibre type composition of human muscles has been thoroughly studied (ERIKSSON et al. 1981; 1982; ERIKSSON 1982; ERIKSSON and THORNELL 1983). Using immunocytochemistry, it has been shown that the fibres with intermediate ATPase activity contain both slow-twitch and fast-twitch myosins, whereas type IIA and type I fibres react with fast- or slow-twitch myosin antibodies only (THORNELL et al. 1984).

The jaw-closing muscles of cat and dog are rich in type II fibres which, however, fail to lose their reactivity for ATPase at pH 9.4 after acid preincubation (pH 4.3). The posterior temporalis muscle consists exclusively of these fibres; the masseter and anterior temporalis muscles may contain up to 50% type I fibres, depending on the muscle region. These peculiar type II fibres resemble IIC fibres; they have been termed II M (M for masticatory) fibres (BOSLEY and ROWLERSON 1980). The myosin of IIM fibres isolated from the posterior temporalis muscle has a much higher ATPase activity than myosin from a fast hindlimb muscle, and also the molecular structure of the light and heavy chains is different. Therefore, antibodies against "fast-red" masticatory myosin of cat and dog do not bind to fast fibres in hindlimb muscles (ROWLERSON et al. 1981). In guinea pig, however, the fast "red" fibres (IIA) of masticatory and limb muscles contain similar myosin isoenzymes (DALLA LIBERA et al. 1980; PIEROBON-BORMIOLI et al. 1981) (Figs. 81, 82).

The isometric twitch contraction times of the jaw-closing muscles of cat are 11–13 ms, with fusion frequencies of 300–500 Hz. These muscles are faster than fast limb muscles. Surprisingly, in spite of their high mitochondrial content, the fibres fatigue rather rapidly (TAYLOR et al. 1973).

VII. Facial Muscles

Little is known about facial muscles. The human platysma contains only 20% fibres rich in mitochondria (supposed to be slow-twitch fibres) and the isometric twitch contraction time is shorter (48 ms) than that of a hand muscle (adductor pollicis, 65 ms); the fibres are thinner than those in limb muscles (BUCHTHAL and SCHMALBRUCH 1970; KRARUP 1977).

In cat, the orbicularis oculi and orbicularis oris muscles contain 85%–90% type II fibres, which occupy 95% of the cross-sectional area of the muscles. The fibres of the orbicularis oris muscle react more intensely for mitochondrial enzymes and are more resistant to fatigue than those of the orbicularis oculi muscle. The isometric twitch contraction times are 27–35 ms in the orbicularis oris and less than 10 ms in the orbicularis oculi muscle. This difference does not appear in the histochemical fibre type pattern; it probably reflects the role of the orbicularis oculi during protective reflexes (LINDQUIST 1973; EDSTRÖM and LINDQUIST 1973).

G. Development, Regeneration, Growth

I. An Overview

The understanding of myogenesis and muscle fibre regeneration has benefitted immensely from the methodological advancements made during the past 20 years, and deeply rooted errors could be corrected.

In 1839, SCHWANN described how the multinucleated muscle fibres were formed by coalescence of aligned mononuclear cells. Some authors contested this finding, and in 1861 WEISMANN claimed to have shown how each myofibre originated from one embryonic cell which underwent nuclear divisions without cytokinesis. The number of myofibres then increased by longitudinal splitting within the muscle spindles. KÖLLIKER, who published several authoritative handbooks on histology, initially agreed with SCHWANN (KÖLLIKER 1850), but later (KÖLLIKER 1863) criticized SCHWANN'S views, and strongly supported WEISMANN. Despite contrasting observations, for more than a century, textbooks taught that muscle fibre was a plasmodium originating from one cell, rather than a syncytium formed by a fusion of cells (for review, see HÄGGQUIST 1931, 1956; BOYD 1960; MURRAY 1960). SCHWANN'S (1839) findings were eventually confirmed (LASH et al. 1957; KONIGSBERG et al. 1960; CAPERS 1960; COOPER and KONIGSBERG 1961) and are now undisputed. Most impressive have been the in vitro studies of myogenesis already initiated by LEWIS and LEWIS (1917). KONIGSBERG (1963), SIMPSON and COX (1967, 1972), and YAFFE (1968) cloned myogenic cells of chicken, lizard, and rat muscles, and conclusively showed that single cells divide mitotically, then stop to proliferate, and fuse.

Mitotic figures had only been found during the early stages of fetal myogenesis, and never with certainty in multinucleated cells. Amitosis was the deus ex machina explaining how the number of myonuclei could increase inside a muscle fibre, and amitosis was widely accepted until about 1960 (e.g. ALTSCHUL and LEE 1960) or even longer (HESS and ROSNER 1970) (for review, see STOCKDALE and HOLTZER 1961). Myonuclei are without exception diploid (LASH et al. 1957; STREHLER et al. 1963) and do not synthesize DNA (LASH et al. 1957; HOLTZER et al. 1957, 1958; KONIGSBERG et al. 1960; COOPER and KONIGSBERG 1961; OKAZAKI and HOLTZER 1965, 1966; BISCHOFF and HOLTZER 1969). These two findings are incompatible with amitosis. Long-term observations of living fibres revealed that the dumb-bell-shaped myonuclei believed to indicate amitosis were moving myonuclei which regained their normal shape without dividing (CAPERS 1960; COOPER and KONIGSBERG 1961).

The temporal and causal relation between cell fusion, "switching off" of DNA synthesis, and "switching on" of synthesis of muscle specific proteins is a matter of dispute, but that these three events are somehow coincident is unquestioned (HOLTZER and SANGER 1972; HOLTZER et al. 1974, 1977; DIENSTMAN and HOLTZER 1975; BUCKLEY and KONIGSBERG 1974, 1977; KONIGSBERG et al. 1978).

Myogenesis in situ is less synchronized than in tissue culture. The myogenic cells are arranged in strands which in cross-sections appear as clusters. The fusing myoblasts form multinucleated myotubes which in contrast to myofibres have centrally placed myonuclei. The myotubes occur in generations, and each cluster contains one primary myotube which serves as backbone along which the secondary generation myotubes develop. The diameter of a cluster is about 20 μm, and the cells are ensheathed by a common basal lamina and separated from each other by membrane-bound clefts about 20 nm wide. It is impossible to identify individual cells using light microscopy. The clusters break up as the young fibres mature, and this process has been mistaken for splitting of muscle fibres. The nature of "splitting" during myogenesis was clarified by KELLY and ZACKS (1969a) and KELLY and SCHOTLAND (1972); its quantitative aspects have been studied by ONTELL (1977, 1979) and ONTELL and DUNN (1978). The number of myonuclei increases when daughter cells of dividing myogenic cells fuse with the growing fibres. This process accompanies fibre growth (MOSS and LEBLOND 1970, 1971). The mitotically active myogenic cells in mature muscles are the satellite cells described by MAURO (1961). Satellite cells are closely attached to muscle fibres; using light microscopy they can usually not be distinguished from myonuclei.

The uncertainty about myogenesis led to erroneous views on muscle regeneration, and also hindered a sensible interpretation of the histopathological changes in muscle tissue. KÖLLIKER (1863) maintained that muscle fibres are unable to regenerate. It is not without irony that the strongest evidence against KÖLLIKER'S views on myogenesis and regeneration came from the observation of diseased muscles (e.g. ZENKER 1864), and that now, after the syncytium theory is firmly established, remnants of the plasmodium theory still may be found in studies of muscle pathology (for discussion, see SCHMALBRUCH 1976a, b, 1979b, 1982b).

Regeneration of muscle fibres recapitulates fetal myogenesis, but in contrast to developing fibres which determine the array of the endomysium, fibre regenerates have to accommodate in a pre-existing framework of connective tissue. Hence, in most cases, a regenerated muscle appears histologically abnormal (SCHMALBRUCH 1976a, b, 1979b; ONTELL et al. 1982). When the muscle is completely destroyed, no regeneration occurs because then the myogenic stem cells are lost as well. In several lower vertebrates, epimorphic regeneration is possible and entire tissue systems regenerate (an amputated limb or a lost tail) (CARLSON 1973).

Originally, it was believed that injured fibres regenerated per continuum by budding from surviving fibre fragments (VOLKMANN 1893). When it became clear that myonuclei did not divide, and that continuous regeneration was not possible, it was proposed that the myogenic stem cells arose from dedifferentiation and fragmentation of the mature muscle fibres, i.e. that the myonuclei gathered cytoplasm and a plasma membrane, became mononuclear cells, and started dividing (HAY 1959, 1970, 1979; BINTLIFF and WALKER 1960; LEE 1965; LENTZ 1969; REZNIK 1969a, b; TERÄVÄINEN 1970; HESS and ROSNER 1970; BARBERIE 1970; HAY and DOYLE 1973). Others believed that myogenic cells were blood-borne (SLOPER et al. 1970). Nevertheless, the satellite cells (MAURO 1961) which have been found in all skeletal muscles of all species are resting myoblasts which are induced to proliferate after injury. Roughly one satellite cell is found for every 20 myonuclei, which suffices to replace all myonuclei within a few days; the generation time of myogenic cells is less than 24 h. Conclusive evidence for the satellite cell concept was presented by BISCHOFF (1974) and NAG and FOSTER (1981), who isolated myogenic cells from mature normal rat muscles under conditions preventing formation of new cell membranes. A prerequisite was that basal lamina material was dissolved, which agrees with the localization of the satellite cells beneath the basal lamina of the muscle fibres. BISCHOFF (1975) placed individual muscle fibres in a culture dish, and demonstrated how they necrotized, and how only the satellite cells survived and started to divide. No new evidence against the satellite cell concept has been presented.

Transverse growth of mature muscle during the late stages of development and also in response to training, is due to the increase of the diameter of the individual fibres and not due to hyperplasia. Only few papers claim formation of new fibres during training (e.g. SCHIAFFINO et al. 1979; HO et al. 1980). The new fibres usually resemble regenerated fibres; probably some of the overloaded fibres had necrotized in these studies.

Longitudinal growth of muscle fibres during normal development, and also under experimental conditions, is usually explained by the formation of additional sarcomeres which are added to the myofibrils at the myotendinous junctions (GOLDSPINK 1972). This is in harmony with the concept that "myocommata" (sarcomeres) are indivisible (HÄGGQUIST 1931), a view already disputed by HEIDENHAIN (1919). The matter has not yet been properly investigated, but there are experimental results suggesting that the fibres grow along the entire length.

In the following parts of this chapter the results pertinent to the problems raised shall be reviewed. Exhaustive reviews of the literature of the 1960s are found in FISCHMAN (1972) and MURRAY (1972).

II. Myogenic Cells

1. The Origin of Myogenic Cells

It is well established that the trunk musculature originates from the myotomic portions of the paired and segmentally arranged somites. The experimental evidence for this rule and its exceptions has been reviewed by HÄGGQUIST (1931, 1956) and BOYD (1960). Since then, there has been relatively little research on the embryonic aspects of muscle development. The musculature of the limbs was assumed to originate from the lateral plate (somatopleuric) mesoderm, i.e. to differentiate in situ from the limb bud mesenchyme. This problem was reinvestigated when a reliable marker for embryonic cells became available. All cells of quails differ from chicken cells by a prominent condensation of the chromatin, which allows identification of quail cells growing in chicken tissue. CHRIST and co-workers (1974, 1977) and CHEVALLIER and co-workers (1976, 1977a) produced chicken-quail chimeras. By grafting quail somites into chick embryos, it was shown that the myogenic cells of the limb, like those of the trunk, come from the somites; the skeletal elements, as well as the intramuscular connective tissue and the tendons originate from the somatopleura. The myogenic cells migrate, during early development, from the somites into the limb buds (JACOB et al. 1978, 1979; NEWMAN et al. 1981); the somatic plate mesoderm does not give rise to any cross-striated muscle fibres (CHRIST et al. 1978, 1979). Also the ventrolateral abdominal muscles of chicken are of somitic origin whereas the connective tissue derives from the somatopleura (CHRIST et al. 1983). A "double origin" has been found for extrinsic ocular muscles as well. In chicken, the connective tissue of the muscles is derived from the neural crest whereas the muscle blastema develops from the prechordal mesoderm (JACOB et al. 1984).

2. Myoblasts

a) Myoblasts In Vitro

The mononuclear cells that proliferate and then fuse to become part of a multinucleated myotube are traditionally called myoblasts. Myogenesis is characterized as much by the failure of the cells to enter the DNA synthetic phase of the cell cycle as it is by cell fusion and contractile protein synthesis. These potentialities must be present in the proliferating myoblasts although they are not yet phenotypically expressed.

The events during myogenesis were originally studied in vivo in fetal or regenerating muscles. Tissues are heterogeneous with respect to cell types, and myogenesis is an asynchronous process. Both these difficulties may be circumvented by using tissue culture systems and studying myogenesis in vitro. The usual way is to dissect muscle tissue from the animal, and to dissociate the cells by incubating small pieces in a solution of trypsin and collagenase (HOLTZER et al. 1957; KONIGSBERG 1963). The tissue is removed from the enzyme solution by centrifugation, and a cell suspension is plated out in an appropriate culture medium. The percentage of cells capable of undergoing myogenic differentiation depends on the age of the donor and on the tissue. Myoblasts from rat, mouse, chick, quail and lizard are readily cultured. Several authors have established cultures from human muscles as well (for reviews, see ASKANAS and ENGEL 1975a; WITKOWSKI and DUBOWITZ 1976; HAUSCHKA 1982). Muscle cultures with very pronounced differentiation may be obtained from neonatal rats and mice, and from chick embryos at 11 days, and from quail embryos at 9 days incubation. Primary cultures consist of myogenic and fibrogenic cells, the term fibrogenic probably covering a variety of cells which have in common only the fact that they are not myogenic (Fig. 101).

It is often desirable to increase the number of myogenic cells, relative to the non-myogenic ones, prior to culturing. This may be achieved by using a tissue particularly rich in myogenic cells (somites of 3-day chick embryos, HOLTZER and SANGER 1972), by cloning (KONIGSBERG 1963), or by separating myogenic and fibrogenic cells. The myogenic cells tend to be less adhesive than fibrogenic cells, and when a culture is briefly treated with trypsin, the myogenic cells detach more quickly from the surface of the disc. Myogenic cultures up to 90% pure have been obtained by repeated reculturing (YAFFE 1968; RICHLER and YAFFE 1970).

Chick myoblasts maintained under standard conditions start to fuse about 40 h after plating, and within 10 h 80% of the cells have fused (O'NEILL and STOCKDALE 1972a). Rat myoblasts start to fuse after 50 h and fusion proceeds over several days (HOLTZER et al. 1975c). Fusion of rat myoblasts is prevented and the cells continue to proliferate when they are grown in a medium containing high concentrations of fetal calf serum and embryonic extract. When the cells are then transferred to a standard medium, after about 18 h, a burst of fusion occurs (YAFFE 1971; YAFFE et al. 1972). "Growth" and "fusion" media have been described for lizard myoblasts as well; the delay after switching from growth to fusion medium is 24 h (SIMPSON and COX 1972). Fusion is also prevented by reducing the Ca^{2+} concentration of the medium; the Ca^{2+} concentration needed for fusion is higher than the Ca^{2+} concentration for replication (SHAINBERG et al. 1969; PATERSON and STROHMAN 1972). Fusion starts within 2 h after Ca^{2+} is added. Cytochalasin B or D inhibits cytokinesis of the proliferating cells, and a large number of binucleated cells occur which do not fuse. Also this effect is reversible (HOLTZER et al. 1971, 1975a, b; MIRANDA and GODMAN 1973; CROOP and HOLTZER 1975).

KONIGSBERG (1963) and YAFFE (1968) maintained myogenic cells for a long time in culture; the cells continued to proliferate and did not lose their capacity to differentiate. A certain number of cells was continuously "lost" because they became postmitotic and fused. Hence, the cell line was maintained by obviously differentiated precursor cells which, however, did not express their differentiation potentiality. The differentiation in the embryo proceeds from a mesenchymal cell with many potentialities to the committed muscle cell, and it is reasonable to assume that this process involves intermediate stages.

b) Stages of Differentiation

HOLTZER and co-workers (BISCHOFF and HOLTZER 1969; BISCHOFF 1970; HOLTZER and SANGER 1972; HOLTZER et al. 1974, 1977; DIENSTMAN and HOLTZER 1975, 1977) proposed and elaborated a provocative hypothesis for the progressive differentiation of cells. This concept had great impact on the research on myogenesis, independent of whether it was accepted or contested. The core of this hypothesis is the "quantal cell cycle" (Fig. 102); differentiating cells are in developmental compartments, and a cell must pass through a mitotic

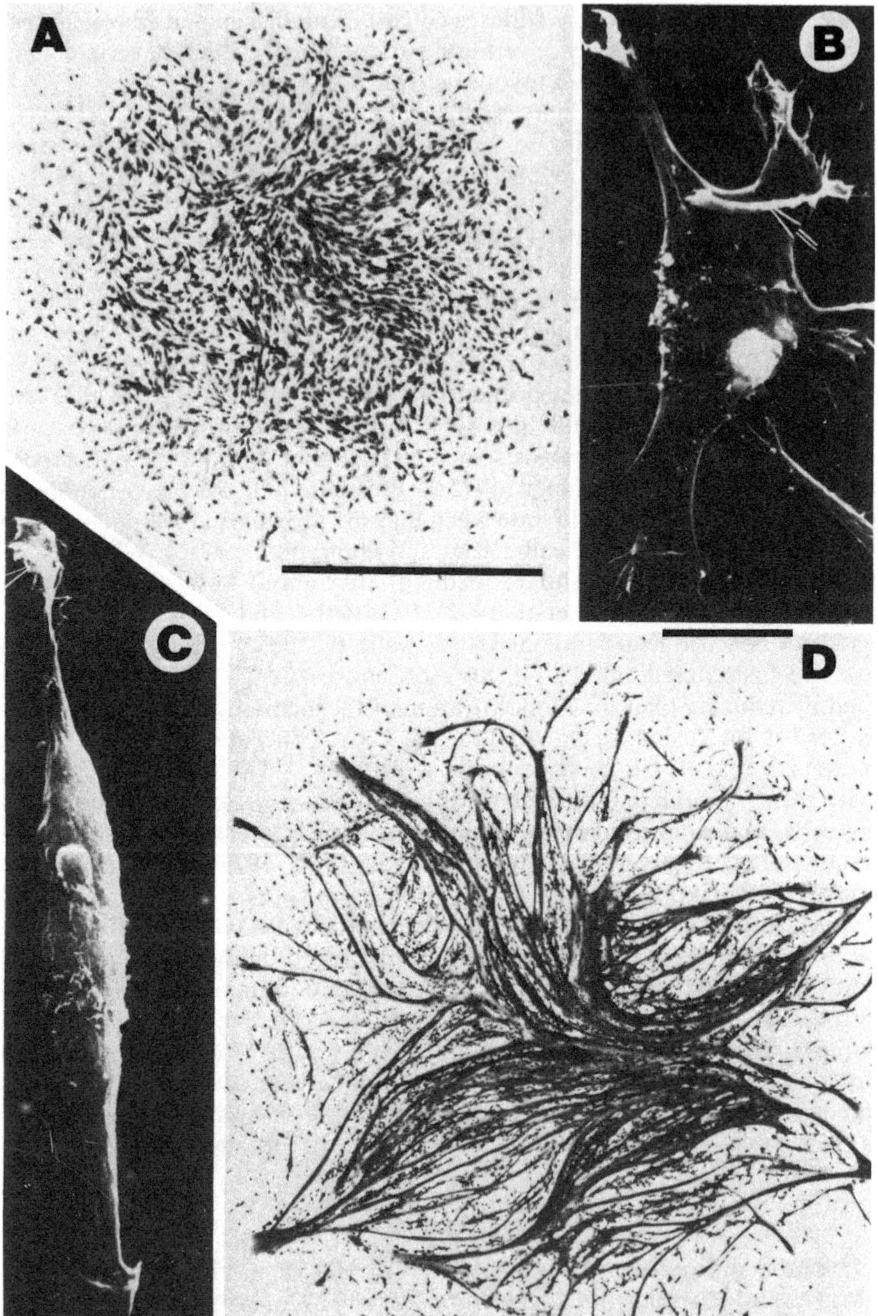

Fig. 101. **A** Chick fibroblast-like cell colony grown for 7 days in clonal density culture. The colony appears "smooth" because numerous cells are flattened. **B** Scanning electron micrograph of a fibroblast-like cell. The cell is flattened, has several projections, and multiple sites are ruffled (*arrow*).

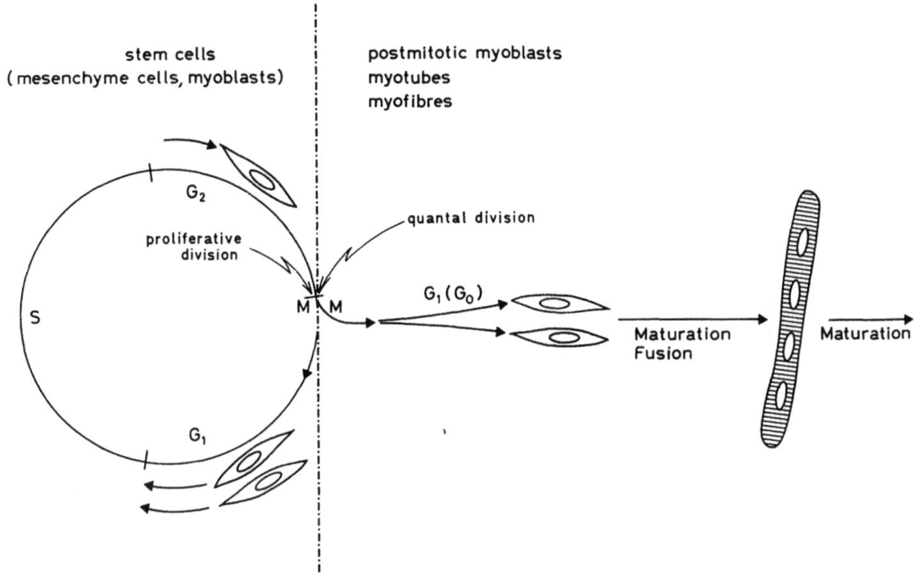

Fig. 102. Scheme to explain the "quantal division concept" (see text). As long as the progeny of a dividing cell remains in the same developmental compartment, the cells go through a proliferative division cycle; a new synthetic program can only be initiated as a consequence of a quantal division which moves the cell into the next developmental compartment. The cells have to pass through a fixed but unknown number of compartments on the way from the mesenchyme cell to the fusion-capable myoblast. The proliferating and postmitotic myoblasts only represent the penultimate and ultimate steps of myogenesis. (Adapted from BISCHOFF 1970)

cycle in order to initiate totally new biosynthetic activities, and to enter a new developmental compartment. This cell cycle is a quantal cell cycle, in contrast to proliferative cell cycles which produce identical progeny remaining in the same compartment. According to the quantal division concept, there are no undifferentiated cells. All embryonic cells exist as differentiated cells, each committed to a unique and limited program of synthesis reflected by a unique repertoire of messenger RNAs. All cells are members of functional lineages. The sequence of compartments in a lineage is obligatory and unidirectional, and a terminal cell type appears only as the result of the appearance of an invariable line of precursors. Because quantal division is the condition for moving from one compartment to the next, no cell has the option of producing more than two new cell lines as progeny. Hence, differentiation is the integration of a small and fixed number of quantal cell cycles interspersed with variable

C Scanning electron micrograph of a typical spindle-shaped myoblast. Such cells do not occur in cultures as shown in **A** and invariably give rise to cultures as shown in **D**. **D** Differentiating muscle clone derived from a quail myoblast grown for 6 days in vitro. The fibrous texture of the colony reflects the presence of cross-striated myotubes and myofibres containing numerous nuclei. In addition there are many mononucleated myoblasts. Bars, 1 mm (**A**, **D**), 10 µm (**B**, **C**). (From LIPTON 1977a, with copyright permission of Academic Press) numbers of proliferative cell cycles. The precursor cell is not pluripotential, but the system has many options depending on the number of quantal divisions. Applied to myogenesis, this means that the myoblasts that proliferate are basically different from those that fuse. Fusion-capable myoblasts become postmitotic and able to produce muscle-specific proteins during the last, quantal, cell cycle that transfers them from the presumptive to the postmitotic myoblast compartment.

The fusing myoblasts are in a prolonged G_1 phase (OKAZAKI and HOLTZER 1966); whether they are able to re-enter the mitotic cycle and start an S-phase is controversial (see below). According to Holtzer's system of differentiation, the postmitotic and presumptive myoblasts represent only the ultimate and penultimate compartments of myoblast differentiation. Primary cultures may contain ancestral cells, which may become myoblasts, but may as well exercise the option of differentiating into fibrogenic cells. This may be one of the reasons why myoblast cultures often are overgrown by fibrogenic cells (or "dedifferentiate") (DIENSTMAN et al. 1974). Even clones of myogenic cells may ultimately consist of a heterogeneous cell population, if the cloned cell was not a presumptive myoblast, but a cell from an earlier compartment of the myogenic lineage (ABBOT et al. 1974).

These views have been contested, and it was mainly KONIGSBERG and coworkers who maintained that (a) all myoblasts can re-enter the mitotic cycle until they fuse; (b) fusion depends on signal substances accumulated in the culture medium and is initiated by a high plating density; and (c) fusion is the signal for the synthesis of muscle-specific proteins.

The following findings support the notion that a quantal mitotic cycle is necessary for the formation of postmitotic myoblasts, that fusion is not the regulating mechanism that represses DNA synthesis, and that there are at least two populations of myoblasts in a culture. Cells that were forced to proliferate by keeping them in a "growth" medium had to go through one mitotic cycle after they had been brought into a standard medium, and hence fused with a delay of 18–24 h (YAFFE 1971; YAFFE et al. 1972; SIMPSON and COX 1972). Some of the myoblasts in cytochalasin B-treated cultures ceased to synthesize DNA and started synthesis of myosin and actin filaments. These cells were assumed to be postmitotic fusion-arrested myoblasts. When cytochalasin B was removed and replaced by colcemid, only those cells that were devoid of muscle-specific proteins, i.e. the presumptive myoblasts, assembled massive, meandering cables of 10-nm filaments (HOLTZER et al. 1975a, b, 1977; CROOP and HOLTZER 1975). When cells that had been fusion arrested by Ca^{2+} deprivation were incubated with normal Ca^{2+} and ^3H-thymidine to label those cells that still synthesized DNA, they fused immediately; only the mononuclear cells were labelled, but no labelled nuclei occurred in the myotubes. This indicated that the fusing cells had been born and were "selected" to fuse in the primary culture. The rate of fusion was independent of the plating density (HOLTZER et al. 1977). Inhibition of DNA synthesis by cytosine arabinoside did not delay fusion (YEOH and HOLTZER 1977). TURNER (1978), in a fusion-permissive medium, re-cultured myoblasts that had previously been fusion arrested, and added cytosine arabinoside to the new cultures to block proliferation. The number of nuclei in myotubes increased at the same rate as in cultures without cytosine arabinoside, but the myoblasts did not proliferate. The formation of myotubes levelled off in the cytosine arabinoside-treated cultures although there were still mononuclear myoblasts available. This indicated that only some of the myoblasts were capable of fusion. 5-Bromodeoxyuridine, a thymidine analogue, does not interfere with cell division but prevents terminal myoblast differentiation (STOCKDALE et al. 1964; COLEMAN et al. 1969; BISCHOFF and HOLTZER 1970; LOUGH and BISCHOFF 1976). The initial rate of myotube formation was the same in cultures with 5-bromodeoxyuridine, in cultures with arabinoside, and in standard cultures, but the formation of myotubes eventually stopped in both drug-treated cultures although the mononuclear cells continued to proliferate with 5-bromodeoxyuridine (TURNER 1978).

The following results speak against the quantal division concept of myogenesis, and against the assumption that there are two populations of myoblasts. KONIGSBERG (1971), O'NEILL and STOCKDALE (1972b), BUCKLEY and KONIGSBERG (1974, 1977), and DOERING and FISCHMAN (1974) observed that the rate of fusion depended on the plating density. The generation cycle increased in older cultures from 10–20 h. This was due to changes in the medium, and young cultures transferred to a "conditioned" medium behaved like older cultures. Conditioned medium increased the percentage of myogenic cells differentiating from muscles of chick embryos from 60% to over 90% (HAUSCHKA and WHITE 1972). All mononuclear cells in a differentiating muscle could be labelled with ^{3}H-thymidine both in vivo and in vitro, and no cells had withdrawn from the cell cycle before they fused. This was assumed to indicate that synthesis of DNA was not repressed before fusion, but that fusion itself blocked DNA synthesis (BUCKLEY and KONIGSBERG 1974, 1977). The activity of DNA polymerase decreased by >90% at the time of fusion (STOCKDALE 1970). KONIGSBERG et al. (1978), in mature cultures, measured a G_1 phase duration of 7 h, as compared with 3 h in young cultures. The minimum time myoblasts had to spend in G_1 before they fused was 4 h. The authors reasoned that a prolongation of the G_1 phase increased the probability of fusion, in particular in densely populated cultures, and proposed that the fusion rate was controlled through the regulation of the cell cycle by the medium.

Whether fusion or a terminal (quantal) mitotic cycle initiates the synthesis of muscle-specific proteins is controversial as well. HOLTZER et al. (1957), OKAZAKI and HOLTZER (1965), and HOLTZER and SANGER (1972) observed mononucleated cells containing cross-striated myofibrils, and SANGER (1974), CROOP and HOLTZER (1975), and HOLTZER et al. (1975a, b) demonstrated that cells that had been fusion arrested with cytochalasin B formed thick and thin filaments in an hexagonal array, and were labelled by antibodies against myosin. Also myoblasts that were fusion arrested by Ca^{2+} deprivation initiated myosin synthesis (DIENSTMAN and HOLTZER 1975). The actin and myosin in 1-day cultures were of the cytoplasmic type occurring in chondroblasts or fibroblasts as well, but in 4-day cultures of fusion-arrested cells typical muscle proteins were present (CHI et al. 1975a; HOLTZER et al. 1977). Nevertheless, PATERSON and STROHMAN (1972) found that myoblasts at low Ca^{2+} concentrations did not synthesize myosin, but did so immediately after fusion was initiated by a normal Ca^{2+} concentration. PRZYBYLA and STROHMAN (1974) and DEVLIN and EMERSON (1979) did not find messenger RNA coding for any of the contractile proteins before fusion had commenced, indicating that the specific mRNA species begin to coordinately accumulate at fusion. According to others, coding for contractile proteins precedes fusion. BUCKINGHAM et al. (1974) reported that in cultures from fetal calf muscles the synthesis of mRNA for myosin took place during the phase of proliferation, but that the half-lives of several RNAs increased with fusion; possibly the already transcribed mRNA was stabilized at this stage of development. Actinomycin-D treatment of cultures approaching fusion did not impair fusion; the synthesis of muscle-specific proteins continued for 6–8 h, indicating that mRNA had been transcribed before actinomyosin D was applied (SHAINBERG et al. 1971; YAFFE et al. 1972; YAFFE and DYM 1972). The specific sequences of RNA were found to accumulate a few hours before fusion (PATERSON et al. 1974; YABLONKA and YAFFE 1977; ZEVIN-SONKIN and YAFFE 1980). A dissociation of myosin synthesis and fusion was also observed when myoblasts were treated with low doses of diazepam. Diazepam was supposed to restrict the availability of Ca^{2+}. In contrast to Ca^{2+} deprivation, low-dose diazepam blocked myosin synthesis and did not interfere with cell fusion (BANDMAN et al. 1978).

The concentration of muscle-specific metabolic enzymes increases with fusion; this increase may be controlled by lowering the Ca^{2+} concentration, i.e. it is coupled to fusion (SHAINBERG et al. 1971). By contrast, the normal occurrence of acetylcholine receptors and of acetylcholinesterase at the time of myotube formation (FAMBROUGH and RASH 1971; OH and JOHNSON 1972) is not prevented when fusion is blocked by a low Ca^{2+} concentration; the synthesis of these proteins seems to follow a developmental scheme not dependent on fusion (PRIVES and PATERSON 1974). In concert with this observation, SARTORE et al. (1979) found that a probe for membrane proteins (2,4,6-trinitrobenzene sulphonate) was bound to myotubes and fusion-arrested mature myoblasts, but not to young myoblasts. Fibroblast growth factor (FGF) maintains the proliferation of mouse myoblasts and prevents fusion; removal of this mitogen initiates fusion after a 14-h delay. In cultures that were depleted for 4 h for FGF and then re-fed with FGF-rich medium, 10% of the mononuclear cells started to synthesize acetylcholine receptors and ceased to synthesize DNA, i.e. they had withdrawn from the mitotic cycle during the previous 4 h (HAUSCHKA et al. 1982). Proliferating mouse myoblasts have receptors for epidermal growth factor; irreversible loss of these receptors precedes fusion (LIM and HAUSCHKA 1984). These results support the hypothesis that there are proliferating and non-proliferating myoblasts in the same culture, and that cells that irreversibly withdraw from the cell cycle are committed cells.

YAFFE and GERSHON (1967) obtained results suggesting that the withdrawal of myotube nuclei from the mitotic cycle might be reversible. Polyoma virus induces in rat myotubes incorporation of ^3H-thymidine into DNA, and the formation of mitotic figures in metaphase. Nevertheless, the nuclei do not complete mitosis and do not divide, but many fuse into giant nuclei. HOLTZER and SANGER (1972) suggested that unscheduled DNA synthesis might lead to the death of the myotube.

Many questions with respect to myoblast differentiation are still open, but there is increasing evidence that the terminal differentiation starts before, and is independent of fusion. Although it is appealing to link the change of the pattern of gene activity to cell replication, this link has not been proven to be causal or obligatory. Despite this reservation, Holtzer's quantal division concept has had the indisputable merit, for regenerating muscle tissue, of discrediting speculations about "dedifferentiation" or "nuclei gathering cytoplasm", and has focussed the interest on testable events, such as DNA replication and RNA transcription.

The effect of "growth" and "fusion" media indicates that mitogenic substances exist which induce proliferation of myoblasts; possibly other factors induce myoblasts to differentiate. Whether the substances present in "conditioned" media, or the specific components of collagen that influence differentiation in vitro (HAUSCHKA and WHITE 1972) are operative in vivo is obscure (YAFFE et al. 1972). NAMEROFF and HOLTZER (1969, 1970) observed that myotubes inhibit the proliferation of myoblasts, but only when they are in direct contact, not through a millipore filter. The mechanism is not understood; it might be related to the regulation of myoblast proliferation in vivo (BISCHOFF 1970; BISCHOFF and LOWE 1974). A similar action of myofibres could explain why the satellite cells in adult muscles remain inactive as long as the muscle fibre is intact, but proliferate in injured muscles (for references, see MAURO et al. 1970; MAURO 1979). Also tissue hormones, such as chalones (BULLOUGH 1962, 1975) or prostaglandins (ZALIN 1977) might be involved.

c) Transdifferentiation

There are contrasting reports on the differentiation of myoblasts from and into other cell types. LENNON et al. (1979) observed that cells from a rat glial cell line differentiated into multinucleated skeletal muscle cells. They speculated that neuroectodermal cells of the cephalic neural crest might have a myogenic potentiality. Similarly, multinucleated muscle fibres were found in cultured optic nerves of fetal rats (WIER and LENNON 1981). Both observations were tentatively connected with the occasional occurrence of dystopic skeletal muscle fibres in the central nervous system. It is indeed possible, by hybridization with differentiated chick muscle cells, to induce synthesis of rat muscle-specific proteins in neural cells of rat brain (WRIGHT 1984a). ABBOTT et al. (1974) found fibrogenic cells in subcultures of myogenic clones and hypothesized that the cloned cells were common ancestors of myoblasts and fibroblasts. Because the cells could not be induced to produce cartilage-specific substances, the authors assumed that in an earlier compartment the cell lineage had bifurcated into chondrogenic and myogenic-fibrogenic precursors. LIPTON (1977b) contested this interpretation and reported that pure myoblast cultures produce a soluble collagenous protein. NATHANSON and HAY (1980a, b) cultured minced muscles from fetal rats, and reversibly transformed myoblasts into fibroblasts or chondroblasts, depending on whether the cells were plated on a bone or collagen matrix. Based on a diametrically different philosophy, also ABBOTT et al. (1974) and DIENSTMAN et al. (1974) associated chondrogenic and myogenic cell lineages (see above). This seemed natural as long as the mesenchymal cell of the limb bud was assumed to be the common ancestor of myogenic and chondrogenic cell lineages. Strong evidence has since been presented that the myoblasts, but not the chondroblasts, of the limb originate from the somites (see Sect. G.II.1).

d) Myoblasts In Vivo

Recent research development in myogenesis shows two main lines. YAFFE and his associates brought a molecular biology approach into the field, and several groups are now concerned with gene expression during myogenesis, for experimental reasons sometimes resorting to rather exotic systems, such as nematode or *Drosophila* muscles (for references, see HARRIS 1980; SCHOTLAND 1982; KONIENCZNY et al. 1983; WRIGHT 1984a, b). These studies are beyond the scope of this handbook. Other workers try to relate the events in culture to myogenesis in vivo.

HAUSCHKA (1974) and HAUSCHKA et al. (1979, 1982) plated the mononuclear cells obtained from dissociated muscle tissue of mouse, chicken, or man at low density, and maintained the cultures for several weeks. Those cells that were successfully plated gave rise to clones, and some of these clones were myogenic. The plating efficiency and the percentage of myogenic clones were determined. Cells giving rise to a muscle clone were termed MCF (muscle colony forming) cells. The earliest MCF cells in man were obtained at 33 days fetal age. The plating efficiency was low, and only 10% of the colonies were myogenic. Whether the other stem cells were non-myogenic, or whether they were earlier forms of myogenic stem cells that could not differentiate in culture remained open. From a fetal age of 90 days on and also in adults, 90% of the successfully plated cells were MCF cells. The MCF cells from adult muscles were probably the satellite cells one sees in sections. The fetal MCF cells replicated immediately, and had a generation time of 15 h. The cells were able to go through about 70 mitotic cycles, giving rise to a large progeny. The MCF cells from adult muscles started replicating after a 24-h delay, had a generation time of 21 h, and went through up to 30 mitotic cycles. Although the proliferative capacity of the adult MCF cells was smaller than that of the fetal ones, probably because they had already gone through many generation cycles in vivo, they could still give rise to a huge progeny. Some of the subclones were unlimited with respect to the number of replication cycles, and these cells allowed the establishment of permanent cell lines, which might be useful for the investigation of inherited metabolic disorders (HAUSCHKA 1982). The first MCF cells of chick embryos were obtained at the third day of development, and, as in man, during development the percentage of MCF cells increased from 10% to 90%. The plating efficiency increased from 1% to 25%. Hence, the number of MCF cells increased from at least 1 to at least 200 cells per 1,000 mononuclear cells in muscle tissue (BONNER and HAUSCHKA 1974; HAUSCHKA et al. 1982), i.e. at least 20% of the mononuclear cells in a late fetal muscle are myoblasts. The incidence of satellite cells in relation to other mononuclear cells in chick muscles is unknown; in adult rat and human muscles, 4%–5% of the mononuclear cells are satellite cells (calculated from data in SCHMALBRUCH and HELL-

HAMMER 1976, 1977; Tables 4, 11). The earliest MCF cells that occurred during ontogenesis were in the proximal parts of the limb buds; a proximo-distal gradient was still observed in 5-day chick embryos and in 42-day human fetuses (HAUSCHKA et al. 1982; RUTZ et al. 1982). Serial cross-sections of leg buds of 5-day chick embryos were cultured; the MCF cells were found to be concentrated in discrete dorsal and ventral regions. Two central regions provided mainly cartilage-forming cells, and the most central and the most superficial part of the cross-sections contained cells producing fibroblastic clones. Thus, the distribution of stem cells reflected the future array of the bones and the flexor and extensor muscles.

e) The Morphology of Myoblasts in Culture

Cultured mononuclear cells with myogenic potentialities are fusiform, and have a small ruffled area on one of the tips; by contrast, cells with fibrogenic potentialities are flattened, have multiple extensions and large parts of the surface are ruffled (KONIGSBERG 1963) (Fig. 101). Cloned myoblasts kept under optimum culture conditions are filled with many free ribosomes, one Golgi complex, and sparse rough endoplasmic reticulum (Fig. 103). Fibrogenic cells show an extensive elaboration of rough endoplasmic reticulum, multiple Golgi complexes, and dense accumulations of 10-nm filaments. Myoblasts grown in the presence of 5-bromodeoxyuridine acquire morphological criteria of fibrogenic cells; they become flattened and accumulate 10-nm filaments (LIPTON 1977a). Twenty percent of the myoblasts of a culture accumulate thick and thin filaments in the presence of 5-bromodeoxyuridine (HOLTZER et al. 1975a, b; LIPTON 1977a). The myoblasts stop to differentiate and to proliferate in an hypoxic environment; they accumulate giant lipid droplets, and closely resemble adipocytes (Fig. 104). The changes induced by 5-bromodeoxyuridine and hypoxia are reversible; when returned to standard conditions the cells actively divide and differentiate into myotubes. The phenotypical modulations by an altered environment do not reflect a change of the genetic program. Interestingly, ISHIKAWA et al. (1968) first described the ubiquitous 10-nm filaments in cultured muscle cells that had been treated with colchicine.

LIPTON'S (1977a) description of cloned myoblasts is in agreement with the morphology of cells which, on the basis of their localization beneath the sarcolemma, in regenerating muscles are identified as myoblasts (MENDELL et al. 1972; SCHMALBRUCH 1976b). The phenotypic modulation by hypoxia may explain why, in human dystrophic muscles, presumed myoblasts often are loaded with large triglyceride droplets (SCHMALBRUCH, unpublished). One may be tempted to speculate that the numerous fat cells in dystrophic human muscles in reality are "degenerated" myoblasts.

f) Satellite Cells

The myogenic cells in mature muscles, the satellite cells (MAURO 1961; for review CAMPION 1984), have a characteristic localization. They are wedged between the basal lamina and the plasma membrane of the myofibres, usually

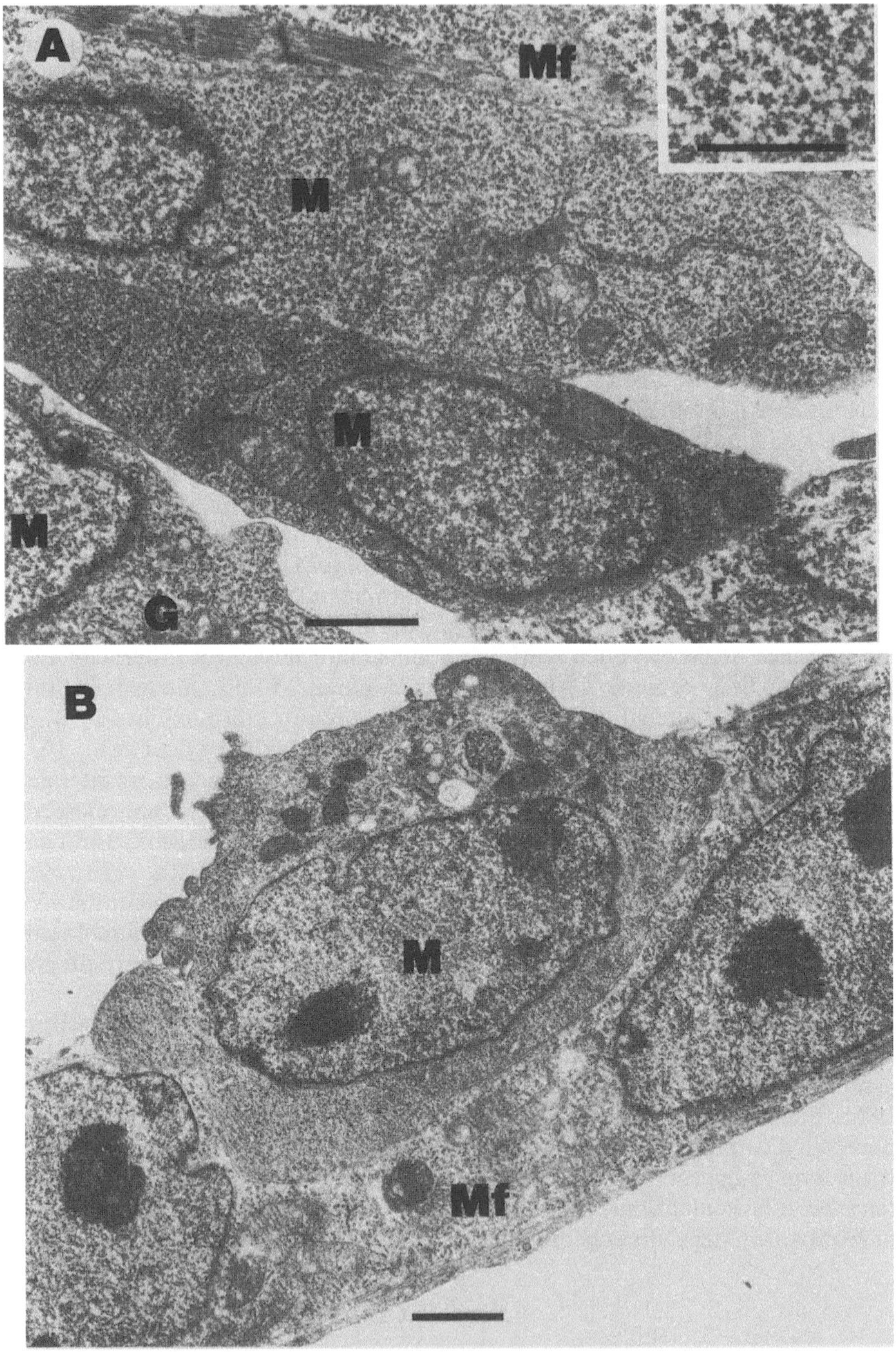

Fig. 103. A, B. Myogenic cells in culture. Mononucleated myoblasts (*M*) in close association with immature multinucleated myofibres (*Mf*). The cytoplasm of the myoblasts is stuffed with ribosomes (*inset*). *G*, Golgi zone; bars, 1 µm, 0.5 µm (inset). (From Lipton 1977a, with copyright permission of Academic Press)

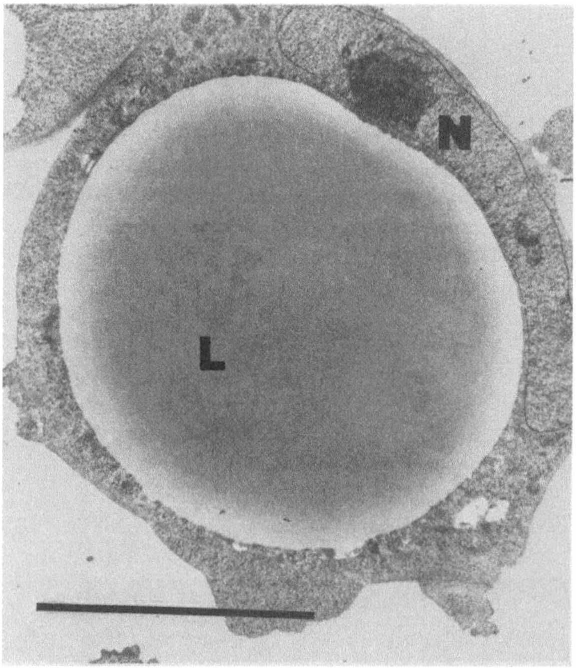

Fig. 104. Chick myoblast in culture following prolonged exposure to hypoxia. The cell contains a large lipid droplet (*L*) and has become virtually indistinguishable from an adipocyte. The nucleus is flattened (*N*) and lies in the periphery of the cell. When returned to normal conditions, these cells differentiate into myotubes. Bar, 5 μm. (From LIPTON 1977a, with copyright permission of Academic Press)

in depressions of the fibre surface. The gap between the two plasma membranes is about 20-nm wide. The cytoplasm is scanty, glycogen granules are lacking, and the nuclei are heterochromatic and thus differ from myonuclei which are usually euchromatic (Fig. 105).

It is impossible to distinguish satellite cells from myonuclei in routine sections for light microscopy. ONTELL (1974) described a staining procedure which could be applied to semithin plastic sections; the method is mainly based on the different chromatin pattern. Nevertheless, the incidence of satellite cells in rat muscles was smaller than when the cells were counted in electron micrographs (Table 11). This suggests that not all satellite cells are reliably identified. Furthermore, recently incorporated myonuclei are heterochromatic (MOSS and LEBLOND 1971), which makes the staining procedure still more unreliable when regenerating or fast-growing muscles are to be studied.

Satellite cells are usually described as fusiform. Freeze fractures provide en face views of the fibre surface (Fig. 106), and reveal that the satellite cells of rat muscles have numerous slender projections embedded in grooves of the muscle fibre membrane. The plasma membrane of the satellite cells has fewer intramembrane particles and fewer caveolae than the plasma membrane of the muscle fibre. Square arrays (see Sect. C. IV. 3 c γ) are absent. Membrane junctions between muscle fibre and satellite cell have not been found (SCHMALBRUCH 1978a). Also the satellite cells of shark muscles have several long, often bifurcating projections (KRYVI 1975). MAZANET et al. (1982) found, using scanning elec-

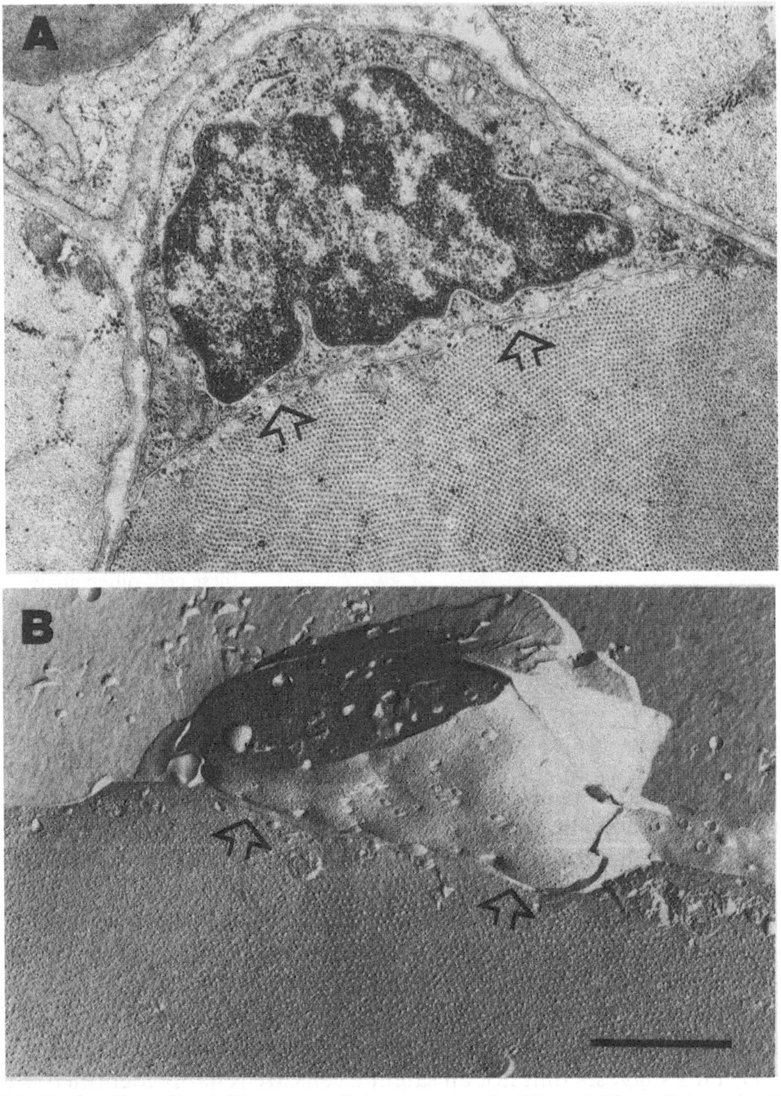

Fig. 105 A, B. Satellite cells. **A** Human medial vastus muscle. The satellite cell is wedged between plasma membrane and basal lamina of the muscle fibre. The 20-nm-wide membrane-bounded gap between the two cells (*arrows*) is beyond the resolution of the light microscope and the satellite cell will be mistaken as myonucleus. **B** Rat soleus muscle. Freeze fracture replica showing a cross fracture through a muscle fibre (*bottom*). The satellite cell is partly extending above the fracture face. The P face of the satellite cell and the E face of the muscle fibre plasma membrane, and the gap between the two cells (*arrows*) are visible. The interstitium (*top*) contains cross-fractured collagen fibrils. Bar, 1 μm

---

Fig. 106 A, B. Freeze fractures of the surface of muscle fibres of the rat soleus muscle with attached satellite cells (*S*). The satellite cells have slender projections and are embedded in grooves of the fibre surface. Note the different intramembrane particle density of the P faces of muscle fibre and satellite cell membrane. Bars, 0.5 μm

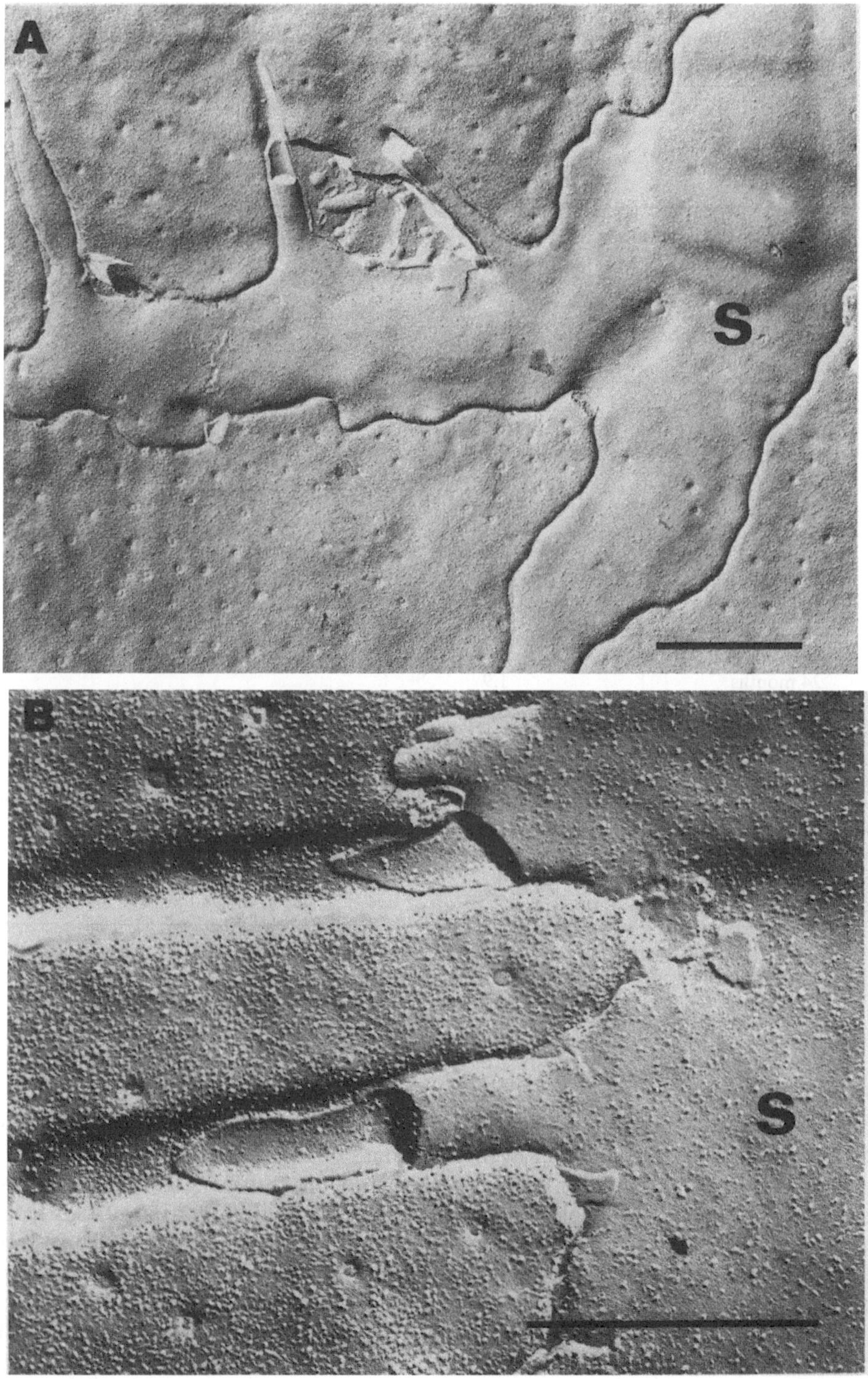
A
S
B
S

Table 11. The incidence of satellite cells in relation to all nuclei within the basal lamina of the muscle fibres (myonuclei + satellite cell nuclei). For some muscles, the absolute number of satellite cells per mg muscle can be computed from the data given here and those given in Table 4. *EM*, electron microscopy; *LM*, light microscopy

Species	Age or weight	Muscle	% of intrasarcolemmal nuclei	Method	References
Mouse	1 day	Peroneus long.	35	EM	ALLBROOK et al. (1971a)
	35 days		6		
	70 days		5		
	140 days		4		
	7 days	Lumbricalis	27	EM	SCHULTZ (1974)
	30 days		6		
Rat	1 day	Subclavius	32	EM	ALLBROOK et al. (1971b)
	35 days		9		
	70 days		5		
	170 days		4		
	600–800 g	Soleus	2–3	LM	ONTELL (1974)
	200–250 g	Soleus	11	EM	SCHMALBRUCH and HELLHAMMER (1977)
	1 month	Soleus	9.6	EM	GIBSON and SCHULTZ (1983)
	12 months		6.6		
	24 months		4.7		
	1 month	Extensor digitorum longus	7.0		
	12 months		2.9		
	24 months		1.9		
	200–250 g	Anterior tibial superfic. part	4	EM	SCHMALBRUCH and HELLHAMMER (1977)
		Diaphragm	8		
	21 days	Intrafusal fibres	8	EM	RUMPELT and SCHMALBRUCH (1969)
Fruit bat	Adult	Web muscle	12–20	LM	CHURCH (1970a)
Man	7–73 years	Various	4	EM	SCHMALBRUCH and HELLHAMMER (1976)
	7 years	Intrafusal fibres	17	EM	RUMPELT and SCHMALBRUCH (1969)
Frog	Adult	Gastrocnemius	3	EM	TRUPIN (1976)
Lizard	?	Tail muscles	5–7	EM	KAHN and SIMPSON (1974)

tron microscopy, fusiform cells on frog muscle fibres and assumed that these cells were satellite cells; they showed only small changes after muscle injury.

The incidence of satellite cells in relation to the number of myonuclei, assessed in cross-sections for electron microscopy, is much higher in muscles from very young rats and mice than in adult animals (ALLBROOK et al. 1971b; SCHULTZ 1974) (Table 11). These counts have been criticized; immature muscles still contain secondary myotubes, which in some cross-sections are devoid of myofilaments and hence may be mistaken for satellite cells (ONTELL 1979). On the other hand, there is little doubt that the incidence of satellite cells decreases during development; there is also a decrease in old age although this is difficult to ascertain by determining the percentage of satellite cells (ALLBROOK et al. 1971b; SCHMALBRUCH and HELLHAMMER 1976; SNOW 1977b). The number of satellite cells per mm^3 tissue in rat is about 5,000 in the diaphragm and soleus muscle, and about 1,000 in the superficial (white) part of the anterior tibial muscle (SCHMALBRUCH and HELLHAMMER 1977). The density in human muscles is roughly $800/mm^3$ (SCHMALBRUCH, unpublished). GIBSON and SCHULTZ (1983) determined the number of nuclei in the soleus and extensor digitorum longus muscles of rats 1, 12, and 24 months of age, using biochemical and morphological techniques. At 1 month, the soleus muscle contains 5.2×10^5 satellite cells; there is a modest increase to 7.3×10^5 at 12 months, while at 24 months the 1-month level is reached again. The extensor digitorum longus muscle at 1 month contains 3.1×10^5 satellite cells, the number decreases continuously to 1.3×10^5 at 24 months. Between 1 and 12 months, the number of myonuclei increases by 100% in the soleus, and by 60% in the extensor digitorum longus muscle; then it remains constant. The number of satellite cells per mg muscle at 1, 12, and 24 months of age is 12,600, 3,300, and 2,400 in the soleus muscle; it is 7,700, 1,150, and 700 in the extensor digitorum muscle (calculated from the data in Tables 4 and 11, and the muscle weights given by GIBSON and SCHULTZ (1983)). These data compare reasonably well with those obtained by SCHMALBRUCH and HELLHAMMER (1977) in 2-month-old rats, using a morphological technique. GIBSON and SCHULTZ (1983) relate the higher incidence of satellite cells in "red" muscles to the fact that these muscles are more frequently activated (see Chap. H) and thus need a larger reservoir of myoblasts for maintenance and repair of the wear and tear of daily activity.

The satellite cells show pronounced mitotic activity in growing, denervated, or injured muscles (SHAFIQ et al. 1968; CHURCH 1970a; MOSS and LEBLOND 1970, 1971; ALLBROOK et al. 1971b; SCHIAFFINO et al. 1972; ALOISI et al. 1973; ONTELL 1974, 1975; SCHULTZ 1974, 1976, 1979; HANZLIKOVA et al. 1975; SCHMALBRUCH 1976b; SNOW 1977a, 1979, 1983; MCGEACHIE and ALLBROOK 1978; KELLY 1979; KLEIN-OGUS and HARRIS 1983). Denervation of rat muscles shortly after birth stops the growth of the muscle, and the mitotic activity of the satellite cells declines (KELLY 1979). In harmony with this observation, BONNER (1978, 1980) found that embryonic chick muscles, which, after 5–6 days development, had been denervated or curare-treated, yielded reduced proportions of clonable myoblasts.

Encroachments of basal lamina material into the space between satellite cell and myofibre, such that in places the satellite cell is detached from the myofibre, are seen in growing, regenerating, diseased, and also in senile muscles (VENABLE 1966b; ONTELL 1975; SCHMALBRUCH 1976b; SNOW 1977b; SCHULTZ 1974, 1976, 1978; KLEIN-OGUS and HARRIS 1983). This configuration may suggest that the satellite cells are released from the underlying muscle fibre. HANSEN-SMITH et al. (1979) found in severely malnourished children a reduced incidence of satellite cells; the number of satellite cells increased during recovery and many satellite cells were partly detached from the myofibre by basal lamina intrusions.

Two closely associated heterochromatic nuclei, of which one is a myonucleus and one a satellite cell nucleus, are quite often found in growing (MOSS and

LEBLOND 1971) and diseased muscles. The two nuclei are probably the result of satellite cell mitosis; one of the daughter cells fuses with the myofibre and one remains a proliferating satellite cell. According to MOSS and LEBLOND (1971), half of the progeny of satellite cells in growing muscles fuse with the underlying myofibres. It is unknown whether this is obligatory, but statistically, it must be the rule: the total number of satellite cells of a muscle remains more or less constant during postnatal growth (HELLMUTH and ALLBROOK 1973), whereas the number of myonuclei increases (ENESCO and LEBLOND 1962). Satellite cells are not evenly distributed but show a concentration around the neuromuscular junction (KELLY 1978b).

g) Fusion of Myoblasts

The initial step of myotube formation is the fusion of two myoblasts, subsequently other myoblasts follow. Also two myotubes can fuse with each other. Myoblast fusion is tissue specific, i.e. liver cells or fibroblasts from the same animal do not fuse with a myoblast, but myoblasts from different genetic origin may fuse (YAFFE and FELDMAN 1965; for references, see FISCHMAN 1972; HAUSCHKA et al. 1982). Isolated and re-implanted satellite cells from rats or mice are incorporated into muscle fibres of the same animal (JONES 1979; LIPTON and SCHULTZ 1979) or of an animal from a genetically different strain (PARTRIDGE et al. 1978; WATT et al. 1982). It is obvious that fusion-capable myoblasts and myotubes must have markers to recognize each other. WAKSHULL et al. (1983) describe a plasma membrane glycoprotein with a mol.wt. of 38,000 daltons which is present in chicken myoblasts, immature myotubes and satellite cells, but not in mature muscle fibres and not in fibroblasts. The molecule is exposed on the surface of the cells and can be labelled with a monoclonal antibody and visualized using immunocytochemistry. The function is enigmatic and the molecule is not a marker of the myogenic lineage because it occurs in glia cells as well. Nevertheless, this observation is interesting because it supports the view that satellite cells are resting myoblasts, and because the molecule is present in cells that are active during myogenesis (myoblasts and immature myotubes) and also in the resting but division capable satellite cells.

Chick myoblasts in vitro move continuously and spend long times in close association with other myoblasts or with myotubes (Fig. 103). The cells have to be in contact for several hours prior to fusion (YAFFE et al. 1972; BISCHOFF and LOWE 1974). Myoblasts form linearly arranged aggregates, the cells being in contact with each other either end-to-end or side-by-side. Replicating cells are usually adhering to myotubes, the daughter cells move apart in opposite directions, but do not lose contact with the aggregate (FEAR 1977; BACHMAN 1980). Cells that approach each other make the first contacts through thin filopodia. Fusion-arrested cells growing in a Ca^{2+}-depleted medium lack these filopodia (CHIQUET et al. 1975) and do not assemble (PAPADIMITRIOU and DAWKINS 1973). Fusion-capable but not proliferating cells are covered with a ruthenium-red-staining surface coat (SHIMADA 1972). EDTA binds Ca^{2+}, suppresses fusion, and induces release of unknown substances into the culture medium. These substances immediately restore the adhesion of myoblasts transferred to normal medium; normal medium alone is ineffective in restoring adhesion for several hours (BISCHOFF and LOWE 1974). Mucopolysaccharides and glycoproteins, i.e. components of the cell coat have been implicated in cell adhesion (ROTH et al. 1971; PESSAC and DEFENDI 1972; TURNER and BURGER 1973; BISCHOFF and LOWE 1974; DEN et al. 1975; WINAND and LUZZATI 1975; CHOW and POO 1982).

The orientation of the myoblasts in strands, and the shape of the resulting myotubes, depend on the ability of the cells to adhere to the substrate. Cultures kept in suspension form spherical multinucleated "myoballs" 20–150 µm in diameter. The myoballs contain non-aligned myofibrils, and the plasma membrane develops normal muscle membrane properties. Myoballs are convenient objects for electrophysiological studies (FISCHBACH and LASS 1978). Similar formations, myosacs, have been found in regenerating tail muscles of tadpoles treated with colchicine (WARREN 1968), known to disrupt microtubules.

The attached myoblasts and myotubes are coupled by gap junctions, both in vivo (KELLY and ZACKS 1969a; KEETER et al. 1975; RASH and STAEHELIN 1974) and in vitro (RASH and FAMBROUGH 1973; KALDERON et al. 1977). Gap junctions between young muscle fibres of rats persist until the early postnatal period (SCHMALBRUCH 1982c) (Fig. 107). KEETER et al. (1975) and HAYES (1975) found, in amphibia, gap junctions between the muscle cells of the same (and also between fibres of adjacent) myotomes, until the cells and their motor innervation had fully matured. There are several reports on gap junctions and electrical coupling between developing invertebrate muscles (for review, see RHEUBEN and KAMMER 1982).

According to RASH and FAMBROUGH (1973), gap junctions occur immediately prior to myoblast fusion. The electrical coupling between pairs of cells is initially weak, but suddenly it increases marking the onset of cytoplasmic fusion. The entire fusion process requires less than 15 min. The separating membranes disappear at a rate of more than 1 $\mu m^2/s$. Remnants of gap junctions are found in the cytoplasmic bridge between the cells. The morphology of membrane fusion is difficult to study. The membranes of adjacent myoblasts interdigitate, and in many places putative pores may be seen; nevertheless, tilting the section in the electron microscope discloses that these pores are simulated by oblique projection of the membranes. LIPTON and KONIGSBERG (1972), in a study of fusion in cultured myoblasts, identified only seven pairs of cells connected by cytoplasmic bridges. An important feature of initial stages of fusion is that the interconnecting bridge is contained within a pair of culs de sac formed by the plasma membranes. In contrast to RASH and FAMBROUGH (1973) and FISCHMAN (1972), LIPTON and KONIGSBERG (1972) did not find membrane remnants within the bridge and assumed that fusion was initiated at a single site. The rapid fusion process explains why only few micrographs showing unequivocal membrane fusion have been published (Fig. 108).

Gap junction formation between the myoblasts is not sufficient to initiate fusion; also fusion-arrested myoblasts are connected by gap junctions (KALDERON et al. 1977). Their role in myogenesis may be twofold. Gap junctions are permeable for small molecules, and during the phase of fusion the junctions may serve as communication pathways. KEETER et al. (1975), HAYES (1975),

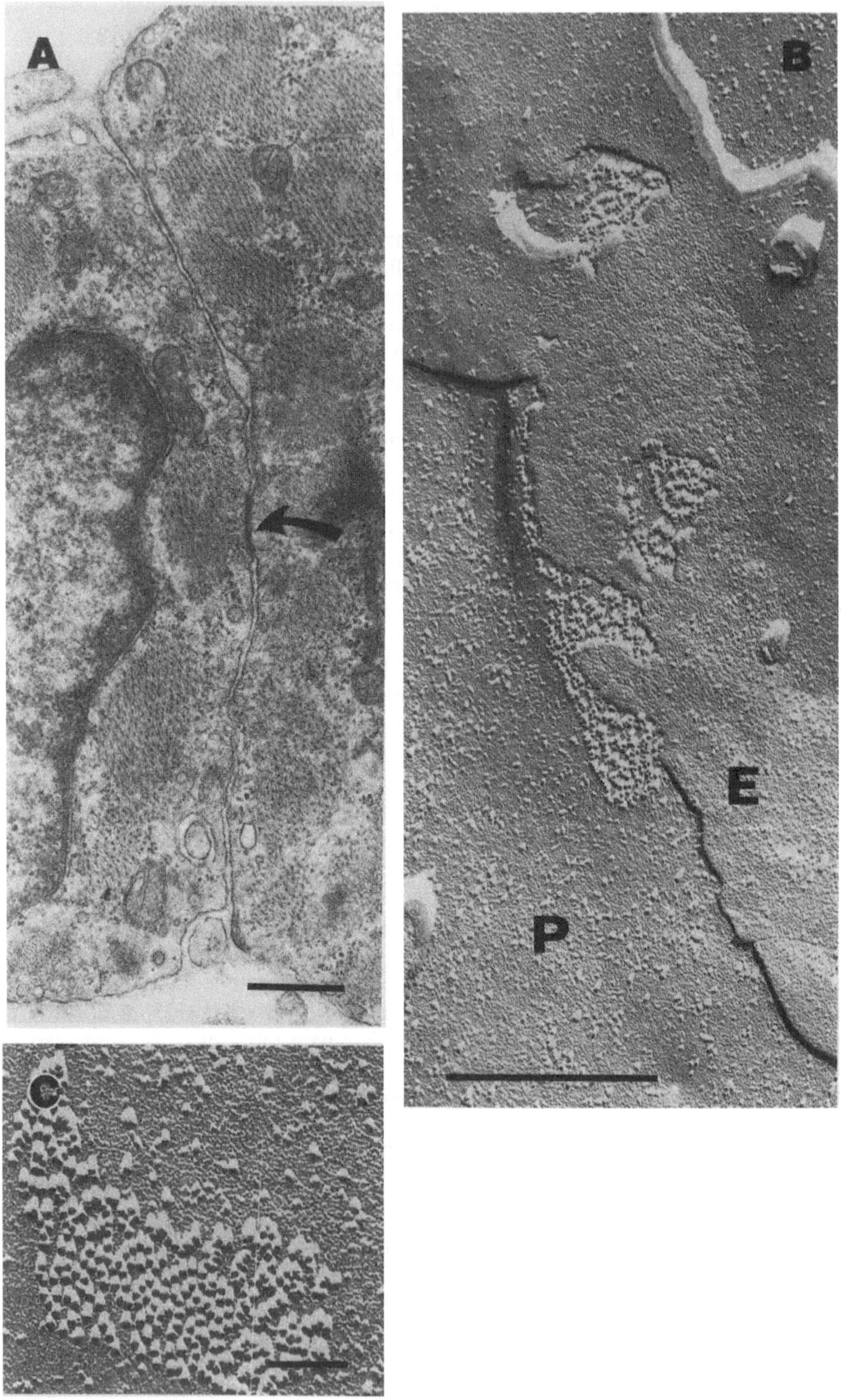
A
B
E
P
C

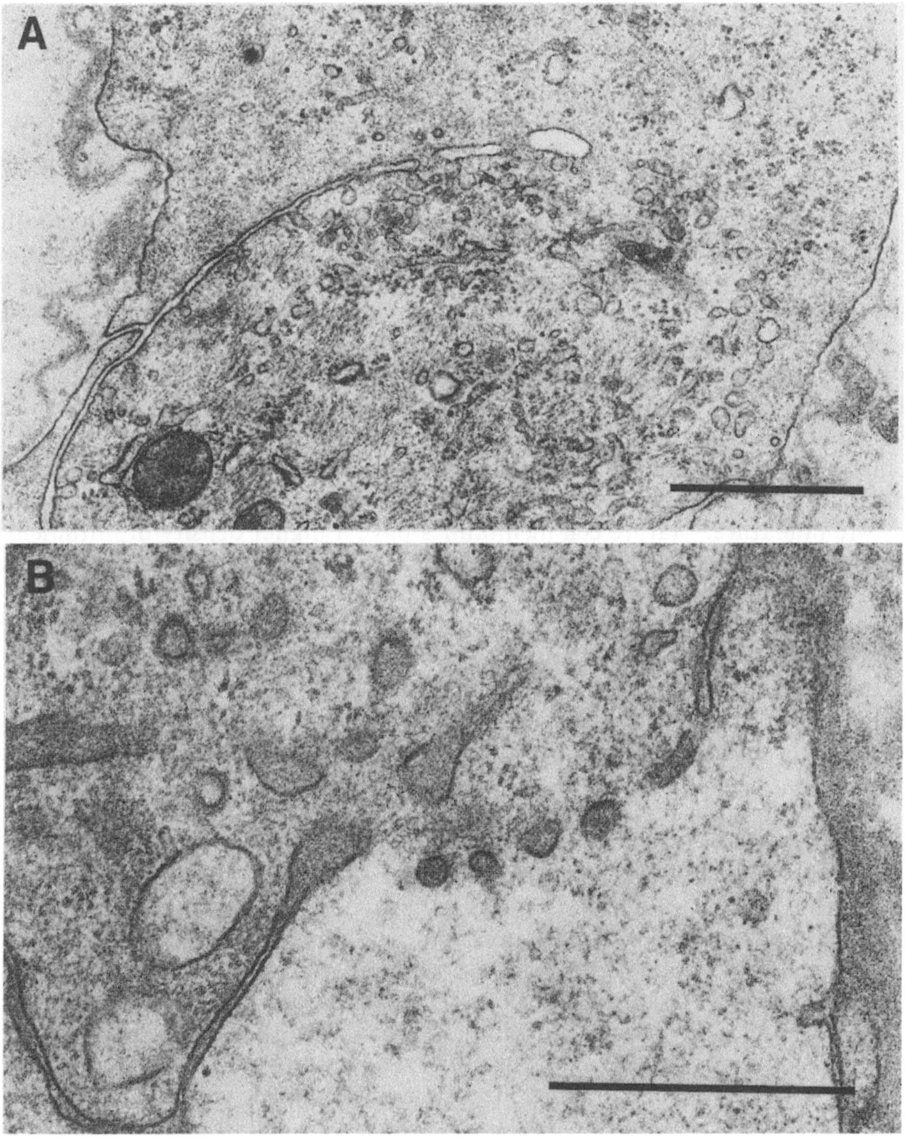

Fig. 108 A, B. Rat soleus muscle regenerating 5 days after injury by hot Ringer's solution. Presumed fusion sites of myotubes (**A**) and myoblasts (**B**) which are contained in the basal lamina of a necrotic fibre (seen in **A**). The cytoplasmic bridges are betweeen membrane pockets. The multiple bridges suggest that fusion commences at several sites. Bars, 1 μm

---

Fig. 107 A–C. Flexor digitorum brevis muscle, newborn rat. The fibres are immature and are coupled by gap junctions. **A** Thin section showing two clustered fibres separated by a narrow membrane-bounded gap without interposed basal lamina, and coupled by a gap junction (*arrow*). Bar, 0.5 μm. **B** Freeze fracture showing the P face of one fibre and the E face of another fibre membrane (*P*, *E*). The gap junction particles are partly exposed. Bar, 0.5 μm. **C** High magnification of the P face of a gap junction showing that the junctional particles are in an irregular array; 10-nm particles are also seen. Bar, 0.1 μm and SCHMALBRUCH (1982c), who found gap junctions between young myofibres as well, proposed that the junctions may provide means for the spread of excitation to fibres not yet innervated. This could explain how tadpoles are able to swim before the motor innervation of their muscles is completed (HAYES 1975).

The very process of membrane fusion is poorly understood, although fusion of membranes is one of the basic events in a cell. The role of the membrane lipids during fusion has been studied in model membranes consisting of lipids only (liposomes) (for review, see TYRRELL et al. 1976). Their ability to fuse depends on the presence of Ca^{2+}, on the charge of the lipids involved, and on the fluidity of the membrane which is temperature related. The same factors influence myoblast fusion in vitro (VAN DER BOSCH et al. 1973; DAHL et al. 1978; GRATZEL et al. 1978). Natural cell membranes contain proteins in addition to the lipid bilayer. Both intact myoblasts and natural cell membranes isolated from myoblasts fuse more readily when the myoblasts are "mature", than when they are immature. Mature cell membranes have a higher density of intramembrane particles; this suggests that also the proteins influence the fusion capacity of natural membranes (DAHL et al. 1978).

The fluidity of the cell membrane of myoblasts increases shortly before they fuse. The subsequent postfusional differentiation is accompanied by the regeneration of membrane rigidity (PRIVES and SHINITZKY 1977). A frequently used model to investigate membrane fusion is exocytosis. Several studies have shown that, prior to fusion, the site of fusion is cleared of intramembrane particles (LAGUNOFF 1973; LAWSON et al. 1977; ORCI et al. 1977).

This observation was in agreement with the following model for membrane fusion (LUCY 1970; POOLE et al. 1970; AHKONG et al. 1975): the bilayer structure of the lipids becomes locally perturbed, and the fluidity increases which, in the extreme, causes micelle formation; the protein particles aggregate, and the disturbed lipid molecules of both apposed membranes intermix within the denuded area. Restoration of the bilayer structure will result in a common membrane (see also CULLIS and DE KRUIJFF 1979). Because liposomes fuse with intact cells, and also induce cell fusion (PAGANO et al. 1974; PAPAHADJOPOULOS et al. 1974; MARTIN and MACDONALD 1974; POSTE and PAPAHADJOPOULOS 1974) it made sense to search for pure lipid domains as prospective fusion sites in the membranes of fusing myoblasts. KALDERON and GILULA (1979) described intracellular particle-free lipid vesicles that fused with the plasma membrane and thus induced particle-free patches; the authors assumed that cell fusion might commence at these sites. SCHMALBRUCH (1982a) observed, in regenerating muscles, extracellular particle-free vesicles that apparently fused with the myotubes and also formed bridges between adjacent cells. These vesicles were possibly remnants of cell debris. The notion that extracellular lipids might be involved in myoblast fusion is supported by the observation that the fusiogenic effect of Sendai virus on erythrocytes is linked to its haemolytic action (KNUTTON and BÄCHI 1980). Haemolysis certainly produces many extracellular particle-free membrane vesicles.

The concept that fusion of membranes requires protein-free membrane domains has been invalidated by the observation that the particle-free areas preceding exocytosis are artefacts, and that membrane fusion apparently can take place without lateral displacement of intramembrane particles (CHANDLER and HEUSER 1979). Nevertheless, an increased lateral mobility of intramembrane particles seems to be related to membrane fusion which may account for the artefactual clearing of fusion sites. The difference between fusion-capable and fusion-resistant fibroblast lines is that the former, but not the latter, respond to polyethylene glycol with intramembrane particle aggregation (ROOS et al. 1983).

FUMAGALLI et al. (1982) proposed a different fusion mechanism for myoblasts. The membranes become closely opposed with a 7-nm gap; then the two outer layers of the lipid bilayers merge and disappear in such a way that, locally, the plasma membrane becomes a diaphragm with two cytoplasmic faces. Finally, "membrane fission" at these sites gives rise to small cytoplasmic bridges, which by membrane endocytosis and retraction are enlarged. Ca^{2+} depletion does not prevent membrane opposition and fusion of the outer lipid layers, but it prevents fission and thereby the formation of cytoplasmic bridges.

III. Myotubes and Muscle Fibres

1. Muscle Fibres as Multinucleated Cells

The large skeletal muscle fibre of vertebrates and its immature form, the myotube, are multinucleated syncytia. It has been proposed that the nuclear-to-cytoplasmic volume ratio is the limiting factor for the size of a cell because one diploid nucleus can sustain only the protein synthetic demands of a specific cytoplasmic volume (SZARSKI 1976). It is interesting to note that cardiac muscle cells, which are widely considered to be mononucleated, in dog and pig often are binucleated. This is due to mitosis without cytokinesis. The chromosomes may fuse giving rise to nuclei with abnormal DNA content (PFITZER 1971; BISHOP and HINE 1975). In cardiac myocytes of man, pig, and monkey, but not of rat, rabbit, cat, and cow, polyploid nuclei are frequent (SANDRITTER and SCOMAZZONI 1964; ADLER 1971; PFITZER 1972; ADLER and COSTABEL 1975). The degree of polyploidy increases considerably in hypertrophying human hearts (PFITZER 1972; ADLER and COSTABEL 1975). The DNA to protein ratio remains fairly constant throughout postnatal growth, such that the gene dosage per volume cytoplasm remains constant (for review, see FISCHMAN 1979).

In contrast to cardiac myocytes, skeletal muscle fibres are mosaic cells containing nuclei with the paternal and maternal genome. This was first shown using chimeric mice, which at the same time was an elegant method of demonstrating the syncytial nature of the myofibre. MINTZ and BAKER (1967) aggregated blastomeres from genetically different mice strains in vitro, and implanted the embryos into a mouse uterus to complete development. Each of the strains was homozygous for an isocitrate dehydrogenase isoenzyme. The muscles of chimeric mice contained a third isoenzyme, which was a hybrid enzyme formed in cells that contained both genomes in a common cytoplasm. This finding was confirmed by BAKER and MINTZ (1969), GEARHART and MINTZ (1972), and FRAIR et al. (1979) for other marker enzymes. The mice strains used by FRAIR et al. (1979) to produce chimeras had electrophoretic variants of a nuclear-coded mitochondrial protein, malic enzyme, which is a polymer of several subunits. The muscles of the chimeric mice contained heteropolymers of this enzyme composed of subunits encoded by nuclei of different origin. This indicates that individual mitochondria accumulate products encoded by several nuclei, and that a myofibre is not subdividable into nuclear territories corresponding to the fused myoblasts. This result may be regarded as the final blow to the fragmentation or dedifferentiation theory for the origin of myoblasts in regenerating muscles.

2. Myotube Differentiation

a) Myofilaments

The synthesis of the myofibrillar proteins in muscles of chick embryos starts synchronously at HAMBURGER and HAMILTON (1951) stage 13–14 (MASAKI and YOSHIZAKI 1972). Thick and thin filaments occur at stage 18–22 (fetal age 3–4 days) and sarcomeres with Z and M lines are present at stage 24 (ALLEN 1973). The myotubes contract spontaneously. This seems to enhance protein synthesis, because cultured chick myotubes produce more contractile proteins when they are electrically stimulated (BREVET et al. 1976). The first myofibrils occur in the periphery of the myotubes, the centre is occupied by nuclei and glycogen granules. The formation of Z bands coincides with the back-to-back apposition of thin filaments, but neither Z band material nor M bridges are essential for the hexagonal array of thick and thin filaments (FIRKET 1967; FISCHMAN 1967; KELLY 1969).

The early development of myofibrils is the same in vivo and in vitro; it is not nerve-dependent (SHIMADA et al. 1967). The contractile proteins are synthesized by free polysomes (Figs. 103, 109). LARSON et al. (1973) observed rows of ribosomes alongside developing myofibrils and proposed that the filaments are synthesized in situ. In view of the self-assembling property of actin and myosin (FISCHMAN 1967), it appears more likely that the structures found by LARSON et al. (1973) were messenger RNA threads together with helically arranged ribosomes (GALAVAZI 1971). The filaments are ordered before they become parts of myofibrils. SHIMADA and OBINATA (1977) determined the polarity of the non-assembled thin filaments by decorating them with heavy meromyosin (arrowhead complexes, ISHIKAWA et al. 1969). An equal number of actin filaments pointed towards each myotube end corresponding to their orientation in the final sarcomeres. The relation to a nearby thick filament was in half of the cases "correct", i.e. cross-bridge action would have pulled the thin filament towards the middle of the thick filament. The rest of the filament pairs were "undetermined", and in only 2% the positioning was "wrong". The authors suggest that the rotational symmetry of the thick and thin filaments and

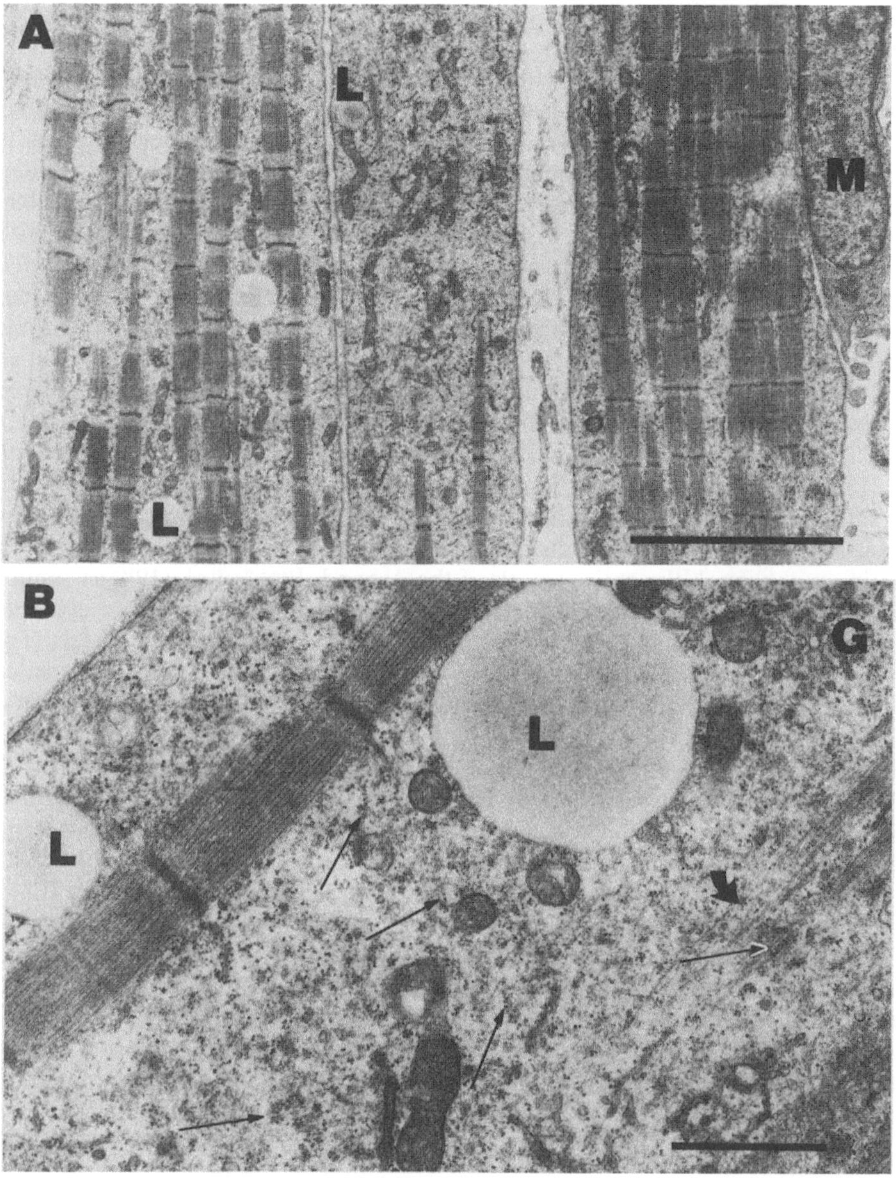

Fig. 109 A, B. Lumbrical muscle, newborn rat. **A** Low-power view of three immature muscle fibres and one presumed myoblast (*M*). The fibres are rich in aqueous sarcoplasm, mitochondria are scarce, and in places large triglyceride droplets (*L*) are seen. Bar, 5 µm. **B** Higher magnification. The sarcoplasm contains ribosomes (*straight arrows*), large lipid droplets (*L*), small mitochondria, and Golgi vesicles (*G*). Regular triadic junctions are absent. *Bent arrow*, longitudinally arranged bundle of microtubules; bar, 1 µm

the predetermined correct polarity disposes the filaments into the sixfold sarcomere pattern such that each thick filament is surrounded by six thin filaments.

Cultured myogenic cells of rat contain two types of myosin. Cytoplasmic myosin is present in myoblasts and myotubes, but skeletal muscle myosin is in the multinucleated cells only. The cytoplasmic myosin of the myotubes is in a submembranous rim all along the cell border and associated with so-called stress fibres at the ends of the cells. The skeletal muscle myosin is in the central part of the myotubes with and without sarcomere pattern. Double-staining with myosin antibodies against cytoplasmic and skeletal myosin reveals that the distribution of the two myosins is complementary. Cytochalasin B causes a diffuse distribution of both myosins; this effect is reversible (FALLON and NACHMIAS 1980).

The stress fibres consist of bundles of 6- to 8-nm filaments apparently linked to the incomplete sarcomeres of the ends of the myofibrils (PENG et al. 1981 a). The stress-fibre filaments contain actin (ISHIKAWA et al. 1969), and PENG et al. (1981 a) suggest that the thick filaments become positioned in a hexagonal fashion into the actin filament bundles of the stress fibres, which thus act as myofibril precursors. The thin filaments of the sarcomeres in young cells appear to be continuous through the entire length of the A band. PENG et al. (1981 a) assume that they eventually break in the middle of the A band and thus give rise to the sarcomeric I filaments. Several points speak against this hypothesis. The ability of myotubes and myoblasts to move and to adhere to a surface is related to the presence of cytoplasmic contractile proteins (FALLON and NACHMIAS 1980), and stress fibres have been shown to be contractile (KREIS and BIRCHMEIER 1980). Actin in non-muscle cells is not of the sarcomeric α-type (WHALEN et al. 1976; GARRELS and GIBSON 1976; STORTI and RICH 1976) and it is unlikely that the actin in stress fibres is identical with sarcomeric actin. Hence, the proposed mechanism of sarcomere formation which contrasts with SHIMADA's and OBINATA's (1977) observations (see above) appears rather speculative (PENG et al. 1981 a). Troponin and its components form the adult 38.5-nm periodicity already in developing sarcomeres, indicating that the regulatory proteins are assembled at their characteristic position, from the intial phases of myofibrillogenesis (OBINATA et al. 1979). Moreover, it is difficult to imagine how a continuous actin thread can split into fragments, each with opposite polarity. On the other hand, it is interesting to propose that the stress fibres constitute the matrix on which the sarcomeres develop in a directed and oriented fashion.

b) Intermediate Filaments

Intermediate filaments are prominent constituents of immature muscle cells (ISHIKAWA et al. 1968). Fibroblast-specific 10-nm filaments run longitudinally throughout the cytoplasm of early myotubes. This protein type disappears with maturation and, soon after sarcomere formation, smooth muscle-specific filaments occur and run transversely at the I–Z region of the myofibrils (BENNETT et al. 1979). The filaments are connected to the Z discs and linked to filamentous structures of the cytoplasm (PENG et al. 1981 a). WALKER et al. (1968) and WALKER and EDGE (1971) found links between the developing Z discs and

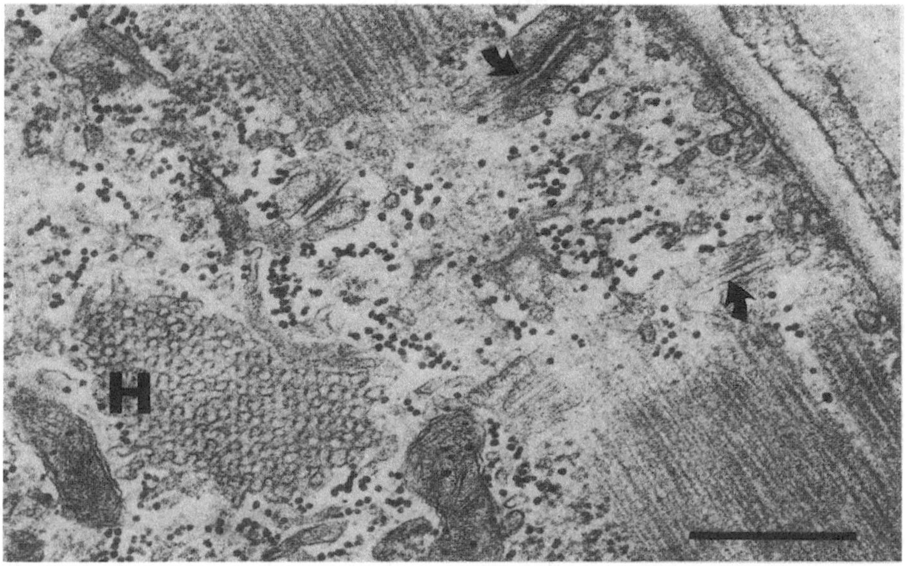

Fig. 110. Regenerating muscle fibre in a "free" graft of the rat soleus muscle. Beneath the plasma membrane (*left*) are two T tubules forming triadic junctions (*arrows*). The crystalline honey-comb structure (*H*) is assumed to originate from excessively growing T tubules. Bar, 0.5 μm

elements of the SR and concluded that the SR was involved in Z disc formation; WARREN (1973) showed that these filamentous links occur after the formation of the Z discs. (With respect to the array of 10-nm filaments in mature muscle, see Sect. C.II.)

c) Sarcoplasmic Reticulum and T System

The SR originates from the rough endoplasmic reticulum of the myotubes, and the T system originates from invaginations of the plasma membrane (EZERMAN and ISHIKAWA 1967). The formation of both systems in cultured chicken muscles starts at the early myotube stage, and labelling with ferritin shows that only the future T system, but not the future SR, is accessible. The junctional structures (triads, diads) are formed concomitantly with the occurrence of myofibrils (WALKER et al. 1975). The orientation is initially random (WALKER and SCHRODT 1968). The T tubules develop bulbous outpocketings which resemble caveolae of the plasma membrane, and which give rise to new T-tubule branches. These outpocketings often become abundant and form elaborate. three-dimensional structures of an almost crystalline pattern. The array is hexagonal or tetragonal (ISHIKAWA 1968). In muscle pathology these networks are well known as "honeycomb structures" (MAIR and TOMÉ 1972); they are found in regenerating muscles as well (Fig. 110). The honeycomb structures indicate high-speed or unbalanced growth, and they are inconspicuous in normal rat muscle fibres developing in vivo (KELLY and ZACKS 1969a; SCHIAFFINO and MARGRETH 1969; KELLY 1971). The formation of the SR in fetal rat muscles precedes the development of the T system, and the SR transiently forms peripheral couplings with the plasma membrane (KELLY 1971) (Fig. 54C). Cultured rat muscle cells behave like cultured chicken muscle cells, and also here labyrinthine T-system networks are formed, suggesting that the T system formation in vitro is unco-ordinated (SCHIAFFINO et al. 1977). Human muscle fibres growing in tissue culture develop T systems and SR, but the maturation is delayed, compared with cultured avian muscles (HEIMAN-PATTERSON 1983).

Muscle fibres of neonatal rats contain numerous T tubules running both transversely and longitudinally; the system has many blind-ending transverse elements (KELLY 1980). In embryonic chicken fibres, only 12% of the T tubules are oriented transversely. The volume of the T system increases before hatching from 0.02% to 0.3% of the fibre volume, and the total SR volume increases from 0.07% to 4.1% of the fibre volume. Initially, there is a large amount of vesicular reticulum that comprises 9.7% of the cell volume, but these vesicles disappear during later development (CROWE and BASKIN 1977). The volume fraction of the T system in the extensor digitorum longus of mice increases from 0.08% to 0.4% during postnatal development. The increase in the soleus muscle is from 0.04% to 0.22%. The ratio of T-membrane area to surface membrane area increases in the extensor digitorum longus muscle from 0.6 to 3.1, and in the soleus muscle from 0.15 to 1.8. The increase in SR volume is from 1.1% to 5.5% and from 1.7% to 2.9%, respectively (see also Table 3). The development of the SR and T system in mice is almost complete 2 weeks after birth; it appears to be more rapid in the soleus than in the extensor digitorum longus muscle (LUFF and ATWOOD 1971). The adult array of the SR in rat psoas muscle is attained 2 weeks after birth (SCHIAFFINO and MARGRETH 1969).

The membranes of the SR pass through remarkable developmental changes. The activity of the Ca^{2+}-transport ATPase in chicken muscles increases by a factor of 20 from day 10 of gestation to full maturation; correspondingly, the density of 8-nm intramembrane particles assumed to represent 4 ATPase molecules each (see Sect. C.III), increases from 200/μm^2 to 4,000/μm^2 (BASKIN 1974; BOLAND et al. 1974; TILLACK et al. 1974; MARTONOSI et al. 1977; MARTONOSI 1982a, b). Cultured myoblasts of rat or chick accumulate Ca^{2+}-transport ATPase during fusion, but the activity in cultured muscle cells does not reach the same level as in muscle cells differentiating in vivo (MARTONOSI et al. 1977; MARTONOSI 1982b). The Ca^{2+}-transport ATPase is synthesized by membrane-bound polysomes of the rough endoplasmic reticulum from which the SR evolves (CHYN et al. 1979).

d) Innervation

α) Acetylcholine Receptors and Acetylcholinesterase

The differentiation of myoblasts into myotubes is associated with the occurrence of specific membrane proteins. The most thoroughly studied ones are the acetylcholine receptors (AChRs). They are lacking in dividing myoblasts. Early myotubes are characterized by a uniform distribution of AChRs along the entire surface (FAMBROUGH and RASH 1971; VOGEL et al. 1972). This localization resembles that in denervated (MILEDI 1960) or embryonic muscle fibres (DIAMOND and MILEDI 1962). AChRs have been visualized as 15-nm intramembrane particles in cultured chick myotubes (COHEN and PUMPLIN 1979) (see also Sect. C.IV. 3b δ, and Fig. 63).

As the myotubes develop further, in culture, the receptor density becomes non-uniform, with randomly distributed "hot spots" of clustered AChRs. This process is nerve independent (SYTKOWSKI et al. 1973; PRIVES et al. 1976; BEKOFF and BETZ 1976; JACOB and LENTZ 1979; FAMBROUGH 1979; FRANKLIN et al. 1980; WALSH and RITTER 1981), but muscle activity increases both total AChR and the number of AChR clusters (COHEN and FISCHBACH 1973; SHAINBERG et al. 1976).

The effect of innervation of the AChR distribution is often studied in cocultures of nerve and muscle cells. Functional synapses may form across species lines. Myoblasts have been combined with dissociated spinal cord cells, with fetal spinal cord explants, or with ciliary ganglia (CRAIN 1964, 1970; BORNSTEIN et al. 1968; JAMES and TRESMAN 1968; VENERONI and MURRAY 1969; SHIMADA et al. 1969; PETERSON and CRAIN 1970; CRAIN et al. 1970; FISCHBACH 1970, 1972, 1982; HARRIS et al. 1971; SHIMADA and KANO 1971; SHIMADA and FISCHMAN 1973; FISCHBACH and COHEN 1973; BIRD and JAMES 1974; BETZ 1976a, b; COHEN and FISCHBACH 1977; RUBIN et al. 1980; ZISKIND-COHNHAIM and DENNIS 1981).

After functional contact with the motor axons is made (FRANK and FISCHBACH 1979; DENNIS et al. 1981; FISCHBACH 1982; CHOW and COHEN 1983), clusters of AChRs packed at high density appear at sites of neuromuscular contacs (BEVAN and STEINBACH 1977; BRAITHWAITE and HARRIS 1979). During further development, the AChRs at the endplate increase in number, and the density of receptors at the extrajunctional membrane gradually declines (DIAMOND and MILEDI 1962; BEVAN and STEINBACH 1977; ANDERSON and COHEN 1977; ANDERSON et al. 1977).

The source of the initial AChR clusters at the endplate is unknown. The site of biosynthesis of AChRs is the rough endoplasmic reticulum (MERLIE et al. 1981). Newly synthesized AChRs are localized in the Golgi apparatus from where they are transported to and incorporated into the plasma membrane (FAMBROUGH and DEVREOTES 1978; FAMBROUGH 1979; ROTUNDO and FAMBROUGH 1980; ATSUMI 1981) via coated vesicles (BURSZTAIN and FISCHBACH 1984). AChRs are degraded in secondary lysosomes (FAMBROUGH 1979; PUMPLIN and FAMBROUGH 1982). The initial junctional receptors are either newly synthesized, or they migrate within the membrane towards the nerve contact area (EDIDIN and FAMBROUGH 1973; AXELROD et al. 1976).

There is ample evidence that it is not the electrical activity of the muscle fibre, or synaptic transmission, that induces the AChR accumulation at the developing endplate, but possibly a water-soluble diffusible factor (PODLESKI et al. 1978; CHRISTIAN et al. 1978; FISCHBACH et al. 1979; JESSELL et al. 1979; AXELROD et al. 1981; BAUER et al. 1981; FISCHBACH 1982; CONOLLY et al. 1982). Contact with positively charged Latex beads induces accumulation of AChRs as well (PENG et al. 1981b). Blockade of nerve and muscle activity by a high K^+ concentration and by tetrodotoxin does not inhibit the accumulation of

AChRs in rat myotubes co-cultured with neuroblastoma cells (STEINBACH 1974). On the other hand, rat myotubes that are in vitro functionally innervated by rat or *Xenopus* neurons do not show clusters of AChRs at their endplates (KIDOKORO 1980; KIDOKORO and EH 1981). Myotubes co-cultured with spinal cord develop increased numbers of AChRs even when they remain uninnervated (FISCHBACH and COHEN 1973; BETZ and OSBORNE 1977). These observations show that different capabilities are required of the nerve to cause AChR accumulation and to form functional synapses.

The muscle fibres of rat intercostal muscles occur at the 13th and 14th days of fetal age. The motor nerves arrive and establish functional contacts at day 14–15; junctional AChR clusters occur at day 16 (DENNIS et al. 1981). The extrajunctional AChR density remains high until 3 days after birth, and then declines during the first postnatal week (DIAMOND and MILEDI 1962; BERG et al. 1972; FAMBROUGH and HARTZELL 1972; BEVAN and STEINBACH 1977; BRAITHWAITE and HARRIS 1979; ZISKIND-COHNHAIM and DENNIS 1981). Tetrodotoxin applied at day 17 of fetal life blocks the electrical activity of the nerve but not the aggregation of AChRs at the endplate zone. According to ZISKIND-COHNHAIM and BENNETT (1982), tetrodotoxin blockade induces sprouting of the terminal axons and formation of multiple extrajunctional AChR clusters not topically related to the axonal sprouts. These authors conclude that electrical activity is necessary to prevent the occurrence of extrajunctional AChR clusters once the muscle fibre is innervated.

The AChRs clustering at the developing endplate are of the extrajunctional type. Mature synaptic receptors have a longer life time (10 days vs 30 h) than extrasynaptic embryonic AChRs, and the channels they regulate have shorter open times (REINESS and WEINBERG 1981; SAKMANN and BRENNER 1978; MICHLER and SAKMANN 1980). The protein molecules are different, and mature and embryonic AChRs are distinguishable by immunofluorescence (REINESS and HALL 1981). The conversion of junctional AChRs occurs late during the development of the neuromuscular junction (second postnatal week in rat, FISCHBACH and SCHUETZE 1980; third week after hatching in chicken, BETZ et al. 1980). The conversion is induced by the motor axon, but when a rat muscle is denervated before the conversion it nonetheless takes place. This indicates that the neural signal must act at the earliest stages of junctional development (BRENNER et al. 1983).

Whereas AChRs accumulate at the endplate in response to an unknown humoral factor, the appearance of acetylcholinesterase during endplate formation in vitro and in vivo depends on muscle activity. Acetylcholinesterase is lacking when muscle activity is prevented by curare, α-bungarotoxin, or chronic nerve conduction block (GIACOBINI et al. 1973; RUBIN et al. 1980; CANGIANO et al. 1980).

β) Neuromuscular Contacts

It is unknown how the axons are guided towards the muscle fibres. Originally it was assumed that the AChRs attracted the axon and that a hot spot was the site at which the future endplate was formed (KATZ and MILEDI 1964).

Nevertheless, the nerves do not grow towards a hot spot (FRANK and FISCHBACH 1979). Innervation takes place when muscle activity is blocked by curare (COHEN 1972), and denervated rat muscle fibres are also re-innervated when the AChRs are blocked by α-bungarotoxin (JANSEN and VAN ESSEN 1975). The anticholinesterase agent, neostigmin, lowers the number of AChRs, but in duck embryos it does not impair the development of normal neuromuscular junctions; in adult animals neostigmin causes degeneration of the endplates (SOHAL and BOYDSTON 1982). In analogy to experiments with mature muscles (GLICKSMAN and SANES 1983) (see Sect. C.IV.2), it is conceivable that basal lamina-attached molecules are involved both in nerve contact and in AChR clustering. Small aggregates of basal lamina occur on myotubes shortly after myoblast fusion, they tend to be associated with clusters of AChRs and with cytoplasmic filamentous material (JACOB and LENTZ 1979). A basal lamina is regularly found in the synaptic cleft of newly formed neuromuscular junctions (KELLY and ZACKS 1969b; SHIMADA and KANO 1971; BENNETT et al. 1974; KULLBERG et al. 1977; BENNETT 1983) (Fig. 111). BAILEY et al. (1979) failed to find, in chick embryos, collagen type V which is exclusive to the basal lamina, before innervation took place; this may be due to the fact that the immunofluorescence technique is not sensitive enough to demonstrate the initial formation of the basal lamina (BENNETT 1983). According to CHIU and SANES (1984) the basal lamina of muscle fibres in rat intercostal muscles is formed between embryonic day 15 and 17, whereas synapses are formed on embryonic day 14. Initially extrasynaptic and synaptic regions have the same antigens, subsequently the synaptic basal lamina differentiates by loosing those antigens that are specific for the extrasynaptic basal lamina. GRUMET et al. (1982) and RUTISHAUSER et al. (1983) describe a cell surface glycoprotein involved in adhesion between neurons; the same glycoprotein is present on chick embryo skeletal muscle cells. Blockade of the molecule by an antibody prevents adhesion of axons to myotubes, whereas binding of neurites to fibroblasts or a collagen substrate is unimpaired. Hence, binding involving this specific glycoprotein may be an early step in synaptogenesis. Fetal rat muscles produce a factor that induces neurite outgrowth and enhances acetylcholine synthesis. This factor influences only anterior horn cells but not posterior horn cells (SMITH and APPEL 1983).

Rat muscle cells co-cultured with dissociated fetal spinal cord cells become innervated after about 3 weeks. The axon terminals are filled with synaptic vesicles, and the postsynaptic membrane is thickened. Secondary synaptic folds do not develop. Nerve stimulation elicits action potentials and synchronized twitches, both of which are blocked by curare. The spontaneous contractions of the myotubes, and their fibrillation potentials continue after curare block (CRAIN et al. 1970; BIRD and JAMES 1974). BIRD and JAMES (1974) describe junctions with several apparently independent axon terminals. This is in agreement with the observation by BETZ (1976a) that chick muscle cells co-cultured with several ciliary ganglia become polyneurally innervated.

The nerve-muscle contacts formed in vitro on rat myotubes have been randomly located along the myotubes, in contrast to the specific localization of endplates in the middle of the fibre which is characteristic of normal developing muscles (DIAMOND and MILEDI 1962; BEVAN and STEINBACH 1977; BENNETT and PETTIGREW 1974a). The synapse formation in organ cultures of the entire body wall of rat embryos is similar to that in vivo (ZISKIND-COHNHAIM and DENNIS 1981).

The first nerve-muscle contacts in rat intercostal muscles in vivo consist of clusters of axons apposed to groups of myotubes. The muscle cell membrane is covered with a rudimentary basal lamina. At day 18 of fetal life, large primary myotubes are surrounded by groups of small secondary myotubes. Clusters of axonal sprouts mutually innervate large and small myotubes. The postsynaptic membrane of the large myotubes is thickened. The myotubes of intercostal muscles have detached at birth, and bundles of axons contained within a single Schwann cell innervate each myofibre. Secondary synaptic clefts occur during the second postnatal week, i.e. concomitantly with the conversion of embryonic into mature AChRs (see above). Acetylcholinesterase may, using histochemistry, be shown in rats from day 18 of fetal life (KELLY and ZACKS 1969b). In proximal limb muscles of mouse, it is present at day 15 and in distal muscles at day 17. Electron microscopic signs of synapse formation are present at the same time (CARRY et al. 1983). Clusters of axonal sprouts each 0.1–0.3 µm in diameter and loosely covered by a Schwann cell projection are characteristic of early stages of endplate formation (SEINSCH 1974; SISTO DANEO and FILOGAMO 1974, 1975; FIDZIANSKA 1980b; PILAR et al. 1981; ONTELL 1977; RUBINSTEIN and KELLY 1981; CARRY et al. 1983) (Fig. 111). This picture may be found in limb muscles of rat until the second postnatal week (KORNELIUSSEN and JANSEN 1976; SCHMALBRUCH, unpublished). The physiological implications of the fact that the motor axons are in direct contact with each other and not separated by Schwann cells – not only at the primitive endplate but also in the nerves – remain unknown.

Newly formed neuromuscular junctions respond to nerve stimulation with subthreshold endplate potentials only; later, the endplate potentials become large enough to evoke action potentials of the muscle fibre. The neuromuscular transmission during repetitive stimulation is susceptible to fatigue, probably because the axon terminal fails to sustain adequate levels of transmitter release (DIAMOND and MILEDI 1962; BENNETT et al. 1973; BENNETT and PETTIGREW 1974a, b). In the (cross-striated) chick iris muscle, the synthesis of acetylcholine increases steeply just before hatching, and 1 day after hatching transmission is far more reliable than 2 days before hatching. The morphological maturation of the junctions is completed 10 days after birth (PILAR et al. 1981).

γ) Polyneural Innervation

It is physiologically well established that muscle fibres developing in vivo are innervated by several motoneurons which all terminate at the same endplate (REDFERN 1970; BAGUST et al. 1973; BENNETT and PETTIGREW 1974a; JANSEN et al. 1976; BROWN et al. 1976; O'BRIEN et al. 1978; BETZ et al. 1979). The redundant synaptic inputs are eliminated during the first 2 postnatal weeks in rat, and during the first 4 postnatal weeks in chicken. The fact that a muscle fibre is innervated by more than one motoneuron is difficult to demonstrate using electron microscopy, because different terminal axons may ultimately con-

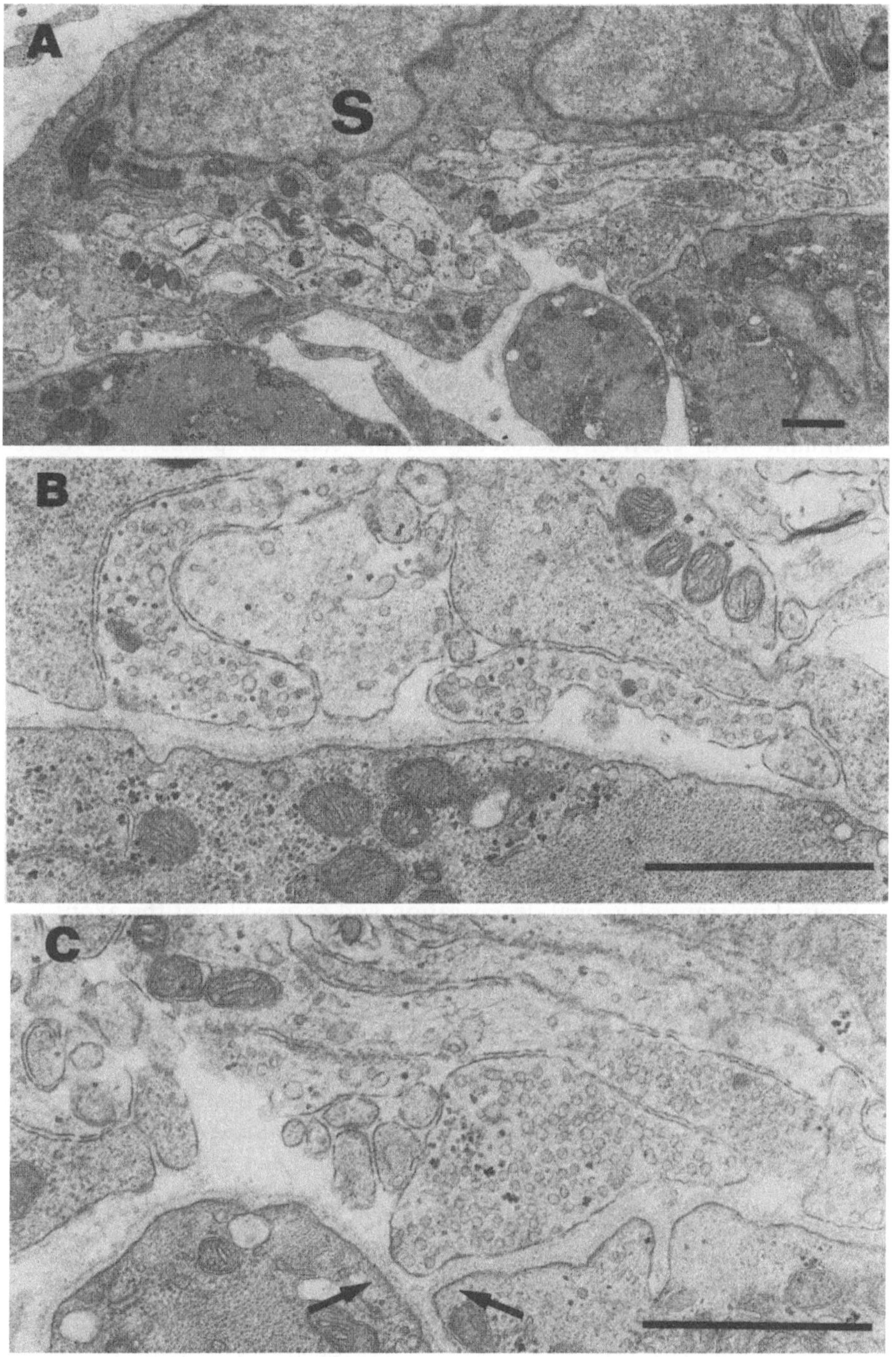

Fig. 111 A–C. Lumbrical muscle, newborn rat. Developing neuromuscular junction. **A** A group of muscle fibres some of which are part of clusters (not shown) is approached by a large number of non-myelinated axons covered by Schwann cells (*S*). **B, C** Higher magnifications of the same area. The axon terminals contain synaptic vesicles, and are not separated from each other by Schwann cell projections. The postsynaptic membrane is thickened but junctional folds have not yet formed. One of the terminal axons seems to approach two muscle fibres at the same time (*arrows*). Bars, 1 μm

nect to the same motor axon. Light micrographs of immature muscle impregnated with metal salts have been published, showing that several axons approach the same endplate (e.g. BROWN et al. 1976; O'BRIEN et al. 1978). These micrographs are ambiguous; this is not surprising considering the diameter of the still unmyelinated axons (0.1–0.3 µm), their grouping in the same Schwann cell, the diameter of the still-clustered myofibres (about 5 µm) (Figs. 111–113), and the unspecificity of this histological technique. The results of a combined nitroblue tetrazolium-acetylcholinesterase staining procedure (LETINSKY and DECINO 1980; LETINSKY and MORRISON-GRAHAM 1980) appear more assuring, but one is still faced with the problem that the structures to be shown are at or beyond the theoretical limit of the light microscope. MORRISON-GRAHAM (1983) studied synapse elimination in focally innervated frog muscle fibres, and supplemented this technique with electron microscopy of the same endplates. Half of the fibres are singly innervated 3 weeks after metamorphosis, but 20% of the endplates are still dually innervated after 1–2 years. This ratio falls to less than 10% during the subsequent 2–3 years.

According to BROWN et al. (1976), the number of motoneurons remains constant after birth, and each motoneuron loses peripheral branches until each muscle fibre receives synaptic inputs from one and only one motoneuron. Thus, the size of the motor units decreases during maturation. Nevertheless, there is evidence, that at least in rat, the period of natural motoneuron death extends into the postnatal phase; death of motoneurons may account for, or at least contribute to, the loss of polyneural innervation (ROOTMAN et al. 1981; BENNETT et al. 1983b; SCHMALBRUCH 1984). The period of natural motoneuron death in chicken precedes the period during which the polyneural innervation is eliminated (OPPENHEIM and MAJORS-WILLARD 1978). The cellular mechanism by which redundant axons are removed is controversial. KORNELIUSSEN and JANSEN (1976) found no signs of axonal degeneration and assumed that the axons "retracted". ROSENTHAL and TARASKEVICH (1977) and SCHMALBRUCH (unpublished) found degenerating axons within developing rat muscles.

The redundant axon terminals do not abruptly cease to function but continue to elicit small subthreshold endplate potentials which occasionally are present even in one-month-old rats, i.e. 2 weeks after the muscle fibres have become mononeurally innervated (TAXT et al. 1983). This indicates, in harmony with the findings in frog muscles (see above), that the anatomical removal of redundant axons, whether by degeneration or "retraction", is a prolonged process.

3. Histogenesis

Myogenesis in vitro produces elongated myotubes which run in all directions and even branch (HOLTZER et al. 1975; KONIGSBERG 1979) (Fig. 101D), but development in situ results in a highly ordered tissue. The myotubes and myoblasts in fetal rats are longitudinally oriented and clustered in groups. This is evident at day 16 of gestation. The groups of clearly myogenic cells are surrounded by undifferentiated cells which frequently extend pseudopodia between adjacent myotubes. Adjacent myotubes are coupled by gap junctions; in places, patches of basal lamina material are present. At 18 days gestation, each cluster of myogenic cells is dominated by one relatively large myotube with central glycogen and peripheral myofilaments. The entire cluster is ensheathed by a common basal lamina. The one large myotube represents the most mature cell in the cluster (primary myotube) along which myoblasts in linear sequence fuse to form later-generation (secondary) myotubes. The muscle cells within a cluster mature, detach, and form their own basal lamina sheaths. The break-up of clusters occurs in different muscles at different ages. Intercostal muscles at birth consist of individual myofibres, whereas in limb muscles clusters are found throughout the first postnatal week. The distal lumbrical muscles mature later than the soleus or extensor digitorum longus muscles. Because the secondary myotubes initially do not run the entire length of the muscle, the number of muscle cells in a given cross-section is smaller before than after cluster break-up. Fibre counts in immature muscles have to be done using electron microscopy, because light microscopy does not allow identification of all cells in a cluster. There are two morphologically distinct mononuclear cell types within the clusters. Some are undifferentiated and are probably myoblasts and satellite cells, others are fibroblast-like and have a well-developed rough endoplasmic reticulum. They have been termed "interstitial cells" (ONTELL 1977); their nature is obscure but one might speculate that they are fibroblasts trapped between the fusing myoblasts. They are possibly involved in the process of cluster break-up (Figs. 112–115).

Myogenesis in vivo, from initial myotube formation to maturation of myofibres, has most systematically been studied in rat (KELLY and ZACKS 1969a, b; LANDON 1970c, 1971; ISHIKAWA 1970; KELLY and SCHOTLAND 1972; RASH and STAEHELIN 1974; ONTELL 1977; ONTELL and DUNN 1978; KELLY 1978a, 1979; BETZ et al. 1979). The time course may be different in other species, but the sequence of events is the same (rabbit, WITTIG 1968; mouse, PLATZER 1978; lamb, ASHMORE et al. 1972; pig, SWATLAND and CASSENS 1973; BEERMANN and CASSENS 1977; CAMPION et al. 1981; cow, RUSSEL and OTERUELO 1981; man, ISHIKAWA 1970; GAMBLE et al. 1978; FIDZIANSKA 1980a, b). The intrafusal fibres of a muscle spindle are formed by the fusion of myoblasts giving rise to one cluster and successive generations of myotubes (KOZEKA and ONTELL 1981) (Fig. 116).

The mechanisms by which the developing myofibres and also the approaching nerves are organized are poorly understood. An organizing function has been ascribed to fibronectin, which is an external protein synthesized by fibroblasts. It may bind to cytoplasmic actin and thus establish transmembrane connections (SINGER 1979; KESKI-OJA et al. 1980). Premuscle masses in chick embryos are devoid of fibronectin, the splitting into muscle anlagen is associated with the appearance of fibronectin-accumulating cells in the cleavage furrows. Chicken myoblasts and myotubes in culture attach to solid surfaces, but only in the presence of fibronectin derived from serum or from fibroblasts contaminating the culture. The myoblasts align themselves along oriented streaks of purified fibronectin, according to CHIQUET et al. (1981), who suggest that the formation of myotubes in a certain spatial order during morphogenesis is regulated by a fibronectin-containing matrix produced by connective tissue cells.

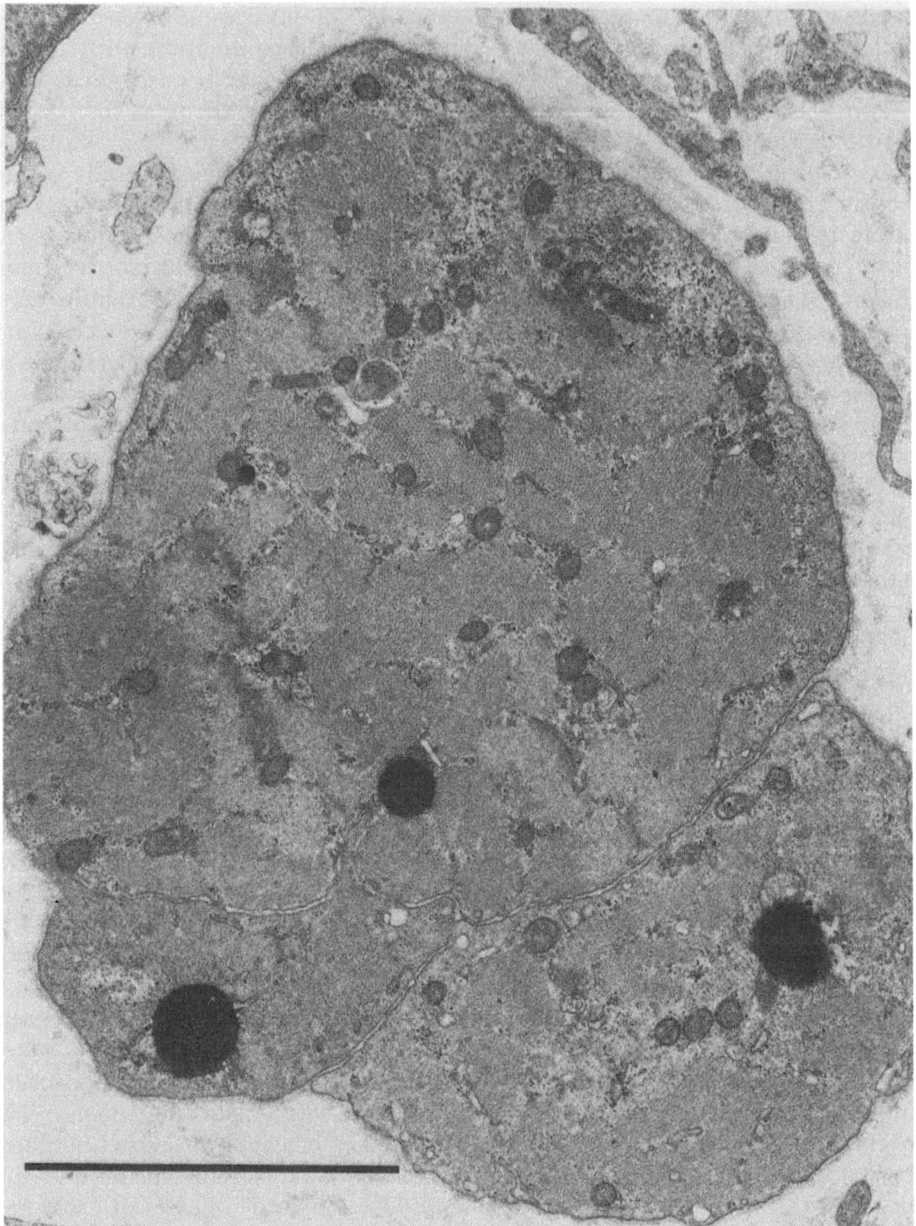

Fig. 112. Lumbrical muscle, newborn rat. Three myofibres form a cluster and are enclosed by a common basal lamina. Note three very electron-dense lipid droplets (see also Fig. 113). Bar, 5 µm

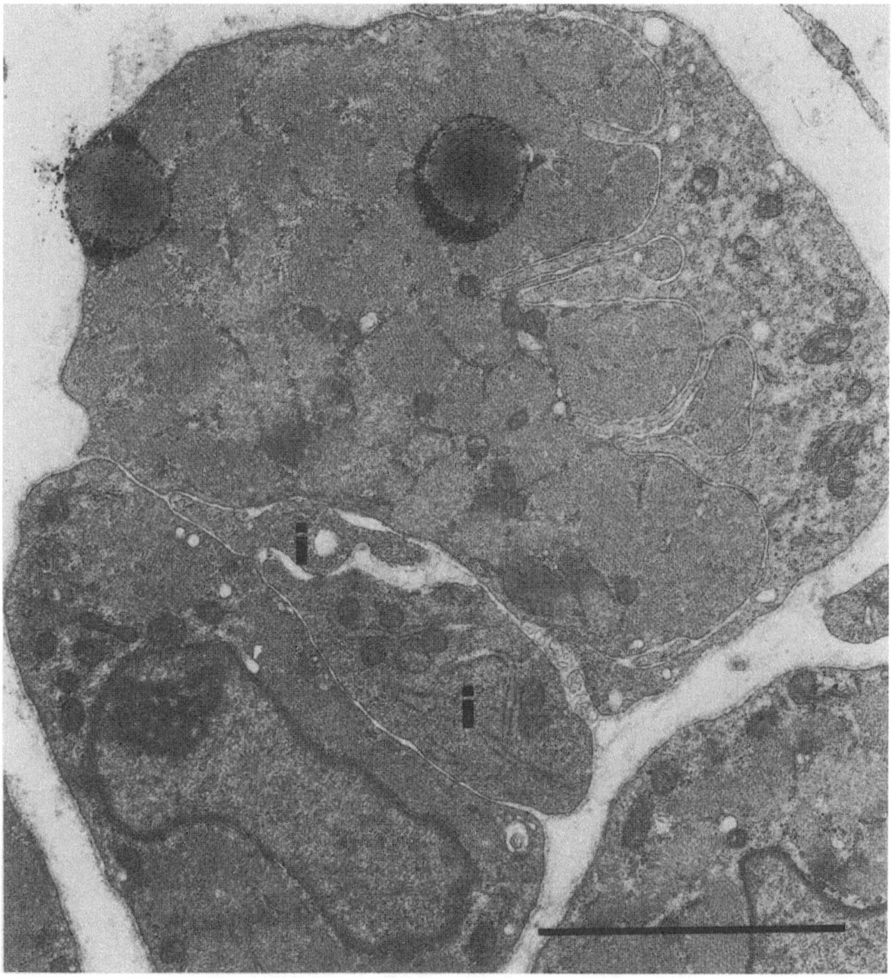

Fig. 113. Lumbrical muscle, newborn rat. A cluster of three muscle fibres apparently in the process of break-up. The cells are separated by projections of fibroblast-like "interstitial cells" (*i*). Two large lipid droplets are seen. The lipid was not completely dissolved during the glutaraldehyde fixation step, but the osmium-lipid complex, during later stages of preparation, has diffused into the surrounding cytoplasm. Bar, 5 μm

It is tempting to speculate along this line of thought about the fibroblast-like cells within the fetal myotube clusters (Figs. 113, 115) (see above).

The pathways taken by the large nerves to the muscles are preprogrammed. No evidence for competition between adjacent segmental nerves has been found in chick embryos (LANCE-JONES and LANDMESSER 1980a); when the spinal cord is excised and reimplanted upside-down, or when the limb buds are dorsoventrally rotated, the motoneurons still try to reach their predetermined targets (LANCE-JONES and LANDMESSER 1980b; FERGUSON 1983). Limb buds without muscles become normally innervated, but the muscle nerves proper do not devel-

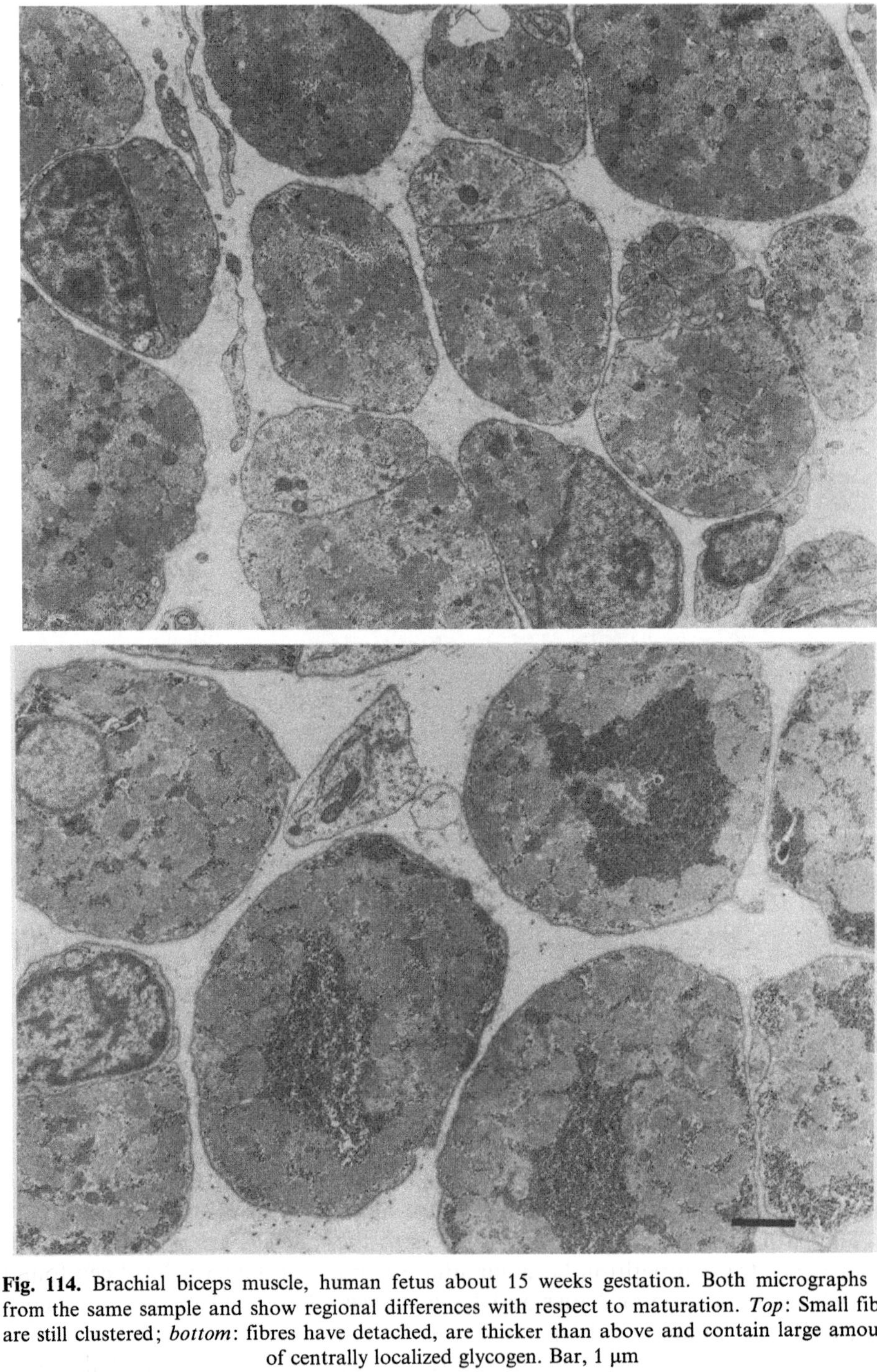

Fig. 114. Brachial biceps muscle, human fetus about 15 weeks gestation. Both micrographs are from the same sample and show regional differences with respect to maturation. *Top*: Small fibres are still clustered; *bottom*: fibres have detached, are thicker than above and contain large amounts of centrally localized glycogen. Bar, 1 μm

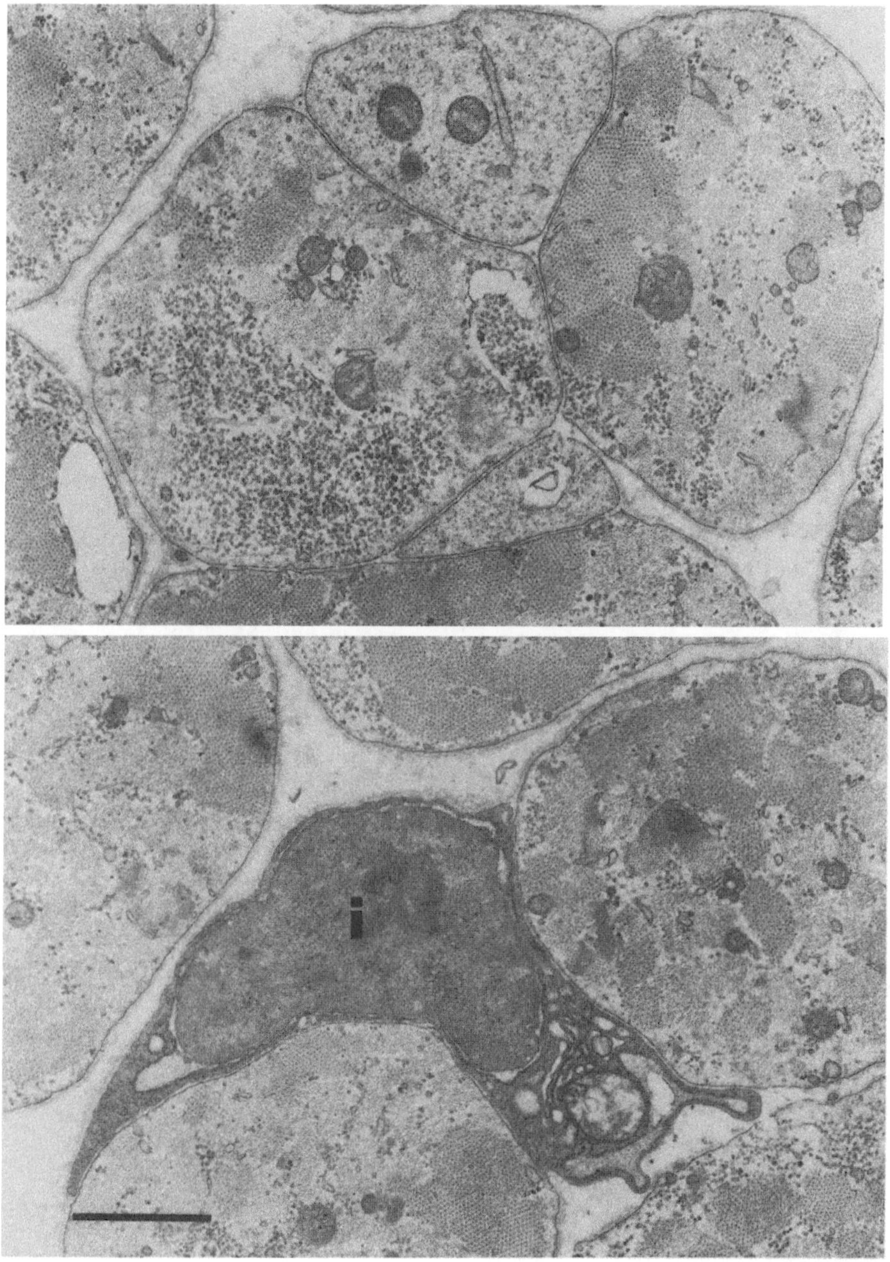

Fig. 115. Brachial biceps muscle, human fetus about 15 weeks gestation. Clusters of immature muscle fibres are together with undifferentiated cells enclosed by a common basal lamina. In one place, the myofibres are separated by a fibroblast-like "interstitial cell" (*i*) which may be involved in the break-up of the clusters. Bar, 1 μm

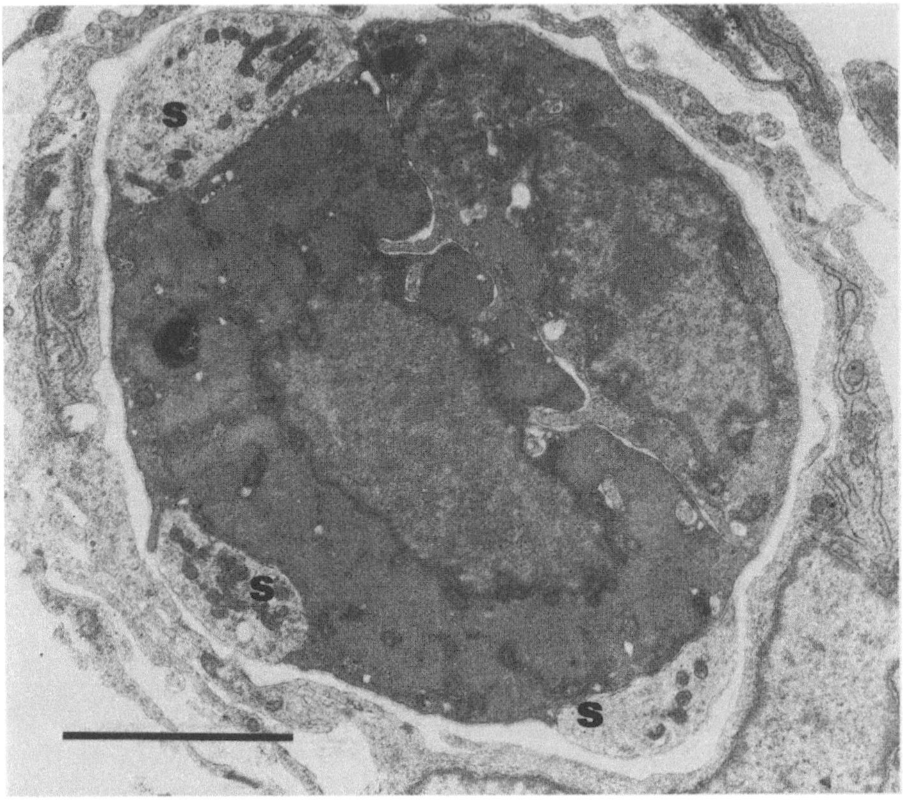

Fig. 116. Lumbrical muscle, newborn rat, muscle spindle. Two intrafusal fibres are seen which still are clustered. The cluster is encircled by a sensory ending three profiles of which are visible (*S*). The developing spindle capsule is seen as well. Bar, 5 μm

op (LEWIS et al. 1981; BENNETT et al. 1983a). HO et al. (1983) observed, in grasshoppers, large mesodermal cells that arise early during development and form a scaffolding for later developing muscles and motoneuron growth cones. Whether this observation is relevant for the understanding of vertebrate morphogenesis remains to be investigated.

IV. Regeneration

1. Epimorphic and Tissue Modes

Regeneration of muscle fibres takes place either in the epimorphic mode or in the tissue mode (CARLSON 1973). Classical examples for the epimorphic mode are the regeneration of a lost limb in urodele amphibians and larval anurans, and the regeneration of the lost tail in lizards. It is essential to recognize that muscle is not regenerating alone, but as part of a process which includes all components of the limb. Epimorphic regeneration repeats embryonic development; the formation of muscle anlagen in the new limb is nerve-independent (GRIM and CARLSON 1974, 1979). Despite controversies concerning the cellular events, it is generally agreed upon that a population of morphologically undifferentiated cells (blastema) gives rise to the different tissues within the new limb or tail (for reviews, see CARLSON 1973; MAURO 1979).

The tissue mode of regeneration is the reparative process which follows damage to the muscle alone. It is the only type of muscle regeneration in mammals. It is interesting to note that the same muscles may be capable of participating in either type of regenerative process if the animal is able to regenerate amputated limbs. This suggests that the cellular events are basically identical and that muscle fibres are not formed in different ways in amphibia and mammals.

In this review I wish to concentrate on tissue mode regeneration in mammals. The field is covered by several informative review articles (CARLSON 1973; ALLBROOK 1981) and proceedings of symposia (MAURO et al. 1970; MAURO 1979).

2. Muscle Fibre Necrosis

Muscle regeneration follows necrosis of muscle fibres; the stimulus for the onset of regeneration is unknown. Necrosis is a spontaneous event during the course of many neuromuscular disorders; it occurs in normal muscles after microtraumata or after extraordinary efforts (migrating birds, SCHUMANN and BERGER 1974; marathon runners, HIKIDA et al. 1983; exhaustive running of mice, VIHKO et al. 1978). The fact that certain types of exercise induce muscle fibre necrosis is also evident from the increase in specific muscle enzymes in blood. The serum creatine kinase level increases grossly in untrained subjects after eccentric contractions, and the muscle fibres are damaged (NEWHAM et al. 1983a, b; FRIDÉN et al. 1981, 1983; FRIDÉN 1983). "Eccentric contraction" means that the contracted muscle is stretched by a load; this may occur during stepping or jumping.

Experimental damage to muscles has been inflicted mechanically (crush, cut), by myotoxic substances (local anaesthetics, e.g. bupivacaine), by ischemia, by heating or cooling, or by acute overload (removal or denervation of synergists). Denervation, however, does not induce fibre necrosis in the first place. Interestingly, incandescent light at an environmental temperature of 39° C causes necrosis of extraocular muscles of rats (O'STEEN et al. 1975).

A physiological form of muscle necrosis has been reported for fetal development. Fetal sarcolysis is assumed to be involved in remodelling the muscles during ontogenesis. Muscle break-down during ontogenesis of insects is beyond doubt, but the histological evidence presented for this phenomenon in mammals (for review, see HÄGGQUIST 1931, 1956) has for a long time been ambiguous at best. Fetal muscle is extremely susceptible to damage, and the morphological reaction to improper handling during preparation is the same as to a damaging factor in vivo. This criticism also applies to electron microscopic studies of fetal sarcolysis (e.g. WEBB 1972). GRIM (1978), however, reported an increase in acid phosphatase activity and phagocytosis in early human fetuses during the period preceding the formation of muscle anlagen of the hand. SEINSCH and SCHWEICHEL (1974) demonstrated, in embryonic mice, by light microscopy using intravital Nile blue staining and by electron microscopy, necrosis and phagocytosis of muscle cells in the back and paraxial musculature. This form of muscle-cell death was observed only between day 9 and day 13 of gestation, before the formation of myotube clusters and of innervation.

Necrosis of mature muscle fibres is usually segmental; the large number of nuclei and the size of the cell assure that fragments of the fibre survive. These fragments may participate in regeneration. One of the earliest events during necrosis is the occurrence of defects of the plasma membrane (SCHMALBRUCH 1975). It is believed that the influx of extracellular calcium is the ultimate cause of cell death. The fibre contracts locally, and contraction clots are formed. Phagocytes invade the basal lamina tube and remove the cell debris. The satellite cells proliferate within the persisting basal lamina tube and fuse with each other and the surviving fibre fragments.

The time course depends on whether the blood supply remains intact or not. Phagocytosis of rat muscle fibres may be completed in <48 h. If the blood supply is interrupted, for instance by complete devascularization or heating the muscle to 80° C, the phagocytotic response is greatly delayed. One week after the lesion, muscles fibres may (by light microscopy) still appear "normal": the cross-striation is preserved and only the absence of stainable nuclei indicates that the cell, in fact, is dead (CLARK 1946; SCHMALBRUCH 1976b).

3. Regeneration In Situ

Many workers have performed experiments with mechanical and thermal lesions of a muscle, and have found satellite cell activation, proliferation of mononuclear cells, myotubes, and maturation of young myofibres in the previously injured muscle region (REZNIK 1969b; CHURCH 1970a, b; REZNIK and ENGEL 1970; TERÄVÄINEN 1970; VRACKO and BENDITT 1972; HALL-CRAGGS 1971, 1972, 1974; RILEY 1973). CLARK (1946) published the most thorough light microscopical study before the satellite cells were discovered; the interpretation of the findings still follows VOLKMANN'S (1893) concept of regeneration per continuum which holds that regeneration is by outgrowth of surviving fibre fragments (see Sect. G.I).

Transverse division of the muscle fibres by a cut or segmental crush divides the fibres into an innervated central and non-innervated peripheral part. The non-innervated peripheral fibre segments survive and are assumed to be reconnected with the central part (RILEY 1973). Denervation and re-innervation alters the normal checkerboard pattern of fibre types of the muscle, and the fibre types become grouped. HALL-CRAGGS (1971, 1972) finds the same reaction in the peripheral and central parts of a previously segmentally crushed sternohyoideus muscle of rats, and concludes that the fibres must have reunited. The continuity of interrupted muscle fibres may indeed be restored under favourable conditions. This has beautifully been shown for the web muscle of bats. This muscle is very thin, and segmental necrosis of the muscle fibres without damage of the basal lamina and the endomysium may be produced by gentle transcutaneous squeezing. Only an increased density of muscle and fibrocyte nuclei marks the site of the lesion after completed regeneration (CHURCH 1970a, b). Regeneration is less perfect when the muscle is severely damaged because then also the endomysial framework is destroyed, and the proliferating fibrotic tissue interferes with the growing myotubes. The difficulty with a rat muscle (HALL-CRAGGS 1972), which is much thicker than the web muscle of bat, is that one cannot be sure that all fibres are divided; after a cut the fibres do not reunite because a scar is formed (VOLKMANN 1893). SCHMALBRUCH (1976b) crushed the rat soleus muscle; the peripheral part contained areas with many small new fibres but also apparently normal regions. The central part did not contain young fibres. The total number of fibre cross-sections had greatly increased in the peripheral but not in the central part of the muscle. This suggests that division by the crush was incomplete and that new fibres had been formed in the peripheral part only. There was no evidence for reunion of fibres.

Experiments with local muscle injury are hampered by the fact that the lesions are not reproducible, and that there is always a gradient from non-regenerating destroyed muscle through a zone of damaged muscle with regeneration to normal muscle tissue. The myotubes may be interpreted as projections of surviving fibre stumps along the ancient plasmodium-budding concept (HALL-CRAGGS 1974). This contrasts with the information available about myogenesis and the role of the satellite cells in injured muscles, but morphology alone does not conclusively disprove budding because there are indeed connections between old and new fibres.

Devascularization, which is part of grafting (see below), may be used to produce homogeneous fibre necrosis in situ (ALLBROOK et al. 1971b; HALL-CRAGGS 1978). The regenerative response again shows a gradient because it depends on the ingrowth of capillaries. Permanent devascularization prevents regeneration and the muscle is replaced by fibrotic tissue. In large muscles, regeneration is restricted to the periphery, and the core of the muscle becomes fibrotic.

Several myotoxic drugs have become useful in experimental muscle research. Local anaesthetics, the most widely employed one being bupivacaine, at clinically used concentrations, damage the muscle fibres (BENOIT and BELT 1970). Intramuscular nerve branches, satellite cells, connective tissue, and blood vessels are left intact. Soleus muscle fibres of rat develop plasma membrane defects already 15 min after injection, and the fibres contract locally. Phagocytes are abundant after 12 h, and after 4 days new myofibres are prominent. The fibres mature and become innervated during the next 2–3 weeks (JIRMANOVA and THESLEFF 1972; HALL-CRAGGS and SEYAN 1975; JIRMANOVA 1975; SEIBEL et al. 1978; BRADLEY 1979; HALL-CRAGGS 1980; JONES 1982; NONAKA et al. 1983). These substances possibly initiate selective myofibre death by releasing Ca^{2+} from the sarcoplasmic reticulum (NONAKA et al. 1983).

Local anaesthetics prevent fusion of chick myoblasts. When bupivacaine is added to older cultures, the differentiating myotubes are disrupted within 60 min; 120 min after treatment, all myotubes have disappeared (STYGALL et al. 1979). Traces of the drug might affect myoblast fusion and myotube maturation in situ as well. SCHMALBRUCH (1976b) used an alternative model: Ringer's solution of 60°–70° C injected beneath the gastrocnemius muscle of rats caused complete necrosis of the soleus muscle, but left the blood supply largely intact. The peak of phagocytosis was 1–2 days after the lesion. Regeneration was pronounced and the time course was similar to that after bupivacaine-induced necrosis (JIRMANOVA and THESLEFF 1972). Innervation, however, was delayed, probably because bupivacaine, but not the hot Ringer's solution, spared the intramuscular nerve branches.

4. Autografts

The capability of muscle tissue to regenerate, as well as the cellular mechanisms and controlling factors, have been studied in muscles that had been excised, and minced or implanted intact into the same animal (autografted) or into another animal of the same species (homografted). The muscle necrotizes completely and then regenerates by proliferation of its surviving satellite cells. Homografts have been used to evaluate whether a muscle disease, e.g. murine or chicken muscular dystrophy, is myogenic or induced by the environment. Homografting may be complicated by an immune response; it shall not be discussed in this review.

Muscle grafting, also with clinical-therapeutical intentions, became rather popular after CARLSON (1970a, 1972, 1973) reviewed the original work from Studitsky's laboratory, which is published exclusively in Russian. Studitsky and his co-workers denervated or traumatized the muscle to be grafted several weeks before the operation, excised the muscle and minced it, and then placed the mince either into the former muscle bed or into another site. New muscle fibres were found to grow from the mince. Several others had previously transplanted pieces of muscle tissue and observed regeneration as well (e.g. ELSON 1929; CLARK 1946).

STUDITSKY presented his results in three lectures published in English (1963, 1964, 1974). The background of the experiments, scientific and otherwise, deserves interest. Denervation or trauma are thought to induce a "plastic state" which, however, has never been defined in understandable terms (see discussion of the 1974 lecture). Not only a fresh muscle mince but also trypan blue-soaked totally necrotic muscle induced muscle fibre formation, but only with a now unavailable batch of trypan blue (see CARLSON 1979). STUDITSKY's most recent theoretical concept, presented in 1974, seems to favour de-differentiation of myonuclei (see Sect. G.I), but no suggestion is made how nuclei or cells give rise to new fibres. In this context, it is interesting to note that the aim of the mincing experiments was to prove in "Lysenko-era Russia" a "new cell theory which held that new cells arise in living organisms from an un-identified

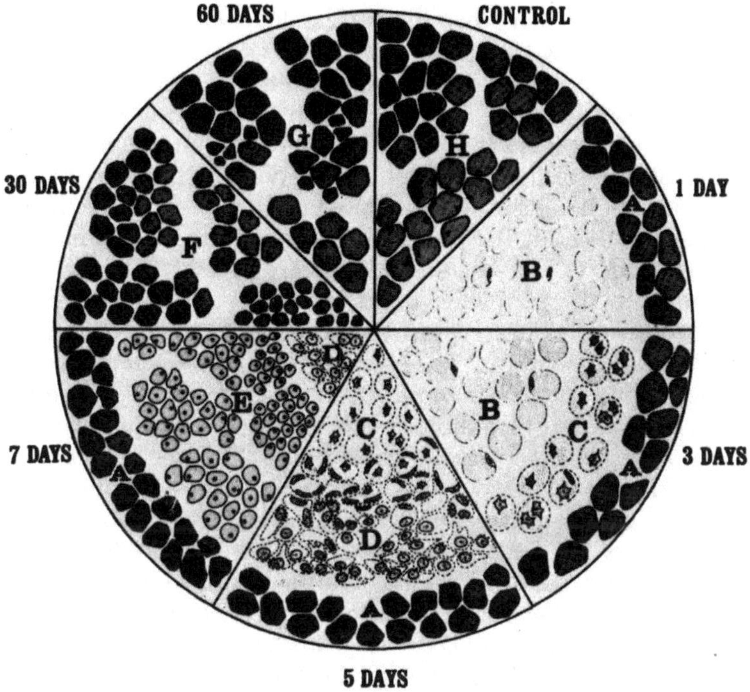

Fig. 117. Schematic representation of cellular reactions within a cross-section of a grafted normal extensor digitorum longus muscle of rat. The diagram is divided into segments representing the histological appearance of grafts at various days after transplantation. The letters refer to groups of cells showing similar histological reactions. (*A*) Surviving muscle fibers; (*B*) original muscle fibers in a state of ischemic necrosis; (*C*) muscle fibers invaded by macrophages that are phagocytozing the necrotic cytoplasm; (*D*) myoblasts and early myotubes within the basal laminae of the original muscle fibers; (*E*) early cross-striated muscle fibers; (*F*) maturing regenerating muscle fibers; (*G*) mature regenerated muscle fibers; (*H*) normal control muscle fibers. (From CARLSON et al. 1979b, with permission of the author and Raven Press)

non-cellular living substance" (quoted from *Editorial Science News* 1975, based on an interview with B.M. CARLSON).

The functional result of regeneration of minced muscles in rat is rather poor. CARLSON and GUTMANN (1972) obtained ≤0.5% of the normal tension in regenerates of the minced rat gastrocnemius muscle. The regenerates macroscopically resemble the muscles they replace, but they are very fibrotic. The anterior tibial muscle of mice, however, regenerates very well and the force may reach 80% of normal (SALAFSKY 1971). The structure and function of the regenerates is much better when the muscles are grafted intact (CARLSON and GUTMANN 1975, 1976; ALLBROOK 1975; SCHMALBRUCH 1977; MUFTI et al. 1977; GORNIAK et al. 1979; FAULKNER et al. 1979, 1983; COAN and TOMANEK 1979, 1981), and denervation or damage to the muscle prior to grafting has no influence on the final result (CARLSON 1973; SCHMALBRUCH 1977; MUFTI et al. 1977; HALL-CRAGGS 1979). Some authors, however, report that the initial rate of regeneration is faster when the graft is denervated about 2 weeks before the operation (ALLBROOK 1975; MUFTI et al. 1977).

The main difference between intact and minced grafts is that the tubes formed by the basal laminae are preserved in intact grafts and thus serve as scaffolding for the regenerating myotubes (VRACKO and BENDITT 1972). The basal lamina of the former endplates probably attracts the sprouting motor axons; re-innervation fails when the endplate region of an otherwise intact graft is excised (BADER 1980). Spontaneous innervation is incomplete or it may fail (CARLSON et al. 1979a, b); the results of grafting are improved when a nerve is sutured to the peripheral nerve stump of the graft (FAULKNER et al. 1983). HALL-CRAGGS (1979) crushed the nerve destined to innervate the graft 2 weeks before the operation and assumed that this enhanced the capability of the severed motor axons to sprout. PRENDERGAST et al. (1977) anastomosed nerves and blood vessels of grafted dog muscles, and saw excellent and rapid recovery. Also human limb muscles have been grafted, occasionally with good functional result (MANKTELOW 1984). It appears that these muscles had more or less survived and not gone through a necrosis-regeneration cycle. Survival of peripheral fibres has been observed in free grafts as well, in particular in those that had been denervated prior to grafting (SCHIAFFINO et al. 1975; CARLSON 1976; HALL-CRAGGS 1979; CARLSON et al. 1979b; MAGON et al. 1981) (Fig. 117). Several authors soaked the grafts in bupivacaine, to exclude the contribution of surviving fibres and probably also to speed up phagocytosis.

The limiting factor for the size of a free graft apparently is the distance over which satellite cells can be kept alive until the blood supply is reestablished. The core of a large graft always becomes fibrotic. FAULKNER et al. (1979) assume that 6 g muscle is the upper limit for a free graft; they obtained good results with cat muscles weighing about 3 g. The force was 12%–85% of that of the normal muscle.

5. Muscle Fibre Regeneration

The cellular events during regeneration are the same in all muscles and in all experimental models. The basic events are the activation and proliferation of satellite cells, i.e. myoblasts, within the basal lamina of a necrotic fibre. This has most beautifully been shown in adult muscle fibres in tissue culture (Fig. 118) (BISCHOFF 1975). The cellular process of regeneration in the tissue may differ, depending on how much of the original muscle cells is preserved.

The time course of regeneration is possibly faster in red than in white muscles, because satellite cells and capillaries are more frequent than in white muscles. Nevertheless, the histology of intact grafts of the rat extensor digitorum longus muscle (CARLSON 1979) and the rat soleus muscle (SCHMALBRUCH 1977) suggests that the grafted extensor digitorum longus muscle regenerates better. CARLSON et al. (1979c) report that levator ani grafts are more apt to be innervated in the bed of the extensor digitorum longus than in the bed of the soleus muscle.

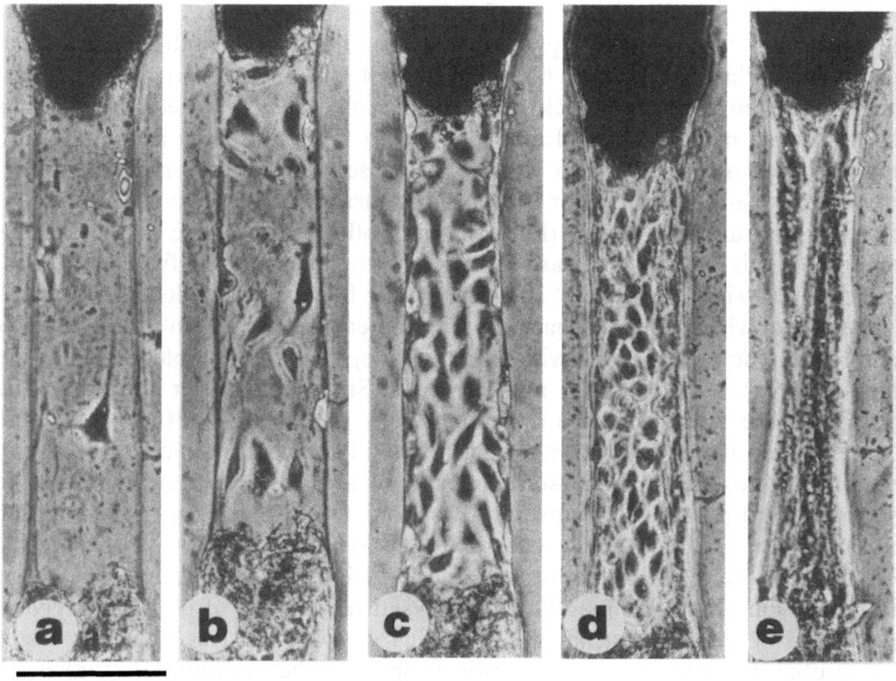

Fig. 118 a–e. Phase-contrast micrographs of the same living rat muscle fibre in vitro at successive stages of regeneration. The fibre has formed retraction clots, a short segment of the sarcolemmal tube is depicted. The microscope has been focussed on the upper surface of the fibre for all exposures; additional cells are out of focus. **a** 2 days in culture, 2 mononuclear cells (myoblasts) are present; **b** 3 days in vitro, 11 cells; **c** 4 days in vitro, 45 cells; **d** 5 days in vitro; **e** 6 days in vitro. The cells have fused to form myotubes; and cross-striations are visible. Bar, 100 µm. (From BISCHOFF 1975, with permission of the author and the Wistar Institute Press)

Necrosis of muscle fibres induces the invasion of phagocytes which enter the basal lamina tubes and digest the cell debris. Pigment-laden histiocytes may persist in the connective tissue for months after the lesion. Without phagocytosis, the onset of regeneration is delayed (ALOISI 1970). The number of phagocytes in the necrotic rat soleus muscle decreases already 2 days after the lesion when the blood supply is intact. In the centre of a free graft of the soleus muscle, phagocytes are still numerous 1 week after the operation (SCHMALBRUCH 1976b, 1977). The mitotic activity of satellite cells peaks about 2 days after the lesion. These cells grow like an endothelium on the inner face of the old basal lamina tubes (MASTAGLIA 1975) and, using light microscopy, have been identified as "basophilic cuffs" (CARLSON 1970b). In contrast to adjacent histiocytes, they lack secondary lysosomes, the nuclei are large and euchromatic, and triglyceride droplets may be prominent. Four to five days after the lesion, many satellite cells have fused into myotubes which start to grow towards the centre of the basal lamina tubes. The myotubes contain many ribosomes, and myofilaments in a hexagonal array. The central "lumen" of the basal lamina tubes may still contain remnants of the necrotic fibres and also histiocytes, but most cells containing phagolysosomes are now in the interstitium close to capillaries. Plasma cells may be found as well, but in rats neutrophilic leucocytes are uncommon. One week later, many cells have matured and contain densely packed myofilaments. Light microscopy does not show basophilia any more. The young myofibres are clustered and separated by 20-nm-wide membrane-bound clefts. Each cluster is eventually surrounded by two basal laminae. The outer one undulates, and the inner one follows the course of the plasma membrane. This is the new basal lamina of the regenerated fibres (VRACKO and BENDITT 1972) (Figs. 119, 120). The young fibres start to detach; the gaps between them widen and new basal lamina material is laid down on the plasma membranes. The opening of a widening cleft is spanned by the old basal lamina which disintegrates during the next 2 weeks. Six months after the lesion the myofibres have a normal mean diameter, but the scatter of fibre diameters is larger than normal. Central myonuclei are still frequent. Adjacent fibres may be molded onto each other, sometimes they are connected by cytoplasmic bridges. Teased preparations and serial sections reveal that many fibres branch and often form complicated syncytia (SCHMALBRUCH 1976b; ONTELL et al. 1982). This is the result of incomplete lateral fusion of adjacent myotubes born within the same basal lamina. The endomysial framework of 6-month regenerates is coarser than in normal muscles. The growing clusters of young fibres have probably compressed the empty meshes of the endomysial framework which have condensed into fibrotic sheaths enclosing groups of fibres (Fig. 121).

The change in the basal lamina pattern during regeneration is reflected in the distribution of fibronectin. Fibronectin is an extracellular protein which possibly governs the histogenesis of muscle fibres during development (see Sect. G.3). A thin sheath of fibronectin surrounds each individual muscle fibre of a normal rat muscle. These sheaths disappear when the fibres necrotize. New fibronectin occurs when myotubes are formed and it surrounds "large myotubes". Later, thin faintly stained strands of fibronectin bisect these myotubes (GULATI et al. 1982). The regenerating muscles have not been studied using electron microscopy; inspection of the micrographs suggests that the "large myotubes", in fact, were clusters of myotubes that had developed in a common basal lamina. The dividing strands of fibronectin corresponded to the widening clefts. Fibronectin acts as a barrier to myoblast fusion (FURCHT et al. 1978; PODLESKI et al. 1979) and it is reasonable to suspect that the late and probably unco-ordinated occurrence of fibronectin between the attached myotubes is the cause of the formation of branching and incompletely separated fibres in regenerating muscles.

Viable fragments of the necrotic fibres fuse with developing myotubes, and probably also with myoblasts. This is best seen during the first week after a lesion, when the immature cytoplasm of new fibres makes them distinguishable from surviving mature fibres. In places, groups of myotubes are beneath the basal lamina of surviving fibre fragments. This indicates that the lesion of the fibre was severe enough to initiate myoblast proliferation and fusion, but that the original fibre survived (Fig. 122). The events during fibre regeneration are summarized in Figs. 123, 124.

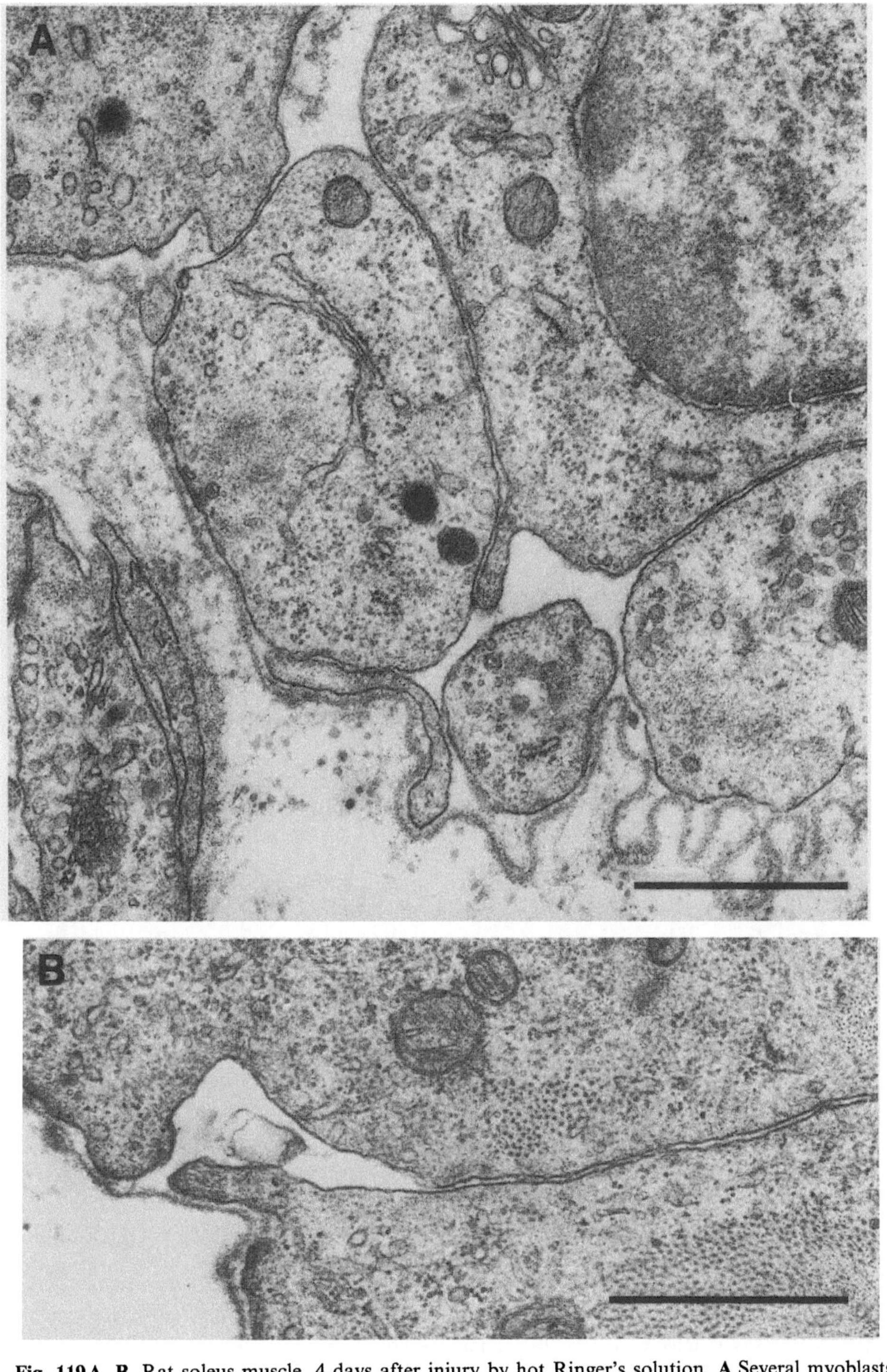

Fig. 119 A, B. Rat soleus muscle, 4 days after injury by hot Ringer's solution. **A** Several myoblasts or myotubes are contained in the collapsed basal lamina of a necrotic muscle fibre. **B** Two new myotubes within the old basal lamina, separated by a 20-nm-wide membrane-bounded cleft. Bars, 1 μm

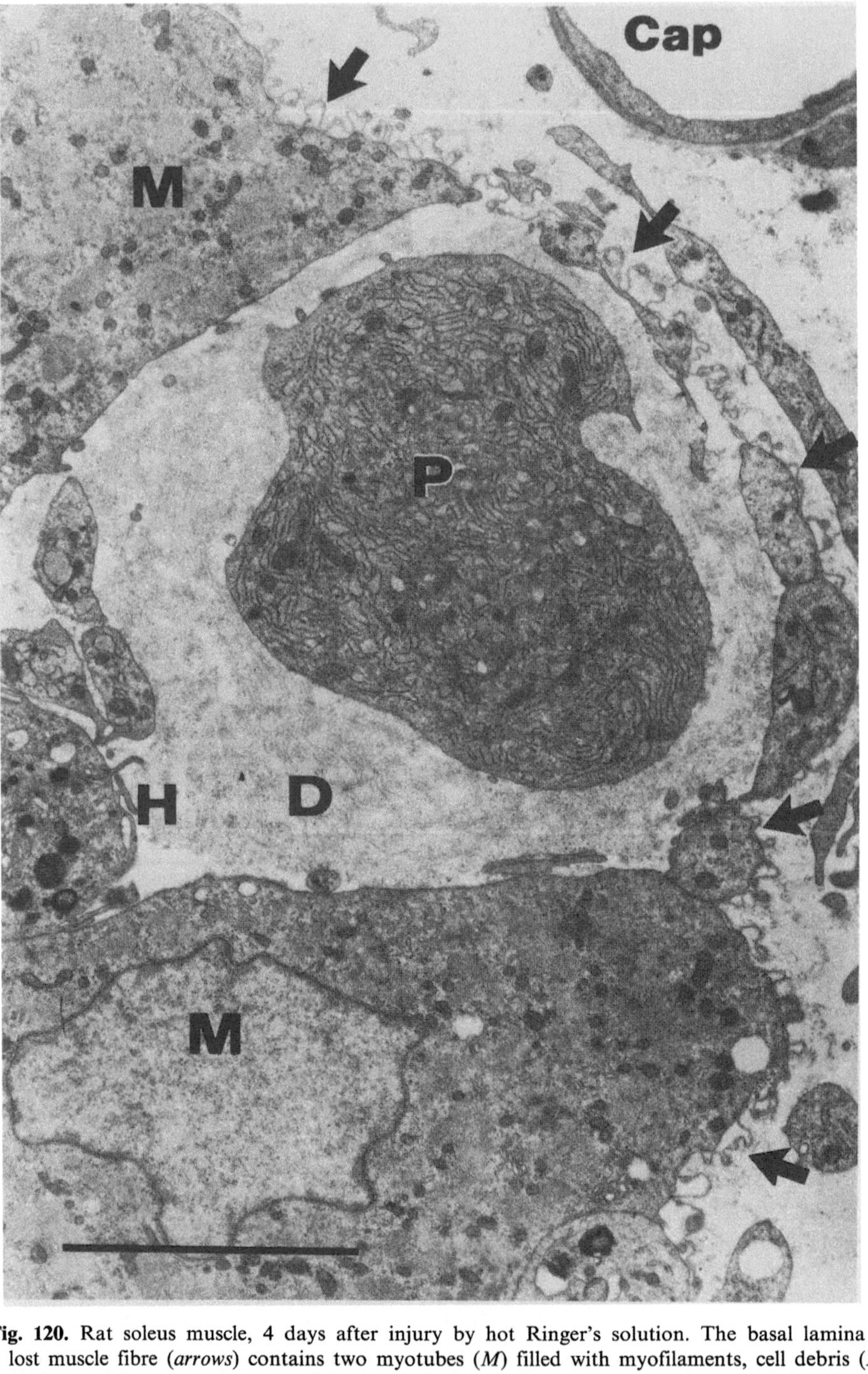

Fig. 120. Rat soleus muscle, 4 days after injury by hot Ringer's solution. The basal lamina of a lost muscle fibre (*arrows*) contains two myotubes (*M*) filled with myofilaments, cell debris (*D*), a histiocyte (*H*) apparently engaged in removal of the cell debris, and a plasma cell (*P*). On the inner face of the old basal lamina tube are projections of undifferentiated myoblasts. *Cap*, capillary. Bar, 5 μm

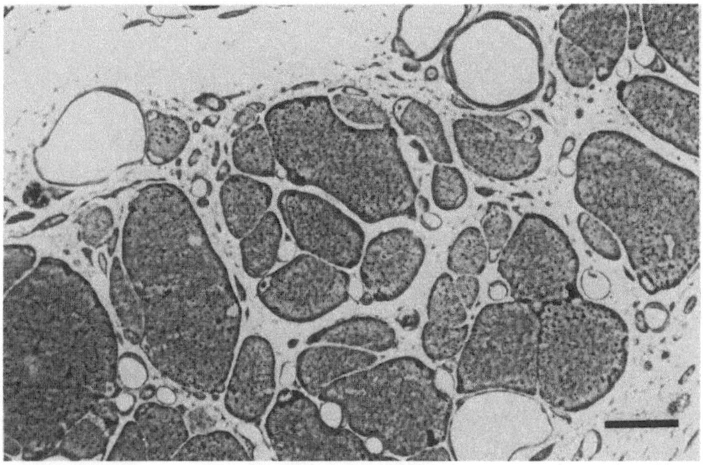

Fig. 121. Rat soleus muscle, 17 days after injury by hot Ringer's solution; 3-μm plastic section stained with p-phenylenediamine and photographed with phase optics. All fibres shown are regenerates. Their origin from clusters is apparent (for explanation, see Fig. 123). Only few capillaries (widened by perfusion fixation) have invaded the clefts between the formerly clustered fibres. The muscle fibres vary considerably in diameter, and some are connected by cytoplasmic bridges indicating incomplete lateral fusion during the myotube stage. Bar, 25 μm

The result of regeneration after injury is seldom a histologically normal muscle. The sequels (central myonuclei, branching and "splitting" fibres, large fibre diameter variation, fibrosis) correspond to what is known as "myopathic changes" in diseased muscles (Figs. 121, 125A). The view that these structural abnormalities indicate regeneration rather than "degeneration" of mature fibres is supported by the facts that "myopathic changes" may be produced experimentally under conditions excluding survival of fibres (SCHMALBRUCH 1979) and that dystrophic mouse muscles do not look "dystrophic" after X-irradiation to prevent regeneration, although the disease process continues (WIRTZ et al. 1982).

In a grafted muscle the regenerative response is delayed, and a gradient is formed corresponding to the gradual vascularization from the periphery (Fig. 117). This gradient is best seen in a 7-day graft (Fig. 126). In the central part regeneration has barely begun, but in the periphery the myotubes already fill the old basal lamina tubes. The myelinated nerves growing into the graft either follow the old intramuscular nerve branches and grow along bands of empty Schwann cells, or they accompany ingrowing blood vessels (Fig. 125). Many muscle fibres remain non-innervated, and eventually atrophy and degenerate. Innervated muscle fibres may even hypertrophy. Innervation of grafted rat muscles starts during the third week (CARLSON et al. 1979b), in grafted cat muscles endplates have been found after 5 weeks (HÅKELIUS and NYSTRÖM 1975). Fat cells are frequent in old muscle grafts. Satellite cells are regularly present in regenerated grafts indicating that, like in fetal myogenesis, some myoblasts withdraw to become reserve myoblasts. The branching myotubes

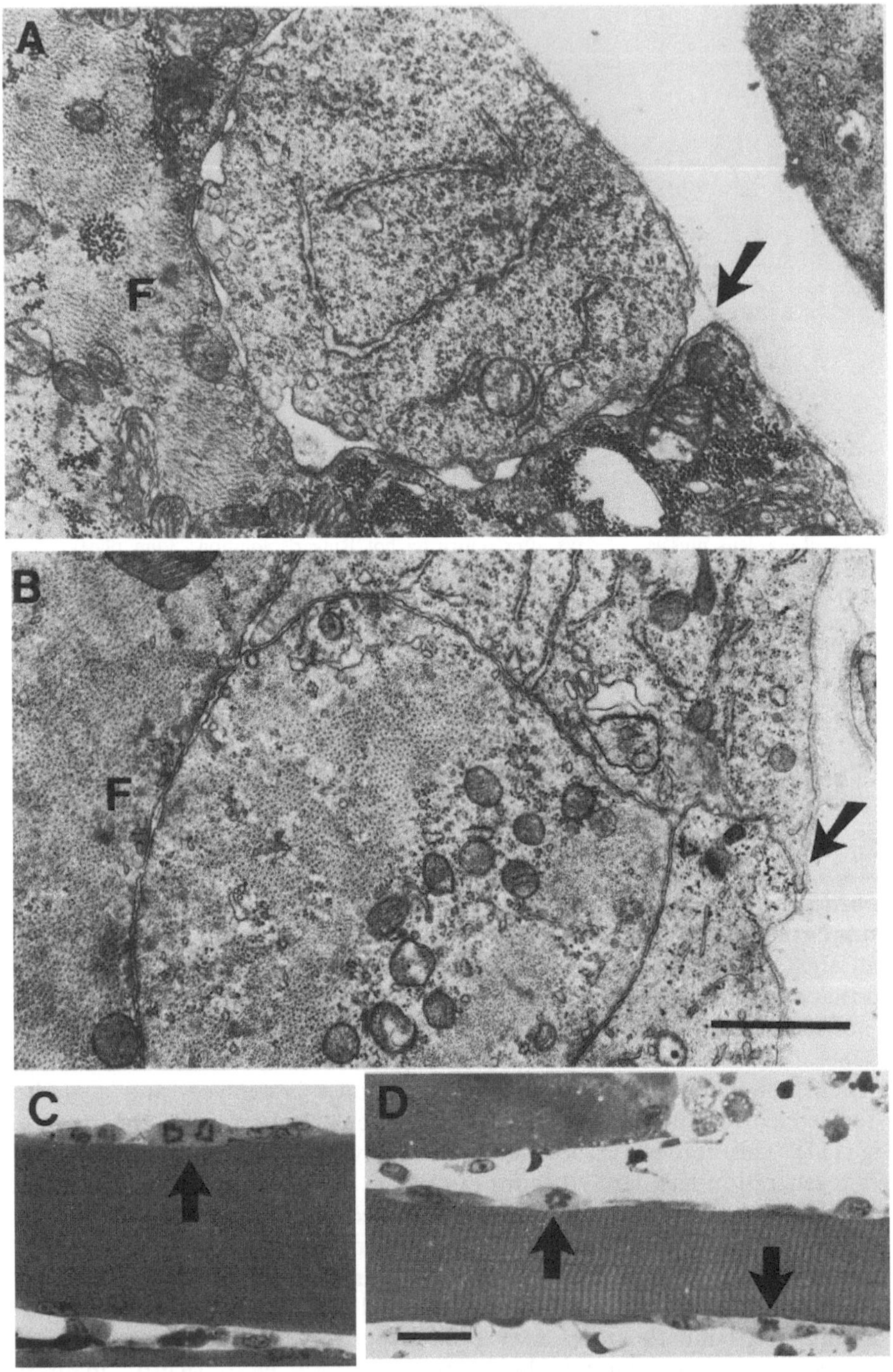

Fig. 122 A–D. Rat soleus muscle injured by hot Ringer's solution. **A** After 1 day. Damaged but viable muscle fibre (*F*) with satellite cell. The satellite cell is still beneath the basal lamina (*arrow*) but is "activated" and contains many free and also membrane-bound ribosomes. **B** After 4 days. A satellite cell has proliferated and several myotubes have arisen beneath the basal lamina (*arrow*) of a surviving muscle fibre (*F*). Bar, 1 μm. **C, D** After 2 days, and 2 h after i.p. injection of colchicine. Light micrographs of surviving fibres showing arrested mitotic figures (*arrows*) in satellite cells. Bar, 25 μm

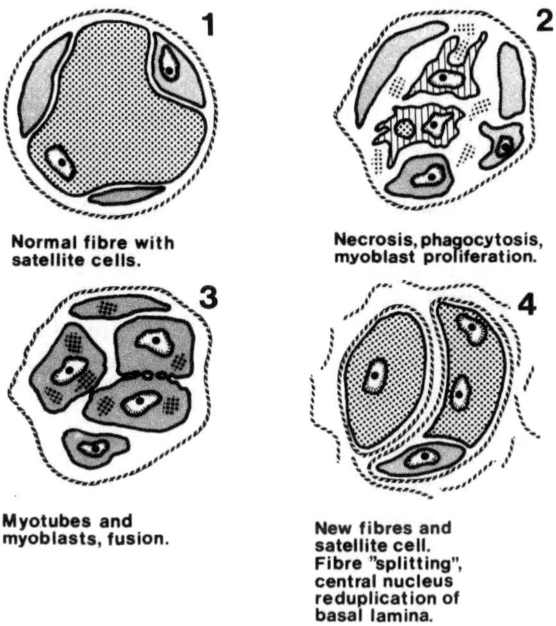

Fig. 123. Diagram illustrating the formation of several new myofibres within the persisting basal lamina tube of a necrotic muscle fibre, and the development of clusters (see Figs. 120, 121)

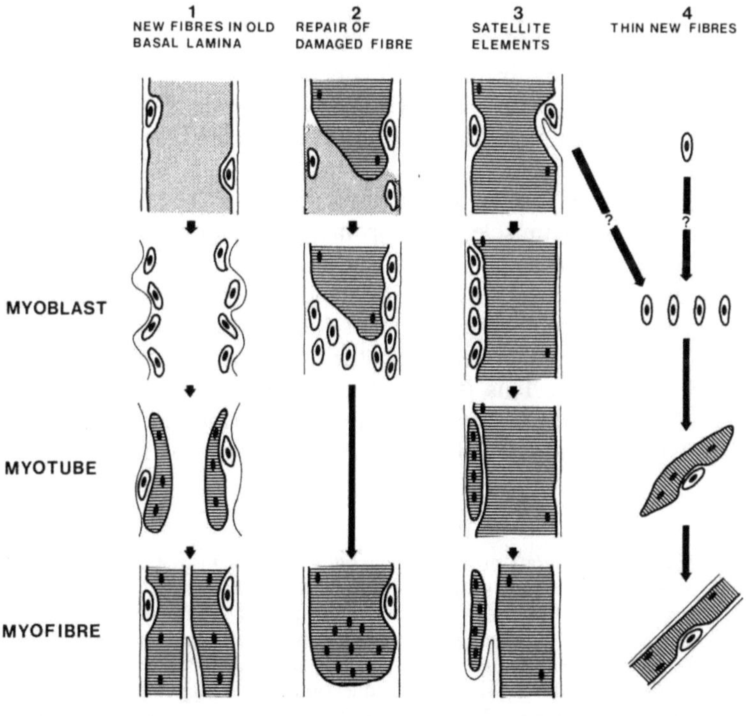

Fig. 124. Schematic illustration of the cellular mechanisms of muscle regeneration. *Thin lines*, basal laminae; *thicker lines*, plasma membranes; *stippled areas*, necrotic cell substance

found in early grafts (ONTELL et al. 1982) persist and become extensively branching mature myofibres which in cross-sections appear divided by incomplete and complete clefts (BOURKE and ONTELL 1984).

Muscle spindles are regularly found. The intrafusal fibres apparently regenerate like the extrafusal ones; efferent or afferent innervation of soleus muscle spindles has not been observed using electron microscopy (SCHMALBRUCH 1977), but CARLSON et al. (1979b) find argyrophilic fibres presumed to be nerve fibres in spindles of the grafted rat extensor digitorum longus muscle. Numerous fibres with ringbinden are characteristic of regenerates of minced muscles, but they occur also in "free" grafts (Fig. 125). Ringbinden are a normal finding in muscles not connected to the skeleton (LINDNER 1968), and they are regularly formed in non-attached muscle fibres regenerating in culture (BISCHOFF 1979). This suggests that ringbinden are a developmental phenomenon caused by lack of directed tension.

Young regenerated muscle fibres, like fetal muscle fibres, react intensely for ATPase at pH 9.4 both after alkaline and acid preincubation (RILEY 1973); the myosin isoenzyme pattern before innervation takes place is that of developing myotubes (CARRARO et al. 1983). A monoclonal antibody against fetal human muscle stains the regenerating fibres in diseased muscles selectively (HURKO and WALSH 1983). Rat muscle grafts do not show fibre type differentiation until 30 days after grafting; from day 40 on the innervated fibres are differentiated (CARLSON et al. 1979b). There is pronounced type grouping (COAN and TOMANEK 1979). In mammals, the properties of the innervating axons probably determine the fibre type differentiation (Fig. 125).

Evidence for myogenic influences has been obtained in grafted pigeon muscles. Fast muscles regenerate better than slow-tonic muscles, and the ultrastructural and histochemical properties of fibres regenerated from a mince are those of fast fibres independent of the source (HIKIDA and LOMBARDO 1974; HIKIDA 1976). Intact grafts of fast muscles implanted into the bed, and innervated by the nerve of a slow muscle, develop fast properties; slow fibres first occur after 4 months (WALRO et al. 1982). The levator ani muscle of rats maintains its testosterone-dependence, whether it regenerates in situ (GUTMANN and CARLSEN 1978) or as a graft replacing the extensor digitorum longus muscle (CARLSON et al. 1979c). These results suggest that the hormone-sensitive muscles originate from special myoblasts. This is at variance with the fact that denervation of adult levator ani muscles in situ abolishes the hormone sensitivity (BURESOVA et al. 1972). No explanation can be given for this contradiction.

Fig. 125A–C. Sixty-day autograft of a rat soleus muscle replacing the entire contralateral triceps surae muscle; 3-μm plastic sections stained with p-phenylenediamine and photographed with phase contrast (**A, B**) or bright field (**C**). **A** Cross-section showing rather large muscle fibres with increased diameter variation (a normal rat soleus muscle prepared with the same technique is shown in Fig. 13). The fibres differ in staining intensity indicating that the mitochondrial content is different. This probably reflects fibre type differentiation and innervation by motoneurons of different type from the previous gastrocnemius nerve. That the fibres are innervated is evident from their size; in non-innervated grafts the fibre diameter never reaches 30 μm and after 60 days they atrophy again. One fibre shows *ringbinde*, indicating aberrant myofibrillogenesis (see text). The capillaries are local-

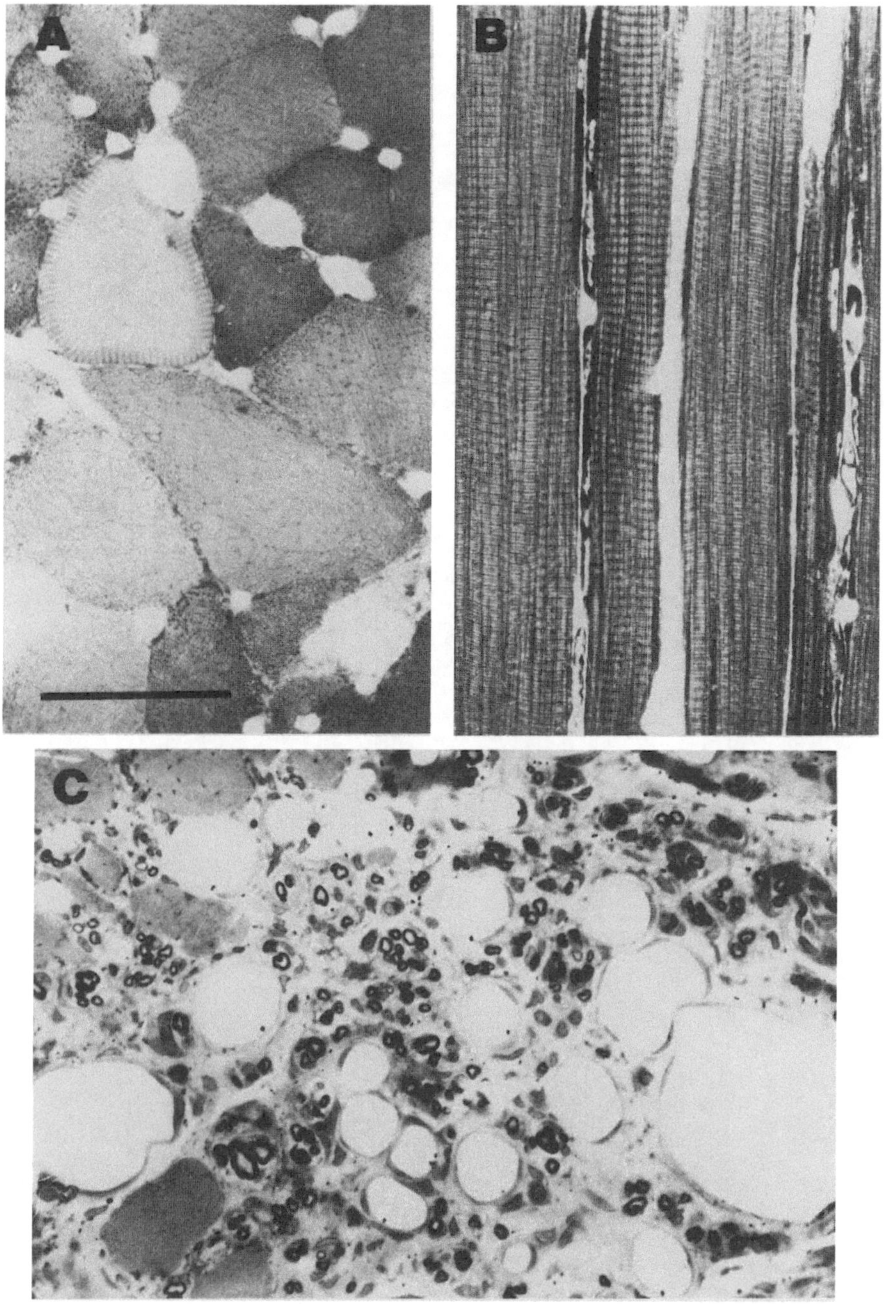

ized preferentially around fibres rich in mitochondria. **B** Longitudinal section showing almost perfect cross-striations. **C** Cross-section through the muscle hilus. A bundle of blood vessels intermingled with myelinated nerve fibres invades the muscle. The sprouting proximal nerve fibres have not grown into the nerve stump of the graft; within the graft, they follow the blood vessels but also enter the former intramuscular nerve branches. Bar, 50 μm. (From SCHMALBRUCH 1977)

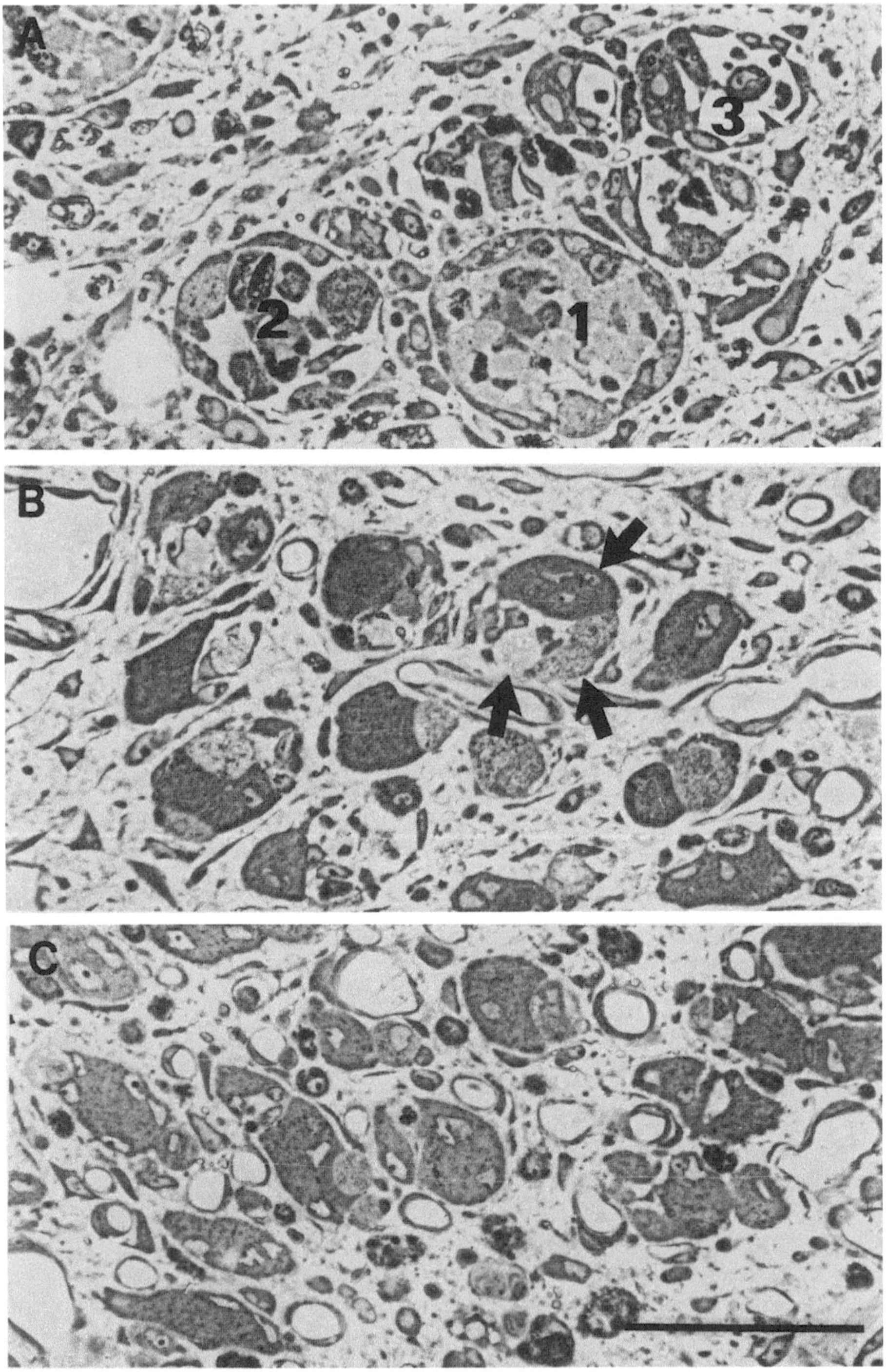

Fig. 126 A–C. Seven-day regenerate of a freely grafted rat soleus muscle. **A** Close to centre. Remnants of necrotic fibres are shown. In fibre *1*, cell debris is still present and the inner face of the basal

V. Muscle Fibre Growth

1. Transverse Growth

The size of a muscle increases during development, and in response to an increased workload (training). Moreover, in some systems a distinct increase in muscle weight has been observed after denervation before atrophy commences, and also when a muscle is immobilized in a stretched position.

The postnatal growth of human muscles is exclusively due to the growth of the individual fibres; the number of fibres remains constant (MONTGOMERY 1962; STICKLAND 1981). In mice and rats, however, new fibres occur until about 3 weeks after birth; the bulk of the increase is during the first week. Thereafter, the number of fibres passing the widest girth of a muscle remains stable (GOLDSPINK 1962; CHIAKULAS and PAULY 1965; BETZ et al. 1979) (Fig. 127). The final number of muscle fibres is apparently predetermined; retardation of postnatal growth of mice by increasing the litter size does not alter the number of muscle fibres (TIMSON 1982). Immobilization of the limb at birth induces muscle atrophy, but a normal number of fibres develops (GARDINER 1981). By contrast, in fish muscles the number of fibres increases continuously during growth (STICKLAND 1983).

The total number of myonuclei in the human sartorius muscle increases during fetal development, but at a body weight of 2500 g a constant level is reached (2×10^8 nuclei, STICKLAND 1981). Rat soleus muscles contain more satellite cells than extensor digitorum longus muscles (Table 11). Postnatal growth of the soleus muscle is associated with pronounced mitotic activity, and numerous new myonuclei occur; this is less pronounced in the extensor digitorum longus muscle. The amount of cytoplasm per myonucleus is smaller in the soleus than in the extensor digitorum longus muscle, although the fibres are larger (KELLY 1978a) (see also Sect. G.II.2f).

Both the fibre diameter and the relative contribution of the contractile filaments to the total fibre volume increase during postnatal development; hence, there is a tremendous increase in the number of myofilaments. G. GOLDSPINK and co-workers (SHEAR and GOLDSPINK 1971; GOLDSPINK 1970; PATTERSON and GOLDSPINK 1976) have proposed a mechanism for how the increase in myofilament mass is translated into increased myofibril number. The thin filaments of the I band take a slightly oblique course to reach their places within

lamina tube is covered by myoblasts. In *2*, the cell debris has disappeared and the centre of the former fibre is taken up by histiocytes. In *3*, the basal lamina has partly collapsed and myotubes are growing towards the centre of the old basal lamina tube. There are no open blood vessels in this region. **B** Intermediate zone. Some blood vessels are present, and the myotubes have increased in diameter. Those that are most mature, i.e. contain most myofilaments, stain darkest. Myotubes of different maturity (*arrows*) are present in the same basal lamina tube. **C** Periphery. This area is richly vascularized, and the myotubes are rather mature. Several of them have peripheral nuclei. The histiocytes are now in the interstitium rather than inside the basal lamina tubes as in **A**. Bar, 50 μm. (From SCHMALBRUCH 1977)

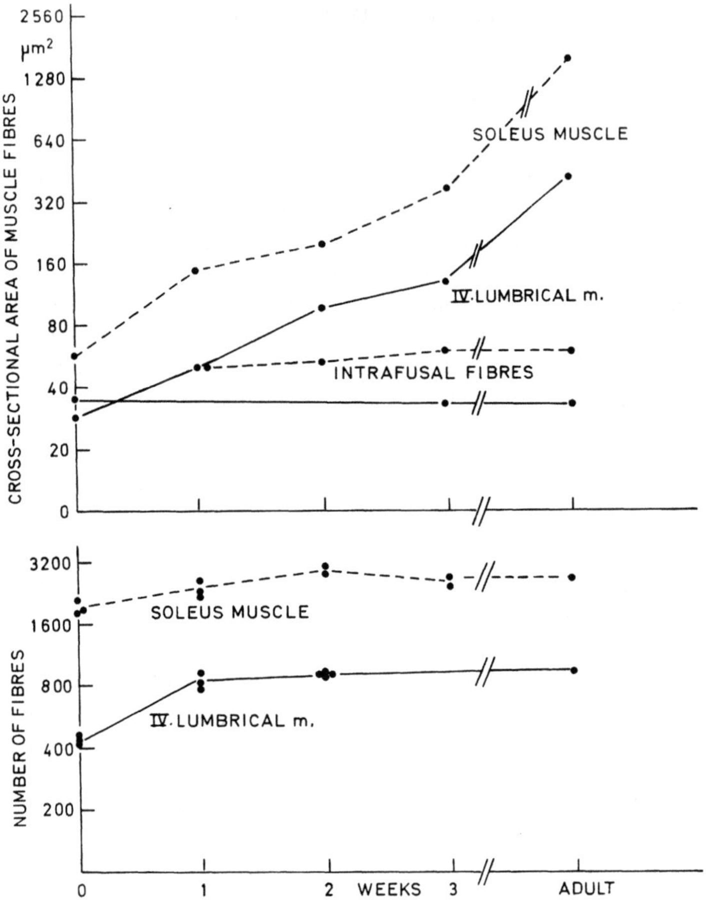

Fig. 127. Diagrams showing the postnatal increase in fibre size and fibre number in the soleus (*dotted lines*) and lumbrical (*solid lines*) muscles of rat. The size of the extrafusal fibres increases considerably, but that of the intrafusal fibres remains constant. Note logarithmic scales. (From SCHMALBRUCH, unpublished)

the hexagonal pattern in the A band. This is due to the fact that the thick filament spacing in the A band is larger than would be required to transform the tetragonal lattice of the thin filaments at the Z disc into a hexagonal lattice in the A band. For geometrical reasons, the deviation from the direction of pull is greater in the periphery than in the centre of a myofibril. This deviation increases when the myofibril increases in diameter; the Z disc disrupts such that new myofibrils arise by longitudinal splitting. Signs of presumed Z disc splitting are more distinct in the periphery of the fibre than in its centre; also an autoradiographic study in fish muscles (PATTERSON and GOLDSPINK 1976) suggests that the new filaments are laid down predominantly in the fibre periphery. The poor definition of the myofibrils in avian slow tonic (Felderstruktur), fibres, compared with that in fast (Fibrillenstruktur) fibres, is attributed to the less rapid pull during contraction.

SOLA and MARTIN (1953) observed that denervated hemidiaphragm of rats transiently increases in weight. This is attributed to the rhythmic passive stretching by the contractile activity of the innervated counterpart (STEWART et al. 1972). The degree of stretch in the diaphragm is difficult to control, and others have investigated the effect of stretch in limb muscles. In young rats (50 g weight) the soleus muscle immobilized in a shortened position decreases in wet weight by about 40% within 1 week; the weight of the simultaneously stretched extensor digitorum longus muscle increases by 10% within 1 day and then remains constant. RNA and protein synthesis of both muscles show corresponding changes (GOLDSPINK 1977a). Hypertrophy and atrophy are reversible within 1 week when the leg is remobilized. Blockade of de novo synthesis of RNA, but not DNA, during the remobilization phase restricts the rapid enhancement of protein synthesis in the soleus muscle (GOLDSPINK 1977b). Denervated limb muscles atrophy, but whether there is a short initial phase of hypertrophy is controversial. This may be due to different experimental conditions; some rats drag the denervated leg and others hold it close to the body (for references, see GOLDSPINK 1977a; LAURENT and MILLWARD 1980). Denervation hypertrophy may only occur when the denervated muscle is stretched (JIRMANOVA and ZELENA 1970; GOLDSPINK 1978). It has to be noted that also the method of immobilization itself may affect the muscle weight. Plaster casts are difficult to apply to hindlimbs of rats, and they may be slack and allow movement, or they may be too tight and interfere with the blood supply and cause muscle degeneration. It is not certain that hypertrophy as determined by weighing and protein analysis reflects increase in girth, it may also reflect increase in length. Histology is of little help because a change in fibre diameter corresponding to a 10% increase in cross-sectional area cannot be distinguished from artefactual swelling or shrinking. The results with respect to change in length after immobilization are inconsistent.

The number of sarcomeres in cat soleus muscles immobilized in a stretched position increases by 20%; immobilization in a shortened position decreases the number of sarcomeres by 40% (TABARY et al. 1972). The soleus muscle of young rabbits immobilized under stretch does not form new sarcomeres, only the length of the tendon increases. When the muscles are denervated and stretched, the number of sarcomeres increases by 20% (TABARY and TARDIEU 1983). The stretched anterior tibial muscle of young rabbits shows a 20% decrease in sarcomere number when it is immobilized, not by a plaster cast but by a wire driven through the joint into the tibia (CRAWFORD 1973).

It is difficult by training to induce muscle hypertrophy in small laboratory animals. Endurance training with isotonic contractions has little or no effect on muscle fibre diameter. GOLDSPINK (1964b) and GOLDSPINK and WARD (1979) designed an experimental set-up to train hamsters in weight-lifting. These authors find, after 2 months training, an increase of the cross-sectional area by 50% for fast-twitch and by 25% for slow-twitch fibres, without any change in fibre number. GONYEA and ERICSON (1976) and GONYEA et al. (1977) built a weight-lifting apparatus for cats and HO et al. (1980) designed one for rats.

Both groups describe a 20% increase in weight of the trained muscle, and both find small regenerating fibres and branched fibres which they interpret as signs of compensatory hyperplasia.

Functional isolation of a muscle may be achieved by denervation or tenotomy of its synergists (VAN LINGE 1962; HAMOSH et al. 1967; GOLDBERG 1968; ROWE and GOLDSPINK 1968; HALL-CRAGGS 1970; MACKOVA and HNIK 1971; JAMES 1973; JABLECKI et al. 1973; HOFMANN 1980). In these experiments the muscle has to carry a heavy load, and during locomotion it may be stretched by the weight of the animal, i.e. many contractions will be eccentric. It is essential to note that eccentric contractions in untrained human subjects give rise to fibre damage (see Sect. G.IV.2). Functionally isolated rat muscles contain either outright necrotic fibres (VAN LINGE 1962; JABLECKI et al. 1973) or they show histological changes which have to be interpreted as sequels of muscle fibre regeneration (VAN LINGE 1962; JAMES 1973; HALL-CRAGGS 1970). JABLECKI et al. (1973) find in the rat soleus muscle, studied using autoradiography 2 days after functional isolation, that RNA synthesis is restricted to the connective tissue; the published micrographs show massive fibre necrosis and macrophage infiltration. The hypertrophy of the muscle fibres produced by acute overload is modest compared with that produced in weight-lifting hamsters (see above). The wet weight of the extensor digitorum longus muscle after 2 months had increased by 10%–20% and the mean fibre diameter had increased by 2%–12% (JAMES 1973). It is disputable whether acute overload by functional isolation simulates natural work-induced hypertrophy. Biochemical changes assumed to reflect myofibre growth may be influenced by the synthesis of hydrolytic enzymes in macrophages, by fibroblast activity, or by myoblast proliferation.

Stretch-induced hypertrophy is prominent in the wing musculature of chicken. These muscles may be stretched by attaching a weight to the wing or by implanting a spring-loaded bar (JIRMANOVA and ZELENA 1970; SOLA et al. 1973; LAURENT and MILLWARD 1980; ASHMORE and SUMMERS 1981). The results are not consistent. ASHMORE and SUMMERS (1981) find, after 1 week, a 55% increase in the cross-sectional area of stretched innervated fast fibres. The tonic anterior latissimus dorsi muscle, 3 weeks after plain denervation, increases 60% in weight and 30% in fibre diameter. Simultaneous denervation and tenotomy cause atrophy. The denervated fast posterior latissimus dorsi muscle atrophies after denervation, whether it is tenotomized or not (JIRMANOVA and ZELENA 1970).

It is obvious that stretch is an important factor regulating muscle fibre growth, both longitudinally and transversely. It may well be that it is the key factor during normal development by stimulating coordinately DNA synthesis and mitosis of satellite cells, RNA synthesis in muscle fibres, and also fibroblast activity (LAURENT and MILLWARD 1980). The growing bones induce stretch during normal development, but actively elicited stretch reflexes may enhance the effect.

2. Longitudinal Growth

The mechanism of longitudinal growth of muscle fibres in vertebrates is unknown. ARONSON (1961) reports that, in an embryonic insect muscle which is only three sarcomeres long, growth is entirely due to lengthening of the A band and to a decrease in the A–I overlap. On the basis of this observation in a muscle very much different from vertebrate muscles, GOLDSPINK (1964a, 1968) reports that also in mouse brachial biceps muscle, lengthening of the sarcomeres accounts for 25% of fibre growth. Confirming for mouse HUXLEY's and PEACHEY's (1961) observation in frog, he shows that the sarcomeres are shorter at the ends than in the middle of a fibre. The shorter terminal sarcomeres appear less extensible in young than in adult muscles, and GOLDSPINK regards them as immature non-functioning sarcomeres. Maximum tetanic force is in adult muscles attained at full extension of the limb, in young mice the optimal length is 10% above the maximum in vivo length. This indicates that there is an excess of sarcomeres in young mice, and that the sarcomere length adapts as the animal grows. By contrast, MACKAY and HARROP (1969) do not find a systematic postnatal increase in sarcomere length in growing rat muscles. GOLDSPINK (1968) assumes that new sarcomeres are added to the myofibrils at the fibre ends, and GRIFFIN et al. (1971) present light microscopical autoradiographs (also published in GOLDSPINK 1972), which they interpret as showing an increased incorporation of adenosine into the fibre ends.

This mechanism is difficult to accept, and there are indeed observations suggesting that a muscle fibre grows along its entire length. In very early myotubes, the myofibrils do not extend throughout the cell and new sarcomeres are added at their ends (see Sect. G.III.2a). Nevertheless, a young myotube is short in comparison with a mature myofibre the length of which may be 200–2,000 times greater than its diameter. This prohibits the free flow of proteins along the entire fibre, and only a negligible fraction of the protein-synthesizing machinery would be involved in growth. The contractile proteins are constantly renewed (see Sect. C.I.16) and a sarcomere to be replaced might be replaced by two instead of one if the fibre is growing. There are roughly 30 myonuclei per mm fibre length (i.e. 300–3,000 in all), and it does not seem to make sense that new sarcomeres are added one by one at the fibre ends only.

Wire markers implanted into the anterior tibial (CRAWFORD 1954) and sternomastoid and gracilis muscles (MACKAY and HARROP 1969) of young rabbits and rats, and regularly located using X-rays, retain their position in relation to each other and to the myotendinous junctions. During the period studied, the muscle belly increases almost twofold in length. Transferring the tendon of the rabbit anterior tibial muscle to in front of the crural ligament induces an increase in muscle length at the expense of the tendon. This probably compensates for the increased degree of shortening necessary to raise the foot. Also in these muscles, wire markers retain their relative positions (CRAWFORD 1954). GINSBORG and MACKAY (1961) demonstrate that the distances between the approximately 80 endplates on individual fibres of the chicken anterior latissimus dorsi muscle increase proportionally with the increase in length of the fibre. Despite these seemingly unequivocal findings, MACKAY and HARROP (1969) state that "the presence of slender bundles of poorly aligned filaments would tend to support the hypothesis that growth in length of the skeletal muscle myofibrils takes place at the muscle-tendon junction". Considering their own results (see above), this conclusion is surprising, because the evidence for

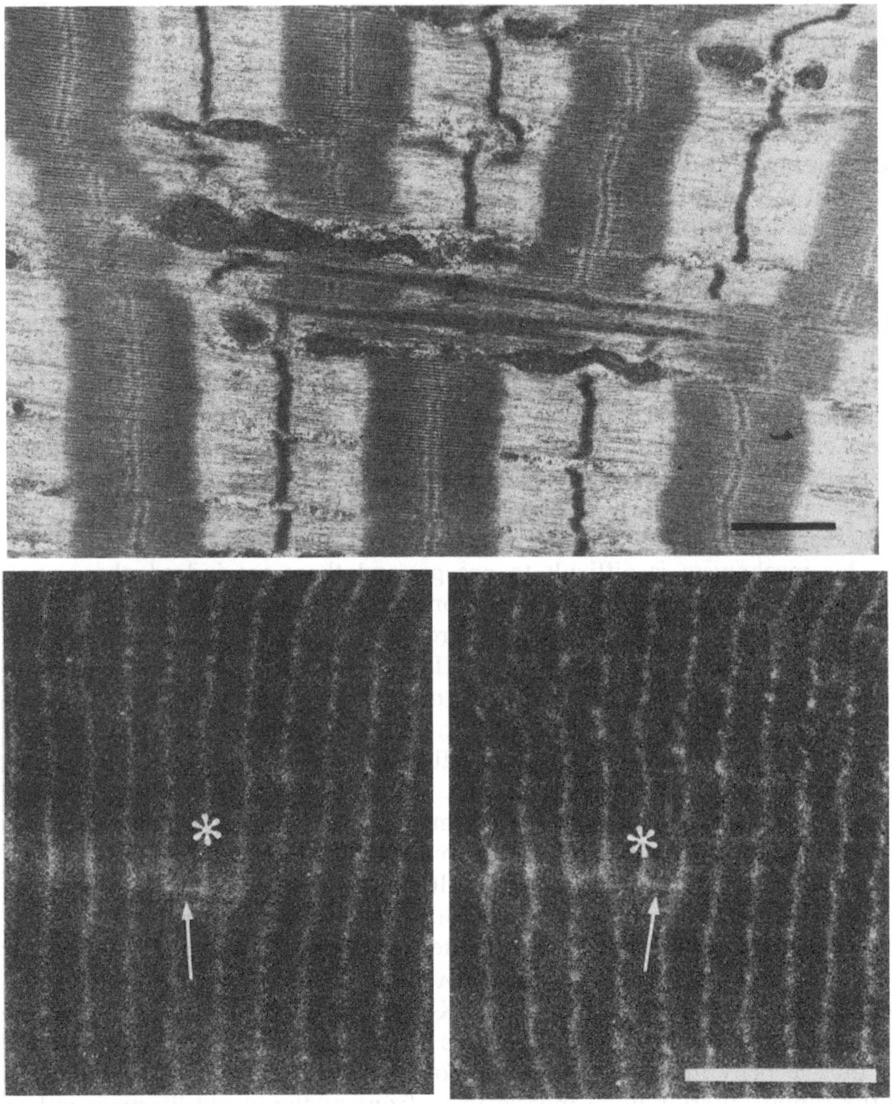

Fig. 128. *Top*: Medial vastus muscle, 2-year-old boy. Centre of a "sphenode" or Vernier-type shift of the cross-striation. This micrograph may be interpreted to mean that, within a sarcomere plane, one sarcomere is broken down and replaced by two new sarcomeres, and that this process continues in the transverse direction and in fact represents the mechanism of longitudinal growth of the muscle fibre. Bar, 1 μm. *Bottom*: Human muscle fibre, stained with fluorescent antibodies against skeletin, a protein of the cytoskeletal system. The two micrographs represent different focal settings. The reactive sites are at the Z line level. One sarcomere plane is branching, giving rise to a supernumerous sarcomere (*asterisk*). At the site of branching, a longitudinal fluorescent strand extends along the sarcomere between consecutive Z discs (*arrow*). Bar, 10 μm. (Light micrographs courtesy of Dr. THORNELL, Umeå)

sarcomere formation derived from the presence of unidentified filaments and of ribosomes (MacKay et al. 1969) is circumstantial at best.

Schmalbruch (1968) proposed a mechanism for sarcomere formation based on a study of Vernier-type band shifts (see Sect. C.I.17) in the diaphragm of young rats and in muscles of children. Some of the shifts ("sphenoden", Heidenhain 1919) originated from a focus of disintegrating Z-disc material. One or two sarcomere planes on one side of the focus corresponded to two to three sarcomere planes on the other side, i.e. there was an incomplete sarcomere plane (Fig. 128). This was interpreted to mean that a sarcomere progressively was broken down and replaced by two new sarcomeres. In that way new sarcomere planes would be added without interruption of the continuity of the fibre. Once an incomplete sarcomere plane had been formed, the oblique pull of the fibre would stress the dividing Z-disc and possibly propagate the process. Foci of Z disintegration with unequal numbers of sarcomeres on both sides have been seen in human muscle fibres damaged by eccentric contractions (Newham, personal communication; Fridén 1984). It is conceivable that overstretched muscles respond with an increase in length, which would explain why the damage to the muscle becomes less severe when the eccentric exercise is continued. A similar process for longitudinal growth has been inferred from electron micrographs of growing crab muscles (Jahromi and Charlton 1979). (With respect to Vernier shifts and the presumed helicoidal structure of the sarcomeres, see Sect. C.I.17.)

H. Muscle Fibres as Members of Motor Units

I. Definition

Muscle fibres function only when they are connected to motoneurons. Each motoneuron innervates many muscle fibres which together with the nerve cell constitute a "motor unit". Motor units are the final common pathways of motor activity onto which central and peripheral influences converge. Motor units differ with respect to force, and contractile properties (speed of contraction, fatiguability). All muscle fibres of a given motor unit are of the same histochemical and physiological type. Originally, the definition of a motor unit was a physiological one: all its muscle fibres discharge and contract synchronously. First the glycogen depletion method (Sect. D.II.2) made it possible to identify the fibres of a physiologically characterized motor unit using histology.

II. The Size of a Motor Unit

The size of a motor unit is defined by its twitch or tetanic force or by the number of muscle fibres it comprises. The number of motor units of a muscle is given by the number of large efferent axons in the nerve branch to the muscle (Sect. B.VI.2). It is also possible to calculate the number of units from the average force of individual units and the total force of the muscle; this determination in some muscles is hampered by a large range of motor unit forces. The average size of the motor units of a muscle may be computed from the number of motor units and the total number of muscle fibres (Sect. B.III). The number of fibres of an individual unit is obtained by the glycogen depletion method; it may also be calculated from the force of the unit in relation to the force of the entire muscle and the total number of muscle fibres, provided the force produced by each muscle fibre is identical.

Reliable data for the number and size of motor units are available for several muscles of cat, rat, and mouse (Table 5, last column). In the baboon, the average motor unit size is 220 fibres in the abductor pollicis brevis muscle (total 43 motor units), and it is 790 fibres in the medial gastrocnemius muscle (total 292 motor units) (Wray 1969). Fibre counts in individual motor units by the glycogen depletion method are listed in Table 12.

The determination of motor unit number and size in human muscles fully depends on morphological methods. The muscle fibre counts are reasonably

Table 12. Number of muscle fibres in individual motor units determined by glycogen depletion. *FR*, fast fatigue resistant, *FF*, fast fatiguable, *S*, slow fatigue resistant; *I*, weakly stained for ATPase at pH 9.4, presumably slow-twitch, *II*, intensely stained for ATPase, presumably fast-twitch; "*white*", poor in mitochondrial enzymes

Species and muscle	Number of fibres per individual motor unit	Number and type of units examined	References
Rat			
Anterior tibial	55–172, mean 118	36 unclassified	Brandstater and Lambert (1973)
	80–178, mean 132	7 "white"	Edström and Kugelberg (1968)
	93–160, mean 132	6 intermediate	
Soleus	50–120, mean 83	18	Kugelberg (1976)
Cat			
Anterior tibial	469-1,323, mean 775	3 type I	Doyle and Mayer (1969)
Soleus[a]	20–170, mean 90 calculated for same specific force as in gastrocnemius: 50–427, mean 229	6	Burke et al. (1974)
Gastrocnemius[a]	167–695, mean 380	9 FF	Burke and Tsairis (1973)
	170–338, mean 200	4 FR	
	14–164	3 S	
	Calculated for percentage of units of different type:		
	570–760	FF	
	420–560	FR	
	?	S	
Extensor digit. long.	522–1,323	4 type I	Doyle and Mayer (1969)
	100	1 type II	

[a] The fibre counts were recalculated, because the authors assumed that not all fibres had been depleted (soleus) and that not all fibres passed through the plane of the cross-section. The figures given first are the fibre counts.

consistent; only with respect to the sartorius muscle disagreement exists (see Table 1). This muscle is special in that its fibres are arranged in series. For several reasons, the nerve fibre counts, however, are dubious. In this context, it is important to note that the number and size of the motor units in the inner ear muscles of cat, determined using morphological methods (Blevins 1963, 1964), deviated by almost a factor 10 from the results of later physiological studies (Teig 1972b) (Sect. F.V). The mean size of motor units in human muscles has been determined by Feinstein et al. (1955) using autopsy samples of adults and by Christensen (1959) using stillborn infants. My attempts to repeat Christensen's (1959) nerve fibre counts in the musculo-cutaneous nerve of a stillborn infant with improved technique (osmium-glutaraldehyde, semithin sections) did not give interpretable results. At variance with Christensen's (1959)

Table 13. Number of muscle fibres, of motor units (60% of all large nerve fibres), and of muscle fibres per motor unit in various human muscles. The number of infant muscles assessed is not given. (See also text)

Material	Muscle	Total number of muscle fibres	Total number of motor units	Number of muscle fibres per motor unit	References
Male 22y	Platysma	27,100	1,096	25	FEINSTEIN et al. (1955)
Male 40y	Brachioradialis	>129,200 (right)	315 350 (left)	>410	
Male 22y	Dorsal interosseus I	40,500	119	340	
Male 54y	Lumbricalis I	10,038	93	108	
Female 29y		10,500	98	107	
Male 40y	Anterior tibial	250,200	445	562	
Male 22y		292,500		657	
Male 28y	Med. gastrocnemius	1,120,000	579	1,934	
Male 22y		946,000		1,634	
Stillborn infants	Superior rectus	42,441	1,779	23	CHRISTENSEN (1959)
	Opponens pollicis	79,080	6,047	13	
	Brachial biceps	580,000	3,552	163	
	Sartorius	222,424	740	300	
	Rectus femoris	186,292	609	305	
	Gracilis	144,933	275	527	
	Semitendinosus	508,219	712	713	
	Gastrocnemius	1,505,538	778	2,037	

statements, the diameters of the myelinated nerve fibres were unimodally distributed, and there was no possibility to estimate the proportion of γ-axons. FEINSTEIN et al. (1955) and CHRISTENSEN (1959) calculated the number of motor units by substracting the thin γ-fibres, and by assuming that 60% of the large fibres were motor axons. The original results – without BUCHTHAL'S (1961) recalculations, some of which also are contained in BUCHTHAL and SCHMALBRUCH (1980) – are compiled in Table 13. None of these data should be taken at face value; it is obvious that the problem needs to be reinvestigated. Data exist also for laryngeal and masticatory muscles (for references, see BUCHTHAL 1961), but in these muscles it is still more difficult to count the nerve fibres supplying the muscle.

It is frequently stated that muscles involved in very accurate movements, like the extraocular muscles, contain few muscle fibres per motor unit. Experimental evidence for that notion is scanty. The results of single motor unit studies in the inferior oblique (LENNERSTRAND 1974a), gastrocnemius and soleus muscles of cat (BURKE et al. 1971, 1974; BURKE and TSAIRIS 1973) allow a rough estimate of the motor unit size in the extraocular muscle. The maximal tetanic tensions of twitch motor units of the inferior oblique muscle range from 40 to 400 mg, the mean being about 200 mg. The mean tetanic tension of slow-twitch gastrocnemius units is 6 g, and that of the slow-twitch soleus units is

11 g. The tension of fast-twitch units of the gastrocnemius muscle ranges from 5 to 130 g. Three identified fast-twitch units of the gastrocnemius muscle contained 300, 500, and 750 fibres with mean cross-sectional areas of 8,100, 2,700 and 8,100 μm^2; the tetanic tensions were 38, 38, and 120 g. For a mean fibre diameter of 15 µm in the extraocular muscle, i.e. a cross-sectional area of 177 μm^2, the number of fibres per average motor unit with 200 mg force is calculated to be 72, 40, or 120, depending on which gastrocnemius motor units is used as reference. For this calculation, it is assumed that the specific force is identical in both muscles. Nevertheless, the force per unit cross-sectional area in the eye muscles is only one-third to one-half of that in limb muscles (rat, CLOSE and LUFF 1974), and a mean unit size of 100–200 fibres in the extraocular muscle is probably correct. Motor units of cat limb muscles comprising more than 100 muscle fibres would develop >10 g tetanic force and would be within the normal range of size. Hence, the motor units of extraocular muscles do not seem to comprise unusually few muscle fibres. Compared with the units of limb muscles, they appear better-suited for force gradation because the tension is small due to the smallness of the muscle fibres. The frequency modulation of force is probably more effective in fast fibres with high than in relatively slow fibres with low tetanic fusion frequency (Sect. F.I.5; see also Fig. 129).

An electrophysiological method for estimating the number of muscle fibres of a human motor unit has been described by STÅLBERG and GATH (1977). They measure, with a multilead electrode, the total area of the muscle which is occupied by synchronously discharging muscle fibres of a motor unit, and the mean density of the fibres of that unit within the known pick-up area of a "single-fibre" electrode. For the human biceps muscle, they find 225 fibres per unit.

III. The Array of the Muscle Fibres of a Motor Unit

The numerous motor units of a moderately or fully activated muscle discharge asynchronously. A concentric needle electrode records an interference pattern indicating that many different units are represented within the limited pick-up area (<1 mm diameter) of this electrode type (BUCHTHAL 1958, 1961). This suggests that the muscle fibres of different units are intermingled.

At very weak effort, individual motor unit potentials are recorded because only few units are active within the part of the muscle "seen" by the electrode. The potentials recorded from different units differ in size and shape because these parameters depend on the spatial distribution of the generators, i.e. the nearest muscle fibres of the unit, in relation to the electrode. The fibres of a unit discharge synchronously, but not simultaneously, because the lengths of the terminal nerve branches and the distances between endplates and recording site vary. The size of an extracellular potential decreases rapidly with increasing distance between generator and electrode. Therefore, the motor unit potentials recorded with an extracellular electrode consist of large and small components with temporal dispersion, and each motor unit potential, as long as the electrode is not moved, has a characteristic shape. This allows to decide whether one or two or three units are active within the pick-up area of the electrode. In normal muscles, most motor unit potentials have one narrow peak originating from the one generator that is nearest to the electrode; the number of small components and the duration of the potential increases with the pick-up area of the electrode (for references, see BUCHTHAL and SCHMALBRUCH 1980).

Originally, it was believed that single muscle fibres do not give rise to sufficiently strong extracellular potentials, and that in man the fibres of a motor unit are clustered and form "subunits" which discharge more or less simultaneously. A reassessment of that problem resulted in the conclusion that the motor unit potentials that are recorded in human muscles at weak effort with an intramuscular electrode indeed arise from individual muscle fibres, and that there is no need to postulate "subunits" (ROSENFALCK and BUCHTHAL 1970).

For animal muscles, the glycogen depletion method revealed that only rarely two fibres of the same motor unit are adjacent to each other (rat, KUGELBERG and EDSTRÖM 1968; EDSTRÖM and KUGELBERG 1968; cat, BURKE and TSAIRIS 1973; BURKE et al. 1974). A given region of the cat gastrocnemius muscle may be shared by as many as 50 different motor units. Primary fascicles contain 15–90 muscle fibres and only one or two belong to the same unit. The average over-all density of fibres of one unit is 4.5 per 100 total fibres (BURKE and TSAIRIS 1973). The density is even lower in the cat soleus muscle, with often only one muscle fibre of a given motor unit among 100 fibres from other motor units. Thus, the fibres of each motor unit are spread over a large part of the cross-section of the muscle. This probably smoothens weak contractions with only few activated motor units.

The cross-sectional area of human muscles over which the fibres of a motor unit are spread has been assessed with an electrode that records the electrical potentials of the fibres at multiple sites of the muscle. The synchronously discharging fibres of a unit are scattered over a circular area 5–10 mm in diameter (BUCHTHAL 1961). The pick-up area of a "single-fibre electrode", a semicircle 270 μm in diameter, may contain 10 fibres 50 μm in diameter. Only an average of 1.5 muscle fibres belongs to the same motor unit, i.e. the pick-up area contains fibres of six different motor units (STÅLBERG and THIELE 1975; STÅLBERG et al. 1976). It appears that the fibre density of a given motor unit is somewhat higher in human than in cat muscles (15 per 100 fibres, compared with 1–4.5 per 100 fibres) but this may be due to the different experimental approaches.

ENGLISH and WEEKS (1984) identified individual motor units in the lateral gastrocnemius muscle of cats using stimulation of single motor axons in ventral roots and the glycogen depletion technique. They found that all fast-twitch motor units localized in the same regions of the muscle were innervated by the same primary branches of the nerve to the muscle. This means that the muscle is subdividable into compartments each compartment being supplied by a primary branch of the nerve, and containing a unique population of motor units. A primary muscle nerve branch is defined as the naturally-occurring branches of the muscle nerve as it enters the muscle at its hilus.

IV. How Are Motor Units of Different Types Used?

Motor units differ in size and contractile properties, and it is obvious that their use during voluntary activity must be related to the type of movement required. In 1965, HENNEMAN et al. reported that, during reflex contractions, the threshold for the activation of motoneurons was lower in small and slowly contracting motor units than in large and fast motor units. Hence, small slow-twitch units are first recruited. "HENNEMAN'S size principle" has been extended to voluntary contractions. Intuitively, this appears sensible because small slow-twitch motor units are fatigue-resistant; thus, fatiguable fast motor units would rest unless extraordinary forces were needed. Nevertheless, BURKE et al. (1970) observed that the threshold of the motoneurons changed with inputs from skin receptors or through the rubrospinal tract. Therefore, not only the synaptic organization and the membrane properties of the motoneuron, but also "external" factors, might influence the sequence of recruitment. Most authors have found a stereotyped sequence of recruitment during voluntary movements according to "Henneman's size principle" (for references, see BUCHTHAL and SCHMALBRUCH 1980); recently, however, evidence has been presented that the order may be reversed, both in cat and man (KANDA et al. 1977; STEPHENS et al. 1978; SMITH et al. 1981; GARNETT and STEPHENS 1980). The present state of the concept is discussed by ENOKA and STUART (1984).

Voluntary contractions have mostly been examined in man because the cooperation of the subject is essential. Motor unit studies during voluntary activity entail two main problems which shall be briefly outlined.

(a) It is difficult to classify the motor units that are activated. The isometric twitch contraction time can only reliably be recorded during very weak contractions when the discharge rate of the motoneurons is low, i.e. before the muscle fibres develop tetanus (Sect. D.IV.2). The electrical potentials of the motor units cannot tell whether a unit is fast or slow, or large or small, because the distance between electrode tip and muscle fibre influences the shape and size of the potential (see above). The conduction velocity of the motor axons in the nerve to a muscle relates to the contractile properties of the muscle fibres they innervate; large motor units consisting of fast fibres have larger motoneurons and the axons conduct faster than those of small and slow motor units. Nevertheless, this property can only be assessed in nerves innervating very small muscles. The glycogen depletion method which would allow identification and typing of the activated muscle fibres using histochemistry is of little value for voluntary contractions (see Sect. D.II.2). The pattern of discharge during voluntary activity is the commonly used marker of the motor unit type (continuously or in bursts, high or low frequency). A classification system established by that criterion need not be identical with the fast- or slow-twitch motor units or with the histochemical muscle fibre types.

(b) Ordinary concentric needle electrodes record an interference pattern from rather low levels of activation on, and the potentials of the one unit one wants to study "drown" among the many additional units being activated. Hence, electrodes have to be "selective", which means that they must have a small pick-up area such that they record only the potentials of one or a few motor units. These are then identified throughout the experiment by the shape of their action potentials. Small electrode displacements, which are difficult to avoid during a moderate or strong contraction, change the shape of the potential, and the unit under study is lost. Probably these difficulties have not been overcome in any of the studies confirming the fixed recruitment order (for details see BUCHTHAL and SCHMALBRUCH 1980).

GRIMBY (1984) recorded, in man, the motor unit activity in the short extensor muscles of the foot; these muscles contain only few motor units which can be identified by the axonal conduction velocity. The number of motor units was intentionally reduced by partial denervation, which in addition has the advantage that the surviving motor axons sprout such that the density of muscle fibres of a given motor unit increases. The resulting "type grouping" (Sect. D.V) makes the selectivity of the electrode less critical. During walking, only low threshold units with low axonal conduction velocity are activated. They fire with a frequency of 15–20 Hz and are presumably slow-twitch motor units. Presumed fast-twitch motor units have a high axonal conduction velocity; they fire in short bursts with frequencies of up to 100 Hz during corrective movements only (for previous studies on this muscle from the same laboratory, see GRIMBY 1984).

Most interesting results have recently been reported by HODGSON (1983). He implanted electrodes and force transducers into the cat gastrocnemius and soleus muscles and determined the relative and absolute contribution of either muscle to different types of locomotion. The two synergistic muscles are predominantly fast and slow, respectively, and their function may model the behaviour of fast- and slow-twitch units in a mixed muscle. Posture is the only activity where the soleus muscle is active alone. In locomotion, the soleus-gastrocnemius activity ratio decreases with increasing speed. The activity of the soleus never exceeds 80% of maximum. HODGSON proposes that the "fast" motoneurons during fast locomotion are driven by excitatory inputs, whereas at the same time inhibitory inputs operate on the "slow" soleus motoneurons. This ascertains that not all slow motor units are activated during "maximum" effort, and maintains the possibility that segmental regulatory inputs still can act on the slow motor unit pool. In conclusion, the activation of motor units according to the size of the motoneurons seems to be a basic property of the motoneuron pool, but the order of recruitment is modified by synaptic inputs to the motoneuron and by the suprasegmental flow of impulses.

The force of a muscle may be adjusted either by recruiting or silencing entire motor units, or by a change of the innervation rate of already active units. Both mechanisms are operative, but it is not clear which one dominates during voluntary activity (MARSDEN et al. 1971; PERSON and KUDINA 1972; MILNER-BROWN et al. 1973b; TANJI and KATO 1973; MATON 1976). Also these studies are hampered by the difficulty of recording from individual motor units during moderate and strong contractions. HENNIG and LØMO (1985) recorded chronically from hindlimb muscles of freely moving rats. The most common motor unit firing rate in the extensor digitorum longus muscle was about 70 Hz,

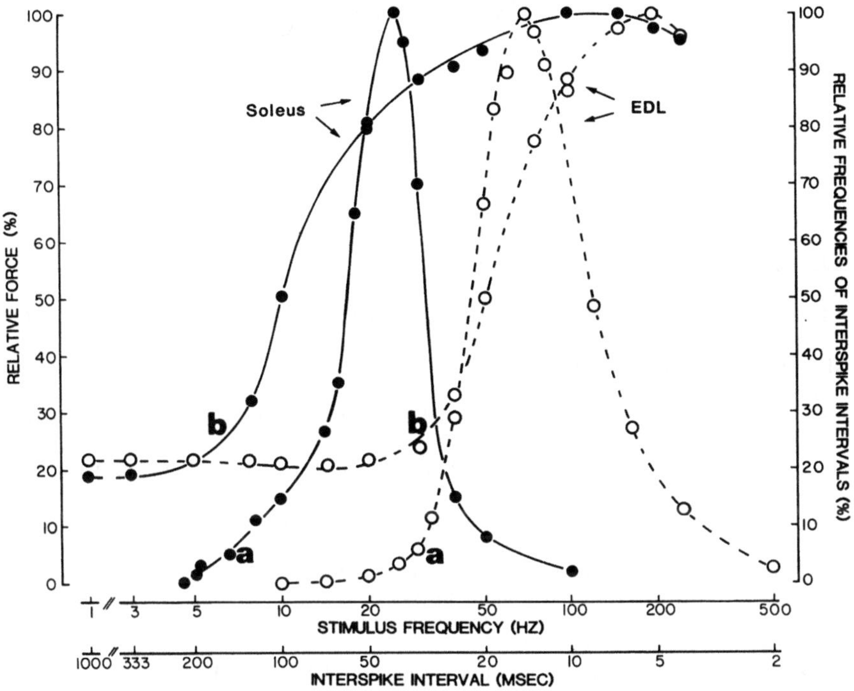

Fig. 129. Motor unit firing rates (*a*) in the soleus and extensor digitorum longus (*EDL*) muscles of freely moving rats, and the tension frequency curves (*b*) of these muscles. Soleus, *filled circles*; EDL, *open circles*. The curves for the motor unit firing rates during voluntary activity (*a*) (expressed as interspike interval) were obtained from pooled histograms for 6 soleus and 10 EDL motor units, respectively; the peak of the histograms representing the incidence of the most commonly occurring interspike interval in each muscle is set at 100% (*right*). The tension frequency curves (*b*) were determined in acute experiments by stimulating normal muscles in situ at 35° C with l-s long trains of supramaximal stimuli at different stimulus frequencies. The isometric tension was recorded at optimal length, and expressed in % of maximal force (*left*). Each symbol is the median of 8 (soleus) or 6 (EDL) muscles. (Figure kindly provided by Dr. LØMO, Oslo; the results are contained in HENNIG and LØMO 1985)

in the soleus muscle it was about 20 Hz. When the muscles were electrically stimulated, the twitches started to fuse above 5 Hz in the soleus and above 20 Hz in the extensor digitorum longus muscle, the maximal tension was reached at 100 and 200 Hz, respectively (Fig. 129). The most common natural firing rate in each muscle coincided with the rise phase of the tension-frequency curve. In the fast extensor muscle, a small change in firing rate would result in a large change in force output. These experiments also show that, under natural conditions, the motor units are mostly innervated at a rate producing only 70%–80% of their maximum force.

References

Abbott J, Schiltz J, Dienstman S, Holtzer H (1974) The phenotypic complexity of myogenic clones. Proc Natl Acad Sci USA 71:1506–1510

Achatz I (1968) Ultrastructure of the striated muscle and moire patterns. Acta Biochim Biophys Acad Sci Hung 3:183–193

Adal MN (1969) The fine structure of the sensory region of cat muscle spindles. J Ultrastruct Res 26:332–354

Adal MN (1973) The fine structure of the intrafusal muscle fibres of muscle spindles in the domestic fowl. J Anat 115:407–413

Adler C-P (1971) Polyploidisierung und Zellzahl im menschlichen Herzen. Thesis, Universität Freiburg

Adler C-P, Costabel U (1975) Cell number in human heart in atrophy, hypertrophy, and under the influence of cytostatics. In: Fleckenstein A, Rona G (eds) Recent advances in studies on cardiac structure and metabolism, vol 6: Pathophysiology and morphology of myocardial cell alteration. University Park Press, Baltimore, pp 343–355

Adrian ED (1925) The spread of activity in the tenuissimus muscle of the cat and in other complex muscles. J Physiol (Lond) 60:301–315

Adrian RH, Peachey LD (1965) The membrane capacity of frog twitch and slow muscle fibres. J Physiol (Lond) 181:324–336

Adrian RH, Costantin LL, Peachey LD (1969) Radial spread of contraction in frog muscle fibres. J Physiol (Lond) 204:231–257

Aebi U, Fowler WE, Isenberg G, Pollard TD, Smith PR (1981) Crystalline actin sheets: their structure and polymorphism. J Cell Biol 91:340–351

Agostini B, Hasselbach W (1971) Electron cytochemistry of calcium uptake in the fragmented sarcoplasmic reticulum. Histochemie 28:55–67

Ahkong QF, Fisher D, Tampion W, Lucy JA (1975) Mechanisms of cell fusion. Nature 253:194–195

Al-Amood WS, Buller AJ, Pope R (1973) Long-term stimulation of cat fast-twitch skeletal muscle. Nature 244:225–227

Albuquerque EX, Thesleff S (1968) A comparative study of membrane properties of innervated and chronically denervated fast and slow skeletal muscles of the rat. Acta Physiol Scand 73:471–480

Allbrook D (1975) Transplantation and regeneration of striated muscle. Ann R Coll Surg Engl 56:312–324

Allbrook D (1981) Skeletal muscle regeneration. Muscle Nerve 4:234–245

Allbrook D, Batalin NJ, James D, Brearley R (1971a) Morphogenetic or cellular control mechanisms in mammalian striated muscle regeneration and reconstitution. In: Kakulas BA (ed) Clinical studies in myology. Excerpta Medica, Amsterdam, pp 686–694

Allbrook DB, Han MF, Hellmuth AE (1971b) Population of muscle satellite cells in relation to age and mitotic activity. Pathology 3:233–243

Allen ER (1973) Sarcomere formation in chick striated muscle. Z Zellforsch Mikrosk Anat 145:167–170

Almers W, Best PM (1976) Effects of tetracaine on displacement currents and contraction of frog skeletal muscle. J Physiol (Lond) 262:583–611

Almers W, Fink R, Shepherd N (1982) Lateral distribution of ionic channels in the cell membrane of skeletal muscle. In: Schotland DL (ed) Disorders of the motor unit. Wiley, New York, pp 349–366

Aloisi M (1970) Patterns of muscle regeneration. In: Mauro A, Shafiq SA, Milhorat AT (eds) Regeneration of striated muscle, and myogenesis. Excerpta Medica, Amsterdam, pp 180–193

Aloisi M, Mussini I, Schiaffino S (1973) Activation of muscle nuclei in denervation and hypertrophy. In: Kakulas BA (ed) Basic research in myology. Excerpta Medica Amsterdam, pp 338–342
Altschul R, Lee JC (1960) Nuclear reaction in denervated and in injured skeletal muscle (Abstr). Anat Rec 136:153
Alvarado-Mallart R-M (1972) Ultrastructure of muscle fibers of an extraocular muscle of the pigeon. Tissue Cell 4:327–339
Amos LA, Huxley HE, Holmes KC, Goody RS, Taylor KA (1982) Structural evidence that myosin heads may interact with two sites on F-actin. Nature 299:467–469
Andersen P (1975) Capillary density in skeletal muscle of man. Acta Physiol Scand 95:203–205
Andersen P, Henriksson J (1977a) Training induced changes in the subgroups of human type II skeletal muscle fibres. Acta Physiol Scand 99:123–125
Andersen P, Henriksson J (1977b) Capillary supply of the quadriceps femoris muscle of man: adaptive response to exercise. J Physiol (Lond) 270:677–690
Andersen P, Sears TA (1964) The mechanical properties and innervation of fast and slow motor units in the intercostal muscles of the cat. J Physiol (Lond) 173:114–129
Anderson MJ, Cohen MW (1974) Fluorescent staining of acetylcholine receptors in vertebrate skeletal muscle. J Physiol (Lond) 237:385–400
Anderson MJ, Cohen MW (1977) Nerve-induced and spontaneous redistribution of acetylcholine receptors on cultured muscle cells. J Physiol (Lond) 268:757–773
Anderson MJ, Cohen MW, Zorychta E (1977) Effects of innervation on the distribution of acetylcholine receptors on cultured muscle cells. J Physiol (Lond) 268:731–756
Andersson-Cedergren E (1959) Ultrastructure of motor endplate and sarcoplasmic components of mouse skeletal muscle fiber as revealed by three-dimensional reconstructions from serial sections. J Ultrastruct Res [Suppl] 1:1–191
Änggård L, Ottoson D (1963) Observations on the functional development of the neuromuscular apparatus in fetal sheep. Exp Neurol 7:294–304
Ängqvist K-A, Sjöström M (1980) Intermittent claudication and muscle fibre fine structure. Ultrastruct Pathol 1:461–470
Aniansson A, Grimby G, Nygaard E, Saltin B (1980) Muscle fiber composition and fiber area in various age groups (Abstr) Muscle Nerve 3:271–272
Anzenbacher H, Zenker W (1962) Über die cholinesterase-aktiven Formelemente des M. thyreoarytaenoideus und ihre Beziehungen zur Struktur dieses Muskels. Z Anat Entwicklungsgesch 123:221–245
Anzenbacher H, Zenker W (1963) Über die Grössenbeziehung der Muskelfasern zu ihren motorischen Endplatten und Nerven. Z Zellforsch Mikrosk Anat 60:860–871
Aquilonius S-M, Askmark H, Gillberg P-G, Nandedkar S, Olsson Y, Stålberg E (1984) Topographical localization of motor endplates in cryosections of whole human muscles. Muscle Nerve 7:287–293
Arangio GA, Hagstrom JWC (1969) The histochemical classification of rabbit hindlimb striated muscle. J Histochem Cytochem 17:127–128
Arendt K-W, Asmussen G (1974a) Die Anzahl und die Verteilung von Muskelspindeln im M. triceps surae der Ratte. Anat Anz 136:207–216
Arendt K-W, Asmussen G (1974b) Enzymhistochemische Untersuchungen an Muskelspindeln verschiedener Spezies. Anat Anz 136:217–228
Ariano MA, Armstrong RB, Edgerton VR (1973) Hindlimb muscle fiber populations of five mammals. J Histochem Cytochem 21:51–55
Armitage PM, Tregear RT, Miller A (1975) Effect of activation by calcium on the X-ray diffraction pattern from insect flight muscle. J Mol Biol 92:39–53
Armstrong RB, Saubert CW, IV, Seeherman HJ, Taylor CR (1982) Distribution of fiber types in locomotory muscles of dogs. Am J Anat 163:87–98
Arndt I, Pepe FA (1975) Antigenic specificity of red and white muscle myosin. J Histochem Cytochem 23:159–168
Arnold H, Pette D (1968) Binding of glycolytic enzymes to structure proteins of the muscle. Eur J Biochem 6:163–171
Arnold H, Pette D (1970) Binding of aldolase and triosephosphate dehydrogenase to F-actin and modification of catalytic properties of aldolase. Eur J Biochem 15:360–366
Arnold H, Nolte J, Pette D (1969) Quantitative and histochemical studies on the desorption and readsorption of aldolase in cross-striated muscle. J Histochem Cytochem 17:314–320

Arnold H, Henning R, Pette D (1971) Quantitative comparison of the binding of various glycolytic enzymes to F-actin and the interaction of aldolase with G-actin. Eur J Biochem 22:121–126

Aronson J (1961) Sarcomere size in developing muscles of a tarsonemid mite. J Biophys Biochem Cytol 11:147–156

Ashhurst DE (1967) Z-line of the flight muscle of belostomatid water bugs. J Mol Biol 27:385–389

Ashley CC, Ridgway EB (1970) On the relationships between membrane potential, calcium transient and tension in single barnacle muscle fibres. J Physiol (Lond) 209:105–130

Ashmore CR, Summers PJ (1981) Stretch-induced growth in chicken wing muscles: Myofibrillar proliferation. Am J Physiol 51:C93–C97

Ashmore CR, Robinson DW, Rattray P, Doerr L (1972) Biphasic development of muscle fibers in the fetal lamb. Exp Neurol 37:241–255

Askanas V, Engel WK (1975a) A new program for investigating adult human skeletal muscle grown aneurally in tissue culture. Neurology (NY) 25:58–67

Askanas V, Engel WK (1975b) Distinct subtypes of type I fibers of human skeletal muscle. Neurology 25:879–887

Asmussen G (1974) Über das Vorkommen von zwei motorischen Systemen im okulomotorischen Apparat der Säuger (eine Literaturübersicht). Wiss Z Karl-Marx Univ Leipzig, Math-Naturwiss R 23:125–141

Asmussen G (1978a) The properties of the extraocular muscles of the frog. I. Mechanical properties of the isolated superior oblique and superior rectus muscles. Acta Biol Med Ger 37:301–312

Asmussen G (1978b) The properties of the extraocular muscles of the frog. II. Pharmacological properties of the isolated superior oblique and superior rectus muscles. Acta Biol Med Ger 37:313–321

Asmussen G (1978c) The properties of the extraocular muscles of the frog. III. Morphological, mechanical and pharmacological properties of the isolated retractor bulbi muscle. Acta Biol Med Ger 37:323–333

Asmussen G, Gaunitz U (1978) Kontraktile Eigenschaften der quergestreiften Muskelfasern des Ösophagus im Vergleich mit ausgewählten Skelettmuskeln der Ratte. Acta Biol Med Ger 37:335–346

Asmussen G, Gaunitz U (1981) Mechanical properties of the isolated inferior oblique muscle of the rabbit. Pflugers Arch 392:183–190

Asmussen G, Kiessling A (1970) Die Muskelfasersorten des Frosches: Ihre Identifikation und die Gesetzmäßigkeiten ihrer Anordnung in der Skelettmuskulatur. Acta Biol Med Ger 24:871–889

Asmussen G, Kiessling A (1974) Charakterisierung von besonderen Muskelfasergruppen in der Skeletmuskulatur des Frosches durch ihre motorische Innervation und ihre Gefäßversorgung. Acta Anat (Basel) 90:226–242

Asmussen G, Wohlrab F (1971) Enzymhistochemische Differenzierung von Muskelfasertypen im M. tensor tympani von Kaninchen und Katzen. Z Mikrosk Anat Forsch 83:507–517

Asmussen G, Wohlrab F (1974) Gemeinsamkeiten im Aufbau der äußeren Augenmuskulatur der Wirbeltiere. Wiss Z Karl-Marx-Univ, Leipzig, Math-Naturwiss R 23:143–154

Asmussen G, Kiessling A, Wohlrab F (1971) Histochemische Charakterisierung der verschiedenen Muskelfasertypen in den äußeren Augenmuskeln von Säugetieren. Acta Anat (Basel) 79:526–545

Atsumi S (1981) Localization of surface and internal acetylcholine receptors in developing fast and slow muscles of the chick embryo. Dev Biol 86:122–135

Auber J, Couteaux R (1963) Ultrastructure de la strie Z dans des muscles de diptères. J Microsc 2:309–324

Axelrod D, Ravdin P, Koppel DE, Schlessinger J, Webb WW, Elson EL, Podleski TR (1976) Lateral motion of fluorescently labeled acteylcholine receptors in membranes of developing muscle fibers. Proc Natl Acad Sci USA 73:4594–4598

Axelrod D, Bauer HC, Stya M, Christian CN (1981) A factor from neurons induces partial immobilization of nonclustered acetylcholine receptors on cultured muscle cells. J Cell Biol 88:459–462

Axelsson J, Thesleff S (1959) A study of supersensitivity in denervated mammalian skeletal muscle. J Physiol (Lond) 147:178–193

Bach-y-Rita P, Ito F (1966) In vivo studies of fast and slow muscle fibers in cat extraocular muscles. J Gen Physiol 49:1177–1198

Bach-y-Rita P, Lennerstrand G (1975) Absence of polyneuronal innervation in cat extraocular muscles. J Physiol (Lond) 244:613–624

Bachmann P (1980) Motility, linear arrangement and cell-to-cell contact of myogenic cells prior to fusion. Cell Tissue Res 206:431–440

Bacou F, Vigneron P, Massoulie J (1982) Acetylcholinesterase forms in fast and slow rabbit muscle. Nature 296:661–664

Bader D (1980) Reinnervation of motor endplate-containing and motor endplate-less muscle grafts. Dev Biol 77:315–327

Bader D, Masaki T, Fischman DA (1982) Immunochemical analysis of myosin heavy chain during avian myogenesis in vivo and in vitro. J Cell Biol 95:763–770

Bagust J, Lewis DM, Westerman RA (1973) Polyneuronal innervation of kitten skeletal muscle. J Physiol (Lond) 229:241–255

Bailey AJ, Shellswell GB, Duance VC (1979) Identification and change of collagen types in differentiating myoblasts and developing chick muscle. Nature 278:67–68

Bailey K (1942) Myosin and adenosinetriphosphatase. Biochem J 36:121–139

Baker AJ, Lewis DM (1983) The effects of denervation on isotonic shortening velocity of rat fast and slow muscle (Abstr). J Physiol (Lond) 345:56P

Baker WW, Mintz B (1969) Subunit structure and gene control of mouse NADP-malate dehydrogenase. Biochem Genet 2:351–360

Bakker GJ, Richmond FJR (1981) Two types of muscle spindles in cat neck muscles: a histochemical study of intrafusal fiber composition. J Neurophysiol 45:973–986

Bandman E, Walker CR, Strohman RC (1978) Diazepam inhibits myoblast fusion and expression of muscle specific protein synthesis. Science 200:559–561

Banker BQ, Girvin JP (1971) The ultrastructural features of the mammalian muscle spindle. J Neuropathol Exp Neurol 30:155–195

Banker BQ, Girvin JP (1972) The ultrastructural features of the normal and de-efferented mammalian muscle spindle. In: Banker BQ, Przybylski RJ, Van Der Meulen JP, Victor M (eds) Research in muscle development and the muscle spindle. Excerpta Medica, Amsterdam, pp 267–298

Banker BQ, Kelly SS, Robbins N (1983) Neuromuscular transmission and correlative morphology in young and old mice. J Physiol (Lond) 339:355–375

Banks RW, James NT (1973) The fine structure of the guinea-pig muscle spindle. Z Zellforsch Mikrosk Anat 140:357–368

Banks RW, James NT (1974) Quantitative studies on the distribution of myofilaments in intrafusal muscle fibres. Cell Tissue Res 150:521–527

Banks RW, Barker D, Harker DW, Stacey MJ (1975) Correlation between ultrastructure and histochemistry of mammalian intrafusal muscle fibres (Abstr). J Physiol (Lond) 252:16P–17P

Banks RW, Harker DW, Stacey MJ (1977) A study of mammalian intrafusal muscle fibres using a combined histochemical and ultrastructural technique. J Anat 123:783–796

Barany M (1967) ATPase activity of myosin correlated with speed of muscle shortening. J Gen Physiol [Suppl] (6, Part 2) 50:197–218

Barany M, Close RI (1971) The transformation of myosin in cross-innervated rat muscles. J Physiol (Lond) 213:455–474

Barany M, Barany K, Reckard T, Volpe A (1965) Myosin of fast and slow muscles of the rabbit. Arch Biochem Biophys 109:185–191

Barberie ME (1970) The formation of presumptive myoblasts in injured amphibian skeletal muscle. In: Mauro A, Shafiq SA, Milhorat AT (eds) Regeneration of striated muscle, and myogenesis. Excerpta Medica, Amsterdam, pp 267–270

Barcroft H, Millen JLE (1939) The blood flow through muscle during sustained contraction. J Physiol (Lond) 97:17–31

Barin-Baum DE (1963) Über die Faserzahl im Musculus biceps brachii des Menschen unter Berücksichtigung des Körperbautypus. Acta Anat (Basel) 55:224–254

Barker D (1959) Some results of a quantitative histological investigation of stretch receptors in limb muscles of the cat (Abstr). J Physiol (Lond) 149:7P–9P

Barker D (1962) The structure and distribution of muscle receptors. In: Barker D (ed) Symposium on muscle receptors. Hong Kong University Press, Hong Kong, pp 227–240

Barker D (1966) Three types of motor ending in cat spindles (Abstr). J Physiol (Lond) 186:27P–28P

Barker D, Gidumal JL (1960) Some observations on the morphology of the intrafusal muscle fibre (Abstr). J Physiol (Lond) 153:28P–29P

Barker D, Gidumal JL (1961) The morphology of intrafusal muscle fibres in the cat. J Physiol (Lond) 157:513–528
Barker D, Harker DW (1972) Two types of multiply-innervated muscle fibre in the superior rectus muscle of the sheep (Abstr). J Physiol (Lond) 222:74P–75P
Barker D, Hunt JP (1964) Mammalian intrafusal muscle fibres. Nature 203:1193
Barker D, Harker D, Stacey MJ, Smith CR (1972) Fusimotor innervation. In: Banker BQ, Przybylski RJ, Van Der Meulen JP, Victor M (eds) Research in muscle development and the muscle spindle. Excerpta Medica, Amsterdam, pp 227–253
Barker D, Emonet-Denand F, Laporte Y, Proske U, Stacey MJ (1973) Morphological identification and intrafusal distribution of the endings of static fusimotor axons in the cat. J Physiol (Lond) 230:405–427
Barker D, Banks RW, Harker DW, Milburn A, Stacey MJ (1976a) Studies of the histochemistry, ultrastructure, motor innervation, and regeneration of mammalian intrafusal muscle fibres. Prog Brain Res 44:67–88
Barker D, Emonet-Denand F, Harker DW, Jami L, Laporte Y (1976b) Distribution of fusimotor axons to intrafusal muscle fibres in cat tenuissimus spindles as determined by the glycogen-depletion method. J Physiol (Lond) 261:49–69
Barker D, Emonet-Denand F, Harker DW, Jami L, Laporte Y (1977) Types of intra- and extrafusal muscle fibre innervated by dynamic skeleto-fusimotor axons in cat peroneus brevis and tenuissimus muscles, as determined by the glycogen-depletion method. J Physiol (Lond) 266:713–726
Barker D, Bessou P, Jankowska E, Pages B, Stacey MJ (1978) Identification of intrafusal muscle fibres activated by single fusimotor axons and injected with fluorescent dye in cat tenuissimus spindles. J Physiol (Lond) 275:149–165
Barnard EA, Rogers AW (1967) Determination of the number, distribution, and some *in situ* properties of cholinesterase molecules in the motor end plate, using labeled inhibitor methods. Ann NY Acad Sci 144:584–612
Barrnett RJ (1962) The fine structural localization of acetylcholinesterase at the myoneural junction. J Cell Biol 12:247–262
Baskin RJ (1974) Ultrastructure and calcium transport in microsomes from developing muscle. J Ultrastruct Res 49:348–371
Bateson DS, Parry DJ (1983) Motor units in a fast-twitch muscle of normal and dystrophic mice. J Physiol (Lond) 345:515–523
Bauer HC, Daniels MP, Pudimat PA, Jacques L, Sugiyama H, Christian CN (1981) Characterization and partial purification of a neuronal factor which increases acetylcholine receptor aggregation on cultured muscle cells. Brain Res 209:395–404
Baylor SM, Oetliker H (1975) Birefringence experiments on isolated skeletal muscle fibres suggest a possible signal from the sarcoplasmic reticulum. Nature 253:97–101
Baylor SM, Oetliker H (1977) Birefringence signals from surface and T-system membranes of frog single muscle fibres. J Physiol (Lond) 264:199–213
Beatty CH, Basinger GM, Bocek RM (1967) Differentiation of red and white fibers in muscle from fetal, neonatal and infant rhesus monkeys. J Histochem Cytochem 15:93–103
Beermann DH, Cassens RG (1977) Indirect fluorescence of primary and secondary myofibers in developing porcine muscle. J Histochem Cytochem 25:439–442
Bekoff A, Betz WJ (1976) Acetylcholine hot spots: Development on myotubes cultured from aneural limb buds. Science 193:915–917
Bekoff A, Betz WJ (1977) Physiological properties of dissociated muscle fibres obtained from innervated and denervated adult rat muscle. J Physiol (Lond) 271:25–40
Bell RD, MacDougall JD, Billeter R, Howald H (1980) Muscle fiber types and morphometric analysis of skeletal muscle in six-year-old children. Med Sci Sports Exerc 12:28–31
Bendall JR (1952) Effect of the "Marsh factor" on the shortening of muscle fibre models in the presence of adenosine. Nature 170:1058–1060
Bender AN, Ringel SP, Engel WK (1976) The acetylcholine receptor in normal and pathologic states. Immunoperoxidase visualization of alpha-bungarotoxin binding at a light and electron-microscopic level. Neurology (NY) 26:477–483
Bennett GS, Fellini SA, Toyama Y, Holtzer H (1979) Redistribution of intermediate filament subunits during skeletal myogenesis and maturation in vitro. J Cell Biol 82:577–584
Bennett HS, Porter KR (1953) An electron microscope study of sectioned breast muscle of the domestic fowl. Am J Anat 93:61–106

Bennett MR (1983) Development of neuromuscular synapses. Physiol Rev 63:915–1048
Bennett MR, Pettigrew AG (1974a) The formation of synapses in striated muscle during development. J Physiol (Lond) 241:515–545
Bennett MR, Pettigrew AG (1974b) The formation of synapses in reinnervated and cross-reinnervated striated muscle during development. J Physiol (Lond) 241:547–573
Bennett MR, McLachlan EM, Taylor RS (1973) The formation of synapses in reinnervated mammalian striated muscle. J Physiol (Lond) 233:481–500
Bennett MR, Florin T, Woog R (1974) The formation of synapses in regenerating mammalian striated muscle. J Physiol (Lond) 238:79–92
Bennett MR, Davey DF, Marshall JJ (1983a) The growth of nerves in relation to the formation of premuscle cell masses in the developing chick forelimb. J Comp Neurol 215:217–227
Bennett MR, McGrath PA, Davey DF, Hutchinson I (1983b) Death of motorneurons during the postnatal loss of polyneural innervation of rat muscles. J Comp Neurol 218:351–363
Bennett MVL (1970) Comparative physiology: Electric organs. Annu Rev Physiol 32:471–528
Benoit PW, Belt WD (1970) Destruction and regeneration of skeletal muscle after treatment with a local anaesthetic, bupivacaine (Marcaine). J Anat 107:547–556
Berg DK, Kelly RB, Sargent PB, Williamson P, Hall ZW (1972) Binding of α-bungarotoxin to acetylcholine receptors in mammalian muscle. Proc Natl Acad Sci USA 69:147–151
Beringer T (1976) A freeze-fracture study of sarcoplasmic reticulum from fast and slow muscle of the mouse. Anat Rec 184:647–664
Beringer T, Koenig E (1975) The sarcolemmal membrane: Evacuation of short myofiber bundles. Exp Neurol 46:20–31
Bertaud WS, Rayns DG, Simpson FO (1970) Freeze-etch studies on fish skeletal muscle. J Cell Sci 6:537–557
Bessou P, Emonet-Denand F, Laporte Y (1965) Motor fibres innervating extrafusal and intrafusal muscle fibres in the cat. J Physiol (Lond) 180:649–672
Betz H, Bourgeois J-P, Changeux J-P (1980) Evolution of cholinergic proteins in developing slow and fast skeletal muscles in chick embryo. J Physiol (Lond) 302:197–218
Betz W (1976a) The formation of synapses between chick embryo skeletal muscle and ciliary ganglia grown *in vitro*. J Physiol (Lond) 254:63–73
Betz W (1976b) Functional and non-functional contacts between ciliary neurones and muscle grown *in vitro*. J Physiol (Lond) 254:75–86
Betz W, Osborne M (1977) Effects of innervation on acetylcholine sensitivity of developing muscle *in vitro*. J Physiol (Lond) 270:75–88
Betz W, Sakmann B (1973) Effects of proteolytic enzymes on function and structure of frog neuromuscular junctions. J Physiol (Lond) 230:673–688
Betz WJ, Caldwell JH, Ribchester RR (1979) The size of motor units during post-natal development of rat lumbrical muscle. J Physiol (Lond) 297:463–478
Bevan S, Steinbach JH (1977) The distribution of α-bungarotoxin binding sites on mammalian skeletal muscle developing *in vivo*. J Physiol (Lond) 267:195–213
Bezanilla F, Horowicz P (1975) Fluorescence intensity changes associated with contractile activation in frog muscle stained with Nile blue A. J Physiol (Lond) 246:709–735
Billeter R, Weber H, Lutz H, Howald H, Eppenberger HM, Jenny E (1980) Myosin types in human skeletal muscle fibers. Histochemistry 65:249–259
Bintliff S, Walker BE (1960) Radioautographic study of skeletal muscle regeneration. Am J Anat 106:233–245
Bird JWC, Carter JH, Triemer RE, Brooks RM, Spanier AM (1980) Proteinases in cardiac and skeletal muscle. Fed Proc 39:20–25
Bird MM, James DW (1974) An ultrastructural and electrophysiological study of the development of neuro-muscular junctions between previously dissociated foetal rat cells *in vitro*. Cell Tissue Res 155:269–282
Birks RI (1965) The sarcoplasmic reticulum of twitch fibres in the frog sartorius muscle. In: Paul WW, Daniel EE, Kay CM, Monckton G (eds) Muscle. Pergamon, Oxford, pp 199–216
Birks RI, Davey DF (1969) Osmotic responses demonstrating the extracellular character of the sarcoplasmic reticulum. J Physiol (Lond) 202:171–188
Birks RI, Huxley HE, Katz B (1960) The fine structure of the neuromuscular junction of the frog. J Physiol (Lond) 150:134–144
Bischoff R (1970) The myogenic stem cell in development of skeletal muscle. In: Mauro A, Shafiq

SA, Milhorat AT (eds) Regeneration of striated muscle, and myogenesis. Excerpta Medica, Amsterdam, pp 218–231
Bischoff R (1974) Enzymatic liberation of myogenic cells from adult rat muscle. Anat Rec 180:645–662
Bischoff R (1975) Regeneration of single skeletal muscle fibers in vitro. Anat Rec 182:215–236
Bischoff R (1979) Tissue culture studies on the origin of myogenic cells during muscle regeneration in the rat. In: Mauro A (ed) Muscle regeneration. Raven Press, New York, pp 13–29
Bischoff R, Holtzer H (1969) Mitosis and the processes of differentiation of myogenic cells in vitro. J Cell Biol 41:188–200
Bischoff R, Holtzer H (1970) Inhibition of myoblast fusion after one round of DNA synthesis in 5-bromodeoxyuridine. J Cell Biol 44:134–150
Bischoff R, Lowe M (1974) Cell surface components and the interaction of myogenic cells. In: Milhorat AT (ed) Exploratory concepts in muscular dystrophy II. Excerpta Medica, Amsterdam, pp 17–31
Bishop SP, Hine P (1975) Cardiac muscle cytoplasmic and nuclear development during canine neonatal growth. In: Roy P-E, Harris P (eds) Recent advances in studies on cardiac structure and metabolism. Vol 8. The cardiac sarcoplasm. University Park Press, Baltimore, pp 77–98
Bleisch WV, Harrelson AL, Luine VN (1982) Testosterone increases acetylcholine receptor number in the "levator ani" muscle of the rat. J Neurobiol 13:153–161
Blevins CE (1963) Innervation of the tensor tympani muscle of the cat. Am J Anat 113:287–301
Blevins CE (1964) Studies on the innervation of the stapedius muscle of the cat. Anat Rec 149:157–172
Blinks JR, Rudel R, Taylor SR (1978) Calcium transients in isolated amphibian skeletal muscle fibres: Detection with aequorin. J Physiol (Lond) 277:291–323
Blinks JR, Weir WG, Hess P, Prendergast FG (1982) Measurement of Ca^{2+} concentrations in living cells. Prog Biophys Mol Biol 40:1–114
Boland R, Martonosi A, Tillack TW (1974) Developmental changes in the composition and function of sarcoplasmic reticulum. J Biol Chem 249:612–623
Bondani A, Karler R (1970) Interaction of calcium and local anesthetics with skeletal muscle microsomes. J Cell Physiol 75:199–212
Bonde-Petersen F, Robertson CH Jr (1981) Blood flow in "red" and "white" calf muscles in cats during isometric and isotonic exercise. Acta Physiol Scand 112:243–251
Bonilla E (1977) Staining of transverse tubular system of skeletal muscle by tannic acid-glutaraldehyde fixation. J Ultrastruct Res 58:162–165
Bonilla E (1983) Ultrastructural study of the muscle cell surface. J Ultrastruct Res 82:341–345
Bonilla E, Fischbeck K, Schotland DL (1981) Freeze-fracture studies of muscle caveolae in human muscular dystrophy. Am J Pathol 104:167–173
Bonner PH (1978) Nerve-dependent changes in clonable myoblast populations. Dev Biol 66:207–219
Bonner PH (1980) Differentiation of chick embryo myoblasts is transiently sensitive to functional denervation. Dev Biol 76:79–86
Bonner PH, Hauschka SD (1974) Clonal analysis of vertebrate myogenesis. I. Early developmental events in the chick limb. Dev Biol 37:317–328
Booth FW, Kelso JR (1973) Effect of hind-limb immobilization on contractile and histochemical properties of skeletal muscle. Pflugers Arch 342:231–238
Bornstein MB, Iwanami H, Lehrer GM, Breitbart L (1968) Observations on the appearance of neuromuscular relationships in cultured mouse tissues. Z Zellforsch Mikrosk Anat 92:197–206
Borys HK, Karler R (1971) Effects of caffeine on the intracellular distribution of calcium in frog sartorius muscle. J Cell Physiol 78:387–404
Bosley M, Rowlerson A (1980) A new myosin associated with the fast fibre type in the cat jaw-closer muscles? (Abstr). J Physiol (Lond) 306:18P–19P
Bosley MA, Cody FWJ, Taylor A (1972) The correlation of histochemistry and speed of contraction in cat jaw muscles (Abstr). J Physiol (Lond) 224:92P–94P
Bourke DL, Ontell M (1984) Branched myofibers in long-term whole muscle transplants: a quantitative study. Anat Rec 209:281–288
Bourne G (1943) Distribution of alkaline phosphatase in various tissues. Q J Exp Physiol 32:1–19
Bowden REM, Duchen LW (1976) The anatomy and pathology of the neuromuscular junction. In: Zaimis E (ed) Neuromuscular junction. Springer: Berlin Heidelberg New York, pp 23–97

Bowman W (1840) On the minute structure and movements of voluntary muscle. Philos Trans R Soc Lond [Biol] 130:457–501
Boyd IA (1956) The tenuissimus muscle of the cat (Abstr). J Physiol (Lond) 133:35P–36P
Boyd IA (1960) The diameter and distribution of the nuclear bag and nuclear chain muscle fibres in the muscle spindles of the cat (Abstr). J Physiol (Lond) 153:23P–24P
Boyd IA (1962a) The structure and innervation of the nuclear bag muscle fibre system and the nuclear chain muscle fibre system in mammalian muscle spindles. Philos Trans R Soc Lond [Biol] 245:81–136
Boyd IA (1962b) The nuclear-bag fibre and nuclear-chain fibre systems in the muscle spindles of the cat. In: Barker D (ed) Muscle receptors. Hong Kong University Press, Hong Kong, pp 185–190
Boyd IA, Davey MR (1966) The composition of peripheral nerves. In: Andrew BL (ed) Control and innervation of skeletal muscle. University of St. Andrews, Dundee, pp 35–52
Boyd IA, Davey MR (1968) Composition of peripheral nerves. Livingstone Ltd, Edinburgh London
Boyd IA, Ward J (1975) Motor control of nuclear bag and nuclear chain intrafusal fibres in isolated living muscle spindles from the cat. J Physiol (Lond) 244:83–112
Boyd IA, Gladden MH, McWilliam PN, Ward J (1977) Control of dynamic and static nuclear bag fibres and nuclear chain fibres by gamma and beta axons in isolated cat muscle spindles. J Physiol (Lond) 265:133–162
Boyd JD (1960) Development of striated muscle. In: Bourne GH (ed) The structure and function of muscle, vol 1. Academic Press, New York, pp 63–85
Boyde A, Williams JCP (1968) Surface morphology of frog striated muscle as prepared for and examined in the scanning electron microscope (Abstr) J Physiol (Lond) 197:10P–11P
Bradley WG (1979) Muscle fiber splitting. In: Mauro A (ed) Muscle regeneration. Raven Press, New York, pp 215–232
Braithwaite AW, Harris AJ (1979) Neural influence on acetylcholine receptor clusters in embryonic development of skeletal muscles. Nature 279:549–551
Brandstater ME, Lambert EH (1969) A histological study of the spatial arrangement of muscle fibers in single motor units within rat tibialis anterior muscle. Bull Am Ass Electromyogr Electrodiagn 82:15–16
Brandstater ME, Lambert EH (1973) Motor unit anatomy. Type and spatial arrangement of muscle fibers. In: Desmedt JE (ed) New developments in electromyography and clinical neurophysiology, vol 1. Karger, Basel, pp 14–22
Brandt PW, Lopez E, Reuben JP, Grundfest H (1967) The relationship between myofilament packing density and sarcomere length in frog striated muscle. J Cell Biol 33:255–263
Brandt PW, Reuben JP, Grundfest H (1968) Correlated morphological and physiological studies on isolated single muscle fibers. II. The properties of the crayfish transverse tubular system: Localization of the sites of reversible swelling. J Cell Biol 38:115–129
Branton D, Cohen CM, Tyler J (1981) Interaction of cytoskeletal proteins on the human erythrocyte membrane. Cell 24:24–32
Bray DF, Rayns DG (1976) A comparative freeze-etch study of the sarcoplasmic reticulum of avian fast and slow muscle fibers. J Ultrastruct Res 57:251–259
Bray JJ, Forrest JW, Hubbard JI (1982) Evidence for the role of non-quantal acetylcholine in the maintenance of the membrane potential of rat skeletal muscle. J Physiol (Lond) 326:285–296
Brenner HR, Meier Th, Widmer B (1983) Early action of nerve determines motor endplate differentiation in rat muscle. Nature 305:536–537
Brevet A, Pinto E, Peacock J, Stockdale FE (1976) Myosin synthesis increased by electrical stimulation of skeletal muscle cell cultures. Science 193:1152–1154
Briggs FN, Fuchs F (1960) The biosynthesis of a muscle-relaxing substance. Biochim Biophys Acta 42:519–527
Briggs FN, Poland JL, Solaro RJ (1977) Relative capabilities of sarcoplasmic reticulum in fast and slow mammalian skeletal muscles. J Physiol (Lond) 266:587–594
Briskey EJ, Seraydarian K, Mommaerts WFHM (1967) The modification of actomyosin by α-actinin. II. The interaction between α-actinin and actin. Biochim Biophys Acta 133:424–434
Brodal P, Ingjer F, Hermansen L (1977) Capillary supply of skeletal muscle fibers in untrained and endurance-trained men. Am J Physiol 232:H705–H712
Brooke MH, Engel WK (1969) The histographic analysis of human muscle biopsies with regard to fiber types. I. Adult male and female. Neurology 19:221–233

Brooke MH, Kaiser KK (1970a) Muscle fiber types: How many and what kind? Arch Neurol 23:369–379
Brooke MH, Kaiser KK (1970b) Three "myosin adenosine triphosphatase" systems: The nature of their pH lability and sulfhydryl dependence. J Histochem Cytochem 18:670–672
Brooke MH, Williamson E, Kaiser KK (1971) The behavior of the four fiber types in developing and reinnervated muscle. Arch Neurol 25:360–366
Brosemer RW (1972) Immunofluorescent localization of two dehydrogenases in honeybee flight muscle. J Histochem Cytochem 20:266–271
Brown GL, Dale HH, Feldberg W (1936) Reactions of the normal mammalian muscle to acetylcholine and to eserine. J Physiol (Lond) 87:394–424
Brown LM, Gonzalez-Serratos H, Huxley AF (1970) Electron microscopy of frog muscle fibres in extreme passive shortening (Abstr). J Physiol (Lond) 208:86P–88P
Brown MC, Jansen JKS, van Essen D (1976) Polyneuronal innervation of skeletal muscle in new-born rats and its elimination during maturation. J Physiol (Lond) 261:387–422
Brown MD, Cotter MA, Hudlicka O, Vrbova G (1976) The effects of different patterns of muscle activity on capillary density, mechanical properties and structure of slow and fast rabbit muscles. Pflugers Arch 361:241–250
Brunner R, Zimmermann P, Klussmann FW (1980) Localization and neurophysiological properties of motoneurones of the M. triceps surae of the rat after retrograde labelling with Evans blue. Cell Tissue Res 212:73–81
Brust M, Cosla HW (1967) Contractility of isolated human skeletal muscle. Arch Phys Med Rehabil 48:543–555
Bubenzer H-J (1964) Ein neues Konzept der interfibrillären Raumaufteilung in Muskelfasern auf Grund von Befunden am Rattenzwerchfell. Third European Regional Conference on Electron Microscopy. Czechoslovak Academy of Sciences Prague, pp 577–578
Bubenzer H-J (1966) Die dünnen und die dicken Muskelfasern des Zwerchfells der Ratte. Z Zellforsch Mikrosk Anat 69:520–550
Buchthal F (1942) The mechanical properties of the single striated muscle fibre at rest and during contraction and their structural interpretation. Dan Biol Medd 17:1–138
Buchthal F (1958) Einführung in die Elektromyographie. Urban & Schwarzenberg, München Berlin, pp 1–39
Buchthal F (1961) The general concept of the motor unit. Res Publ Assoc Nerv Ment Dis 38:3–30
Buchthal F, Kaiser E (1951) The rheology of the cross striated muscle fibre with particular reference to isotonic conditions. Dan Biol Medd 21/7:1–318
Buchthal F, Knappeis GG (1940) Diffraction spectra and minute structure of the cross-striated muscle fibre. Skand Arch Physiol 83:281–307
Buchthal F, Lindhard J (1937) Direct application of acetylcholine to motor end plates of voluntary muscle fibres (Abstr). J Physiol (Lond) 90:82P–83P
Buchthal F, Lindhard J (1942) Transmission of impulses from nerve to muscle fibre. Acta Physiol Scand 4:136–148
Buchthal F, Schmalbruch H (1969) Spectrum of contraction times of different fibre bundles in the brachial biceps and triceps muscles of man. Nature 222:89
Buchthal F, Schmalbruch H (1970) Contraction times and fibre types in intact human muscle. Acta Physiol Scand 79:435–452
Buchthal F, Schmalbruch H (1980) Motor unit of mammalian muscle. Physiol Rev 60:90–142
Buchthal F, Kamieniecka Z, Schmalbruch H (1974) Fibre types in normal and diseased human muscles and their physiological correlates. In: Milhorat AT (ed) Exploratory concepts in muscular dystrophy II. Excerpta Medica Amsterdam, pp 526–551
Buckingham ME, Caput D, Cohen A, Whalen RG, Gros F (1974) The synthesis and stability of cytoplasmic messenger RNA during myoblast differentiation in culture. Proc Natl Acad Sci USA 71:1466–1470
Buckley PA, Konigsberg IR (1974) Myogenic fusion and the duration of the post-mitotic gap (G_1). Dev Biol 37:193–212
Buckley PA, Konigsberg IR (1977) Do myoblasts *in vivo* withdraw from the cell cycle? A reexamination. Proc Natl Acad Sci USA 74:2031–2035
Bullard B, Hammond KS, Luke BM (1977) The site of paramyosin in insect flight muscle and

the presence of an unidentified protein between myosin filaments and Z-line. J Mol Biol 115:417–440
Bullard HH (1919) Histological as related to physiological and chemical differences in certain muscles of the cat. Johns Hopkins Hosp Rep 18:323–328
Buller AJ, Lewis DM (1965a) Some observations on the effects of tenotomy in the rabbit. J Physiol (Lond) 178:326–342
Buller AJ, Lewis DM (1965b) Further observations on the differentiation of skeletal muscles in the kitten hind limb. J Physiol (Lond) 176:355–370
Buller AJ, Dornhorst AC, Edwards R, Kerr D, Whelan RF (1959) Fast and slow muscles in mammals. Nature, 183:1516–1517
Buller AJ, Eccles JC, Eccles RM (1960a) Differentiation of fast and slow muscles in the cat hind limb. J Physiol (Lond) 150:399–416
Buller AJ, Eccles JC, Eccles RM (1960b) Interactions between motorneurons and muscles in respect of the characteristic speeds of their responses. J Physiol (Lond) 150:417–439
Buller AJ, Mommaerts WFHM, Seraydarian K (1969) Enzymic properties of myosin in fast and slow twitch muscles of the cat following cross-innervation. J Physiol (Lond) 205:581–597
Bullough WS (1962) The control of mitotic activity in adult mammalian tissues. Biol Rev 37:307–342
Bullough WS (1975) Mitotic control in adult mammalian tissues. Biol Rev 50:99–127
Burden SJ, Sargent PB, McMahan UJ (1979) Acetylcholine receptors in regenerating muscle accumulate at original synaptic sites in the absence of the nerve. J Cell Biol 82:412–425
Buresova M, Gutmann E, Hanzlikova V (1972) Differential effects of castration and denervation on protein synthesis in the levator ani muscle of the rat. J Endocrinol 54:3–14
Burgener J, Mayr R (1980) Guinea pig stapedius muscle. A histochemical, light and electron microscopic study. Anat Embryol 161:65–81
Burke RE, Tsairis P (1973) Anatomy and innervation ratios in motor units of cat gastrocnemius. J Physiol (Lond) 234:749–765
Burke RE, Jankowska E, Ten Bruggencate G (1970) A comparison of peripheral and rubrospinal synaptic input to slow and fast twitch motor units of triceps surae. J Physiol (Lond) 207:709–732
Burke RE, Levine DN, Zajac FE, III, Tsairis P, Engel WK (1971) Mammalian motor units: Physiological-histochemical correlation in three types in cat gastrocnemius. Science 174:709–712
Burke RE, Levine DN, Tsairis P, Zajac FE, III (1973) Physiological types and histochemical profiles in motor units of the cat gastrocnemius. J Physiol (Lond) 234:723–748
Burke RE, Levine DN, Salcman M, Tsairis P (1974) Motor units in cat soleus muscle: physiological, histochemical and morphological characteristics. J Physiol (Lond) 238:503–514
Burke RE, Rymer WZ, Walsh JV (1976) Relative strength of synaptic input from short-latency pathways to motor units of defined type in cat medial gastrocnemius. J Neurophysiol 39:447–458
Burke RE, Strick PL, Kanda K, Kim CC, Walmsley B (1977) Anatomy of medial gastrocnemius and soleus motor nuclei in cat spinal cord. J Neurophysiol 40:667–680
Burleigh IG (1977) Observations on the number of nuclei within the fibres of some red and white muscles. J Cell Sci 23:269–284
Bursztajn S, Fischbach GD (1984) Evidence that coated vesicles transport acetylcholine receptors to the surface membrane of chick myotubes. J Cell Biol 98:498–506
Bursztajn S, Libby P (1981) Morphological changes in cultured myotubes treated with agents that interfere with lysosomal function. Cell Tissue Res 220:573–588
Busch WA, Stromer MH, Goll DE, Suzuki A (1972) Ca^{2+}-specific removal of Z lines from rabbit skeletal muscle. J Cell Biol 52:367–381
Butler-Browne GS, Bugaisky LB, Cuenoud S, Schwartz K, Whalen RG (1982) Denervation of newborn rat muscles does not block the appearance of adult fast myosin heavy chain. Nature 299:830–833
Campion DR (1984) The muscle satellite cell: A review. Int Rev Cytol 87:225–251
Campion DR, Fowler SP, Hausman GJ, Reagan JO (1981) Ultrastructural analysis of skeletal muscle development in the fetal pig. Acta Anat (Basel) 110:277–284
Cangiano A, Lømo T, Lutzemberger L, Sveen O (1980) Effects of chronic nerve conduction block on formation of neuromuscular junctions and junctional AChE in the rat. Acta Physiol Scand 109:283–296
Cantini M, Sartore S, Schiaffino S (1980) Myosin types in cultured muscle cells. J Cell Biol 85:903–909

Capers CR (1960) Multinucleation of skeletal muscle *in vitro*. J Biophys Biochem Cytol 7:559–566

Caputo C, Vergara J, Bezanilla F (1979) Local anaesthetics inhibit tension development and Nile blue fluorescence signals in frog muscle fibres. Nature 277:400–402

Carafoli E, Crompton M (1978) The regulation of intracellular calcium. Curr Top Membr Transport 10:151–216

Cardasis CA, Padykula HA (1981) Ultrastructural evidence indicating reorganization at the neuromuscular junction in the normal rat soleus muscle. Anat Rec 200:41–59

Carlsen F, Knappeis GG, Buchthal F (1961) Ultrastructure of the resting and contracted striated muscle fiber at different degrees of stretch. J Biophys Biochem Cytol 11:95–117

Carlson BM (1970a) The regeneration of entire muscles from minced fragments. In: Mauro A, Shafiq SA, Milhorat AT (eds) Regeneration of striated muscle, and myogenesis. Excerpta Medica, Amsterdam, pp 25–37

Carlson BM (1970b) Histological observations on the regeneration of mammalian and amphibian muscle. In: Mauro A, Shafiq SA, Milhorat AT (eds) Regeneration of striated muscle, and myogenesis. Excerpta Medica, Amsterdam, pp 38–72

Carlson BM (1972) The regeneration of minced muscles. Karger, Basel

Carlson BM (1973) The regeneration of skeletal muscle – a review. Am J Anat 137:119–150

Carlson BM (1976) A quantitative study of muscle fiber survival and regeneration in normal, predenervated, and Marcaine-treated free muscle grafts in the rat. Exp Neurol 52:421–432

Carlson BM (1979) Relationship between tissue and epimorphic regeneration of skeletal muscle. In: Mauro A (ed) Muscle regeneration. Raven Press, New York, pp 57–71

Carlson BM, Gutmann E (1972) Development of contractile properties of minced muscle regenerates in the rat. Exp Neurol 36:239–249

Carlson BM, Gutmann E (1975) Regeneration in free grafts of normal and denervated muscles in the rat: Morphology and histochemistry. Anat Rec 183:47–62

Carlson BM, Gutmann E (1976) Free grafting of the extensor digitorum longus muscle in the rat after Marcaine pretreatment. Exp Neurol 53:82–93

Carlson BM, Wagner KR, Max SR (1979a) Reinnervation of rat extensor digitorum longus muscles after free grafting. Muscle Nerve 2:304–307

Carlson BM, Hansen-Smith FM, Magon DK (1979b) The life history of a free muscle graft. In: Mauro A (ed) Muscle regeneration. Raven Press, New York, pp 493–507

Carlson BM, Herbrychova A, Gutmann E (1979c) Retention of hormonal sensitivity in free grafts of the levator ani muscle. Exp Neurol 63:94–107

Carnay LD, Barry WH (1969) Turbidity, birefringence, and fluorescence changes in skeletal muscle coincident with the action potential. Science 165:608–609

Carpenter S, Karpati G (1984) Pathology of skeletal muscle. Churchill Livingstone, New York, pp 1–754

Carraro U, Dalla Libera L, Catani C, Danieli-Betto D (1982) Chronic denervation of rat diaphragm: Selective maintenance of adult fast myosin heavy chains. Muscle Nerve 5:515–524

Carraro U, Dalla Libera L, Catani C (1983) Myosin light and heavy chains in muscle regenerating in absence of nerve: Transient appearance of the embryonic light chain. Exp Neurol 79:106–117

Carry MR, Morita M, Nornes HO (1983) Morphogenesis of motor endplates along the proximodistal axis of the mouse hindlimb. Anat Rec 207:473–485

Carvalho AP (1968a) Effects of potentiators of muscular contraction on binding of cations by sarcoplasmic reticulum. J Gen Physiol 51:427–442

Carvalho AP (1968b) Calcium-binding properties of sarcoplasmic reticulum as influenced by ATP, caffeine, quinine, and local anesthetics. J Gen Physiol 52:622–642

Carvalho AP, Leo B (1967) Effects of ATP on the interaction of Ca^{++}, Mg^{++}, and K^{+} with fragmented sarcoplasmic reticulum isolated from rabbit skeletal muscle. J Gen Physiol 50:1327–1352

Casella C (1951) Tensile force in total striated muscle, isolated fibre and sarcolemma. Acta Physiol Scand 21:380–401

Caspar DLD, Cohen C, Longley W (1969) Tropomyosin: crystal structure, polymorphism and molecular interactions. J Mol Biol 41:87–107

Ceccarelli B, Grohovaz F, Hurlbut WP (1979a) Freeze-fracture studies of frog neuromuscular junctions during intense release of neurotransmitter. I. Effects of black widow spider venom and Ca^{2+}-free solutions on the structure of the active zone. J Cell Biol 81:163–177

Ceccarelli B, Grohovaz F, Hurlbut WP (1979b) Freeze-fracture studies of frog neuromuscular junctions during intense release of neurotransmitter. II. Effects of electrical stimulation and high potassium. J Cell Biol 81:178–192

Celio MR, Heizmann CW (1982) Calcium-binding protein parvalbumin is associated with fast contracting muscle fibres. Nature 297:504–506

Chan AK, Edgerton VR, Goslow GE, Kurata H, Rasmussen SA, Spector SA (1982) Histochemical and physiological properties of cat motor units after self- and cross-reinnervation. J Physiol (Lond) 332:343–361

Chandler DE, Heuser J (1979) Membrane fusion during secretion. Cortical granule exocytosis in sea urchin eggs as studied by quick-freezing and freeze-fracture. J Cell Biol 83:91–108

Chang CC, Lee CY (1963) Isolation of neurotoxins from the venom of *Bungarus multicinctus* and their modes of neuromuscular blocking action. Arch Int Pharmacodyn Ther 144:241–257

Chaplain RA, Tregear RT (1966) The mass of myosin per cross-bridge in insect fibrillar flight muscle. J Mol Biol 21:275–280

Cheng K, Breinin GM (1965) Fine structure of nerve endings in extraocular muscle. Arch Ophthalmol 74:822–834

Cheng K, Breinin GM (1966) A comparison of the fine structure of extraocular and interosseus muscles in the monkey. Invest Ophthalmol Vis Sci 5:535–549

Cheng-Minoda K, Davidowitz J, Liebowitz A, Breinin GM (1968) Fine structure of extraocular muscle in rabbit. J Cell Biol 39:193–197

Chevallier A, Kieny M, Mauger A (1976) Sur l'origine de la musculature de l'aile chez les oiseaux. CR Seances Acad Sci III 282:309–311

Chevallier A, Kieny M, Mauger A (1977a) Limb-somite relationship: origin of the limb musculature. J Embryol Exp Morphol 41:245–258

Chevallier J, Bonnet J-P, Galante M, Tenu J-P, Gulik-Krzywicki T (1977b) Functional and structural heterogeneity of sarcoplasmic reticulum preparations. Rev Biologie Cellulaire 30:103–110

Chi JC, Fellini SA, Holtzer H (1975a) Differences among myosins synthesized in non-myogenic cells, presumptive myoblasts, and myoblasts. Proc Natl Acad Sci (USA) 72:4999–5003

Chi JCH, Rubinstein N, Strahs K, Holtzer H (1975b) Synthesis of myosin heavy and light chains in muscle cultures. J Cell Biol 67:523–537

Chiakulas JJ, Pauly JE (1965) A study of postnatal growth of skeletal muscle in the rat. Anat Rec 152:55–62

Chiquet M, Eppenberger HM, Moor H, Turner DC (1975) Application of freeze-etching to the study of myogenesis in cell culture. Exp Cell Res 93:498–502

Chiquet M, Eppenberger HM, Turner DC (1981) Muscle morphogenesis: Evidence for an organizing function of exogenous fibronectin. Dev Biol 88:220–235

Chiu AY, Sanes JR (1984) Development of basal lamina in synaptic and extrasynaptic portions of embryonic rat muscle. Dev Biol 103:456–467

Cho Y, Sidie JM, DeBruyn PPH (1972) Electron microscopic studies on a tubulo filamentous fasciculus in the bat cricothyroid muscle. J Ultrastruct Res 41:344–357

Chow I, Cohen MW (1983) Developmental changes in the distribution of acetylcholine receptors in the myotomes of *Xenopus laevis*. J Physiol (Lond) 339:553–571

Chow I, Poo M-M (1982) Redistribution of cell surface receptors induced by cell-cell contact. J Cell Biol 95:510–518

Chowrashi PK, Pepe FA (1979) M-band proteins: evidence for more than one component. In: Pepe FA, Sanger JW, Nachmias VT (eds) Motility in cell function. Academic Press, New York, pp 419–422

Christ B, Jacob HJ, Jacob M (1974) Über den Ursprung der Flügelmuskulatur. Experimentelle Untersuchungen mit Wachtel- und Hühnerembryonen. Experientia 30:1446–1448

Christ B, Jacob HJ, Jacob M (1977) Experimental analysis of the origin of the wing musculature in avian embryos. Anat Embryol 150:171–186

Christ B, Jacob HJ, Jacob M (1978) An experimental study on the relative distribution of the somitic and somatic plate mesoderm to the abdominal wall of avian embryos. Experientia 34:241–242

Christ B, Jacob HJ, Jacob M (1979) Differentiating abilities of avian somatopleural mesoderm. Experientia 35:1376–1378

Christ B, Jacob M, Jacob HJ (1983) On the origin and development of the ventrolateral abdominal muscles in the avian embryo. Anat Embryol 166:87–101

Christensen E (1959) Topography of terminal motor innervation in striated muscles from stillborn infants. Am J Phys Med 38:65–78

Christian CN, Daniels MP, Sugiyama H, Vogel Z, Jacques L, Nelson PG (1978) A factor from neurons increases the number of acetylcholine receptor aggregates on cultured muscle cells. Proc Natl Acad Sci USA 75:4011–4015

Church JCT (1970a) Cell populations in skeletal muscle after regeneration. J Embryol Exp Morphol 23:531–537

Church JCT (1970b) Cell quantitation in regenerating bat web muscle. In: Mauro A, Shafiq SA, Milhorat AT (eds) Regeneration of striated muscle, and myogenesis. Excerpta Medica, Amsterdam, pp 101–117

Chyn TL, Martonosi AN, Morimoto T, Sabatini DD (1979) In vitro synthesis of the Ca^{2+} transport ATPase by ribosomes bound to sarcoplasmic reticulum membranes. Proc Natl Acad Sci USA 76:1241–1245

Cilimbaris PA (1910) Histologische Untersuchungen über die Muskelspindeln der Augenmuskeln. Arch Mikrosk Anat Entwicklungsgesch 75:692–747

Clara M (1961) Was ist das Sarkolemm? Anat Anz 110:41–51

Clark DA (1931) Muscle counts of motor units: a study in innervation ratios. Am J Physiol 96:296–304

Clark RW, Luschei ES (1981) Histochemical characteristics of mandibular muscles of monkeys. Exp Neurol 74:654–672

Clark WELeGros (1946) An experimental study of the regeneration of mammalian striped muscle. J Anat 80:24–36

Clark WELeGros, Blomfield LB (1945) Efficiency of intramuscular anastomoses, with observations on regeneration of devascularized muscle. J Anat 79:15–32

Cleworth DR, Edman KAP (1972) Changes in sarcomere length during isometric tension development in frog skeletal muscle. J Physiol (Lond) 227:1–17

Close R (1964) Dynamic properties of fast and slow skeletal muscles of the rat during development. J Physiol (Lond) 173:74–95

Close R (1965) The relation between intrinsic speed of shortening and duration of the active state of muscle. J Physiol (Lond) 180:542–559

Close R (1967) Properties of motor units in fast and slow skeletal muscles of the rat. J Physiol (Lond) 193:45–55

Close R, Hoh JFY (1969) Post-tetanic potentiation of twitch contractions of cross-innervated rat fast and slow muscles. Nature 221:179–181

Close RI (1972) Dynamic properties of mammalian skeletal muscles. Physiol Rev 52:129–197

Close RI, Luff AR (1974) Dynamic properties of inferior rectus muscle of the rat. J Physiol (Lond) 236:259–270

Coan MR, Tomanek RJ (1979) Regeneration of transplanted muscle with reference to overload. In: Mauro A (ed) Muscle regeneration. Raven Press, New York, pp 509–522

Coan MR, Tomanek RJ (1981) The growth of regenerating soleus muscle transplants after ablation of the gastrocnemius muscle. Exp Neurol 71:278–294

Coërs C (1953) La détection histochimique de la cholinestérase au niveau de la jonction neuromusculaire. Rev Belge Pathol Med Exp 22:306–315

Coërs C (1967) Structure and organization of the myoneural junction. Int Rev Cytol 22:239–267

Coërs C, Telerman-Toppet N (1977) Morphological changes of motor units in Duchenne's muscular dystrophy. Arch Neurol 34:396–402

Coërs C, Woolf AL (1959) The innervation of muscle. A biopsy study. Blackwell, Oxford

Coërs C, Telerman-Toppet N, Gerard J-M (1973a) Terminal innervation ratio in neuromuscular disease. I. Methods and controls. Arch Neurol 29:210–214

Coërs C, Telerman-Toppet N, Gerard J-M (1973b) Terminal innervation ratio in neuromuscular disease. II. Disorders of lower motor neuron, peripheral nerve, and muscle. Arch Neurol 29:215–222

Cohen C, Longley W (1966) Tropomyosin paracrystals formed by divalent cations. Science 152:794–796

Cohen C, Lowey S, Harrison RG, Kendrick-Jones J, Szent-Györgyi AG (1970) Segments from myosin rods. J Mol Biol 47:605–609

Cohen LB, Hille B, Keynes RD, Landowne D, Rojas E (1971) Analysis of the potential-dependent changes in optical retardation in the squid giant axon. J Physiol (Lond) 218:205–237

Cohen MW (1972) The development of neuromuscular connexions in the presence of D-tubocurarine. Brain Res 41:457–463
Cohen SA, Fischbach GD (1973) Regulation of muscle acetylcholine sensitivity by muscle activity in cell culture. Science 181:76–78
Cohen SA, Fischbach GD (1977) Clusters of acetylcholine receptors located at identified nerve-muscle synapses in vitro. Dev Biol 59:24–38
Cohen SA, Pumplin DW (1979) Clusters of intramembrane particles associated with binding sites for α-bungarotoxin in cultured chick myotubes. J Cell Biol 82:494–516
Coleman JR, Coleman AW, Hartline EJH (1969) A clonal study of the reversible inhibition of muscle differentiation by the halogenated thymidine analog 5-bromodeoxyuridine. Dev Biol 19:527–548
Connolly JA, St. John PA, Fischbach GD (1982) Extracts of electric lobe and electric organ from *Torpedo californica* increase the total number as well as the number of aggregates of chick myotube acetylcholine receptors. J Neurosci 2:1207–1213
Cooke P (1976) A filamentous cytoskeleton in vertebrate smooth muscle fibers. J Cell Biol 68:539–556
Cooper S (1966) Muscle spindles and motor units. In: Andrew BL (ed) Control and innervation of skeletal muscle. University of St. Andrews, Dundee, pp 9–17
Cooper S, Daniel PM (1949) Muscle spindles in human extrinsic eye muscles. Brain 72:1–24
Cooper S, Daniel PM (1956) Human muscle spindles (Abstr). J Physiol (Lond) 133:1P–3P
Cooper WG, Konigsberg IR (1961) Dynamics of myogenesis *in vitro*. Anat Rec 140:195–205
Corsi A, Muscatello U, Ronchetti I (1967) Electron microscope observations of the location of actin and tropomyosin in the rabbit myofibril. J Ultrastruct Res 19:260–272
Corvaja N, Pompeiano O (1970) The differentiation of two types of intrafusal fibres in rabbit muscle spindles. An electron microscopic study. Pflugers Arch 317:187–197
Corvaja N, Marinozzi V, Pompeiano O (1967) Close appositions and junctions of plasma membranes of intrafusal fibres in mammalian muscle spindles. Pflugers Arch 296:337–345
Corvaja N, Marinozzi V, Pompeiano O (1969) Muscle spindles in the lumbrical muscle of the adult cat. Electron microscopic observations and functional considerations. Arch Ital Biol 107:365–543
Corvaja N, Magherini PC, Pompeiano O (1971) Ultrastructure of glycogen-membrane complexes in sensory nerve fibres of cat muscle spindles. Z Zellforsch Mikrosk Anat 121:199–217
Costantin LL (1975) Electrical properties of the transverse tubular system. Fed Proc 34:1390–1394
Costantin LL, Podolsky RJ (1967) Depolarization of the internal membrane system in the activation of frog skeletal muscle. J Gen Physiol 50:1101–1124
Costantin LL, Taylor SR (1973) Graded activation in frog muscle fibers. J Gen Physiol 61:424–443
Costantin LL, Podolsky RJ, Tice LW (1967) Calcium activation of frog slow muscle fibres. J Physiol (Lond) 188:261–271
Couteaux R (1953) Particularités histochimiques des zones d'insertion du muscle strié. CR Soc Biol (Paris) 147:1974–1976
Couteaux R (1955) Localization of cholinesterases at neuromuscular junctions. Int Rev Cytol 4:335–375
Couteaux R (1959) Observations sur l'ultrastructure de la jonction musculo-tendineuse. CR Acad Sci (Paris) 249:964–966
Couteaux R (1963) The differentiation of synaptic areas. Proc R Soc Lond [Biol] 158B:457–480
Craig R (1977) Structure of A-segments from frog and rabbit skeletal muscle. J Mol Biol 109:69–81
Craig R (1983) Muscle au naturel. Nature 306:112–113
Craig R, Megerman J (1979) Electron microscope studies on muscle thick filaments. In: Pepe FA, Sanger JW, Nachmias VT (eds) Motility in cell function. Academic Press, New York, pp 91–102
Craig R, Offer G (1976a) Axial arrangement of crossbridges in thick filaments of vertebrate skeletal muscle. J Mol Biol 102:325–332
Craig R, Offer G (1976b) The location of C-protein in rabbit skeletal muscle. Proc R Soc Lond [Biol] 192:451–461
Crain SM (1964) Electrophysiological studies of cord-innervated skeletal muscle in long-term tissue cultures of mouse embryo myotomes (Abstr). Anat Rec 148:273–274
Crain SM (1970) Bioelectric interactions between cultured fetal rodent spinal cord and skeletal muscle after innervation *in vitro*. J Exp Zool 173:353–370
Crain SM, Alfei L, Peterson ER (1970) Neuromuscular transmission in cultures of adult human and rodent skeletal muscle after innervation *in vitro* by fetal rodent spinal cord. J Neurobiol 1:471–489

Crandall WF, Goldberg SJ, Wilson JS, McClung JR (1981) Muscle units divided among retractor bulbi muscle slips and between the lateral rectus and retractor bulbi muscles in cat. Exp Neurol 71:251–260
Crawford GNC (1954) An experimental study of muscle growth in the rabbit. J Bone Joint Surg [Br] 36:294–303
Crawford GNC (1973) The growth of striated muscle immobilized in extension. J Anat 114:165–183
Cronkite AE (1936) The tensile strength of human tendons. Anat Rec 64:173–186
Croop J, Holtzer H (1975) Response of myogenic and fibrogenic cells to cytochalasin B and to colcemid. I. Light microscope observations. J Cell Biol 65:271–285
Crowe LM, Baskin RJ (1977) Stereological analysis of developing sarcotubular membranes. J Ultrastruct Res 58:10–21
Crowther RA, Luther PK (1984) Three-dimensional reconstruction from a single oblique section of fish muscle M-band. Nature 307:569–570
Cull-Candy SG, Miledi R, Uchitel OD (1982) Properties of junctional and extrajunctional acetylcholine-receptor channels in organ cultured human muscle fibres. J Physiol (Lond) 333:251–267
Cullen MJ, Mastaglia FL (1982) Pathological reactions of skeletal muscle. In: Mastaglia FL, Walton Sir John (eds) Skeletal muscle pathology. Churchill Livingstone, Edinburgh, pp 88–139
Cullen MJ, Weightman D (1975) The ultrastructure of normal human muscle in relation to fibre type. J Neurol Sci 25:43–56
Cullen MJ, Hollingworth S, Marshall MW (1984) A comparative study of the transverse tubular system of the rat extensor digitorum longus and soleus muscles. J Anat 138:297–308
Cullis PR, De Kruijff B (1979) Lipid polymorphism and the functional roles of lipids in biological membranes. Biochim Biophys Acta 559:399–420
D'Albis A, Pantaloni C, Bechet J-J (1979) An electrophoretic study of native myosin isozymes and of their subunit content. Eur J Biochem 99:261–272
Dahl G, Schudt C, Gratzl M (1978) Fusion of isolated myoblast plasma membranes. An approach to the mechanism. Biochim Biophys Acta 514:105–116
Dale HH, Feldberg W, Vogt M (1936) Release of acetylcholine at voluntary motor nerve endings. J Physiol (Lond) 86:353–380
Dale MM, Muid R (1978) A mammalian muscle with the pharmacological characteristics of slow tonic muscle. J Pharm Pharmacol 30:297–300
Dalla Libera L (1981) Myosin heavy chains in fast skeletal muscle of chick embryo. Experientia 37:1268–1270
Dalla Libera L, Sartore S, Pierobon-Bormioli S, Schiaffino S (1980) Fast-white and fast-red isomyosins in guinea pig muscles. Biochem Biophys Res Commun 96:1662–1670
Dalrymple RH, Cassens RG, Kastenschmidt LL (1974) Glycolytic enzyme activity in developing red and white muscle. J Cell Physiol 83:251–258
Daniel P (1946) Spiral nerve endings in the extrinsic eye muscles of man. J Anat 80:189–193
Daniels MP, Vogel Z (1975) Immunoperoxidase staining of α-bungarotoxinbinding sites in muscle endplates shows distribution of acetylcholine receptors. Nature 254:339–341
Dauber W (1979) Zur fasertypischen Morphologie und Funktion der Triaden im Skeletmuskel des Frosches (*Rana esculenta*). Z Mikrosk Anat Forsch 93:512–536
Davey DF (1973) The effect of fixative tonicity on the myosin filament lattice volume of frog muscle fixed following exposure to normal or hypertonic Ringer. Histochem J 5:87–104
Davey DF, O'Brien GM (1978) The sarcoplasmic reticulum and T-system of rat extensor digitorum longus muscles exposed to hypertonic solutions. Aust J Exp Biol Med Sci 56:409–419
Davey DF, Wong SYP (1980) Morphometric analysis of rat extensor digitorum longus and soleus muscles. Aust J Exp Biol Med Sci 58:213–230
Davey DF, Mark RF, Marotte LR, Proske U (1975) Structure and innervation of extraocular muscles of Carassius. J Anat 120:131–147
Davey DF, Dulhunty AF, Fatkin D (1980) Glycerol treatment in mammalian skeletal muscle. J Membr Biol 53:223–233
David H, Gerhardt H-J, Uerlings I (1966) Die submikroskopische Struktur des Musculus stapedius des Meerschweinchens. Z Zellforsch Mikrosk Anat 70:334–346
Davidowitz J, Philips G, Breinin GM (1980a) Variation of mitochondrial volume fraction along multiply innervated fibers in rabbit extraocular muscle. Tissue Cell 12:449–457
Davidowitz J, Philips G, Breinin GM (1980b) The distribution of membrane-glycogen complexes in the orbital surface layer of rabbit superior rectus. Tissue Cell 12:459–465

Davidowitz J, Philips G, Breinin GM (1983) Membrane-glycogen complexes in rabbit extraocular muscle. J Ultrastruct Res 82:64–75

Davies AS (1972) Postnatal changes in the histochemical properties of porcine skeletal muscle (Abstr). J Anat 111:487–489

Davis CJF, Montgomery A (1977) The effect of prolonged inactivity upon the contraction characteristics of fast and slow mammalian twitch muscle. J Physiol (Lond) 270:581–594

Davis DA, Wasserkrug HL, Heyman IA, Padmanabhan KC, Seligman GA, Plapinger RE, Seligman AM (1972) Comparison of ultrastructural cholinesterase demonstration in the motor end plate with α-acetylthiol-m-toluenediazonium ion and 3-acetoxy-5-indolediazonium ion. J Histochem Cytochem 20:161–172

Davis R, Koelle GB (1967) Electron microscopic localization of acetylcholinesterase and nonspecific cholinesterase at the neuromuscular junction by the gold-thiocholine and gold-thiolacetic acid methods. J Cell Biol 34:157–171

De Duve C, Wattiaux R (1966) Functions of lysosomes. Annu Rev Physiol 28:435–492

De Harven E, Coërs C (1959) Electron microscope study of the human neuromuscular junction. J Biophys Biochem Cytol 6:7–10

De Meis L, Hasselbach W, Machado RD (1974) Characterization of calcium oxalate and calcium phosphate deposits in sarcoplasmic reticulum vesicles. J Cell Biol 62:505–509

De Robertis E (1958) Submicroscopic morphology and function of the synapse. Exp Cell Res [Suppl] 5:347–369

Deamer DW, Baskin RJ (1969) Ultrastructure of sarcoplasmic reticulum preparations. J Cell Biol 42:296–307

Decorte L, Emonet-Denand F, Harker DW, Jami L, Laporte Y (1984) Glycogen depletion elicited in tenuissimus intrafusal muscle fibres by stimulation of static γ-axons in the cat. J Physiol (Lond) 346:341–352

Delbrück A, Zebe E, Bücher TH (1959) Über Verteilungsmuster von Enzymen des Energie liefernden Stoffwechsels im Flugmuskel, Sprungmuskel und Fettkörper von *Locusta migratoria* und ihre cytologische Zuordnung. Biochem Z 331:273–296

Den H, Malinzak DA, Keating HJ, Rosenberg A (1975) Influence of concavalin A, wheat germ agglutinin, and soybean agglutinin on the fusion of myoblasts in vitro. J Cell Biol 67:826–834

Dennis MJ, Ziskind-Conhaim L, Harris AJ (1981) Development of neuromuscular junctions in rat embryos. Dev Biol 81:266–279

Denny-Brown DE (1929) The histological features of striped muscle in relation to its functional activity. Proc R Soc Lond [Biol] 104:371–411

Dermietzel R (1973) Visualization by freeze-fracture of regular structures in glial cell membranes. Naturwissenschaften 60:208

Deshpande SS, Warnick JE, Guth L, Albuquerque EX (1980) Quantal release of acetylcholine does not regulate the resting membrane potential of mammalian skeletal muscle: Evidence from in vivo experiments. Exp Neurol 70:122–137

Desmedt JE, Godaux E (1977) Fast motor units are not preferentially activated in rapid voluntary contractions in man. Nature 267:717–718

Desmedt JE, Hainaut K (1977) Inhibition of the intracellular release of calcium by dantrolene in barnacle giant muscle fibres. J Physiol (Lond) 265:565–585

Devanandan MS, Eccles RM, Westerman RA (1965) Single motor units of mammalian muscle. J Physiol (Lond) 178:359–367

Devlin RB, Emerson Jr CP (1979) Coordinate accumulation of contractile protein mRNAs during myoblast differentiation. Dev Biol 69:202–216

Dewey MM, Levine RJC, Colflesh DE (1973) Structure of limulus striated muscle. The contractile apparatus at various sarcomere lengths. J Cell Biol 58:574–593

Dhoot GK (1983) Differentiation of skeletal muscle fibre types in the rat fetus (Abstr). J Physiol (Lond) 345:53P

Dhoot GK, Perry SV (1980) The components of the troponin complex and development in skeletal muscle. Exp Cell Res 127:75–87

Diamond J, Miledi R (1962) A study of foetal and new-born rat muscle fibres. J Physiol (Lond) 162:393–408

Dienstman SR, Holtzer H (1975) Myogenesis: A cell lineage interpretation. In: Reinert J, Holtzer H (eds) Results and problems in cell differentiation, vol 7. Springer: Berlin Heidelberg New York, pp 1–25

Dienstman SR, Holtzer H (1977) Skeletal myogenesis. Control of proliferation in a normal cell lineage. Exp Cell Res 107:355–364
Dienstman SR, Biehl J, Holtzer S, Holtzer H (1974) Myogenic and chondrogenic lineages in developing limb buds grown *in vitro*. Dev Biol 39:83–95
Dodd L, Gray SD, Hudlicka O, Renkin EM (1980) Evaluation of capillary density in relation to fibre types in electrically stimulated rabbit fast muscles (Abstr). J Physiol (Lond) 301:11P–12P
Doering JL, Fischman DA (1974) The *in vitro* cell fusion of embryonic chick muscle without DNA synthesis. Dev Biol 36:225–235
Dölken G, Leisner E, Pette D (1975) Immunofluorescent localization of glycogenolytic and glycolytic enzyme proteins and of malate dehydrogenase isozymes in cross-striated skeletal muscle and heart of the rabbit. Histochemistry 43:113–121
Doyle AM, Mayer RF (1969) Studies of the motor unit in the cat. Bulletin School of Medicine University of Maryland 54:11–17
Drachman DB, Johnston DM (1975) Neurotrophic regulation of dynamic properties of skeletal muscle: effects of botulinum toxin and denervation. J Physiol (Lond) 252:657–667
Draper MH, Hodge AJ (1949) Studies on muscle with the electron microscope. 1. The ultrastructure of toad striated muscle. Aust J Exp Biol Med Sci 27:465–503
Drenckhahn D, Lüllmann-Rauch R (1976) Myopathy in rats treated with chlorphentermine or iprindole. Virchows Arch [Cell Pathol] 20:343–346
Drenckhahn D, Lüllmann-Rauch R (1979) Experimental myopathy induced by amphiphilic cationic compounds including several psychotropic drugs. Neuroscience 4:549–562
Duance VC, Restall DJ, Beard H, Bourne FJ, Bailey AJ (1977) The location of three collagen types in skeletal muscle. FEBS Lett 79:248–252
Dubowitz V (1965) Enzyme histochemistry of skeletal muscle. Part II. Developing human muscle. J Neurol Neurosurg Psychiatry 28:519–524
Dubowitz V (1967) Cross-innervated mammalian skeletal muscle: histochemical, physiological and biochemical observations. J Physiol (Lond) 193:481–496
Dubowitz V (1968) Developing and diseased muscle. A histochemical study. Heinemann Medical Books, London
Duchen LW (1971) An electron microscopic comparison of motor end-plates of slow and fast skeletal muscle fibres of the mouse. J Neurol Sci 14:37–45
Dulhunty AF, Franzini-Armstrong C (1975) The relative contributions of the folds and caveolae to the surface membrane of frog skeletal muscle fibres at different sarcomere lengths. J Physiol (Lond) 250:513–539
Dum RP, Kennedy TT (1980) Physiological and histochemical characteristics of motor units in cat tibialis anterior and extensor digitorum longus muscles. J Neurophysiol 43:1615–1630
Düring M von (1967) Über die Feinstruktur der motorischen Endplatte von höheren Wirbeltieren. Z Zellforsch Mikrosk Anat 81:74–90
Düring M von, Andres KH (1969) Zur Feinstruktur der Muskelspindel von Mammalia. Anat Anz 124:566–573
Eastwood AB, Wood DS, Bock KL, Sorenson MM (1979) Chemically skinned mammalian skeletal muscle. I. The structure of skinned rabbit psoas. Tissue Cell 11:553–566
Ebashi F, Ebashi S (1962) Removal of calcium and relaxation in actomyosin systems. Nature 194:378–379
Ebashi S (1960) Calcium binding and relaxation in the actomyosin system. J Biochem (Tokyo) 48:150–151
Ebashi S (1961a) Calcium binding activity of vesicular relaxing factor. J Biochem (Tokyo) 50:236–244
Ebashi S (1961b) The role of "relaxing factor" in contraction-relaxation cycle of muscle. Progr Theor Physics [Suppl] 17:35–40
Ebashi S (1963) Third component participating in the superprecipitation of "natural actomyosin". Nature 200:1010
Ebashi S (1980) Regulation of muscle contraction. Proc R Soc Lond [Biol] 207:259–286
Ebashi S, Ebashi F (1964) A new protein component participating in the superprecipitation of myosin B. J Biochem (Tokyo) 55:604–613
Ebashi S, Ebashi F (1965) Alpha-actinin, a new structural protein from striated muscle. I. Preparation and action on actomyosin-ATP interaction. J Biochem (Tokyo) 58:7–12
Ebashi S, Endo M (1968) Calcium ion and muscle contraction. Progr Biophys Mol Biol 18:123–183

Ebashi S, Kodama A (1966) Native tropomyosin-like action of troponin on trypsin-treated myosin B. J Biochem (Tokyo) 60:733–734
Ebashi S, Lipmann F (1962) Adenosine triphosphate-linked concentration of calcium ions in a particulate fraction of rabbit muscle. J Cell Biol 14:389–400
Ebashi S, Nonomura Y (1973) Proteins of the myofibril. In: Bourne GH (ed) The structure and function of muscle, vol III, 2nd edn. Academic Press, New York, pp 285–362
Ebashi S, Otsuka M, Endo M (1962) Calcium binding of the relaxing factor and the link between excitation and contraction. 22. Int Congr Physiol Sci Leiden. Excerpta Medica Int Congr Series. No 48. Abstract 899
Ebashi S, Endo M, Ohtsuki I (1969) Control of muscle contraction. Q Rev Biophys 2:351–384
Ebashi S, Ohtsuki I, Mihashi K (1972) Regulatory proteins of muscle with special reference to troponin. Cold Spring Harbor Symp Quant Biol 37:215–224
Eberstein A, Goodgold J (1968) Slow and fast twitch fibers in human skeletal muscle. Am J Physiol 215:535–541
Eberstein A, Rosenfalck A (1963) Birefringence of isolated muscle fibres in twitch and tetanus. Acta Physiol Scand 57:144–166
Eccles JC, Sherrington CS (1930) Number and contraction values of individual motor-units examined in some muscles of the limb. Proc R Soc Lond [Biol] 106:326–357
Eccles JC, Eccles RM, Lundberg A (1958) The action potentials of the alpha motoneurones supplying fast and slow muscles. J Physiol (Lond) 142:275–291
Eccles JC, Eccles RM, Kozak W (1962) Further investigations on the influence of motoneurones on the speed of muscle contraction. J Physiol (Lond) 163:324–339
Edds MV (1950) Collateral regeneration of residual motor axons in partially denervated muscles. J Exp Zool 113:507–552
Edgerton VR (1970) Morphology and histochemistry of the soleus muscle from normal and exercised rats. Am J Anat 127:81–88
Edgerton VR, Simpson DR (1969) The intermediate muscle fiber of rats and guinea pigs. J Histochem Cytochem 17:828–838
Edgerton VR, Gerchman L, Carrow R (1969) Histochemical changes in rat skeletal muscle after exercise. Exp Neurol 24:110–123
Edidin M, Fambrough D (1973) Fluidity of the surface of cultured muscle fibers. Rapid lateral diffusion of marked surface antigens. J Cell Biol 57:27–37
Editorial Science News (1975) Successful transplant of cat muscle. Sci News 107:268
Edström L, Kugelberg E (1968) Histochemical composition, distribution of fibres and fatiguability of single motor units. Anterior tibial muscle of the rat. J Neurol Neurosurg Psychiatry 31:424–433
Edström L, Lindquist C (1973) Histochemical fiber composition of some facial muscles in the cat in relation to their contraction properties. Acta Physiol Scand 89:491–503
Edström L, Nyström B (1969) Histochemical types and sizes of fibres in normal human muscles. Acta Neurol Scand 45:257–269
Edwards GA, Ruska H, De Souza Santos P, Vallejo-Freire A (1956) Comparative cytophysiology of striated muscle with special reference to the role of the endoplasmic reticulum. J Biophys Biochem Cytol [Suppl] 2:143–156
Edwards RP (1975) An ultrastructural study of neuromuscular spindles in normal mice: with reference to mice and man infected with *Mycobacterium leprae*. J Anat 120:149–168
Egelman EH, Padron R (1984) X-ray diffraction evidence that actin is a 100 Å filament. Nature 307:56–58
Ehrlich P (1885) Das Sauerstoff-Bedürfniss des Organismus. Eine farbenanalytische Studie. Hirschwald, Berlin
Eichelberg H, Schneider H (1973) Die Feinstruktur der Kehlkopfmuskeln des Laubfrosches, *Hyla arborea arborea* (L.), im Vergleich zu einem Skelettmuskel. Z Zellforsch Mikrosk Anat 141:223–233
Eichelberg H, Schneider H (1974) The fine structure of the larynx muscles in female tree frogs, *Hyla a. arborea* L. (Anura, amphibia). Cell Tissue Res 152:185–191
Eisen A, Karpati G, Carpenter S, Danon J (1974) The motor unit profile of the rat soleus in experimental myopathy and reinnervation. Neurology (Minneap) 24:878–884
Eisen A, Karpati G, Carpenter S (1975) Reserpine induced alteration of physiological properties and histochemical fiber types in rat skeletal muscle. Exp Neurol 46:554–565

Eisenberg B, Eisenberg RS (1968) Selective disruption of the sarcotubular system in frog sartorius muscle. A quantitative study with exogenous peroxidase as a marker. J Cell Biol 39:451–467
Eisenberg BR (1972) Three dimensional branching of the T system in frog sartorius muscle (Abstr). J Cell Biol 55:68a
Eisenberg BR (1983) Quantitative ultrastructure of mammalian skeletal muscle. In: Peachey LD, Adrian RH (eds) Handbook of Physiology, Sect 10: Skeletal muscle. American Physiological Society, Bethesda, pp 73–112
Eisenberg BR, Eisenberg RS (1982) The T-SR junction in contracting single skeletal muscle fibers. J Gen Physiol 79:1–19
Eisenberg BR, Gilai A (1979) Structural changes in single muscle fibers after stimulation at a low frequency. J Gen Physiol 74:1–16
Eisenberg BR, Kuda AM (1975) Stereological analysis of mammalian skeletal muscle. II. White vastus muscle of the adult guinea pig. J Ultrastruct Res 51:176–187
Eisenberg BR, Kuda AM (1976) Discrimination between fiber populations in mammalian skeletal muscle by using ultrastructural parameters. J Ultrastruct Res 54:76–88
Eisenberg BR, Kuda AM (1977) Retrieval of cryostat sections for comparison of histochemistry and quantitative electron microscopy in a muscle fiber. J Histochem Cytochem 25:1169–1177
Eisenberg BR, Peachey LD (1975) The network parameters of the T-system in frog muscle measured with the high voltage electron microscope. In: Bailey GW (ed) 33rd Ann Proc Electron Microscopy Soc Amer. Las Vegas, Nevada, pp 550–551
Eisenberg BR, Salmons S (1981) The reorganization of subcellular structure in muscle undergoing fast-to-slow type transformation. A stereological study. Cell Tissue Res 220:449–471
Eisenberg BR, Kuda AM, Peter JB (1974) Stereological analysis of mammalian skeletal muscle. I. Soleus muscle of the adult guinea pig. J Cell Biol 60:732–754
Eisenberg RS, Gage PW (1967) Frog skeletal muscle fibers: changes in electrical properties after disruption of transverse tubular system. Science 158:1700–1701
Eisenberg RS, Gage PW (1968) The surface and tubular membranes of frog sartorius muscle fibers: continuous membranes with apparently different properties (Abstr). J Cell Biol 39:39a
Elftman H (1944) Skeletal and muscular systems: Structure and function. In: Glasser O (ed) Medical physics. The Year Book Publishers, Chicago, pp 1421–1430
Elizalde A, Huerta M, Stefani E (1983) Selective reinnervation of twitch and tonic muscle fibres of the frog. J Physiol (Lond) 340:513–524
Elliott GF (1964) X-ray diffraction studies on striated and smooth muscles. Proc R Soc Lond [Biol] 160:467–472
Elliott GF (1973) The muscle fiber: liquid-crystalline and hydraulic aspects. Ann NY Acad Sci 204:564–574
Elliott GF, Lowy J, Worthington CR (1963) An X-ray and light-diffraction study of the filament lattice of striated muscle in the living state and in rigor. J Mol Biol 6:295–305
Elliott GF, Lowy J, Millman BM (1965) X-ray diffraction from living striated muscle during contraction. Nature 206:1357–1359
Elliott GF, Rome EM, Spencer M (1970) A type of contraction hypothesis applicable to all muscles. Nature 226:417–420
Ellisman MH, Rash JE (1977) Studies of excitable membranes. III. Freeze-fracture examination of the membrane specializations at the neuromuscular junction and in the non-junctional sarcolemma after denervation. Brain Res 137:197–206
Ellisman MH, Rash JE, Staehelin LA, Porter KR (1976) Studies of excitable membranes. II. A comparison of specializations at neuromuscular junctions and nonjunctional sarcolemmas of mammalian fast and slow twitch muscle fibers. J Cell Biol 68:752–774
Ellisman MH, Brooke MH, Kaiser KK, Rash JE (1978) Appearance in slow muscle sarcolemma of specializations characteristic of fast muscle after reinnervation by a fast muscle nerve. Exp Neurol 58:59–67
Elson J (1929) Auto- and homoiotransplantation of cross-striated muscle tissue in the rat. Am J Pathol 5:425–438
Emonet-Denand F, Jami L, Laporte Y, Tankov N (1980) Glycogen depletion of bag_1 fibres elicited by stimulation of static γ axons in cat peroneus brevis muscle spindles. J Physiol (Lond) 302:311–321
Endo M (1964) Entry of dye into the sarcotubular system of muscle. Nature 202:1115–1116

Endo M (1966) Entry of fluorescent dyes into the sarcotubular system of the frog muscle. J Physiol (Lond) 185:224–238

Enesco M (1961) Increase in the number of nuclei in various striated muscles of the growing rat (Abstr). Anat Rec 139:225–226

Enesco M, Leblond CP (1962) Increase in cell number as a factor in the growth of the organs and tissues of the young male rat. J Embryol Exp Morphol 10:530–562

Engel AG (1966) Recent studies on neuromuscular disease. Late-onset rod myopathy (a new syndrome?): Light and electron microscopic observations in two cases. Mayo Clin Proc 41:713–741

Engel AG, Gomez MR (1967) Nemaline (Z disk) myopathy: Observations on the origin, structure, and solubility properties of the nemaline structures. J Neuropathol Exp Neurol 26:601–619

Engel AG, Santa T (1973) Motor endplate fine structure. Quantitative analysis in disorders of neuromuscular transmission and prostigmine-induced alterations. In: Desmedt JE (ed) New developments in EMG and clinical neurophysiology, vol 1. Karger, Basel, pp 196–228

Engel AG, Santa T, Stonnington HH, Jerusalem F, Tsujihata M, Brownell AKW, Sakakibara H, Banker BQ, Sahashi K, Lambert EH (1979) Morphometric study of skeletal muscle ultrastructure. Muscle Nerve 2:229–237

Engel WK (1962) The essentiality of histo- and cytochemical studies of skeletal muscle in the investigation of neuromuscular disease. Neurology (Minneapolis) 12:778–794

Engelhardt A (1955) Über den helikoiden Bau menschlicher Augenmuskelfasern. Anat Anz 101:233–236

Engelhardt A, Popp O (1953) Ist die Faser des quergestreiften Muskels wendelartig (helikoid) gebaut? Z Biol 105:405–408

Engelhardt WA, Ljubimova MN (1939) Myosine and adenosinetriphosphatase. Nature 144:668–669

English AW, Letbetter WD (1982) A histochemical analysis of identified compartments of cat lateral gastrocnemius muscle. Anat Rec 204:123–130

English AW, Weeks OI (1984) Compartmentalization of single muscle units in cat lateral gastrocnemius. Exp Brain Res 56:361–368

Enoka RM, Stuart DG (1984) Henneman's 'size principle': current issues. Trends in Neuro Sciences 7:226–227

Ericsson JLE (1969) Mechanism of cellular autophagy. In: Dingle JT, Fell HB (eds) Lysosomes in biology and pathology, vol 2. North-Holland Publishing Company, Amsterdam, pp 345–394

Eriksson P-O (1982) Muscle-fibre composition of the human mandibular locomotor system. Enzyme-histochemical and morphological characteristics of functionally different parts (Thesis). Swed Dent J [Suppl] 12:1–44

Eriksson P-O, Thornell L-E (1983) Histochemical and morphological muscle-fibre characteristics of the human masseter, the medial pterygoid and the temporal muscles. Arch Oral Biol 28:781–795

Eriksson P-O, Eriksson A, Ringqvist M, Thornell L-E (1981) Special histochemical muscle-fibre characteristics of the human lateral pterygoid muscle. Arch Oral Biol 26:495–507

Eriksson P-O, Eriksson A, Ringqvist M, Thornell L-E (1982) Histochemical fibre composition of the human digastric muscle. Arch Oral Biol 27:207–215

Erulkar SD, Shelanski ML, Whitsel BL, Ogle P (1964) Studies of muscle fibers of the tensor tympani of the cat. Anat Rec 149:279–298

Essén B, Lindholm A, Thornton J (1980) Histochemical properties of muscle fibre types and enzyme activities in skeletal muscles of standardbred trotters of different ages. Equine Vet J 12:175–180

Etemadi AA, Hosseini F (1968) Frequency and size of muscle fibers in athletic body build. Anat Rec 163:269–273

Eusebi F, Miledi R, Takahashi T (1983) Aequorin-calcium transients in frog twitch muscle fibres. J Physiol (Lond) 340:91–106

Eversole LR, Standish SM (1970) Histochemical demonstration of muscle fiber types. J Histochem Cytochem 18:591–593

Ezerman EB, Ishikawa H (1967) Differentiation of the sarcoplasmic reticulum and T system in developing chick skeletal muscle in vitro. J Cell Biol 35:405–420

Fahim MA, Robbins N (1982) Ultrastructural studies of young and old mouse neuromuscular junctions. J Neurocytol 11:641–656

Fahrenbach WH (1967) The fine structure of fast and slow crustacean muscles. J Cell Biol 35:69–79

Falk G, Fatt P (1964) Linear electrical properties of striated muscle fibres observed with intracellular electrodes. Proc R Soc Lond [Biol] 160:69–123

Fallon JR, Nachmias VT (1980) Localization of cytoplasmic and skeletal myosins in developing muscle cells by double-label immunofluorescence. J Cell Biol 87:237–247
Fambrough D, Rash JE (1971) Development of acetylcholine sensitivity during myogenesis. Dev Biol 26:55–68
Fambrough DM (1979) Control of acetylcholine receptors in skeletal muscle. Physiol Rev 59:165–227
Fambrough DM, Devreotes PN (1978) Newly synthesized acetylcholine receptors are located in the Golgi apparatus. J Cell Biol 76:237–244
Fambrough DM, Hartzell HC (1972) Acetylcholine receptors: Number and distribution at neuromuscular junctions in rat diaphragm. Science 176:189–191
Fambrough DM, Rotundo R, Gardner JM, Bayne EK, Wakshull E, Anderson MJ (1982) Cell surface and secretory proteins of skeletal muscle. In: Schotland DL (ed) Disorders of the motor unit. Wiley, New York, pp 197–212
Fanburg BL, Drachman DB, Moll D, Roth SI (1968) Calcium transport in isolated sarcoplasmic reticulum during muscle maturation. Nature 218:962–964
Fardeau M (1973) Ultrastructure des jonctions neuro-musculaires dans la musculature squelettique du cobaye. In: Fardeau M, Israel M, Manaranche R (eds) La transmission cholinergique de l'excitation. Inserm, Paris, pp 29–50
Faulkner JA, Maxwell LC, Lieberman DA (1972) Histochemical characteristics of muscle fibers from trained and detrained guinea pigs. Am J Physiol 222:836–840
Faulkner JA, Maxwell LC, White TP, Niemeyer JH (1979) Characteristics of autografted mammalian skeletal muscles. In: Mauro A (ed) Muscle regeneration. Raven Press, New York, pp 485–492
Faulkner JA, Markley JM Jr., McCully KK, Watters CR, White TP (1983) Characteristics of cat skeletal muscles grafted with intact nerves or with anastomosed nerves. Exp Neurol 80:682–696
Fawcett DW, Revel JP (1961) The sarcoplasmic reticulum of a fast-acting fish muscle. J Biophys Biochem Cytol [Suppl] 10:89–109
Fawcett DW, Selby CC (1958) Observations on the fine structure of the turtle atrium. J Biophys Biochem Cytol 4:63–72
Fear J (1977) Observations on the fusion of chick embryo myoblasts in culture. J Anat 124:437–444
Feinstein B, Lindegård B, Nyman E, Wohlfart G (1955) Morphological studies of motor units in normal human muscles. Acta Anat (Basel) 23:127–142
Feinstein MB (1963) Inhibition of caffeine rigor and radiocalcium movements by local anesthetics in frog sartorius muscle. J Gen Physiol 47:151–172
Feldberg W, Fessard A (1942) The cholinergic nature of the nerves to the electric organ of the torpedo (*Torpedo marmorata*). J Physiol (Lond) 101:200–216
Feneis H (1935) Über die Anordnung und die Bedeutung des Bindegewebes für die Mechanik der Skelettmuskulatur. Morphol Jahrb 76:161–202
Fenichel GM (1966) A histochemical study of developing human skeletal muscle. Neurology 16:741–745
Ferguson BA (1983) Development of motor innervation of the chick following dorsal-ventral limb bud rotations. J Neurosci 3:1760–1772
Fernand VSV, Hess A (1969) The occurrence, structure and innervation of slow and twitch muscle fibres in the tensor tympani and stapedius of the cat. J Physiol (Lond) 200:547–554
Fernand VSV, Young JZ (1951) Sizes of the nerve fibres of muscle nerves. Proc R Soc Lond [Biol] 139:38–58
Fertuck HC, Salpeter MM (1976) Quantitation of junctional and extrajunctional acetylcholine receptors by electron microscope autoradiography after ^{125}I-α-bungarotoxin binding at mouse neuromuscular junctions. J Cell Biol 69:144–158
Festoff BW, Oliver KL, Reddy NB (1977) In vitro studies of skeletal musle membranes. Effects of denervation on the macromolecular components of cation transport in red and white skeletal muscle. J Membr Biol 32:345–360
Fidzianska A (1980a) Human ontogenesis. I. Ultrastructural characteristics of developing human muscle. J Neuropathol Exp Neurol 39:476–486
Fidzianska A (1980b) Human ontogenesis. II. Development of the human neuromuscular junction. J Neuropathol Exp Neurol 39:606–615
Fiehn W, Peter JB (1973) Lipid composition of muscles of nearly homogeneous fiber type. Exp Neurol 39:372–380
Fields RW (1970) Mechanical properties of the frog sarcolemma. Biophys J 10:462–480

Fields RW, Faber JJ (1970) Biophysical analysis of the mechanical properties of the sarcolemma. Can J Physiol Pharmacol 48:394–404

Fink WJ, Costill DL, Pollock ML (1977) Submaximal and maximal working capacity of elite distance runners. Part II. Muscle fiber composition and enzyme activities. Ann NY Acad Sci 301:323–327

Firket H (1967) Ultrastructural aspects of myofibrils formation in cultured skeletal muscle. Z Zellforsch Mikrosk Anat 78:313–327

Fischbach GD (1970) Synaptic potentials recorded in cell cultures of nerve and muscle. Science 169:1331–1333

Fischbach GD (1972) Synapse formation between dissociated nerve and muscle cells in low density cell cultures. Dev Biol 28:407–429

Fischbach GD (1982) Induction of high acetylcholine receptor density at nerve-muscle synapses in vitro. In: Schotland DL (ed) Disorders of the motor unit. Wiley, New York, pp 137–141

Fischbach GD, Cohen SA (1973) The distribution of acetylcholine sensitivity over uninnervated and innervated muscle fibers grown in cell culture. Dev Biol 31:147–162

Fischbach GD, Lass Y (1978) Acetylcholine noise in cultured chick myoballs: A voltage clamp analysis. J Physiol (Lond) 280:515–526

Fischbach GD, Robbins N (1969) Changes in contractile properties of disused soleus muscles. J Physiol (Lond) 201:305–320

Fischbach GD, Schuetze SM (1980) A post-natal decrease in acetylcholine channel open time at rat end-plates. J Physiol (Lond) 303:125–137

Fischbach GD, Frank E, Jessell TM, Rubin LL, Schuetze SM (1979) Accumulation of acetylcholine receptors and acetylcholinesterase at newly formed nerve-muscle synapses. Pharmacol Rev 30:411–428

Fischbeck KH, Bonilla E, Schotland DL (1982) Freeze-fracture analysis of plasma membrane cholesterol in fast- and slow-twitch muscles. J Ultrastruct Res 81:117–123

Fischer E (1877) Ueber die Endigung der Nerven im quergestreiften Muskel der Wirbelthiere. Arch Mikrosk Anat 13:365–390

Fischman DA (1967) An electron microscope study of myofibril formation in embryonic chick skeletal muscle. J Cell Biol 32:557–575

Fischman DA (1972) Development of striated muscle. In: Bourne GH (ed) The structure and function of muscle, vol I/1, 2nd edn. Academic Press, New York, pp 75–148

Fischman DA (1979) Multinucleation: Comparative aspects and functional implications. In: Mauro A (ed) Muscle regeneration. Raven Press, New York, pp 9–12

Fitts RH, Nagle FJ, Cassens RG (1973) Characteristics of skeletal muscle fiber types in the miniature pig and the effect of training. Can J Physiol Pharmacol 51:825–831

Fitzsimons RB, Hoh JFY (1981) Isomyosins in human type 1 and type 2 skeletal muscle fibres. Biochem J 193:229–233

Fitzsimons RB, Hoh JFY (1983) Myosin isoenzymes in fast-twitch and slow-twitch muscles of normal and dystrophic mice. J Physiol (Lond) 343:539–550

Flood PR (1977) The sarcoplasmic reticulum and associated plasma membrane of trunk muscle lamellae in *Branchiostoma lanceolatum* (Pallas). A transmission and scanning electron microscopic study including freeze-fractures, direct replicas and X-ray microanalysis of calcium oxalate deposits. Cell Tissue Res 181:169–196

Floyd K (1970) Junctions between muscle fibres in cat extraocular muscles. Nature 227:185–186

Floyd K (1973) Cholinesterase activity in sheep oesophageal muscle. J Anat 116:357–373

Floyd K, Morrison JFB (1975) The mechanical properties of oesophageal striated muscle in the cat and sheep. J Physiol (Lond) 248:717–724

Folkow B, Halicka HD (1968) A comparison between "red" and "white" muscle with respect to blood supply, capillary surface area and oxygen uptake during rest and exercise. Microvasc Res 1:1–14

Forbes MS, Sperelakis N (1979) Ruthenium-red staining of skeletal and cardiac muscles. Cell Tissue Res 200:367–382

Forssmann WG, Matter A (1966) Ultrastruktureller Nachweis von zwei Myofibrillentypen in den Muskelfasern des Rattenzwerchfells. Experientia 22:816

Fowler WE, Aebi U (1983) A consistent picture of the actin filament related to the orientation of the actin molecule. J Cell Biol 97:264–269

Frair PM, Strasberg PM, Freeman KB, Peterson AC (1979) Mitochondrial malic enzyme in mosaic skeletal muscle of mouse chimeras. Biochem Genet 17:693–702

Frank E, Fischbach GD (1979) Early events in neuromuscular junction formation in vitro. Induction of acetylcholine receptor clusters in the postsynaptic membrane and morphology of newly formed synapses. J Cell Biol 83:143–158

Frank E, Jansen JKS, Lømo T, Westgaard RH (1975) The interaction between foreign and original motor nerves innervating the soleus muscle of rats. J Physiol (Lond) 247:725–743

Frank G, Weeds AG (1974) The aminoacid sequence of the alkali light chains of rabbit skeletal-muscle myosin. Eur J Biochem 44:317–334

Franke WW, Schinko W (1969) Nuclear shape in muscle cells. J Cell Biol 42:326–331

Franklin GI, Yasin R, Hughes BP, Thompson EJ (1980) Acetylcholine receptors in cultured human muscle cells. J Neurol Sci 47:317–327

Franzini-Armstrong C (1970a) Natural variability in the length of thin and thick filaments in single fibres from a crab, Portunus depurator. J Cell Sci 6:559–592

Franzini-Armstrong C (1970b) Details of the I band structure as revealed by the localization of ferritin. Tissue Cell 2:327–338

Franzini-Armstrong C (1970c) Studies of the triad. I. Structure of the junction in frog twitch fibers. J Cell Biol 47:488–499

Franzini-Armstrong C (1973a) Studies of the triad. IV. Structure of the junction in frog slow fibers. J Cell Biol 56:120–128

Franzini-Armstrong C (1973b) The structure of a simple Z line. J Cell Biol 58:630–642

Franzini-Armstrong C (1973c) Membranous systems in muscle fibers. In: Bourne GH (ed) The structure and function of muscle, vol II, 2nd edn. Academic Press, New York London, pp 531–619

Franzini-Armstrong C (1975) Membrane particles and transmission at the triad. Fed Proc 34:1382–1389

Franzini-Armstrong C (1983) Disposition of Ca ATPase in SR membrane from skeletal muscle. (Abstr). J Cell Biol 97:260a

Franzini-Armstrong C (1984) Freeze-fracture of frog slow tonic fibers. Structure of surface and internal membranes. Tissue Cell 16:647–664

Franzini-Armstrong C, Porter KR (1964a) Sarcolemmal invaginations and the T-system in fish skeletal muscle. Nature 202:355–357

Franzini-Armstrong C, Porter KR (1964b) Sarcolemmal invaginations constituting the T system in fish muscle fibers. J Cell Biol 22:675–696

Franzini-Armstrong C, Porter KR (1964c) The Z disc of skeletal muscle fibrils. Z Zellforsch Mikrosk Anat 61:661–672

Franzini-Armstrong C, Landmesser L, Pilar G (1975) Size and shape of transverse tubule openings in frog twitch muscle fibers. J Cell Biol 64:493–497

Freimann R (1954) Untersuchungen über Zahl und Anordnung der Muskelspindeln in den Kaumuskeln des Menschen. Anat Anz 100:258–264

Freygang WH, Goldstein DA, Hellam DC, Peachey LD (1964) The relation between the late after-potential and the size of the transverse tubular system of frog muscle. J Gen Physiol 48:235–263

Fridén J (1983) Exercise-induced muscle soreness (Thesis). Umeå Univ Med Dissertations, Umeå

Fridén J (1984) Changes in human skeletal muscle induced by long-term eccentric exercise. Cell Tissue Res 236:365–372

Fridén J, Sjöström M, Ekblom B (1981) A morphological study of delayed muscle soreness. Experientia 37:506–507

Fridén J, Sjöström M, Ekblom B (1983) Myofibrillar damage following intense eccentric exercise in man. Int J Sports Med 4:170–176

Fridén J, Sjöström M, Ekblom B (1984) Muscle fibre type characteristics in endurance trained and untrained undividuals. Eur J Applied Physiol 52:266–271

Friedenberg RM, Seligman AM (1972) Acetylcholinesterase at the myoneural junction: cytochemical ultrastructure and some biochemical considerations. J Histochem Cytochem 20:771–792

Frisk-Holmberg M, Jorfeldt L, Juhlin-Dannfelt A, Karlsson J (1981) Leg blood flow during exercise in man in relation to muscle fibre composition. Acta Physiol Scand 112:339–342

Froehner SC, Gulbrandsen V, Hyman C, Jeng AY, Neubig RR, Cohen JB (1981) Immunofluoresence localization at the mammalian neuromuscular junction of the Mr 43,000 protein of *Torpedo* postsynaptic membranes. Proc Natl Acad Sci USA 78:5230–5234

Fuchs AF, Luschei ES (1971) Development of isometric tension in simian extraocular muscle. J Physiol (Lond) 219:155–166
Fuchs F, Briggs FN (1961) Direct isolation of a soluble relaxing system from muscle. Biochim Biophys Acta 51:423–425
Fukami Y (1970) Tonic and phasic muscle spindles in snake. J Neurophysiol 33:28–35
Fukami Y, Hunt CC (1970) Structure of snake muscle spindles. J Neurophysiol 33:9–27
Fumagalli G, Brigonzi A, Tachikawa T, Clementi F (1982) Rat myoblast fusion: Morphological study of membrane apposition, fusion, and fission during controlled myogenesis *in vitro*. J Ultrastruct Res 75:112–125
Furcht LT, Mosher DF, Wendelschafer-Crabb G (1978) Immunocytochemical localization of fibronectin (LETS protein) on the surface of L6 myoblasts: Light and electron microscopic studies. Cell 13:263–271
Gage PW, Eisenberg RS (1967) Action potentials without contraction in frog skeletal muscle fibers with disrupted transverse tubules. Science 158:1702–1703
Gage PW, Eisenberg RS (1969a) Capacitance of the surface and transverse tubular membrane of frog sartorius muscle fibers. J Gen Physiol 53:265–278
Gage PW, Eisenberg RS (1969b) Action potentials, afterpotentials, and excitation-contraction coupling in frog sartorius fibers without transverse tubules. J Gen Physiol 53:298–310
Galavazi G (1971) Identification of helical polyribosomes in sections of mature skeletal muscle fibers. Z Zellforsch Mikrosk Anat 121:531–547
Galavazi G, Szirmai JA (1971a) Cytomorphometry of skeletal muscle: The influence of age and testosterone on the rat M. levator ani. Z Zellforsch Mikrosk Anat 121:507–530
Galavazi G, Szirmai JA (1971b) The influence of age and testosterone on the ribosomal population in the M. levator ani and a thigh muscle of the rat. Z Zellforsch Mikrosk Anat 121:548–560
Galey FR (1964) Local contraction patterns of striated muscle. J Ultrastruct Res 11:389–400
Galvas PE, Gonyea WJ (1980) Motor-end-plate and nerve distribution in a histochemically compartmentalized pennate muscle in the cat. Am J Anat 159:147–156
Galvas PE, Neaves WB, Gonyea WJ (1982) Direct correlation of histochemical profile to the ultrastructure of single myofibers and their neuromuscular junctions from a mixed muscle. Anat Rec 203:1–17
Gamble HJ, Fenton J, Allsopp G (1978) Electron microscope observations on human fetal striated muscle. J Anat 126:567–589
Gard DL, Lazarides E (1979) Specific fluorescent labeling of chicken myofibril Z-line proteins catalyzed by guinea pig liver transglutaminase. J Cell Biol 81:336–347
Gardiner PF (1981) Influence of reduced neuromuscular activity on the development of neonatal rat hindlimb muscle. Dev Neurosci 4:382–388
Garnett R, Stephens JA (1981) Changes in the recruitment threshold of motor units produced by cutaneous stimulation in man. J Physiol (Lond) 311:463–473
Garrels JI, Gibson W (1976) Identification and characterization of multiple forms of actin. Cell 9:793–805
Gauthier GF (1969) On the relationship of ultrastructural and cytochemical features to color in mammalian skeletal muscle. Z Zellforsch Mikrosk Anat 95:462–482
Gauthier GF (1974) Some ultrastructural and cytochemical features of fiber populations in the soleus muscle. Anat Rec 180:551–564
Gauthier GF (1979) Ultrastructure identification of muscle fiber types by immunocytochemistry. J Cell Biol 82:391–400
Gauthier GF, Lowey S (1977) Polymorphism of myosin among skeletal muscle fiber types. J Cell Biol 74:760–779
Gauthier GF, Lowey S (1979) Distribution of myosin isoenzymes among skeletal musle fiber types. J Cell Biol 81:10–25
Gauthier GF, Padykula HA (1965) Cytochemical studies of adenosine triphosphatase activity in the sarcoplasmic reticulum. J Cell Biol 27:252–260
Gauthier GF, Padykula HA (1966) Cytological studies of fiber types in skeletal muscle. A comparative study of the mammalian diaphragm. J Cell Biol 28:333–354
Gauthier GF, Lowey S, Hobbs AW (1978) Fast and slow myosin in developing muscle fibres. Nature 274:25–29
Gauthier GF, Lowey S, Benfield PA, Hobbs AW (1982) Distribution and properties of myosin isozymes in developing avian and mammalian skeletal muscle fibers. J Cell Biol 92:471–484

Gauthier GF, Burke RE, Lowey S, Hobbs AW (1983) Myosin isozymes in normal and cross-reinnervated cat skeletal muscle fibers. J Cell Biol 97:756–771
Gearhart JD, Mintz B (1972) Clonal origins of somites and their muscle derivatives: Evidence from allophenic mice. Dev Biol 29:27–37
Gelber D, Moore DH, Ruska H (1960) Observations of the myo-tendon junction in mammalian skeletal muscle. Z Zellforsch Mikrosk Anat 52:396–400
Gerebtzoff MA (1956) Contribution a la morphologie comparée des appareils cholinestérasiques myo-neural et musculo-tendineux des vertébrés. Annales d'Histochimie 1:26–29
Gerebtzoff MA (1957) L'appareil cholinestérasique musculo-tendineux: Structure, développement, effet de la dénervation et de la ténotomie. Acta Physiol Pharmacol Neerlandica 6:419–427
Gergely J (1959) The relaxing factor of muscle. Ann NY Acad Sci 81:490–504
Gerlach J (1877) Ueber das Verhältniss der nervösen und contractilen Substanz des quergestreiften Muskels. Arch Mikrosk Anat 13:399–414
Giacobini G, Filogamo G, Weber M, Boquet P, Changeux JP (1973) Effects of a snake α-neurotoxin on the development of innervated skeletal muscles in chick embryo. Proc Natl Acad Sci USA 70:1708–1712
Gibson MC, Schultz E (1983) Age-related differences in absolute numbers of skeletal muscle satellite cells. Muscle Nerve 6:574–580
Gillespie CA, Simpson DR, Edgerton VR (1970) High glycogen content of red as opposed to white skeletal muscle fibers of guinea pig. J Histochem Cytochem 18:552–558
Gillespie CA, Simpson DR, Edgerton VR (1974) Motor unit recruitment as reflected by muscle fibre glycogen loss in a prosimian (Bushbaby) after running and jumping. J Neurol Neurosurg Psychiatry 37:817–824
Gilliatt RW, Westgaard RH, Williams IR (1978) Extrajunctional acetylcholine sensitivity of inactive muscle fibres in the baboon during prolonged nerve pressure block. J Physiol (Lond) 280:499–514
Gillis JM, Page SG (1967) Localization of ATPase activity in striated muscle and probable sources of artifact. J Cell Sci 2:113–118
Gillis JM, Piront A, Gosselin-Rey C (1979) Parvalbumins. Distribution and physical state inside the muscle cell. Biochim Biophys Acta 585:444–450
Gilly WF (1975) Slow fibers in the frog cruralis muscle. Tissue Cell 7:203–210
Ginsborg BL (1960) Some properties of avian skeletal muscle fibres with multiple neuromuscular junctions. J Physiol (Lond) 154:581–598
Ginsborg BL, MacKay B (1961) A histochemical demonstration of two types of motor innervation in avian skeletal muscle. Bibl Anat 2:174–181
Girardier L, Reuben JP, Brandt PW, Grundfest H (1963) Evidence for anion-permselective membrane in crayfish muscle fibers and its possible role in excitation-contraction coupling. J Gen Physiol 47:189–214
Glicksman MA, Sanes JR (1983) Differentiation of motor nerve terminals formed in the absence of muscle fibres. J Neurocytol 12:661–671
Goldberg AL (1968) Protein synthesis during work-induced growth of skeletal muscle. J Cell Biol 36:653–658
Goldberg SJ, Lennerstrand G, Hull CD (1976) Motor unit responses in the lateral rectus muscle of the cat: Intracellular current injection of abducens nucleus neurons. Acta Physiol Scand 96:58–63
Golder TK, Nieberg PS, Wilson BW (1977) Ultrastructural localization of acetyl-cholinesterase in cultured cells. II. Cycloheximide-treated embryo muscle. J Histochem Cytochem 25:376–383
Goldspink DF (1977a) The influence of immobilization and stretch on protein turnover of rat skeletal muscle. J Physiol (Lond) 264:267–282
Goldspink DF (1977b) The influence of activity on muscle size and protein turnover. J Physiol (Lond) 264:283–296
Goldspink DF (1978) The influence of passive stretch on the growth and protein turnover of the denervated extensor digitorum longus muscle. Biochem J 174:595–602
Goldspink G (1962) Studies on postembryonic growth and development of skeletal muscle. Proc Roy Irish Acad 62B:135–150
Goldspink G (1964a) Increase in length of skeletal muscle during normal growth. Nature 204:1095–1096

Goldspink G (1964b) The combined effects of exercise and reduced food intake on skeletal muscle fibers. J Cell Comp Physiol 63:209–216
Goldspink G (1968) Sarcomere length during post-natal growth of mammalian muscle fibres. J Cell Sci 3:539–548
Goldspink G (1970) The proliferation of myofibrils during muscle fibre growth. J Cell Sci 6:593–604
Goldspink G (1972) Postembryonic growth and differentiation of striated muscle. In: Bourne GH (ed) The structure and function of muscle, vol I/1, 2nd edn. Academic Press, New York London, pp 179–236
Goldspink G, Ward PS (1979) Changes in rodent muscle fibre types during post-natal growth, undernutrition and exercise. J Physiol (Lond) 296:453–469
Goldstein DJ (1959) Some histochemical observations of human striated muscle. Anat Rec 134:217–237
Goldstein MA (1969a) A morphological and cytochemical study of sarcoplasmic reticulum and T system of fish extraocular muscle. Z Zellforsch Mikrosk Anat 102:31–39
Goldstein MA (1969b) Anionic binding of ruthenium red in fish extraocular muscle. Z Zellforsch Mikrosk Anat 102:459–465
Goll DE, Mommaerts FHM, Reedy MK, Seraydarian K (1969) Studies on α-actinin-like proteins liberated during trypsin digestion of α-actinin and of myofibrils. Biochim Biophys Acta 175:174–194
Goll DE, Suzuki A, Temple J, Holmes GR (1972) Studies on purified α-actinin. I. Effect of temperature and tropomyosin on the α-actinin/F-actin interaction. J Mol Biol 67:469–488
Gollnick PD, King DW (1969) Effect of exercise and training on mitochondria of rat skeletal muscle. Am J Physiol 216:1502–1509
Gollnick PD, Saltin B (1982) Significance of skeletal muscle oxidative enzyme enhancement with endurance training. Clin Physiol 2:1–12
Gollnick PD, Piehl K, Saltin B (1974a) Selective glycogen depletion pattern in human muscle fibres after exercise of varying intensity and at varying pedalling rates. J Physiol (Lond) 241:45–57
Gollnick PD, Karlsson J, Piehl K, Saltin B (1974b) Selective glycogen depletion in skeletal muscle fibres of man following sustained contractions. J Physiol (Lond) 241:59–67
Gonyea WJ (1979) Fiber size distribution in the flexor carpi radialis muscle of the cat. Anat Rec 195:447–454
Gonyea WJ (1980) Role of exercise in inducing increases in skeletal muscle fiber number. J Appl Physiol 48:421–426
Gonyea WJ, Ericson GC (1976) An experimental model for the study of exercise-induced skeletal muscle hypertrophy. J Appl Physiol 40:630–633
Gonyea W, Ericson GC, Bonde-Petersen F (1977) Skeletal muscle fiber splitting induced by weight-lifting exercise in cats. Acta Physiol Scand 99:105–109
Gonyea WJ, Marushia SA, Dixon JA (1981) Morphological organization and contractile properties of the wrist flexor muscles in the cat. Anat Rec 199:321–339
Gonzalez-Serratos H (1966) Inward spread of contraction during a twitch (Abstr). J Physiol (Lond) 185:20P–21P
Gonzalez-Serratos H (1971) Inward spread of activation in vertebrate muscle fibres. J Physiol (Lond) 212:777–799
Gonzalez-Serratos H, Hill L, Valle-Aguilera R (1981) Effects of catecholamines and cyclic amp on excitation-contraction coupling in isolated skeletal muscle fibres of the frog. J Physiol (Lond) 315:267–282
Goodall MC, Szent-Györgyi AG (1953) Relaxing factors in muscle. Nature 172:84–85
Gordon AM, Huxley AF, Julian FJ (1966) The variation in isometric tension with sarcomere length in vertebrate muscle fibres. J Physiol (Lond) 184:170–192
Gordon G, Phillips CG (1953) Slow and rapid components in a flexor muscle. Q J Physiol 38:35–45
Gordon GB, Price HM, Blumberg JM (1967) Electron microscopic localization of phosphatase activities within striated muscle fibers. Lab Invest 16:422–435
Gori Z (1972) Proliferations of the sarcoplasmic reticulum and the T system in denervated muscle fibers. Virchows Arch [Cell Pathol] 11:147–160
Gori Z, Pellegrino C, Pollera M (1967) The castration atrophy of the dorsal bulbocavernosus muscle of rat: An electron microscopic study. Exp Mol Pathol 6:172–198

Gori Z, Pellegrino C, Pollera M (1969) The hypertrophy of levator ani muscle of rat induced by testosterone: An electron microscope study. Exp Mol Pathol 10:199–218
Gorniak GC, Gans C, Faulkner JA (1979) Muscle fiber regeneration after transplantation: Prediction of structure and physiology from electromyograms. Science 204:1085–1087
Gottschall J, Zenker W, Neuhuber W, Mysicka A, Müntener M (1980a) The sternomastoid muscle of the rat and its innervation. Muscle fiber composition, perikarya and axons of efferent and afferent neurons. Anat Embryol 160:285–300
Gottschall J, Neuhuber W, Müntener M, Mysicka A (1980b) The ansa cervicalis and the infrahyoid muscles of the rat. II. Motor and sensory neurons. Anat Embryol 159:59–69
Granger BL, Lazarides E (1978) The existence of an insoluble Z disk scaffold in chicken skeletal muscle. Cell 15:1253–1268
Granger BL, Lazarides E (1979) Desmin and vimentin coexist at the periphery of the myofibril Z disc. Cell 18:1053–1063
Granger BL, Lazarides E (1980) Synemin: a new high molecular weight protein associated with desmin and vimentin filaments in muscle. Cell 22:727–738
Gratzl M, Schudt C, Ekerdt R, Dahl G (1980) Fusion of isolated biological membranes. A tool to investigate basic processes of exocytosis and cell-cell fusion. In: Bittar EE (ed), Membrane structure and function, vol 3. Wiley, New York, pp 60–92
Green HJ, Thomson JA, Daub WD, Houston ME, Ranney DA (1979) Fiber composition, fiber size and enzyme activities in vastus lateralis of elite athletes involved in high intensity exercise. Eur J Applied Physiol 41:109–117
Green HJ, Daub B, Houston ME, Thomson JA, Fraser I, Ranney D (1981) Human vastus lateralis and gastrocnemius muscles. A comparative histochemical and biochemical analysis. J Neurol Sci 52:201–210
Green HJ, Reichmann H, Pette D (1983) Fibre type specific transformations in the enzyme activity pattern of rat vastus lateralis muscle by prolonged endurance training. Pflugers Arch 399:216–222
Green HJ, Klug GA, Reichmann H, Seedorf U, Wiehrer W, Pette D (1984) Exercise-induced fibre type transitions with regard to myosin, parvalbumin, and sarcoplasmic reticulum in muscles of the rat. Pflugers Arch 400:432–438
Greenberg AJ, Parker KK, Trevor AJ (1977) Immunochemical studies of mammalian brain acetylcholinesterase. J Neurochem 29:911–917
Greene T, Jampel R (1966) Muscle spindles in the extraocular muscles of the macaque. J Comp Neurol 126:547–549
Griffin GE, Williams PE, Goldspink G (1971) Region of longitudinal growth in striated muscle fibres. Nature New Biology 232:28–29
Grim M (1978) Acid phosphatase activity in normal and sarcolytic myotubes in muscle anlagen of the human hand. Histochemistry 56:307–316
Grim M, Carlson BM (1974) A comparison of morphogenesis of muscles of the forearm and hand during ontogenesis and regeneration in the axolotl (*Ambystoma mexicanum*). II. The development of the muscular pattern in the embryonic and regenerating limb. Z Anat Entwicklungsgesch 145:149–167
Grim M, Carlson BM (1979) The formation of muscles in regenerating limbs of the newt after denervation of the blastema. J Embryol Exp Morphol 54:99–111
Grimby L (1984) Firing properties of single human motor units during locomotion. J Physiol (Lond) 346:195–202
Gröschel-Stewart U, Doniach D (1969) Immunological evidence for human myosin isoenzymes. Immunology 17:991–994
Grove BK, Kurer V, Lehner C, Doetschman TC, Perriard J-C, Eppenberger HM (1984) A new 185,000-dalton skeletal muscle protein detected by monoclonal antibodies. J Cell Biol 98:518–524
Gruber H (1968) Über Struktur und Innervation der quergestreiften Muskulatur des Oesophagus der Ratte. Z Zellforsch Mikrosk Anat 91:236–247
Gruber H (1978) Motor innervation of the striated oesophagus muscle. J Neurol Sci 36:41–53
Gruber H, Zenker W (1973) Acetylcholinesterase: histochemical differentiation between motor and sensory nerve fibres. Brain Res 51:207–214
Grumet M, Rutishauser U, Edelman GM (1982) Neural cell adhesion molecule is on embryonic muscle cells and mediates adhesion to nerve cells *in vitro*. Nature 295:693–695

Gruner J-E (1961) La structure fine du fuseau neuromusculaire humain. Rev Neurol (Paris) 104:490–507
Guba F, Harsanyi V, Vajda E (1968) Size of the filaments in relaxation and contraction. Acta Biochim Biophys Acad Sci Hung 3:441–448
Gulati AK, Reddi AH, Zalewski AA (1982) Distribution of fibronectin in normal and regenerating skeletal muscle. Anat Rec 204:175–183
Gunn HM, Davies AS (1971) Histochemical characteristics of muscle fibres in the diaphragm (Abstr). Biochem J 125:108–109P
Guth L (1968) "Trophic" influences of nerve on muscle. Physiol Rev 48:645–687
Guth L (1973) Fact and artifact in the histochemical procedure for myofibrillar ATPase. Exp Neurol 41:440–450
Guth L, Albuquerque EX (1978) The neurotrophic regulation of resting membrane potential and extrajunctional acetylcholine sensitivity in mammalian skeletal muscle. Physiol Bohemoslov 27:401–414
Guth L, Albuquerque EX (1979) The neurotrophic regulation of resting membrane potential and extrajunctional acetylcholine sensitivity in mammalian skeletal muscle. In: Mauro A (ed) Muscle regeneration. Raven Press, New York, pp 405–415
Guth L, Samaha FJ (1972) Erroneous interpretations which may result from application of the "myofibrillar ATPase" histochemical procedure to developing muscle. Exp Neurol 34:465–475
Guth L, Wells JB (1972) Physiological and histochemical properties of the soleus muscle after denervation of its antagonists. Exp Neurol 36:463–471
Gutmann E, Carlson BM (1978) The regeneration of a hormone-sensitive muscle (levator ani) in the rat. Exp Neurol 58:535–548
Gutmann E, Hanzlikova V (1966) Motor unit in old age. Nature 209:921–922
Gutmann E, Hanzlikova V (1973) Functional and histochemical properties of the persistent levator ani muscle in female rats. Folia Morph 21:142–143
Guy PS, Snow DH (1977) The effect of training and detraining on muscle composition in the horse. J Physiol (Lond) 269:33–51
Hackenbrock CR (1966) Ultrastructural bases for metabolically linked mechanical activity in mitochondria. I. Reversible ultrastructural changes with change in metabolic steady state in isolated liver mitochondria. J Cell Biol 30:269–297
Häggmark T, Jansson E, Svane B (1978) Cross-sectional area of thigh muscles in man measured by computed tomography. Scand J Clin Lab Invest 38:355–360
Häggmark T, Thorstensson A (1979) Fibre types in human abdominal muscles. Acta Physiol Scand 107:319–325
Häggquist G (1931) Gewebe und Systeme der Muskulatur. In: Von Möllendorff W (ed) Die Gewebe: (Handbuch der Mikroskopischen Anatomie des Menschen, Bd II/3). Springer, Berlin
Häggquist G (1956) Gewebe und Systeme der Muskulatur. Ergänzung zu Bd II/3. In: Bargmann W (ed) Die Gewebe. Springer, Berlin Göttingen Heidelberg (Handbuch der Mikroskopischen Anatomie des Menschen, II/4)
Hagopian M (1966) The myofilament arrangement in the femoral muscle of the cockroach *Leucophea maderae* Fabricius. J Cell Biol 28:545–562
Hagopian M, Spiro D (1968) The filament lattice of cockroach thoracic muscle. J Cell Biol 36:433–442
Håkansson CH (1956) Conduction velocity and amplitude of the action potential as related to circumference in the isolated fibre of frog muscle. Acta Physiol Scand 37:14–34
Håkelius L, Nyström B (1975) Histochemical studies of end-plate formation in free autologous muscle transplants in cats. Scand J Plast Reconstr Surg 9:9–14
Halkjaer-Kristensen J, Ingemann-Hansen T (1977) Microphotometry of single fibres from the quadriceps muscle in man. Scand J Clin Lab Invest 37:461–468
Hall CE, Jakus MA, Schmitt FO (1946) An investigation of cross striations and myosin filaments in muscle. Biol Bull 90:32–50
Hall ZW, Kelly RB (1971) Enzymatic detachment of endplate acetylcholinesterase from muscle. Nature New Biol 232:62–63
Hall ZW, Lubit BW, Schwartz JH (1981) Cytoplasmic actin in postsynaptic structures at the neuromuscular junction. J Cell Biol 90:789–792
Hall-Craggs ECB (1968) The contraction times and enzyme activity of two rabbit laryngeal muscles. J Anat 102:241–255

Hall-Craggs ECB (1970) The longitudinal division of fibres in overloaded rat skeletal muscle. J Anat 107:459–470
Hall-Craggs ECB (1971) Observations on the fate of muscle fibres temporarily isolated by transection of a muscle belly. Z Zellforsch Mikrosk Anat 119:68–76
Hall-Craggs ECB (1972) Non-selective reunion of skeletal muscle fibres in the rat. J Anat 111:151–156
Hall-Craggs ECB (1974) The regeneration of skeletal muscle fibres *per continuum*. J Anat 117:171–178
Hall-Craggs ECB (1978) Ischemic muscle as a model of regeneration. Exp Neurol 60:393–399
Hall-Craggs ECB (1979) The free autogenous muscle transplant. In: Mauro A (ed), Muscle regeneration. Raven Press, New York, pp 541–546
Hall-Craggs ECB (1980) Survival of satellite cells following exposure to the local anesthetic bupivacaine (Marcaine). Cell Tissue Res 209:131–135
Hall-Craggs ECB, Seyan HS (1975) Histochemical changes in innervated and denervated skeletal muscle fibers following treatment with bupivacaine (Marcain). Exp Neurol 46:345–354
Hamburger V, Hamilton HL (1951) A series of normal stages in the development of the chick embryo. J Morphol 88:49–92
Hammarberg C (1974) The histochemical appearance of developing muscle fibres in the gastrocnemius, soleus and anterior tibial muscles of the kitten, as viewed in serial sections stained for lipids and succinic dehydrogenase. Acta Neurol Scand 50:285–301
Hammarberg C (1975) On the postnatal differentiation of motor units in the cat ankle muscles. A histochemical and physiological study (Thesis). Karolinska Institute, Stockholm
Hammersen F (1968) The pattern of the terminal vascular bed and the ultrastructure of capillaries in skeletal muscle. In: Lübbers D-W, Luft UC, Thews G, Witzleb E (eds) Oxygen transport in blood and tissue. Thieme, Stuttgart, pp 184–197, pp 258–261
Hammersen F, Appell H-J (1976) Capillary density and patterns in skeletal muscle. 1. On the reliability of various techniques to demonstrate blood capillaries (Abstr). Pflugers Arch [Suppl] 365:109
Hammersen F, Gidlof A, Larsson J, Lewis DH (1980) The occurrence of paracrystalline mitochondrial inclusions in normal human skeletal muscle. Acta Neuropathol (Berl) 49:35–41
Hamosh M, Lesch M, Baron J, Kaufman S (1967) Enhanced protein synthesis in a cell-free system from hypertrophied skeletal muscle. Science 157:935–937
Handbook of Circulation (1959) National Academy of Sciences, National Research Council. Saunders, Philadelphia London
Hansen-Smith FM, Picou D, Golden MH (1979) Muscle satellite cells in malnourished and nutritionally rehabilitated children. J Neurol Sci 41:207–221
Hanson J, Huxley HE (1955) The structural basis of contraction in striated muscle. Symposia of the Society for Experimental Biology 9:228–264
Hanson J, Huxley HE (1957) Quantitative studies on the structure of cross-striated myofibrils. II. Investigations by biochemical techniques. Biochim Biophys Acta 23:250–260
Hanson J, Lowy J (1963) The structure of F-actin and of actin filaments isolated from muscle. J Mol Biol 6:46–60
Hanson J, Lowy J (1965) Molecular basis of contractility in muscle. Br Med Bull 21:264–271
Hanson J, Lennerstrand G (1977) Contractile and histochemical properties of the inferior oblique muscle in the rat and in the cat. Acta Ophthalmol 55:88–102
Hanson J, Lennerstrand G, Nichols KC (1980) The postnatal development of the inferior oblique muscle of the cat. III. Fiber sizes and histochemical properties. Acta Physiol Scand 108:61–71
Hanzlikova V, Schiaffino S (1977) Mitochondrial changes in ischemic skeletal muscle. J Ultrastruct Res 60:121–133
Hanzlikova V, Mackova EV, Hnik P (1975) Satellite cells of the rat soleus muscle in the process of compensatory hypertrophy combined with denervation. Cell Tissue Res 160:411–421
Harker DW (1972a) The structure and innervation of sheep superior rectus and levator palpebrae extraocular muscles. I. Extrafusal muscle fibers. Invest Ophthalmol 11:956–969
Harker DW (1972b) The structure and innervation of sheep superior rectus and levator palpebrae extraocular muscles. II. Muscle spindles. Invest Ophthalmol 11:970–979
Harker DW, Jami L, Laporte Y, Petit J (1977) Fast-conducting skeletofusimotor axons supplying intrafusal chain fibers in the cat peroneus tertius muscle. J Neurophysiol 40:791–799
Harrington WF, Sutoh K, Chiao Y-CC (1979) Myosin filaments and cross bridge movement. In: Pepe FA, Sanger JW, Nachmias VT (eds) Motility in cell function. Academic Press, New York, pp 69–90

Harris AJ, Heinemann S, Schubert D, Tarakis H (1971) Trophic interaction between cloned tissue culture lines of nerve and muscle. Nature 231:296–301
Harris H (1980) Switching on the muscle genes. Nature 286:758–759
Harris JB, Marshall MW (1973) Tetrodotoxin-resistant action potentials in newborn rat muscle. Nature New Biol 243:191–192
Hartzell HC, Fambrough DM (1972) Acetylcholine receptors. Distribution and extrajunctional density in rat diaphragm after denervation correlated with acetylcholine sensitivity. J Gen Physiol 60:248–262
Haselgrove JC, Huxley HE (1973) X-ray evidence for radial cross-bridge movement and for the sliding filament model in actively contracting skeletal muscle. J Mol Biol 77:549–568
Hasselbach W (1953) Elektronenmikroskopische Untersuchungen an Muskelfibrillen bei totaler und partieller Extraktion des L-Myosins. Z Naturforsch 8b:449–454
Hasselbach W (1960) Zur Cytochemie des Herzmuskels. Acta tertii europaei de cordis scientia conventus. Romae 1:113–123
Hasselbach W (1963) Mechanismen der Muskelkontraktion und ihre intracelluläre Steuerung. Naturwissenschaften 50:249–256
Hasselbach W (1964) Relaxation and the sarcotubular calcium pump. Fed Proc 23:909–912
Hasselbach W (1979) The sarcoplasmic calcium pump. A model of energy transduction in biological membranes. Top Curr Chem 78:1–56
Hasselbach W, Elfvin L-G (1967) Structural and chemical asymmetry of the calcium-transporting membranes of the sarcotubular system as revealed by electron microscopy. J Ultrastruct Res 17:598–622
Hasselbach W, Makinose M (1961) Die Calciumpumpe der "Erschlaffungsgrana" des Muskels und ihre Abhängigkeit von der ATP-Spaltung. Biochem Z 333:518–528
Hasselbach W, Schneider G (1951) Der L-Myosin- und Aktingehalt des Kaninchenmuskels. Biochem Z 321:462–475
Hasselbach W, Weber A (1955) Models for the study of the contraction of muscle and of cell protoplasm. Pharmacol Rev 7:97–117
Hasselbach W, Weber HH (1965) Die intrazelluläre Regulation der Muskelaktivität. Naturwissenschaften 52:121–128
Hauschka SD (1974) Clonal analysis of vertebrate myogenesis. II. Environmental influences upon human muscle differentiation. Dev Biol 37:329–344
Hauschka SD (1982) Muscle cell culture: Future goals for facilitating the investigation of human muscle disease. In: Schotland DL (ed) Disorders of the motor unit. Wiley, New York, pp 925–936
Hauschka SD, White NK (1972) Studies of myogenesis in vitro. In: Banker BQ, Przybylski RJ, Van Der Meulen JP, Victor M (eds) Research in muscle development and the muscle spindle. Excerpta Medica, Amsterdam, pp 53–71
Hauschka SD, Linkhart TA, Clegg C, Merrill G (1979) Clonal studies of human and mouse muscle. In: Mauro A (ed) Muscle regeneration. Raven Press, New York, pp 311–322
Hauschka SD, Rutz R, Linkhart TA, Clegg CH, Merrill GF, Haney CM, Lim RW (1982) Skeletal muscle development. In: Schotland DL (ed) Disorders of the motor unit. Wiley, New York, pp 903–923
Hay ED (1959) Electron microscopic observations of muscle dedifferentiation in regenerating *Amblystoma* limbs. Dev Biol 1:555–585
Hay ED (1970) Regeneration of muscle in the amputated amphibian limb. In: Mauro A, Shafiq SA, Milhorat AT (eds) Regeneration of striated muscle, and myogenesis. Excerpta Medica, Amsterdam, pp 3–24
Hay ED (1979) Reversibility of muscle differentiation in newt limb regeneration. In: Mauro A (ed) Muscle regeneration. Raven Press, New York, pp 73–81
Hay ED, Doyle CM (1973) Absence of reserve cells (satellite cells) in nonregenerating muscle of mature newt limbs. Anat Rec 175:339–340
Hayashida Y, Schmalbruch H (1972) Zur Größe der Fettpartikel in mitochondrienreichen Skelettmuskelfasern der Ratte in Abhängigkeit von der Nahrungsaufnahme. Z Zellforsch Mikrosk Anat 127:374–381
Hayes BP (1975) The distribution of intercellular junctions in the developing myotomes of the clawed toad. Anat Embryol 147:345–354

Hayes D, Huang M, Zobel CR (1971) Electron microscope observations on thick filaments in striated muscle from the lobster *Homarus americanus*. J Ultrastruct Res 37:17–30
Hedberg G, Jansson E (1976) Skeletal muscle fibre distribution, capacity and interest in different physical activities among students in high school (In Swedish). Pedagogiska rapporter Umeå 54:1–40
Hegarty PVJ, Dahlin KJ, Benson ES, Allen CE (1973) Ultrastructural and light microscope studies on rigor-extended sarcomeres in avian and porcine skeletal muscles. J Anat 115:203–219
Heidenhain M (1919) Die Entdeckung der Noniusfelder in der quergestreiften Muskelfaser. Anat Anz 51:49–53
Heiman-Patterson TD (1983) Developmental changes of the T-system in cultured human skeletal muscle (Abstr). J Cell Biol 97:45a
Hellmuth AE, Allbrook D (1973) Satellite cells as the stem cells of skeletal muscle. In: Kakulas BA (ed) Basic research in myology. Excerpta Medica, Amsterdam, pp 343–345
Henneman E, Olson CB (1965) Relations between structure and function in the design of skeletal muscles. J Neurophysiol 28:581–598
Henneman E, Somjen G, Carpenter DO (1965) Functional significance of cell size in spinal motoneurons. J Neurophysiol 28:560–580
Hennig G (1969) Die Nervenendigungen der Rattenmuskelspindel im elektronen- und phasenkontrastmikroskopischen Bild. Z Zellforsch Mikrosk Anat 96:275–294
Hennig R, Lømo T (1985) Firing patterns of motor units in normal rats. Nature 314:164–166
Henriksson J (1976) Human skeletal muscle adaptation to physical activity (Thesis Stockholm). August Krogh Inst, Copenhagen
Henriksson J, Reitman JS (1976) Quantitative measures of enzyme activities in type I and type II muscle fibres of man after training. Acta Physiol Scand 97:392–397
Henriksson J, Jansson E, Schantz P (1980) Increase in myofibrillar ATPase intermediate skeletal muscle fibers with endurance training of extreme duration in man (Abstr). Muscle Nerve 3:274
Hess A (1961a) Structural differences of fast and slow extrafusal muscle fibres and their nerve endings in chickens. J Physiol (Lond) 157:221–231
Hess A (1961b) The structure of slow and fast extrafusal muscle fibers in the extraocular muscles and their nerve endings in guinea pigs. J Cell Comp Physiol 58:63–80
Hess A (1963) Two kinds of extrafusal muscle fibers and their nerve endings in the garter snake. Am J Anat 113:347–363
Hess A (1965) The sarcoplasmic reticulum, the T system, and the motor terminals of slow and twitch muscle fibers in the garter snake. J Cell Biol 26:467–476
Hess A (1967) The structure of vertebrate slow and twitch muscle fibers. Invest Ophthalmol 6:217–228
Hess A (1970) Vertebrate slow muscle fibers. Physiol Rev 50:40–62
Hess A, Pilar G (1963) Slow fibres in the extraocular muscles of the cat. J Physiol (Lond) 169:780–798
Hess A, Rosner S (1970) The satellite cell bud and myoblast in denervated mammalian muscle fibers. Am J Anat 129:21–40
Heuser JE (1983) Structure of the myosin crossbridge lattice in insect flight muscle. J Mol Biol 169:123–154
Heuser JE, Cooke R (1983) Actin-myosin interactions visualized by the quick-freeze, deep-etch replica technique. J Mol Biol 169:97–122
Heuser JE, Kirschner MW (1980) Filament organization in platinum replicas of freeze-dried cytoskeletons. J Cell Biol 86:212–234
Heuser JE, Reese TS (1973) Evidence for recycling of synaptic vesicle membrane during transmitter release at the frog neuromuscular junction. J Cell Biol 57:315–344
Heuser JE, Reese TS, Landis DMD (1974) Functional changes in frog neuromuscular junctions studied with freeze-fracture. J Neurocytol 3:109–131
Heuser JE, Reese TS, Dennis MJ, Jan Y, Jan L, Evans L (1979) Synaptic vesicle exocytosis captured by quick freezing and correlated with quantal transmitter release. J Cell Biol 81:275–300
Higashi S, Ooi T (1968) Crystals of tropomyosin and native tropomyosin. J Mol Biol 34:699–701
Hikida RS (1972) The structure of the sarcotubular system in avian muscle. Am J Anat 134:481–496
Hikida RS (1976) Regeneration of a tonic avian muscle: Transformation to a twitch morphology. Dev Biol 48:67–79
Hikida RS, Lombardo JA (1974) Regeneration of pigeon fast and slow muscle fiber types after partial excision and mincing. J Cell Biol 61:414–426

Hikida RS, Peterson WJ (1983) Skeletal muscle-smooth muscle interaction: An unusual myoelastic system. J Morphol 177:231–243
Hikida RS, Staron RS, Hagerman FC, Sherman WM, Costill DL (1983) Muscle fiber necrosis associated with human marathon runners. J Neurol Sci 59:185–203
Hill AV (1938) The heat of shortening and the dynamic constants of muscle. Proc R Soc Lond [Biol] 126:136–195
Hill AV (1949) The abrupt transition from rest to activity in muscle. Proc R Soc Lond [Biol] 136:399–420
Hill AV (1950a) Mechanics of the contractile element of muscle. Nature 166:415–419
Hill AV (1950b) The dimensions of animals and their muscular dynamics. Sci Prog 38:209–230
Hill DK (1964) The space accessible to albumin within the striated muscle fibre of the toad. J Physiol (Lond) 175:275–294
Hill DK (1968) Tension due to interaction between the sliding filaments in resting striated muscle. The effect of stimulation. J Physiol (Lond) 199:637–684
Hill L (1977) A-band length, striation spacing and tension change on stretch of active muscle. J Physiol (Lond) 266:677–685
Hinrichsen C, Dulhunty A (1982) The contractile properties, histochemistry, ultrastructure and electrophysiology of the cricothyroid and posterior cricoarytenoid muscles in the rat. J Muscle Res Cell Motil 3:169–190
Hirche H, Raff WK, Grün D (1970) The resistance to blood flow in the gastrocnemius of the dog during sustained and rhythmical isometric and isotonic contractions. Pflugers Arch 314:97–112
Hnik P, Zelena J (1961) Atypical spindles in reinnervated rat muscles. J Embryol Exp Morphol 9:456–467
Ho KW, Heusner WW, Van Huss J, Van Huss WD (1983) Postnatal muscle fibre histochemistry in the rat. J Embryol Exp Morphol 76:37–49
Ho KW, Roy RR, Tweedle CD, Heusner WW, Van Huss WD, Carrow RE (1980) Skeletal muscle fiber splitting with weight-lifting exercise in rats. Am J Anat 157:433–440
Ho RK, Ball EE, Goodman CS (1983) Muscle pioneers: Large mesodermal cells that erect a scaffold for developing muscles and motoneurones in grasshopper embryos. Nature 301:66–69
Hodgson JA (1983) The relationship between soleus and gastrocnemius muscle activity in conscious cats – a model for motor unit recruitment? J Physiol (Lond) 337:553–562
Hofmann WW (1980) Mechanisms of muscular hypertrophy. J Neurol Sci 45:205–216
Hofmann WW, Peacock JH (1973) Postjunctional changes induced by partial interruption of axoplasmic flow in motor nerves. Exp Neurol 41:345–356
Hofmann WW, Thesleff S (1972) Studies on the trophic influence of nerve on skeletal muscle. Eur J Pharmacol 20:256–260
Hoh JFY, Yeoh GPS (1979) Rabbit skeletal myosin isoenzymes from fetal, fast-twitch and slow-twitch muscles. Nature 280:321–323
Holley JA, Fahim MA (1983) Scanning electron microscopy of mouse muscle microvasculature. Anat Rec 205:109–117
Holloszy JO (1967) Biochemical adaptations in muscle. Effects of exercise on mitochondrial oxygen uptake and respiratory enzyme activity in skeletal muscle. J Biol Chem 242:2278–2282
Holloszy JO, Booth FW (1976) Biochemical adaptations to endurance exercise in muscle. Annu Rev Physiol 38:273–291
Holtzer H, Sanger JW (1972) Myogenesis: Old views rethought. In: Banker PQ, Przybylski RJ, Van Der Meulen JP, Victor M (eds) Research in muscle development and the muscle spindle. Excerpta Medica, Amsterdam, pp 122–133
Holtzer H, Marshall JM, Finck H (1957) An analysis of myogenesis by use of fluorescent antimyosin. J Biophys Biochem Cytol 3:705–724
Holtzer H, Abbott J, Lash J (1958) On the formation of multinucleated myotubes (Abstr). Anat Rec 131:567
Holtzer H, Sanger J, Ishikawa H (1971) Effects of colcemid and cytochalasin on myogenesis (Abstr). J Gen Physiol 57:245
Holtzer H, Rubinstein N, Chi J, Dienstman S, Biehl J (1974) The phenotypic heterogeneity of myogenic clones and what cells synthesize which molecules. In: Milhorat AT (ed) Exploratory concepts in muscular dystrophy II. Excerpta Medica, Amsterdam, pp 3–16

Holtzer H, Strahs K, Biehl J, Somlyo AP, Ishikawa H (1975a) Thick and thin filaments in postmitotic, mononucleated myoblasts. Science 188:943–945

Holtzer H, Croop J, Dienstman S, Ishikawa H, Somlyo AP (1975b) Effects of cytochalasin B and colcemide on myogenic cultures. Proc Natl Acad Sci USA 72:513–517

Holtzer H, Jones KW, Yaffe D (1975c) Research group on neuromuscular diseases. A report on various aspects of myogenic cell culture with particular reference to studies on the muscular dystrophies. J Neurol Sci 26:115–124

Holtzer H, Yeoh G, Rubinstein N, Chi J, Fellini S, Dienstman S (1977) A review of controversial issues in myogenesis. In: Nichols WW, Murphy DG (eds) Regulation of cell proliferation and differentiation. Plenum, New York, pp 87–104

Høncke P (1947) Investigations on the structure and function of living, isolated, cross striated muscle fibres of mammals. Acta Physiol Scand [Suppl 15] 48:1–230

Hoogenraad TU, Jennekens FGI, Tan KEWP (1979) Histochemical fibre types in human extraocular muscles, an investigation of inferior oblique muscle. Acta Neuropathol (Berl) 45:73–78

Hopkins GA (1975) The ultrastructure of the rabbit inferior oblique muscle. Micron 5:407–415

Hoppeler H, Lüthi P, Claassen H, Weibel ER, Howald H (1973) The ultrastructure of the normal human skeletal muscle. Pflugers Arch 344:217–232

Houk JC, Cornew RW, Stark L (1966) A model of adaption in amphibian spindle receptors. J Theor Biol 12:196–215

Howell JN (1969) A lesion of the transverse tubules of skeletal muscle. J Physiol (Lond) 201:515–533

Howell JN (1974) Intracellular binding of ruthenium red in frog skeletal muscle. J Cell Biol 62:242–247

Howell JN, Jenden DJ (1967) T-tubules of skeletal muscle: morphological alterations which interrupt excitation-contraction coupling (Abstr). Fed Proc 26:553

Hoyer E von (1963) Untersuchungen über Anzahl, Länge und Verteilung der Muskelspindeln in der Unterschenkelmuskulatur. Anat Anz 113:36–44

Hoyle G (1967) Diversity of striated muscle. Am Zoologist 7:435–449

Hoyle G (1969) Comparative aspects of muscle. Annu Rev Physiol 31:43–84

Hoyle G (1970) How is muscle turned on and off? Sci Am 222:85–93

Hoyle G, McAlear JH, Selverston A (1965) Mechanism of supercontraction in a striated muscle. J Cell Biol 26:621–640

Hoyle G, McNeill PA, Walcott B (1966) Nature of invaginating tubules in Felderstruktur muscle fibers of the garter snake. J Cell Biol 30:197–201

Hubbard BD, Lazarides E (1979) Copurification of actin and desmin from chicken smooth muscle and their copolymerization in vitro to intermediate filaments. J Cell Biol 80:166–182

Hubbard JI (1973) Microphysiology of vertebrate neuromuscular transmission. Physiol Rev 53:674–723

Hubbard JI, Kwanbunbumpen S (1968) Evidence for the vesicle hypothesis. J Physiol (Lond) 194:407–420

Hudlicka O, Tyler KR (1980) Importance of different patterns of frequency in the development of contractile properties and histochemical characteristics of fast skeletal muscle (Abstr). J Physiol (Lond) 301:10P–11P

Hudlicka O, Tyler KR, Srihari T, Heilig A, Pette D (1982) The effect of different patterns of long-term stimulation on contractile properties and myosin light chains in rabbit fast muscles. Pflugers Arch 393:164–170

Hughes BJ, Bowen JM, Campion DR, Bradley WE (1983a) Effect of denervation or castration on ultrastructural and histochemical properties of feline bulbocavernosus muscle. Acta Anat (Basel) 115:97–106

Hughes BJ, Bowen JM, Campion DR, Bradley WE (1983b) Effect of prepuberal castration on porcine bulbospongiosus muscle. Anat Embryol 168:51–58

Hui SW, Boni LT (1982) Lipidic particles and cubic phase lipids. Nature 296:175–176

Hulsmann WC, DeJong JW, Van Tol A (1968) Mitochondria with loosely and tightly coupled oxidative phosphorylation in skeletal muscle. Biochim Biophys Acta 162:292–293

Hunt CC, Ridge RMAP (1974) Intrafusal muscle-fibre contraction in long-capsule muscle spindles of snake (Abstr). J Anat 119:191–192

Hurko O, Walsh FS (1983) Human fetal muscle-specific antigen is restricted to regenerating myofibers in diseased adult muscle. Neurology (Cleveland) 33:737–743

Huxley AF (1964) Muscle. Annu Rev Physiol 26:131–152
Huxley AF (1974) Muscular contraction. J Physiol (Lond) 243:1–43
Huxley AF, Niedergerke R (1954) Interference microscopy of living muscle fibres. Nature 173:971–973
Huxley AF, Niedergerke R (1958) Measurement of the striations of isolated muscle fibres with the interference microscope. J Physiol (Lond) 144:403–425
Huxley AF, Peachey LD (1961) The maximum length for contraction in vertebrate striated muscle. J Physiol (Lond) 156:150–165
Huxley AF, Simmons RM (1971) Proposed mechanism of force generation in striated muscle. Nature 233:533–538
Huxley AF, Taylor RE (1958) Local activation of striated muscle fibres. J Physiol (Lond) 144:426–441
Huxley HE (1957) The double array of filaments in cross-striated muscle. J Biophys Biochem Cytol 3:631–648
Huxley HE (1963) Electron microscope studies on the structure of natural and synthetic protein filaments from striated muscle. J Mol Biol 7:281–308
Huxley HE (1964) Evidence for continuity between the central elements of the triads and extracellular space in frog sartorius muscle. Nature 202:1067–1071
Huxley HE (1968) Structural difference between resting and rigor muscle; evidence from intensity changes in the low-angle equatorial X-ray diagram. J Mol Biol 37:507–520
Huxley HE (1969) The mechanism of muscular contraction. Science 164:1356–1366
Huxley HE (1972) Molecular basis of contraction in cross-striated muscles. In: Bourne GH (ed) The structure and function of muscle, vol 1, 2nd edn. Academic Press, New York, pp 301–387
Huxley HE (1979) X-ray diffraction on contracting skeletal muscle. In: Sugi H, Pollack GH (eds) Cross-bridge mechanism in muscle contraction. University of Tokyo Press, Tokyo, pp 391–405
Huxley HE (1982) The mechanism of force production in muscle. In: Schotland DL (ed) Disorders of the motor unit. Wiley, New York, pp 1–11
Huxley HE, Brown W (1967) The low-angle X-ray diagram of vertebrate striated muscle and its behaviour during contraction and rigor. J Mol Biol 30:383–434
Huxley HE, Hanson J (1954) Changes in the cross-striations of muscle during contraction and stretch and their structural interpretation. Nature 173:973–976
Huxley HE, Hanson J (1957) Quantitative studies on the structure of cross-striated myofibrils. I. Investigations by interference microscopy. Biochim Biophys Acta 23:229–249
Huxley HE, Faruqi AR, Bordas J, Koch MHJ, Milch JR (1980) The use of synchrotron radiation in time-resolved X-ray diffraction studies of myosin layer-line reflections during muscle contraction. Nature 284:140–143
Huxley HE, Simmons RM, Faruqi AR, Kress M, Bordas J, Koch MHJ (1981) Millisecond time-resolved changes in X-ray reflections from contracting muscle during rapid mechanical transients, recorded using synchrotron radiation. Proc Natl Acad Sci USA 78:2297–2301
Huxley HE, Simmons RM, Faruqi AR, Kress M, Bordas J, Koch MHJ (1983) Changes in the X-ray reflections from contracting muscle during rapid mechanical transients and their structural implications. J Mol Biol 169:469–506
Ianuzzo D, Patel P, Chen V, O'Brien P, Williams C (1977) Thyroidal trophic influence on skeletal muscle myosin. Nature 270:74–76
Ikai M, Fukunaga T (1968) Calculation of muscle strength per unit cross-sectional area of human muscle by means of ultrasonic measurement. Int Z Angew Physiol 26:26–32
Ikemoto N (1982) Structure and function of the calcium pump protein of sarcoplasmic reticulum. Annu Rev Physiol 44:297–317
Ikemoto N, Sreter FA, Gergely J (1971) Structural features of the surface of the vesicles of FSR – lack of functional role in Ca^{2+} uptake and ATPase activity. Arch Biochem Biophys 147:571–582
Illert M, Fritz N, Aschoff A, Holländer H (1982) Fluorescent compounds as retrograde tracers compared with horseradish peroxidase (HRP). II. A parametric study in the peripheral motor system of the cat. J Neurosci Methods 6:199–218
Inestrosa NC, Fernandez HL (1976) Muscle enzymatic changes induced by blockage of axoplasmic transport. J Neurophysiol 39:1236–1245
Ingels NP, Thompson NP (1966) An elektrokinematic theory of muscle contraction. Nature 211:1032–1035
Ingjer F (1979) Capillary supply and mitochondrial content of different skeletal muscle fiber types

in untrained and endurance-trained men. A histochemical and ultrastructural study. Eur J Applied Physiol 40:197–209
Ingjer F, Brodal P (1978) Capillary supply of skeletal muscle fibers in untrained and endurance-trained women. Eur J Applied Physiol 38:291–299
Ishikawa H (1968) Formation of elaborate networks of T-system tubules in cultured skeletal muscle with special reference to the T-system formation. J Cell Biol 38:51–66
Ishikawa H (1970) Satellite cells in developing muscle and tissue culture. In: Mauro A, Shafiq SA, Milhorat AT (eds) Regeneration of striated muscle, and myogenesis. Excerpta Medica, Amsterdam, pp 167–179
Ishikawa H (1980) Fine structure of the striated myofibril as revealed by high-voltage electron microscopy. Biomed Res 1:15–23
Ishikawa H, Bischoff R, Holtzer H (1968) Mitosis and intermediate-sized filaments in developing skeletal muscle. J Cell Biol 38:538–555
Ishikawa H, Bischoff R, Holtzer H (1969) Formation of arrowhead complexes with heavy meromyosin in a variety of cell types. J Cell Biol 43:312–328
Ito S (1965) The enteric surface coat on cat intestinal microvilli. J Cell Biol 27:475–491
Iwayama T (1966) Electron microscopic study on cholinesterase at the neuromuscular junction. Okajimas Folia Anat Jap 42:329–353
Iwayama T (1969) Relation of regenerating nerve terminals to original endplates. Nature 224:81–82
Jablecki CK, Heuser JE, Kaufman S (1973) Autoradiographic localization of new RNA synthesis in hypertrophying skeletal muscle. J Cell Biol 57:743–759
Jacob M, Christ B, Jacob HJ (1978) On the migration of myogenic stem cells into the prospective wing region of chick embryos. A scanning and transmission electron microscope study. Anat Embryol (Berl) 153:179–193
Jacob M, Christ B, Jacob HJ (1979) The migration of myogenic cells from the somites into the leg region of avian embryos. An ultrastructural study. Anat Embryol (Berl) 157:291–309
Jacob M, Jacob HJ, Wachtler F, Christ B (1984) Ontogeny of avian extrinsic ocular muscles. I. A light- and electron-microscopic study. Cell Tissue Res 237:549–557
Jacob M, Lentz TL (1979) Localization of acetylcholine receptors by means of horseradish peroxidase-α-bungarotoxin during formation and development of the neuromuscular junction in the chick embryo. J Cell Biol 82:195–211
Jagendorf-Elfvin M (1967) Ultrastructure of the contraction-relaxation cycle of glycerinated rabbit psoas muscle. J Ultrastruct Res 17:379–400
Jagendorf-Elfvin M (1970) Localized deposition of lead phosphate precipitate in glycerinated rabbit psoas muscle during ATP-induced contraction. Tissue Cell 2:311–326
Jahromi SS, Charlton MP (1979) Transverse sarcomere splitting. A possible means of longitudinal growth in crab muscles. J Cell Biol 80:736–742
Jakus MA, Hall CE (1947) Studies of actin and myosin. J Biol Chem 167:705–714
James DW, Tresman RL (1968) *De novo* formation of neuro-muscular junctions in tissue culture. Nature 220:384–385
James NT (1971a) The histochemical demonstration of three types of intrafusal fibre in rat muscle spindles. Histochem J 3:457–462
James NT (1971b) The histochemical demonstration of mucopolysaccharide in the lymph space of muscle spindles. J Anat 110:163–164
James NT (1972a) The histochemical properties of muscle fibres and the formation of sub-fasciculi in the tibialis anterior muscle of the rabbit. J Neurol Sci 15:429–437
James NT (1972b) A quantitative study of the clumping of muscle fibre types in skeletal muscles. J Neurol Sci 17:41–44
James NT (1973) Compensatory hypertrophy in the extensor digitorum longus muscle of the rat. J Anat 116:57–65
James NT, Meek GA (1973) An electron microscopical study of avian muscle spindles. J Ultrastruct Res 43:193–204
James NT, Meek GA (1979) Stereological analyses of the structure of mitochondria in pigeon skeletal muscle. Cell Tissue Res 202:493–503
Jami L, Murthy KSK, Petit J (1980) Proportion of skeletofusimotor axons (β-axons) in the motor supply to the cat peroneus tertius muscle. Neurosci Lett [Suppl] 5:S225
Jami L, Murthy KSK, Petit J, Zytnicki D (1982) Distribution of physiological types of motor units in the cat peroneus tertius muscle. Exp Brain Res 48:177–184

Jansen JKS, Van Essen DC (1975) Re-innervation of rat skeletal muscle in the presence of α-bungarotoxin. J Physiol (Lond) 250:651–667

Jansen JKS, Van Essen DC, Brown MC (1976) Formation and elimination of synapses in skeletal muscles of rat. Cold Spring Harbor Symp Quant Biol 40:425–434

Jarcho LW, Berman B, Dowben RM, Lilienthal Jr, JL (1954) Site of origin and velocity of conduction of fibrillary potentials in denervated skeletal muscle. Am J Physiol 178:129–134

Jasper D (1967) Body muscles of the lamprey. J Cell Biol 32:219–227

Jaweed MM, Herbison GJ, Ditunno JF (1977) Myosin ATPase activity after strengthening exercise. J Anat 124:371–381

Jerusalem F, Engel AG, Gomez MR (1974) Duchenne dystrophy. I. Morphometric study of the muscle microvasculature. Brain 97:115–122

Jerusalem F, Engel AG, Peterson HA (1975) Human muscle fiber fine structure: morphometric data on controls. Neurology (Minneapolis) 25:127–134

Jessell TM, Siegel RE, Fischbach GD (1979) Induction of acetylcholine receptors on cultured skeletal muscle by a factor extracted from brain and spinal cord. Proc Natl Acad Sci USA 76:5397–5401

Jilka RL, Martonosi AN, Tillack TW (1975) Effect of purified ($Mg^{2+}+Ca^{2+}$) activated ATPase of sarcoplasmic reticulum upon the passive Ca^{2+} permeability and ultrastructure of phospholipid vesicles. J Biol Chem 250:7511–7524

Jirmanova I (1975) Ultrastructure of motor end-plates during pharmacologically-induced degeneration and subsequent regeneration of skeletal muscle. J Neurocytol 4:141–155

Jirmanova I, Thesleff S (1972) Ultrastructural study of experimental muscle degeneration and regeneration in the adult rat. Z Zellforsch Mikrosk Anat 131:77–97

Jirmanova I, Zelena J (1970) Effect of denervation and tenotomy of slow and fast muscles of the chicken. Z Zellforsch Mikrosk Anat 106:333–347

Jobsis FF, O'Connor MJ (1966) Calcium release and reabsorption in the sartorius muscle of the toad. Biochem Biophys Res Commun 25:246–252

Johnson DJ, Smith LA, Eldred E, Edgerton VR (1982) Exercise-induced changes of biochemical, histochemical, and contractile properties of muscle in cordotomized kittens. Exp Neurol 76:414–427

Johnson MA, Polgar J, Weightman D, Appleton D (1973) Data on the distribution of fibre types in thirty-six human muscles. An autopsy study. J Neurol Sci 18:111–129

Johnson MA, Mastaglia FL, Montgomery AG, Pope B, Weeds AG (1980) Changes in myosin light chains in the rat soleus after thyroidectomy. FEBS Lett 110:230–235

Jolesz F, Sreter FA (1981) Development, innervation, and activity-pattern induced changes in skeletal muscle. Annu Rev Physiol 43:531–552

Jones EG (1966) The innervation of muscle spindles in the Australian opossum, *Trichosurus vulpecula*, with special reference to the motor nerve endings. J Anat 100:733–759

Jones GH (1982) Protein synthesis in bupivacaine (Marcaine)-treated, regenerating skeletal muscle. Muscle Nerve 5:281–290

Jones PH (1979) Implantation of cultured regenerate muscle cells into adult rat muscle. Exp Neurol 66:602–610

Jorgensen AO, Kalnins V, MacLennan DH (1979) Localization of sarcoplasmic reticulum proteins in rat skeletal muscle by immunofluorescence. J Cell Biol 80:372–384

Jorgensen AO, Shen ACY, MacLennan DH, Tokuyasu KT (1982) Ultrastructural localization of the Ca^{2+} and Mg^{2+}-dependent ATPase of sarcoplasmic reticulum in rat skeletal muscle by immunoferritin labeling of ultrathin frozen sections. J Cell Biol 92:409–416

Jorgensen AO, Shen ACY, Campbell KP, MacLennan DH (1983) Ultrastructural localization of calsequestrin in rat skeletal muscle by immunoferritin labeling of ultrathin frozen sections. J Cell Biol 97:1573–1581

Joseph MP, Guinan Jr JJ, Fullerton BC, Norris BE, Kiang NYS (1985) Number and distribution of stapedius motoneurons in cats. J Comp Neurol 232:43–54

Kaczmarski F (1969) The fine structure of extraocular muscles of the lizard, *Lacerta agilis* L. Z Mikrosk Anat Forsch 80:517–531

Kaczmarski F (1970a) The fine structure of extraocular muscles of the tree sparrow, *Passer montanus* L. Z Mikrosk Anat Forsch 82:523–536

Kaczmarski F (1970b) The fine structure of extraocular muscles of the bank vole, *Clethrionomys glareolus* Schr. Acta Anat 77:570–582

Kahn EB, Simpson Jr SB (1974) Satellite cells in mature uninjured skeletal muscle of the lizard tail. Dev Biol 37:219–223
Kalderon N, Gilula NB (1979) Membrane events involved in myoblast fusion. J Cell Biol 81:411–425
Kalderon N, Epstein ML, Gilula NB (1977) Cell-to-cell communication and myogenesis. J Cell Biol 75:788–806
Kamieniecka Z, Schmalbruch H (1980) Neuromuscular disorders with abnormal muscle mitochondria. Int Rev Cytol 65:321–357
Kanda K, Burke RE, Walmsley B (1977) Differential control of fast and slow twitch motor units in the decerebrate cat. Exp Brain Res 29:57–74
Karlsson U, Andersson-Cedergren E (1966) Motor myoneural junctions on frog intrafusal muscle fibers. J Ultrastruct Res 14:191–211
Karlsson U, Andersson-Cedergren E, Ottoson D (1966) Cellular organization of the frog muscle spindle as revealed by serial sections for electron microscopy. J Ultrastruct Res 14:1–35
Karpati G, Engel WK (1968) Correlative histochemical study of skeletal muscle after suprasegmental denervation, peripheral nerve section and skeletal fixation. Neurology (Minneapolis) 18:681–692
Katayama E, Nonomura Y (1979a) Quantitative analysis of the mechanism of negative staining with native collagen fibrils and polar tropomyosin paracrystals. J Biochem (Tokyo) 86:1495–1509
Katayama E, Nonomura Y (1979b) Electron microscopic analysis of tropomyosin paracrystals. J Biochem (Tokyo) 86:1511–1522
Katchburian E, Burgess AMC, Johnson FR (1973) The effect of tilting ultrathin sections on the image of the Z-disc of skeletal muscle. Experientia 29:1020–1022
Katz B, Miledi R (1964) Further observations on the distribution of acetylcholine-reactive sites in skeletal muscle. J Physiol (Lond) 170:379–388
Katz B, Miledi R (1972) The statistical nature of the acetylcholine potential and its molecular components. J Physiol (Lond) 224:665–699
Kaufmann P (1981) Einfluß der Fixationstechnik auf morphometrische Daten (Abstr). Acta Anat (Basel) 111:73
Kean CJC, Lewis DM, McGarrick JD (1974) Dynamic properties of denervated fast and slow twitch muscle of the cat. J Physiol (Lond) 237:103–113
Keene MFL (1961) Muscle spindles in human laryngeal muscles. J Anat 95:25–29
Keeter JS, Pappas GD, Model PG (1975) Inter- and intramyotomal gap junctions in the axolotl embryo. Dev Biol 45:21–33
Keller LR, Emerson CP Jr (1980) Synthesis of adult myosin light chains by embryonic muscle cultures. Proc Natl Acad Sci USA 77:1020–1024
Keller LR, Emerson CP Jr (1982) Development of fiber type specificity in embryonic avian skeletal muscle. In: Schotland DL (ed) Disorders of the motor unit. Wiley, New York, pp 863–875
Kelly AM (1971) Sarcoplasmic reticulum and T tubules in differentiating rat skeletal muscle. J Cell Biol 49:335–344
Kelly AM (1978a) Satellite cells and myofiber growth in the rat soleus and extensor digitorum longus muscles. Dev Biol 65:1–10
Kelly AM (1978b) Perisynaptic satellite cells in the developing and mature rat soleus muscle. Anat Rec 190:891–903
Kelly AM (1979) Variations in satellite cell distribution in developing and mature muscles of the rat. In: Mauro A (ed) Muscle regeneration. Raven Press, New York, pp 167–176
Kelly AM (1980) T tubules in neonatal rat soleus and extensor digitorum longus muscles. Dev Biol 80:501–505
Kelly AM, Rubinstein NA (1980) Why are fetal muscles slow? Nature 288:266–269
Kelly AM, Schotland DL (1972) The evolution of the "checkerboard" in a rat muscle. In: Banker BQ, Przybylski RJ, Van Der Meulen JP, Victor M (eds) Research in muscle development and the muscle spindle. Excerpta Medica, Amsterdam, pp 32–52
Kelly AM, Zacks SI (1969a) The histogenesis of rat intercostal muscle. J Cell Biol 42:135–153
Kelly AM, Zacks SI (1969b) The fine structure of motor endplate morphogenesis. J Cell Biol 42:154–169
Kelly DE (1967) Models of muscle Z-band fine structure based on a looping filament configuration. J Cell Biol 34:827–840
Kelly DE (1969) Myofibrillogenesis and Z-band differentiation. Anat Rec 163:403–425
Kelly DE, Cahill MA (1972) Filamentous and matrix components of skeletal muscle Z-disks. Anat Rec 172:623–642

Kelly DE, Kuda AM (1979) Subunits of the triadic junction in fast skeletal muscle as revealed by freeze-fracture. J Ultrastruct Res 68:220–233
Kemplay S, Stolkin C (1980) End plate classification and spontaneous sprouting in the sternocostalis muscle of the rat: A new whole mount preparation. Cell Tissue Res 212:333–339
Kendrick-Jones J, Lehman W, Szent-Györgyi AG (1970) Regulation in molluscan muscles. J Mol Biol 54:313–326
Kendrick-Jones J, Scholey JM (1981) Myosin-linked regulatory systems. J Muscle Res Cell Motil 2:347–372
Kensler RW, Stewart M (1983) Frog skeletal muscle thick filaments are three-stranded. J Cell Biol 96:1797–1802
Kereshi S, McComas A, Kowalchuk N, Stuart A (1983) Absence of fibrillation after muscle injury. Exp Neurol 80:645–651
Kern R (1965) A comparative pharmacologic-histologic study of slow and twitch fibers in the superior rectus muscle of the rabbit. Invest Ophthalmol 4:901–910
Kernell D, Eerbeek O, Verhey BA (1983a) Motor unit categorization on basis of contractile properties: An experimental analysis of the composition of the cat's m. peroneus longus. Exp Brain Res 50:211–219
Kernell D, Eerbeek O, Verhey BA (1983b) Relation between isometric force and stimulus rate in cat's hindlimb motor units of different twitch contraction time. Exp Brain Res 50:220–227
Keski-Oja J, Sen A, Todaro GJ (1980) Direct association of fibronectin and actin molecules in vitro. J Cell Biol 85:527–533
Ketelsen U-P (1974) The plasma membrane of skeletal muscle cells in the pathological state. Freeze etch studies. In: Sanders JV, Goddchild DJ (eds) Eighth international congress on electron microscopy, vol II. The Australian Academy of Science, Canberra, pp 506–507
Ketelsen U-P (1975) The plasma membrane of human skeletal muscle cells in the pathological state. Freeze-etch studies. In: Bradley WG, Gardner-Medwin D, Walton JN (eds) Recent advances in myology. Excerpta Medica, Amsterdam, pp 446–454
Ketelsen U-P (1980) Quantitative freeze fracture studies of human skeletal muscle cell membranes under normal and pathological conditions. In: Angelini C, Danieli GA, Fontanari A (eds) Muscular dystrophy research: advances and new trends. Excerpta Medica, Amsterdam, pp 79–87
Kidokoro Y (1980) Developmental changes of spontaneous synaptic potential properties in the rat neuromuscular contact formed in culture. Dev Biol 78:231–241
Kidokoro Y, Yeh E (1981) Synaptic contacts between embryonic *Xenopus* neurons and myotubes formed from a rat skeletal muscle cell line. Dev Biol 86:12–18
Kielley WW, Meyerhof O (1948) A new magnesium-activated adenosintriphosphatase from muscle. J Biol Chem 174:387–388
Kiessling A, Wohlrab F (1969) Histochemische Differenzierung von Muskelfasertypen in der Skelettmuskulatur des Frosches. Anat Anz 125:131–144
Kilarski W (1964) The organization of the sarcoplasmic reticulum in skeletal muscles of fishes. Part I. The sarcoplasmic reticulum of striated muscles of the swim-bladder in the burbot (*Lota lota* L.). Acta Biol Cracoviensia, Series Zoologia 7:161–168
Kilarski W (1965) The organization of the sarcoplasmic reticulum in skeletal muscles of fishes. Part II. The perch (*Perca fluviatilis* L.). Acta Biol Cracoviensia, Series Zoologia 8:51–57
Kilarski W (1966) Organization of the sarcoplasmic reticulum in skeletal muscles of fishes. III. Pike (*Esox lucius* L.). Bulletin de l'Academie Polonaise des Sciences Cl. II. Serie des sciences biologiques 14:575–580
Kilarski W (1967) The fine structure of striated muscles in teleosts. Z Zellforsch Mikrosk Anat 79:562–580
Kilarski W, Bigaj J (1969) Organization and fine structure of extraocular muscles in *Carassius* and *Rana*. Z Zellforsch Mikrosk Anat 94:194–204
King MV, Young M (1972) Evidence for flexibility of the helical rod section of the myosin molecule. J Mol Biol 65:519–523
Klein-Ogus C, Harris JB (1983) Preliminary observations of satellite cells in undamaged fibres of the rat soleus muscle assaulted by a snake-venom toxin. Cell Tissue Res 230:617–676
Klug A, Franklin RE, Humphreys-Owen SPF (1959) The crystal structure of tipula iridescent virus as determined by Bragg reflection of visible light. Biochim Biophys Acta 32:203–219
Knappeis GG, Carlsen F (1962) The ultrastructure of the Z disc in skeletal muscle. J Cell Biol 13:323–335

Knappeis GG, Carlsen F (1968) The ultrastructure of the M line in skeletal muscle. J Cell Biol 38:202–211
Knutton S, Bachi T (1980) The role of cell swelling and haemolysis in Sendai virus-induced cell fusion and in the diffusion of incorporated viral antigens. J Cell Sci 42:153–167
Koelle GB, Friedenwald JS (1949) A histochemical method for localizing cholinesterase activity. Proc Soc Exp Biol Med 70:617–622
Koizumi T (1974) Turnover rates of structural proteins of rabbit skeletal muscle. J Biochem (Tokyo) 76:431–439
Kölliker A (1850) Mikroskopische Anatomie oder Gewebelehre des Menschen, Bd II. Engelmann, Leipzig, pp 253–255
Kölliker A (1863) Handbuch der Gewebelehre des Menschen. Engelmann, Leipzig, pp 210–214
Konieczny SF, Lawrence JB, Coleman JR (1983) Analysis of muscle protein expression in polyethylene glycol-induced chicken: rat myoblast heterokaryons. J Cell Biol 97:1348–1355
Konigsberg IR (1963) Clonal analysis of myogenesis. Its relevance to the general problem of the stability of cell-type in cultured animal cell is discussed. Science 140:1273–1284
Konigsberg IR (1971) Diffusion-mediated control of myoblast fusion. Dev Biol 26:133–152
Konigsberg IR (1979) Regeneration of single muscle fibers in culture and in vivo. In: Mauro A (ed) Muscle regeneration. Raven Press, New York, pp 41–56
Konigsberg IR, McElvain N, Tootle M, Herrmann H (1960) The dissociability of deoxyribonucleic acid synthesis from the development of multinuclearity of muscle cells in culture. J Biophys Biochem Cytol 8:333–344
Konigsberg IR, Sollmann PA, Mixter LO (1978) The duration of the terminal G_1 of fusing myoblasts. Dev Biol 63:11–26
Kono T, Kakuma F, Homma M, Fukuda S (1964) The electron-microscopic structure and chemical composition of the isolated sarcolemma of the rat skeletal muscle cell. Biochim Biophys Acta 88:155–176
Kordylewski L (1979a) Morphology of muscle fibres in amphibian submandibular muscle. Z Mikrosk Anat Forsch 93:225–243
Kordylewski L (1979b) Morphology of motor end-plate and its size relation to the muscle fibre size in the amphibian submandibular muscle. Z Mikrosk Anat Forsch 93:1038–1050
Korneliussen H (1972) Ultrastructure of normal and stimulated motor endplates. With comments on the origin and fate of synaptic vesicles. Z Zellforsch Mikrosk Anat 130:28–57
Korneliussen H (1973) Ultrastructure of myotendinous junctions in Myxine and rat. Specializations between the plasma membrane and the lamina densa. Z Anat Entwicklungsgesch 142:91–101
Korneliussen H (1977) Tubules invaginating from the sarcolemma in the subneural region of muscle fibers. Cell Tissue Res 181:73–79
Korneliussen H, Jansen JKS (1976) Morphological aspects of the elimination of polyneuronal innervation of skeletal muscle fibres in newborn rats. J Neurocytol 5:591–604
Korneliussen H, Waerhaug O (1973) Three morphological types of motor nerve terminals in the rat diaphragm, and their possible innervation of different muscle fiber types. Z Anat Entwicklungsgesch 140:73–84
Körner G (1960) Untersuchungen über Zahl, Anordnung und Länge der Muskelspindeln in einigen Schulter-, den Oberarmmuskeln und im Musculus sternalis des Menschen. Anat Anz 108:99–103
Kovacs L, Rios E, Schneider MF (1983) Measurement and modification of free calcium transients in frog skeletal muscle fibres by a metallochromic indicator dye. J Physiol (Lond) 343:161–196
Kozeka K, Ontell M (1981) The three-dimensional cytoarchitecture of developing murine muscle spindles. Dev Biol 87:133–147
Krarup C (1977) Electrical and mechanical responses in the platysma and in the adductor pollicis muscle: in normal subjects. J Neurol Neurosurg Psychiatry 40:234–240
Krause W (1877) Die Entladungshypothese und die motorischen Endplatten. Arch Mikrosk Anat 13:170–179
Kreis TE, Birchmeier W (1980) Stress fiber sarcomeres of fibroblasts are contractile. Cell 22:555–561
Kreutziger GO (1968) Freeze-etching of intercellular junctions of mouse liver. Proceedings of the 26th Meeting of the Electron Microscope Society of America. Baton Rouge, Claitor's Publ Div, p 234
Kristensson K, Olsson Y (1971) Uptake and retrograde axonal transport of peroxidase in hypoglossal neurons. Electron microscopical localization in the neuronal perikaryon. Acta Neuropathol 19:1–9

Krnjevic K, Miledi R (1958) Motor units in the rat diaphragm. J Physiol (Lond) 140:427–439
Krolenko SA (1969) Changes in the T-system of muscle fibres under the influence of influx and efflux of glycerol. Nature 221:966–968
Krüger P (1929) Über einen möglichen Zusammenhang zwischen Struktur, Funktion und chemischer Beschaffenheit der Muskeln. Biol Zentralbl 49:616–622
Krüger P (1950) Untersuchungen am Vogelflügel. Zool Anz 145:445–460
Krüger P (1952) Tetanus und Tonus der quergestreiften Skelettmuskeln der Wirbeltiere und des Menschen. Geest und Portig, Leipzig
Kryvi H (1975) The structure of the myosatellite cells in axial muscles of the shark *Galeus melastomus*. Anat Embryol (Berl) 147:35–44
Kucera J (1977) Intrafusal muscle fiber histochemistry following its motor reinnervation. J Histochem Cytochem 25:1260–1263
Kucera J (1980) Histochemical study of long nuclear chain fibers in the cat muscle spindle. Anat Rec 198:567–580
Kucera J (1981) The occurrence of "mixed" nuclear bag intrafusal fibers in the cat muscle spindle. Histochemistry 72:123–131
Kucera J (1985) Histological study of motor innervation of nuclear bag_1 intrafusal muscle fibers in the cat. J Comp Neurol 232:331–346
Kucera J, Dorovini-Zis K (1979) Types of human intrafusal muscle fibers. Muscle Nerve 2:437–451
Kuffler DP, Thompson W, Jansen JKS (1980) The fate of foreign endplates in cross-innervated rat soleus muscle. Proc R Soc Lond [Biol] 208:189–222
Kuffler SW, Vaughan Williams EM (1953) Small-nerve junctional potentials. The distribution of small motor nerves to frog skeletal muscle, and the membrane characteristics of the fibres they innervate. J Physiol (Lond) 121:289–317
Kuffler SW, Yoshikami D (1975) The distribution of acetylcholine sensitivity at the post-synaptic membrane of vertebrate skeletal twitch muscles: Iontophoretic mapping in the micron range. J Physiol (Lond) 244:703–730
Kugelberg E (1973a) Properties of the rat hind-limb motor units. In: Desmedt JE (ed) New developments in electromyography and clinical neurophysiology, vol 1. Karger, Basel, pp 2–13
Kugelberg E (1973b) Histochemical composition, contraction speed and fatiguability of rat soleus motor units. J Neurol Sci 20:177–198
Kugelberg E (1976) Adaptive transformation of rat soleus motor units during growth (Histochemistry and contraction speed). J Neurol Sci 27:269–289
Kugelberg E, Edström L (1968) Differential histochemical effects of muscle contractions on phosphorylase and glycogen in various types of fibres: relation to fatigue. J Neurol Neurosurg Psychiatry 31:415–423
Kugelberg E, Lindegren B (1979) Transmission and contraction fatigue of rat motor units in relation to succinate dehydrogenase activity of motor unit fibres. J Physiol (Lond) 288:285–300
Kugelberg E, Thornell L-E (1983) Contraction time, histochemical type, and terminal cisternae volume of rat motor units. Muscle Nerve 6:149–153
Kullberg RW, Lentz TL, Cohen MW (1977) Development of the myotomal neuromuscular junction in *Xenopus laevis*: an electrophysiological and fine-structural study. Dev Biol 60:101–129
Kundrat E, Pepe FA (1971) The M band. J Cell Biol 48:340–347
Kuner JM, Beyer RE (1970) An ultrastructural study of isolated rat skeletal muscle mitochondria in various metabolic states. J Membr Biol 2:71–84
Kupfer C (1960) Motor innervation of extraocular muscle. J Physiol (Lond) 153:522–526
Labaw LW, Wyckoff RWG (1957) The structure of southern bean mosaic virus protein crystals. Arch Biochem Biophys 67:225–236
Lagunoff D (1973) Membrane fusion during mast cell secretion. J Cell Biol 57:252–259
Lamvik MK (1978) Muscle thick filament mass measured by electron scattering. J Mol Biol 122:55–68
Lance-Jones C, Landmesser L (1980a) Motoneurone projection patterns in embryonic chick limbs following partial deletions of the spinal cord. J Physiol (Lond) 302:559–580
Lance-Jones C, Landmesser L (1980b) Motoneurone projection patterns in the chick hind limb following early partial reversals of the spinal cord. J Physiol (Lond) 302:581–602
Landis DMD, Reese TS (1974) Arrays of particles in freeze-fractured astrocytic membranes. J Cell Biol 60:316–320
Landon DN (1966) Electron microscopy of muscle spindles. In: Andrew BL (ed) Control and innervation of skeletal muscle. University of St Andrews, Dundee, pp 96–111

Landon DN (1970a) The influence of fixation upon the fine structure of the Z-disk of rat striated muscle. J Cell Sci 6:257–276
Landon DN (1970b) Change in Z-disk structure with muscular contraction (Abstr). J Physiol (Lond) 211:44P–45P
Landon DN (1970c) Observations on the morphogenesis of rat skeletal muscle (Abstr). J Anat 107:385–387
Landon DN (1971) A quantitative study of some fine structural features of developing myotubes in the rat (Abstr). J Anat 110:170–171
Landon DN (1982) Skeletal muscle – normal morphology, development and innervation. In: Mastaglia FL, Walton Sir John (eds) Skeletal muscle pathology. Churchill Livingstone, Edinburgh, pp 1–87
Lännergren J (1975) Structure and function of twitch and slow fibres in amphibian skeletal muscle. In: Lennerstrand G, Bach-Y-Rita P (eds) Basic mechanisms of ocular motility and their clinical implications. Pergamon Press, Oxford New York, pp 63–84
Lännergren J (1979) An intermediate type of muscle fibre in *Xenopus laevis*. Nature 279:254–256
Lännergren J, Hoh JFY (1984) Myosin isoenzymes in single muscle fibres of *Xenopus laevis*: analysis of five different functional types. Proc R Soc Lond [Biol] 222:401–408
Lännergren J, Smith RS (1966) Types of muscle fibres in toad skeletal muscle. Acta Physiol Scand 68:263–274
Larson PF, Fulthorpe JJ, Hudgson P (1973) The alignment of polysomes along myosin filaments in developing myofibrils. J Anat 116:327–334
Larsson L, Forssberg A (1980) Morphological muscle characteristics in rowers. Can J Appl Sport Sci 5:239–244
Lash JW, Holtzer H, Swift H (1957) Regeneration of mature skeletal muscle. Anat Rec 128:679–698
Laurent GJ, Millward DJ (1980) Protein turnover during skeletal muscle hypertrophy. Fed Proc 39:42–47
Lawson D, Raff MC, Gomperts B, Fewtrell C, Gilula NB (1977) Molecular events during membrane fusion. A study of exocytosis in rat peritoneal mast cells. J Cell Biol 72:242–259
Lazarides E (1980) Intermediate filaments as mechanical integrators of cellular space. Nature 283:249–256
Lazarides E (1982) Intermediate filaments: a chemically heterogeneous, developmentally regulated class of proteins. Annu Rev Biochem 51:219–250
Lazarides E, Granger BL (1978) Fluorescent localization of membrane sites in glycerinated chicken skeletal muscle fibers and the relationship of these sites to the protein composition of the Z disc. Proc Natl Acad Sci USA 75:3683–3687
Lazarides E, Hubbard BD (1976) Immunological characterization of the subunit of the 100 Å filaments from muscle cells. Proc Natl Acad Sci USA 73:4344–4348
Lee JC (1965) Electron microscope observations on myogenic free cells of denervated skeletal muscle. Exp Neurol 12:123–135
Lee RE, Rash JE, Poulos AC (1983) Changes in sarcolemmal caveolae in human muscular dystrophy: artifact of fixation (Abstr). J Cell Biol 97:461a
Leeson CR, Leeson TS (1969) Mitochondrial organization in skeletal muscle of the rat soft palate. J Anat 105:363–370
Leinonen H, Juntunen J, Somer H, Rapola J (1979) Capillary circulation and morphology in Duchenne muscular dystrophy. Eur Neurol 18:249–255
Lennerstrand G (1974a) Electrical activity and isometric tension in motor units of the cat's inferior oblique muscle. Acta Physiol Scand 91:458–474
Lennerstrand G (1974b) Mechanical studies on the retractor bulbi muscle and its motor units in the cat. J Physiol (Lond) 236:43–55
Lennerstrand G (1980) Histochemical studies on the inferior oblique muscle of Siamese cats and domestic cats with unilateral lid suture. Exp Eye Res 30:619–629
Lennerstrand G, Hanson J (1978a) The postnatal development of the inferior oblique muscle of the cat. I. Isometric twitch and tetanic properties. Acta Physiol Scand 103:132–143
Lennerstrand G, Hanson J (1978b) The postnatal development of the inferior oblique muscle of the cat. II. Effects of repetitive stimulation on isometric tension responses. Acta Physiol Scand 103:144–153

Lennon VA, Peterson S, Schubert D (1979) Neuroectoderm markers retained in phenotypical skeletal muscle cells arising from a glial cell line. Nature 281:586–588

Lentz TL (1969) Cytological studies of muscle dedifferentiation and differentiation during limb regeneration of the newt *Triturus*. Am J Anat 124:447–480

Lentz TL, Mazurkiewicz JE, Rosenthal J (1977) Cytochemical localization of acetylcholine receptors at the neuromuscular junction by means of horseradish peroxidase-labeled α-bungarotoxin. Brain Res 132:423–442

Letinsky MS, Decino PA (1980) Histological staining of pre- and postsynaptic components of amphibian neuromuscular junctions. J Neurocytol 9:305–320

Letinsky MS, Morrison-Graham K (1980) Structure of developing frog neuromuscular junctions. J Neurocytol 9:321–342

Lewis DM (1972) The effect of denervation on the mechanical and electrical responses of fast and slow mammalian twitch muscle. J Physiol (Lond) 222:51–75

Lewis DM, Parry DJ, Rowlerson A (1982a) Isometric contractions of motor units and immunohistochemistry of mouse soleus muscle. J Physiol (Lond) 325:393–401

Lewis DM, Rowlerson A, Webb SN (1982b) Motor units and immunohistochemistry of cat soleus muscle after long periods of cross-reinnervation. J Physiol (Lond) 325:403–418

Lewis J, Chevallier A, Kieny M, Wolpert L (1981) Muscle nerve branches do not develop in chick wings devoid of muscle. J Embryol Exp Morphol 64:211–232

Lewis WH, Lewis MR (1917) Behavior of cross striated muscle in tissue cultures. Am J Anat 22:169–194

Lexell J, Henriksson-Larsén K, Sjöström M (1983a) Distribution of different fibre types in human skeletal muscles. 2. A study of cross-sections of whole m. vastus lateralis. Acta Physiol Scand 117:115–122

Lexell J, Henriksson-Larsén K, Winblad B, Sjöström M (1983b) Distribution of different fiber types in human skeletal muscles: Effects of aging studied in whole muscle cross sections. Muscle Nerve 6:588–595

Libby P, Goldberg AL (1982) The use of protease inhibitors to investigate pathways of protein breakdown in muscle. In: Schotland DL (ed) Disorders of the motor unit. Wiley, New York, pp 683–695

Libelius R (1975) Evidence for endocytotic uptake of cobra neurotoxin in mouse skeletal muscle. J Neurol Transmission 37:61–71

Libelius R (1978) Endocytosis and the lysosome system in skeletal muscle. (Thesis). University of Lund, Lund

Libelius R, Jirmanova I, Lundquist I, Thesleff S (1978) Increased endocytosis with lysosomal activation in skeletal muscle of dystrophic mouse. J Neuropath Exp Neurol 37:387–400

Libelius R, Josefsson J-O, Lundquist I (1979) Endocytosis in chronically denervated mouse skeletal muscle. A biochemical and ultrastructural study with horseradish peroxidase. Neuroscience 4:283–292

Liddell EGT, Sherrington CS (1925) Recruitment and some other features of reflex inhibition. Proc R Soc Lond [Biol] 97:488–518

Lie SO, Schofield B (1973) Inactivation of lysosomal function in normal cultured human fibroblasts by chloroquine. Biochem Pharmacol 22:3109–3114

Lieber RL, Baskin RJ (1983) Intersarcomere dynamics of single muscle fibers during fixed-end tetani. J Gen Physiol 82:347–364

Lim RW, Hauschka SD (1984) A rapid decrease in epidermal growth factor-binding capacity accompanies the terminal differentiation of mouse myoblasts in vitro. J Cell Biol 98:739–747

Lindhard J (1931) Der Skeletmuskel und seine Funktion. Ergebn Physiol 33:337–557

Lindner E (1968) Submikroskopische Ringbinden – funktionelle Strukturen der "Weichteilmuskulatur". Z Zellforsch Mikrosk Anat 88:370–386

Lindquist C (1973) Contraction properties of cat facial muscles. Acta Physiol Scand 89:482–490

Lipton BH (1977a) A fine-structural analysis of normal and modulated cells in myogenic culture. Dev Biol 60:26–47

Lipton BH (1977b) Collagen synthesis by normal and bromodeoxyuridine-modulated cells in myogenic culture. Dev Biol 61:153–165

Lipton BH, Konigsberg IR (1972) A fine-structural analysis of the fusion of myogenic cells. J Cell Biol 53:348–364

Lipton BH, Schultz E (1979) Developmental fate of skeletal muscle satellite cells. Science 205:1292–1294

Locker RH, Leet NG (1975) Histology of highly-stretched beef muscle. I. The fine structure of grossly stretched single fibers. J Ultrastruct Res 52:64–75

Locker RH, Leet NG (1976a) Histology of highly-stretched beef muscle. II. Further evidence on the location and nature of gap filaments. J Ultrastruct Res 55:157–172

Locker RH, Leet NG (1976b) Histology of highly-stretched beef muscle. IV. Evidence for movement of gap filaments through the Z-line, using the N_2-line and M-line as markers. J Ultrastruct Res 56:31–38

Locker RH, Daines GJ, Leet NG (1976) Histology of highly-stretched beef muscle. III. Abnormal contraction patterns in ox muscle, produced by overstretching during prerigor blending. J Ultrastruct Res 55:173–181

Lockhart RD, Brandt W (1938) Length of striated muscle-fibres (Abstr). J Anat 72:470

Lømo T, Rosenthal J (1972) Control of ACh sensitivity by muscle activity in the rat. J Physiol (Lond) 221:493–513

Lømo T, Slater CR (1978) Control of acetylcholine sensitivity and synapse formation by muscle activity. J Physiol (Lond) 275:391–402

Lømo T, Westgaard RH (1975) Further studies on the control of ACh sensitivity by muscle activity in the rat. J Physiol (Lond) 252:603–626

Lømo T, Westgaard RH, Dahl HA (1974) Contractile properties of muscle: control by pattern of muscle activity in the rat. Proc R Soc Lond [Biol] 187:99–103

Long ME (1947) The development of the muscle-tendon attachment in the rat. Am J Anat 81:159–197

Lough J, Bischoff R (1976) Differential sensitivity to 5-bromodeoxyuridine during the S phase of synchronized myogenic cells. Dev Biol 50:457–475

Lowey S, Risby D (1971) Light chains from fast and slow muscle myosins. Nature 234:81–85

Lowey S, Slayter HS, Weeds AG, Baker H (1969) Substructure of the myosin molecule. I. Subfragments of myosin by enzymic degradation. J Mol Biol 42:1–29

Lowey S, Silberstein L, Gauthier GF, Holt JC (1979) Isolation and distribution of myosin isoenzymes. In: Pepe FA, Sanger JW, Nachmias VT (eds) Motility in cell function. Academic Press, New York, pp 53–67

Lowrie MB, Krishnan S, Vrbova G (1982) Recovery of slow and fast muscles following nerve injury during early post-natal development in the rat. J Physiol (Lond) 331:51–66

Luck WAP (1967) Kristallstrukturen aus nichtmolekularen Bausteinen. Phys Blätter 7:304–313

Lucy JA (1970) The fusion of biological membranes. Nature 227:815–817

Lüllmann H, Lüllmann-Rauch R, Wassermann O (1975) Drug-induced phospholipidoses. CRC Crit Rev Toxicol 4:185–218

Lüttgau HC (1963) The action of calcium ions on potassium contractures of single muscle fibres. J Physiol (Lond) 168:679–697

Luff AR (1975) Dynamic properties of fast and slow skeletal muscles in the cat and rat following cross-reinnervation. J Physiol (Lond) 248:83–96

Luff AR (1981) Dynamic properties of the inferior rectus, extensor digitorum longus, diaphragm and soleus muscles of the mouse. J Physiol (Lond) 313:161–171

Luff AR, Atwood HL (1971) Changes in the sarcoplasmic reticulum and transverse tubular system of fast and slow skeletal muscles of the mouse during postnatal development. J Cell Biol 51:369–383

Luff AR, Proske U (1976) Properties of motor units of the frog sartorius muscle. J Physiol (Lond) 258:673–685

Luther P, Squire J (1978) Three-dimensional structure of the vertebrate muscle M-region. J Mol Biol 125:313–324

Luther PK, Crowther RA (1984) Three-dimensional reconstruction from tilted sections of fish muscle M-band. Nature 307:566–568

Lutz H, Weber H, Billeter R, Jenny E (1979) Fast and slow myosin within single skeletal muscle fibres of adult rabbits. Nature 281:142–144

Lyon MJ (1978) The central location of the motor neurons to the stapedius muscle in the cat. Brain Res 143:437–444

MacDonald RD, Engel AG (1970) Experimental chloroquine myopathy. J Neuropathol Exp Neurol 29:479–499

MacDonald RD, Engel AG (1971) Observations on organization of Z-disk components and on rod-bodies of Z-disk origin. J Cell Biol 48:431–437
MacKay B, Harrop TJ (1969) An experimental study of the longitudinal growth of skeletal muscle in the rat. Acta Anat (Basel) 72:38–49
MacKay B, Harrop TJ, Muir AR (1969) The fine structure of the muscle tendon junction in the rat. Acta Anat (Basel) 73:588–604
Mackova E, Hnik P (1971) "Compensatory" muscle hypertrophy in the rat induced by tenotomy of synergistic muscles. Experientia 27:1039–1040
MacLennan DH, Holland PC (1975) Calcium transport in sarcoplasmic reticulum. Annu Rev Biophys Bioeng 4:377–404
MacLennan DH, Wong PTS (1971) Isolation of a calcium-sequestering protein from sarcoplasmic reticulum. Proc Natl Acad Sci USA 68:1231–1235
MacLennan DH, Seeman P, Iles GH, Yip CC (1971) Membrane formation by the adenosine triphosphatase of sarcoplasmic reticulum. J Biol Chem 246:2702–2710
Maeda Y (1983) The arrangement of myosin heads in relaxed crab muscle. Nature 302:69–72
Magon DK, Basson MD, Carlson BM (1981) Histochemical patterns of dehydrogenase activity in the development of free muscle grafts in the rat. Anat Rec 199:341–347
Mahran ZY, Sakla FB (1965) The pattern of innervation of the extrinsic ocular muscles and the intra-orbital ganglia of the albino mouse. Anat Rec 152:173–183
Maier A, De Santis M, Eldred E (1971) Absence of muscle spindles in avian extraocular muscles. Exp Eye Res 12:251–253
Mair WGP, Tomé FMS (1972) Atlas of the ultrastructure of diseased human muscle. Churchill Livingstone, Edinburgh
Malhotra SK, Tipnis U (1978) Alterations in the structure of sarcolemma in a denervated skeletal muscle. Proc R Soc Lond [Biol] 203:59–68
Malmfors T, Wersäll J (1960a) Innervation of the middle ear muscles in the rabbit with special reference to nerve calibres and motor units. I. Musculus tensor tympani. Acta Morph Neerl Scand 3:157–162
Malmfors T, Wersäll J (1960b) Innervation of the middle ear muscles in the rabbit with special reference to nerve calibres and motor units. II. Musculus stapedius. Acta Morph Neerl Scand 3:163–169
Man-i M, Ito K, Kikuchi K (1967) Histological studies of muscular training. Report I. Effect of training upon skeletal muscle fibers. Res J Physical Education 11:153–165
Maniloff J, Vanderkooi G, Hayashi H, Capaldi RA (1973) Optical analysis of electron micrographs of cytochrome oxidase membranes. Biochim Biophys Acta 298:180–183
Manktelow RT (1984) Functioning muscle transplantation to the upper limb. Clin Plast Surg 11:59–63
Marchesi VT, Tillack TW, Jackson RL, Segrest JP, Scott RE (1972) Chemical characterization and surface orientation of the major glycoprotein of the human erythrocyte membrane. Proc Natl Acad Sci USA 69:1445–1449
Margreth A, Dalla Libera L, Salviati G (1980) Postnatal changes in myosin composition of slow muscle in relation to the differentiation of the motoneurons (Abstr). Muscle Nerve 3:273
Margreth A, Salviati G, Di Mauro S, Turati G (1972) Early biochemical consequences of denervation in fast and slow skeletal muscles and their relationship to neural control over muscle differentiation. Biochem J 126:1099–1110
Marsden CD, Meadows JC, Merton PA (1971) Isolated single motor units in human muscle and their rate of discharge during maximal voluntary effort (Abstr). J Physiol (Lond) 217:12P–13P
Marsh BB (1951) A factor modifying muscle fibre synaeresis. Nature 167:1065–1066
Mårtensson A, Skoglund CR (1964) Contraction properties of intrinsic laryngeal muscles. Acta Physiol Scand 60:318–336
Martin F, MacDonald R (1974) Liposomes can mimic virus membranes. Nature 252:161–163
Martin WD (1979) A study of the effect of pregnancy on muscle fibers of the rectus abdominis muscle of the rat. Anat Rec 195:455–462
Martonosi A (1964) Role of phospholipids in ATPase activity and Ca transport of fragmented sarcoplasmic reticulum. Fed Proc 23:913–921
Martonosi AN (1982a) Regulation of cytoplasmic calcium concentration by sarcoplasmic reticulum. In: Schotland DL (ed) Disorders of the motor unit. Wiley, New York, pp 565–583

Martonosi A (1982b) The development of sarcoplasmic reticulum membranes. Annu Rev Physiol 44:337–355
Martonosi A, Feretos R (1964) Sarcoplasmic reticulum. II. Correlation between adenosine triphosphatase activity and Ca^{++} uptake. J Biol Chem 239:659–668
Martonosi A, Roufa D, Boland R, Reyes E, Tillack TW (1977) Development of sarcoplasmic reticulum in cultured chicken muscle. J Biol Chem 252:318–332
Masaki T, Yoshizaki C (1972) The onset of myofibrillar protein synthesis in chick embryo *in vivo*. J Biochem (Tokyo) 71:755–757
Masaki T, Yoshizaki C (1974) Differentiation of myosin in chick embryos. J Biochem (Tokyo) 76:123–131
Masaki T, Takaiti O, Ebashi S (1968) "M-substance", a new protein constituting the M-line of myofibrils. J Biochem (Tokyo) 64:909–910
Mastaglia FL, Dawkins RL, Papadimitriou JM (1975) Morphological changes in skeletal muscle after transplantation. A light and electron-microscopic study of the initial phases of degeneration and regeneration. J Neurol Sci 25:227–247
Mathers DA, Thesleff S (1978) Studies on neurotrophic regulation of murine skeletal muscle. J Physiol (Lond) 282:105–114
Maton B (1976) Motor unit differentiation and integrated surface EMG in voluntary isometric contraction. Eur J Appl Physiol 35:149–157
Matsubara I, Elliott GF (1972) X-ray diffraction studies on skinned single fibres of frog skeletal muscle. J Mol Biol 72:657–669
Matsubara I, Yagi N (1978) A time-resolved X-ray diffraction study of muscle during twitch. J Physiol (Lond) 278:297–307
Matthews-Bellinger J, Salpeter MM (1978) Distribution of acetylcholine receptors at frog neuromuscular junctions with a discussion of some physiological implications. J Physiol (Lond) 279:197–213
Matyushkin DP (1964) Motor systems in the oculomotor apparatus of higher animals. Fed Proc 23:T1103–T1106
Maugham RJ, Watson JS, Weir J (1983) Strength and cross-sectional area of human skeletal muscle. J Physiol (Lond) 338:37–49
Maunsbach AB, Wirsén C (1966) Ultrastructural changes in kidney, myocardium and skeletal muscle of the dog during excessive mobilization of free fatty acids. J Ultrastruct Res 16:35–54
Mauro A (1961) Satellite cell of skeletal muscle fibers. J Biophys Biochem Cytol 9:493–495
Mauro A (ed) (1979) Muscle regeneration. Raven Press, New York
Mauro A, Adams R (1961) The structure of the sarcolemma of the frog skeletal muscle fiber. J Biophys Biochem Cytol [Suppl] 10:177–185
Mauro A, Shafiq SA, Milhorat AT (eds) (1970) Regeneration of striated muscle, and myogenesis. Excerpta Medica, Amsterdam
Maw MC, Rowe AJ (1980) Fraying of A-filaments into three subfilaments. Nature 286:412–414
Maxwell LC, Carlson DS, McNamara Jr JA, Faulkner JA (1981) Adaptation of the masseter and temporalis muscles following alteration in length, with or without surgical detachment. Anat Rec 200:127–137
Mayer RF (1973) Observations on motor units in cat anterior tibial muscle. In: Desmedt JE (ed) New developments in electromyography and clinical neurophysiology, vol 1. Karger, Basel, pp 31–34
Mayer RF, Doyle AM (1970) Studies of the motor unit in the cat. Histochemistry and topology of anterior tibial and extensor digitorum longus muscles. In: Walton JN, Canal N, Scarlato G (eds) Muscle diseases. Excerpta Medica, Amsterdam, pp 159–163
Mayr R (1969) Untersuchungen an isolierten Muskelspindeln der Ratte nach Cholinesterasedarstellung und Sudanschwarz-Färbung. Ein Beitrag zur Frage der Morphologie und Verteilung motorischer Nervenendigungen in der Muskelspindel der Säuger. Z Zellforsch Mikrosk Anat 93:594–606
Mayr R (1971) Structure and distribution of fibre types in the external eye muscles of the rat. Tissue Cell 3:433–462
Mayr R, Zenker W (1969) Narrowings of the synaptic cleft in myoneural junctions. Experientia 25:1319–1321
Mayr R, Stockinger L, Zenker W (1966) Elektronenmikroskopische Untersuchungen an unterschiedlich innervierten Muskelfasern der äußeren Augenmuskulatur des Rhesusaffen. Z Zellforsch Mikrosk Anat 75:434–452

Mayr R, Zenker W, Gruber H (1967) Zwischensehnenfreie Skeletmuskelfaser-Verbindungen. Z Zellforsch Mikrosk Anat 79:319–325

Mazanet R, Reese BF, Franzini-Armstrong C, Reese TS (1982) Variability in the shapes of satellite cells in normal and injured frog sartorius muscle. Dev Biol 93:22–27

McCloskey DI, Mitchell JH (1972) Reflex cardiovascular and respiratory responses originating in exercising muscle. J Physiol (Lond) 224:173–186

McComas AJ, Fawcett PRW, Campbell MJ, Sica REP (1971) Electrophysiological estimation of the number of motor units within a human muscle. J Neurol Neurosurg Psychiatry 34:121–131

McComas AJ, Kereshi S, Stuart A (1982) Significance of fibrillations in myopathies (Abstr). Vth Int Congr Neuromuscular Dis, Marseille, Pl 18.3

McGeachie J, Allbrook D (1978) Cell proliferation in skeletal muscle following denervation or tenotomy. Cell Tissue Res 193:259–267

McMahan UJ, Sanes JR, Marshall LM (1978) Cholinesterase is associated with the basal lamina at the neuromuscular junction. Nature 271:172–174

McPhedran AM, Wuerker RB, Henneman E (1965) Properties of motor units in a homogeneous red muscle (soleus) of the cat. J Neurophysiol 28:71–84

Meissner G (1975) Isolation and characterization of two types of sarcoplasmic reticulum vesicles. Biochim Biophys Acta 389:51–68

Meissner G, Fleischer S (1971) Characterization of sarcoplasmic reticulum from skeletal muscle. Biochim Biophys Acta 241:356–378

Mendell JR, Roelofs RI, Engel WK (1972) Ultrastructural development of explanted human skeletal muscle in tissue culture. J Neuropath Exp Neurol 31:433–446

Mendelson M (1969) Electrical and mechanical characteristics of a very fast lobster muscle. J Cell Biol 42:548–563

Merlie JP, Hofler JG, Sebbane R (1981) Acetylcholine receptor synthesis from membrane polysomes. J Biol Chem 256:6995–6999

Michler A, Sakmann B (1980) Receptor stability and channel conversion in the subsynaptic membrane of the developing mammalian neuromuscular junction. Dev Biol 80:1–17

Mikawa T (1979) 'Freezing' of Ca-regulated conformation of reconstituted thin filament of skeletal muscle by glutaraldehyde. Nature 278:473–474

Miledi R (1960) The acetylcholine sensitivity of frog muscle fibres after complete or partial denervation. J Physiol (Lond) 151:1–23

Miledi R (1963) An influence of nerve not mediated by impulses. In: Gutmann E, Hnik P (eds) The effect of use and disuse on neuromuscular functions. Elsevier, Amsterdam, pp 35–40

Miledi R (1964) Electron-microscopical localization of products from histochemical reactions used to detect cholinesterase in muscle. Nature 204:293–295

Miledi R, Potter LT (1971) Acetylcholine receptors in muscle fibres. Nature 233:599–603

Miledi R, Slater CR (1968) Some mitochondrial changes in denervated muscle. J Cell Sci 3:49–54

Miledi R, Stefani E (1969) Non-selective re-innervation of slow and fast muscle fibres in the rat. Nature 222:569–571

Miledi R, Uchitel OD (1981) Properties of postsynaptic channels induced by acetylcholine in different frog muscle fibres. Nature 291:162–165

Miledi R, Uchitel OD (1984) A study of the submaxillaris muscle of the frog. J Physiol (Lond) 350:279–291

Miledi R, Zelena J (1966) Sensitivity to acetylcholine in rat slow muscle. Nature 210:855–856

Miledi R, Stefani E, Zelena J (1968) Neural control of acetylcholine-sensitivity in rat muscle fibres. Nature 220:497–498

Miledi R, Parker I, Schalow G (1977) Measurement of calcium transients in frog muscle by the use of arsenazo III. Proc R Soc Lond [Biol] 198:201–210

Miledi R, Parker I, Zhu PH (1982) Calcium transients evoked by action potentials in frog twitch muscle fibres. J Physiol (Lond) 333:655–679

Miledi R, Parker I, Zhu PH (1983a) Calcium transients in frog skeletal muscle fibres following conditioning stimuli. J Physiol (Lond) 339:223–242

Miledi R, Parker I, Zhu PH (1983b) Calcium transients studied under voltage-clamp control in frog twitch muscle fibres. J Physiol (Lond) 340:649–680

Miledi R, Reiser G, Uchitel OD (1984) Characteristics of membrane channels induced by acetylcholine at frog muscle-tendon junctions. J Physiol (Lond) 350:269–277

Miller A, Tregear RT (1971) X-ray studies on the structure and function of vertebrate and invertebrate muscle. In: Podolsky RJ (ed) Contractility of muscle cells and related processes. Prentice-Hall, Inc, Englewood Cliffs, pp 205–228
Miller JE (1967) Cellular organization of rhesus extraocular muscle. Invest Ophthalmol 6:18–39
Miller JE (1971) Recent histologic and electron microscopic findings in extraocular muscle. Trans Am Acad Ophthalmol Otolaryngol 75:1175–1185
Miller TM, Heuser JE (1984) Endocytosis of synaptic vesicle membrane at the frog neuromuscular junction. J Cell Biol 98:685–698
Millman BM, Elliott GF, Lowy J (1967) X-ray diffraction studies. Nature 213:356–358
Milner-Brown HS, Stein RB, Yemm R (1973a) The contractile properties of human motor units during voluntary isometric contractions. J Physiol (Lond) 228:285–306
Milner-Brown HS, Stein RB, Yemm R (1973b) Changes in firing rate of human motor units during linearly changing voluntary contractions. J Physiol (Lond) 230:371–390
Mintz B, Baker WW (1967) Normal mammalian muscle differentiation and gene control of isocitrate dehydrogenase synthesis. Proc Natl Acad Sci USA 58:592–598
Miranda AF, Godman GC (1973) The effects of cytochalasin D on differentiating muscle in culture. Tissue Cell 5:1–22
Mishra SK, Haining JL (1980) Measurement of local skeletal muscle blood flow in normal humans by hydrogen clearance. Muscle Nerve 3:289–292
Mitchell JH, Reardon WC, McCloskey DI, Wildenthal K (1977) Possible role of muscle receptors in the cardiovascular response to exercise. Ann NY Acad Sci 301:232–242
Mobley BA, Eisenberg BR (1975) Sizes of components in frog skeletal muscle measured by methods of stereology. J Gen Physiol 66:31–45
Mommaerts WFHM (1951) The scattering of light in myosin solutions. 1. The angular dissymmetry and the molecular length. J Biol Chem 188:553–557
Montgomery RD (1962) Growth of human striated muscle. Nature 195:194–195
Moore MJ (1983) The dual connective tissue system of rat soleus muscle. Muscle Nerve 6:416–422
Moore PB, Huxley HE, DeRosier DJ (1970) Three-dimensional reconstruction of F-actin, thin filaments and decorated thin filaments. J Mol Biol 50:279–295
Moos C (1981) Fluorescence microscope study of the binding of added C protein to skeletal muscle myofibrils. J Cell Biol 90:25–31
Moos C, Mason CM, Besterman JM, Feng I-NM, Dubin JH (1978) The binding of skeletal muscle C-protein to F-actin, and its relation to the interaction of actin with myosin subfragment-l. J Mol Biol 124:571–586
Morimoto K, Harrington WF (1972) Isolation and physical chemical properties of an M-line protein from skeletal muscle. J Biol Chem 247:3052–3061
Morimoto K, Harrington WF (1973) Isolation and composition of thick filaments from rabbit skeletal muscle. J Mol Biol 77:165–175
Mornet D, Bertrand R, Pantel P, Audemard E, Kassab R (1981) Structure of the actin-myosin interface. Nature 292:301–306
Morrison-Graham K (1983) An anatomical and electrophysiological study of synapse elimination at the developing frog neuromuscular junction. Dev Biol 99:298–311
Moss FP, Leblond CP (1970) Nature of dividing nuclei in skeletal muscle of growing rats. J Cell Biol 44:459–462
Moss FP, Leblond CP (1971) Satellite cells as the source of nuclei in muscles of growing rats. Anat Rec 170:421–436
Moulds RFW, Young A, Jones DA, Edwards RHT (1977) A study of the contractility, biochemistry and morphology of an isolated preparation of human skeletal muscle. Clin Sci Mol Med 52:291–297
Mrak RE, Saito A, Evans OB, Fleischer S (1982) Autophagic degradation in human skeletal muscle target fibers. Muscle Nerve 5:745–753
Müntener M (1982) A rapid and reversible muscle fiber transformation in the rat. Exp Neurol 77:668–678
Müntener M, Gottschall J, Neuhuber W, Mysicka A, Zenker W (1980) The ansa cervicalis and the infrahyoid muscles of the rat. I. Anatomy; distribution, number and diameter of fiber types; motor units. Anat Embryol 159:49–57

Mufti SA, Carlson BM, Maxwell LC, Faulkner JA (1977) The free autografting of entire limb muscles in the cat: Morphology. Anat Rec 188:417–430
Murata F, Ogata T (1969) The ultrastructure of neuromuscular junctions of human red, white and intermediate striated muscle fibers. Tohoku J Exp Med 99:289–301
Murray MR (1960) Skeletal muscle tissue in culture. In: Bourne GH (ed) The structure and function of muscle, vol I. Academic Press, New York, pp 111–136
Murray MR (1972) Skeletal muscle in culture. In: Bourne GH (ed) The structure and function of muscle, 2nd edn, vol I/1. Academic Press, New York, pp 237–299
Murthy KSK, Letbetter WD, Eidelberg E, Cameron WE, Petit J (1982) Histochemical evidence for the existence of skeletofusimotor (β) innervation in the primate. Exp Brain Res 46:186–190
Muscatello U, Andersson-Cedergren E, Azzone GF, Von der Decken A (1961) The sarcotubular system of frog skeletal muscle. A morphological and biochemical study. J Biophys Biochem Cytol [Suppl] 10:201–218
Myrhage R (1977) Microvascular supply of skeletal muscle fibres. A microangiographic, histochemical and intravital microscopic study of hind limb muscles in the rat, rabbit and cat (Thesis). Acta Orthop Scand [Suppl] 168:1–46
Myrhage R (1978) Capillary supply of the muscle fibre population in hindlimb muscles of the cat. Acta Physiol Scand 103:19–30
Myrhage R, Eriksson E (1980) Vascular arrangements in hind limb muscles of the cat. J Anat 131:1–17
Nag AC, Cheng M (1982) Differentiation of fibre types in an extraocular muscle of the rat. J Embryol Exp Morphol 71:171–191
Nag AC, Foster JD (1981) Myogenesis in adult mammalian skeletal muscle *in vitro*. J Anat 132:1–18
Nagai T, Makinose M, Hasselbach W (1960) Der physiologische Erschlaffungsfaktor und die Muskelgrana. Biochim Biophys Acta 43:223–238
Nakajima S, Gilai A (1980) Action potentials of isolated single muscle fibers recorded by potential-sensitive dyes. J Gen Physiol 76:729–750
Nakajima S, Nakajima Y, Peachey LD (1973) Speed of repolarization and morphology of glycerol-treated frog muscle fibres. J Physiol (Lond) 234:465–480
Nakamura T, Namba T, Grob D (1967) Motor end plate reactivity to divalent metal ions. Histochemical studies. J Histochem Cytochem 15:276–284
Nakamura A, Sreter F, Gergely J (1971) Comparative studies on light meromyosin paracrystals derived from red, white, and cardiac muscle myosins. J Cell Biol 49:883–898
Nakao T (1975) Fine structure of the myotendinous junction and "terminal coupling" in the skeletal muscle of the lamprey, *Lampetra japonica*. Anat Rec 182:321–338
Namba T, Grob D (1968) Cholinesterase activity of the motor endplate in isolated muscle membrane. J Neurochem 15:1445–1454
Namba T, Nakamura T, Takahashi A, Grob D (1968) Motor nerve endings in extraocular muscles. J Comp Neurol 134:385–396
Nameroff M, Holtzer H (1969) Contact-mediated reversible suppression of myogenesis. Dev Biol 19:380–396
Nameroff M, Holtzer H (1970) Interference with myogenesis. In: Mauro A, Shafiq SA, Milhorat AT (eds) Regeneration of striated muscle, and myogenesis. Excerpta Medica, Amsterdam, pp 251–266
Nasledov GA (1965) Correlative study of certain morphological and functional features of muscle fibers. Fed Proc 24:T1091–T1095
Nathanson MA, Hay ED (1980a) Analysis of cartilage differentiation from skeletal muscle grown on bone matrix. I. Ultrastructural aspects. Dev Biol 78:301–331
Nathanson MA, Hay ED (1980b) Analysis of cartilage differentiation from skeletal muscle grown on bone matrix. II. Chondroitin sulfate synthesis and reaction to exogenous glycosaminoglycans. Dev Biol 78:332–351
Neher E, Sakmann B (1976) Noise analysis of drug induced voltage clamp currents in denervated frog muscle fibres. J Physiol (Lond) 258:705–729
Nemeth PM, Pette D (1981) The limited correlation of myosin-based and metabolism-based classifications of skeletal muscle fibers. J Histochem Cytochem 29:89–90
Newham DJ, Jones DA, Edwards RHT (1983a) Large delayed plasma creatine kinase changes after stepping exercise. Muscle Nerve 6:380–385

Newham DJ, McPhail G, Mills KR, Edwards RHT (1983b) Ultrastructural changes after concentric and eccentric contractions of human muscle. J Neurol Sci 61:109–122

Newman SA, Pautou M-P, Kieny M (1981) The distal boundary of myogenic primordia in chimeric avian limb buds and its relation to an accessible population of cartilage progenitor cells. Dev Biol 84:440–448

Nicol CJM, Bruce DS (1981) Effect of hyperthyroidism on the contractile and histochemical properties of fast and slow twitch skeletal muscle in the rat. Pflugers Arch 390:73–79

Niedergerke R (1955) Local muscular shortening by intracellularly applied calcium (Abstr). J Physiol (Lond) 128:12P–13P

Ninomiya JG, Echeverria OM, Vazquez-Nin GH (1981) Morphological studies of fiber types of striated muscle fibers of the cremaster in the guinea pig. Acta Anat (Basel) 111:240–246

Nolte J, Pette D (1972a) Microphotometric determination of enzyme activity in single cells in cryostat sections. I. Application of the gel film technique to microphotometry and studies on the intralobular distribution of succinate dehydrogenase and lactate dehydrogenase in rat liver. J Histochem Cytochem 20:567–576

Nolte J, Pette D (1972b) Microphotometric determination of enzyme activity in single cells in cryostat sections. II. Succinate dehydrogenase, lactate dehydrogenase and triosephosphate dehydrogenase activities in red, intermediate and white fibers of soleus and rectus femoris muscles of rat. J Histochem Cytochem 20:577–582

Nonaka I, Takagi A, Ishiura S, Nakase H, Sugita H (1983) Pathophysiology of muscle fiber necrosis induced by bupivacaine hydrochloride (Marcaine). Acta Neuropathol (Berlin) 60:167–174

Novikoff AB (1976) The endoplasmic reticulum: A cytochemist's view (A review). Proc Natl Acad Sci USA 73:2781–2787

Novikoff AB, Essner E, Quintana N (1964) Golgi apparatus and lysosomes. Fed Proc 23:1010–1022

Nunzi MG, Franzini-Armstrong C (1980) Trabecular network in adult skeletal muscle. J Ultrastruct Res 73:21–26

Nygaard E (1980) Number of fibers in skeletal muscle of man (Abstr). Muscle Nerve 3:268

Nygaard E (1981) Skeletal muscle fibre characteristics in young women. Acta Physiol Scand 112:299–304

Nygaard E, Sanchez J (1982) Intramuscular variation of fiber types in the brachial biceps and the lateral vastus muscles of elderly men: how representative is a small biopsy sample? Anat Rec 203:451–459

Nygaard E, Schmalbruch H (1980) Incidence of capillaries in skeletal muscle of man determined by light and electron microscopy (Abstr). Muscle Nerve 3:268

Nygaard E, Houston M, Suzuki Y, Jørgensen K, Saltin B (1983) Morphology of the brachial biceps muscle and elbow flexion in man. Acta Physiol Scand 117:287–292

Nyström B (1968a) Postnatal development of motor nerve terminals in "slow-red" and "fast-white" cat muscles. Acta Neurol Scand 44:363–383

Nyström B (1968b) Histochemistry of developing cat muscles. Acta Neurol Scand 44:405–439

O'Brien RAD, Östberg AJC, Vrbova G (1978) Observations on the elimination of polyneural innervation in developing mammalian skeletal muscle. J Physiol (Lond) 282:571–582

O'Neill MC, Stockdale FE (1972a) A kinetic analysis of myogenesis in vitro. J Cell Biol 52:52–65

O'Neill MC, Stockdale FE (1972b) Differentiation without cell division in cultured skeletal muscle. Dev Biol 29:410–418

O'Steen WK, Shear CR, Anderson KV (1975) Extraocular muscle degeneration and regeneration after exposure of rats to incandescent radiant energy. J Cell Sci 18:157–177

Obinata T, Shimada Y, Matsuda R (1979) Troponin in embryonic chick skeletal muscle cells in vitro. An immunoelectron microscope study. J Cell Biol 81:59–66

Obinata T, Masaki T, Takano H (1980) Types of myosin light chains present during the development of fast skeletal muscle in chick embryo. J Biochem (Tokyo) 87:81–88

Odusote K, Karpati G, Carpenter S (1981) An experimental morphometric study of neutral lipid accumulation in skeletal muscles. Muscle Nerve 4:3–9

Oertel G (1983) Durchmesserbestimmung normaler kindlicher Muskulatur. Wiss Ztschr Friedrich-Schiller-Univ Jena, Math-Naturwiss R 32:665–668

Offer G (1976) The antigenicity of myosin and C-protein. Proc R Soc Lond [Biol] 192:439–449

Offer G, Moos C, Starr R (1973) A new protein of the thick filaments of vertebrate skeletal myofibrils. J Mol Biol 74:653–676

Oh TH, Johnson DD (1972) Effects of acetyl-β-methylcholine on development of acetylcholinesterase and butyrylcholinesterase activities in cultured chick embryonic skeletal muscle. Exp Neurol 37:360–370

Ohtsuki I (1975) Distribution of troponin components in the thin filament studied by immunoelectron microscopy. J Biochem (Tokyo) 77:633–639

Ohtsuki I (1979) Molecular arrangement of troponin-T in the thin filament. J Biochem (Tokyo) 86:491–497

Ohtsuki I, Masaki T, Nonomura Y, Ebashi S (1967) Periodic distribution of troponin along the thin filament. J Biochem (Tokyo) 61:817–819

Ohtsuki I, Wakabayashi T (1972) Optical diffraction studies on the structure of troponin-tropomyosin-actin paracrystals. J Biochem (Tokyo) 72:369–377

Okazaki K, Holtzer H (1965) An analysis of myogenesis *in vitro* using fluorescein-labeled antimyosin. J Histochem Cytochem 13:726–739

Okazaki K, Holtzer H (1966) Myogenesis: fusion, myosin synthesis, and the mitotic cycle. Proc Natl Acad Sci USA 56:1484–1490

Oldfors A, Fardeau M (1983) The permeability of the basal lamina at the neuromuscular junction. An ultrastructural study of rat skeletal muscle using particulate tracers. Neuropathol Appl Neurobiol 9:419–432

Olson CB, Swett CP Jr (1966) A functional and histochemical characterization of motor units in a heterogeneous muscle (flexor digitorum longus) of the cat. J Comp Neurol 128:475–498

Ontell M (1974) Muscle satellite cells: A validated technique for light microscopic identification and a quantitative study of changes in their population following denervation. Anat Rec 178:211–228

Ontell M (1975) Evidence for myoblastic potential of satellite cells in denervated muscle. Cell Tissue Res 160:345–353

Ontell M (1977) Neonatal muscle: An electron microscopic study. Anat Rec 189:669–690

Ontell M (1979) The source of "new" muscle fibers in neonatal muscle. In: Mauro A (ed) Muscle regeneration. Raven Press, New York, pp 137–146

Ontell M, Dunn RF (1978) Neonatal muscle growth: a quantitative study. Am J Anat 152:539–556

Ontell M, Hughes D, Bourke D (1982) Secondary myogenesis of normal muscle produces abnormal myotubes. Anat Rec 204:199–207

Oppenheim RW, Majors-Willard C (1978) Neuronal cell death in the brachial spinal cord of the chick is unrelated to the loss of polyneural innervation in wing muscle. Brain Res 154:148–152

Orci L, Perrelet A, Friend DS (1977) Freeze-fracture of membrane fusions during exocytosis in pancreatic B-cells. J Cell Biol 75:23–30

Orkand RK (1963) A further study of electrical responses in slow and twitch muscle fibres of the frog. J Physiol (Lond) 167:181–191

Osame M, Engel AG, Rebouche CJ, Scott RE (1981) Freeze-fracture electronmicroscopic analysis of plasma membranes of cultured muscle cells in Duchenne dystrophy. Neurology (NY) 31:972–979

Ovalle WK (1971) Fine structure of rat intrafusal muscle fibers. The polar region. J Cell Biol 51:83–103

Ovalle WK (1972) Fine structure of rat intrafusal muscle fibers. The equatorial region. J Cell Biol 52:382–396

Ovalle WK (1978) Histochemical dichotomy of extrafusal and intrafusal fibers in an avian slow muscle. Am J Anat 152:587–597

Ovalle WK, Smith RS (1974) Types of intrafusal muscle fibre in the amphibia (Abstr). J Anat 119:189–191

Padykula HA, Gauthier GF (1970) The ultrastructure of the neuromuscular junctions of mammalian red, white, and intermediate skeletal muscle fibers. J Cell Biol 46:27–41

Padykula HA, Herman E (1955a) Factors affecting the activity of adenosine triphosphatase and other phosphatases as measured by histochemical techniques. J Histochem Cytochem 3:161–169

Padykula HA, Herman E (1955b) The specificity of the histochemical method for adenosine triphosphatase. J Histochem Cytochem 3:170–195

Pagano RE, Huang L, Wey C (1974) Interaction of phospholipid vesicles with cultured mammalian cells. Nature 252:166–167

Page S (1964) Filament lengths in resting and excited muscles. Proc R Soc Lond [Biol] 160:460–466

Page S (1966) Filament lengths in vertebrate striated muscles. Molecular basis of muscle contraction. Information exchange group No. 4. Scientific memo 79:1–6

Page SG (1965) A comparison of the fine structures of frog slow and twitch muscle fibres. J Cell Biol 26:477–497

Page SG (1968) Fine structure of tortoise skeletal muscle. J Physiol (Lond) 197:709–715

Page SG (1969) Structure and some contractile properties of fast and slow muscles of the chicken. J Physiol (Lond) 205:131–145

Page SG, Huxley HE (1963) Filament lengths in striated muscle. J Cell Biol 19:369–390

Palade GE (1954) Electron microscope observations of interneuronal and neuromuscular synapses (Abstr). Anat Rec 118:335–336

Pallot DJ (1974) Structural studies on two types of snake spindle (Abstr). J Anat 119:192–193

Pallot DJ, Ridge RMAP (1972) The fine structure of the long-capsule muscle spindles in the snake *Natrix* sp. J Anat 113:61–74

Palmucci L, Anzil AP, Luh S (1983) Crystalline aggregates of protein-glycogen complexes (alias 'Virus-like particles') in skeletal muscle: Report of a case and review of the literature. Neuropathol Appl Neurobiol 9:61–71

Papadimitriou JM, Dawkins RL (1973) Scanning and transmission electron microscopy and immunofluorescence of myogenesis *in vitro*. Cytobios 8:227–241

Papahadjopoulos D, Mayhew E, Poste G, Smith S (1974) Incorporation of lipid vesicles by mammalian cells provides a potential method for modifying cell behaviour. Nature 252:163–166

Pardo JV, Siliciano JD, Craig SW (1983) A vinculin-containing cortical lattice in skeletal muscle: transverse lattice elements ("costameres") mark sites of attachment between myofibrils and sarcolemma. Proc Natl Acad Sci USA 80:1008–1012

Partridge TA, Grounds M, Sloper JC (1978) Evidence of fusion between host and donor myoblasts in skeletal muscle grafts. Nature 273:306–308

Paterson B, Strohman RC (1972) Myosin synthesis in cultures of differentiating chicken embryo skeletal muscle. Dev Biol 29:113–138

Paterson BM, Roberts BE, Yaffe D (1974) Determination of actin messenger RNA in cultures of differentiating embryonic chick skeletal muscle. Proc Natl Acad Sci USA 71:4467–4471

Patterson S, Goldspink G (1976) Mechanism of myofibril growth and proliferation in fish muscle. J Cell Sci 22:607–616

Payne CM, Stern LZ, Curless RG, Hannapel LK (1975) Ultrastructural fiber typing in normal and diseased human muscle. J Neurol Sci 25:99–108

Peachey LD (1961) Structure of the longitudinal body muscles of amphioxus. J Biophys Biochem Cytol [Suppl] 10:159–176

Peachey LD (1965) The sarcoplasmic reticulum and transverse tubules of the frog's sartorius. J Cell Biol 25:209–231

Peachey LD (1968) Muscle. Annu Rev Physiol 30:401–440

Peachey LD, Eisenberg BR (1978) Helicoids in the T system and striations of frog skeletal muscle fibers seen by high voltage electron microscopy. Biophys J 22:145–154

Peachey LD, Huxley AF (1962) Structural identification of twitch and slow striated muscle fibers of the frog. J Cell Biol 13:177–180

Peachey LD, Porter KR (1959) Intracellular impulse conduction in muscle cells. Science 129:721–722

Peachey LD, Schild RF (1968) The distribution of the T-system along the sarcomeres of frog and toad sartorius muscle. J Physiol (Lond) 194:249–258

Peachey LD, Takeichi M, Nag AC (1974) Muscle fiber types and innervation in adult cat extraocular muscles. In: Milhorat AT (ed) Exploratory concepts in muscular dystrophy II. Excerpta Medica, Amsterdam, pp 246–257

Pease DC, Jenden DJ, Howell JN (1965) Calcium uptake in glycerol-extracted rabbit psoas muscle fibers. II. Electron microscopic localization of uptake sites. J Cell Comp Physiol 65:141–154

Pellegrino C, Franzini C (1963) An electron microscope study of denervation atrophy in red and white skeletal muscle fibers. J Cell Biol 17:327–349

Pelloni-Mueller G, Ermini M, Jenny E (1976) Myosin light chains of developing fast and slow rabbit skeletal muscle. FEBS Lett 67:68–74

Pena SDJ, Gordon BB, Karpati G, Carpenter S (1981) Lectin histochemistry of human skeletal muscle. J Histochem Cytochem 29:542–546

Peng HB, Nakajima Y (1978) Membrane particle aggregates in innervated and noninnervated cultures of Xenopus embryonic muscle cells. Proc Natl Acad Sci USA 75:500–504
Peng HB, Wolosewick JJ, Cheng P-C (1981a) The development of myofibrils in cultured muscle cells: A whole-mount and thin-section electron microscopic study. Dev Biol 88:121–136
Peng HB, Cheng P-C, Luther PW (1981b) Formation of ACh receptor clusters induced by positively charged latex beads. Nature 292:831–834
Pepe FA (1966) Some aspects of the structural organization of the myofibril as revealed by antibody-staining methods. J Cell Biol 28:505–525
Pepe FA (1967a) The myosin filament. I. Structural organization from antibody staining observed in electron microscopy. J Mol Biol 27:203–225
Pepe FA (1967b) The myosin filament. II. Interaction between myosin and actin filaments observed using antibody staining in fluorescent and electron microscopy. J Mol Biol 27:227–236
Pepe FA (1975) Structure of muscle filaments from immunohistochemical and ultrastructural studies. J Histochem Cytochem 23:543–562
Pepe FA (1979) The myosin filament: molecular structure. In: Pepe FA, Sanger JW, Nachmias VT (eds) Motility in cell function. Academic Press, New York, pp 103–116
Pepe FA, Dowben P (1977) The myosin filament. V. Intermediate voltage electron microscopy and optical diffraction studies of the substructure. J Mol Biol 113:199–218
Pepe FA, Drucker B (1972) The myosin filament. IV. Observation of the internal structural arrangement. J Cell Biol 52:255–260
Pepe FA, Drucker B (1975) The myosin filament. III. C-protein. J Mol Biol 99:609–617
Peper K, McMahan UJ (1972) Distribution of acetylcholine receptors in the vicinity of nerve terminals on skeletal muscle of the frog. Proc R Soc Lond [Biol] 181:431–440
Peper K, Dreyer F, Sandri C, Akert K, Moor H (1974) Structure and ultrastructure of the frog motor endplate. A freeze-etching study. Cell Tissue Res 149:437–455
Perry SV, Corsi A (1958) Extraction of proteins other than myosin from the isolated rabbit myofibril. Biochem J 68:5–12
Person RS, Kudina LP (1972) Discharge frequency and discharge pattern of human motor units during voluntary contraction of muscle. Electroencephalogr Clin Neurophysiol 32:471–483
Pessac B, Defendi V (1972) Cell aggregation: Role of acid mucopolysaccharides. Science 175:898–900
Pestronk A, Drachman DB, Stanley EF, Price DL, Griffin JW (1980) Cholinergic transmission regulates extrajunctional acetylcholine receptors. Exp Neurol 70:690–696
Peter JB, Barnard RJ, Edgerton VR, Gillespie CA, Stempel KE (1972) Metabolic profiles of three fiber types of skeletal muscle in guinea pigs and rabbits. Biochemistry 11:2627–2633
Petersen H (1924) Histologie und mikroskopische Anatomie, Bd 3. Bergmann, München
Peterson ER, Crain SM (1970) Innervation of cultures of fetal rodent skeletal muscle by organotypic explants of spinal cord from different animals. Z Zellforsch Mikrosk Anat 106:1–21
Pette D (1975) Some aspects of supramolecular organization of glycogenolytic and glycolytic enzymes in muscle. Acta Histochem [Suppl] (Jena) 14:47–68
Pette D (1984) Activity-induced fast to slow transitions in mammalian muscle. Med Sci Sports Exerc 16:517–528
Pette D, Schnez U (1977) Coexistence of fast and slow type myosin light chains in single muscle fibres during transformation as induced by long term stimulation. FEBS Lett 83:128–130
Pette D, Tyler KR (1983) Response of succinate dehydrogenase activity in fibres of rabbit tibialis anterior muscle to chronic nerve stimulation. J Physiol (Lond) 338:1–9
Pette D, Staudte HW, Vrbova G (1972) Physiological and biochemical changes induced by long-term stimulation of fast muscle. Naturwissenschaften 8:469–470
Pette D, Smith ME, Staudte HW, Vrbova G (1973) Effects of long-term electrical stimulation on some contractile and metabolic characteristics of fast rabbit muscles. Pflugers Arch 338:257–272
Pette D, Müller W, Leisner E, Vrbova G (1976) Time dependent effects on contractile properties, fibre population, myosin light chains and enzymes of energy metabolism in intermittently and continuously stimulated fast twitch muscles of the rabbit. Pflugers Arch 364:103–112
Peyronnard J-M, Lamarre Y (1977) Electrophysiological and anatomical estimation of the number of motor units in the monkey extensor digitorum brevis muscle. J Neurol Neurosurg Psychiatry 40:756–764
Pfitzer P (1971) Polyploide Zellkerne im Herzmuskel des Schweins. Virchows Arch [Cell Pathol] 9:180–186

Pfitzer P (1972) Die karyologischen Grundlagen der Hypertrophie. Verh. Dtsch Ges Herz Kreislaufforsch 38:22–34
Philpott DE, Szent-Györgyi AG (1954) The structure of light-meromyosin: an electron microscopic study. Biochim Biophys Acta 15:165–173
Piehl K (1974) Glycogen storage and depletion in human skeletal muscle fibres. Acta Physiol Scand [Suppl] 402:3–32
Pierobon-Bormioli S (1981) Transverse sarcomere filamentous systems: "Z- and M-cables". J Muscle Res Cell Motil 2:401–413
Pierobon Bormioli S, Schiaffino S (1977) Myomuscular junctions in re-innervated rat skeletal muscle. J Anat 124:359–370
Pierobon Bormioli S, Torresan P, Sartore S, Moschini GB, Schiaffino S (1979) Immunohistochemical identification of slow-tonic fibers in human extrinsic eye muscles. Invest Ophthalmol Vis Sci 18:303–306
Pierobon-Bormioli S, Sartore S, Vitadello M, Schiaffino S (1980) "Slow" myosins in vertebrate skeletal muscle. An immunofluorescence study. J Cell Biol 85:672–681
Pierobon-Bormioli S, Sartore S, Dalla Libera L, Vitadello M, Schiaffino S (1981) "Fast" isomyosins and fiber types in mammalian skeletal muscle. J Histochem Cytochem 29:1178–1188
Pilar G, Hess A (1966) Differences in internal structure and nerve terminals of the slow and twitch muscle fibers in the cat superior oblique. Anat Rec 154:243–252
Pilar G, Tuttle J, Vaca K (1981) Functional maturation of motor nerve terminals in the avian iris: Ultrastructure, transmitter metabolism and synaptic reliability. J Physiol (Lond) 321:175–193
Platzer AC (1978) The ultrastructure of normal myogenesis in the limb of the mouse. Anat Rec 190:639–658
Podleski TR, Axelrod D, Ravdin P, Greenberg I, Johnson MM, Salpeter MM (1978) Nerve extract induces increase and redistribution of acetylcholine receptors in cloned muscle cells. Proc Natl Acad Sci USA 75:2035–2039
Podleski TR, Greenberg I, Schlessinger J, Yamada KM (1979) Fibronectin delays the fusion of L6 myoblasts. Exp Cell Res 122:317–326
Podolsky RJ (1964) The maximum sarcomere length for contraction of isolated myofibrils. J Physiol (Lond) 170:110–123
Podolsky RJ (1968) Deposit formation in muscle fibers following contraction in the presence of lead. J Cell Biol 39:197–201
Podolsky RJ (1975) Muscle activation: the current status. Fed Proc 34:1374–1378
Podolsky RJ, Hall T, Hatchett SL (1970) Identification of oxalate precipitates in striated muscle fibers. J Cell Biol 44:699–702
Polgar J, Johnson MA, Weightman D, Appleton D (1973) Data on fibre size in thirty-six human muscles. An autopsy study. J Neurol Sci 19:307–318
Politoff A, Blitz AL, Rose S (1975) Incorporation of acetylcholinesterase into synaptic vesicles is associated with blockade of synaptic transmission. Nature 256:324–325
Pollack GH (1983) The cross-bridge theory. Physiol Rev 63:1049–1113
Pollack MS, Bird JWC (1968) Distribution and particle properties of acid hydrolase in denervated muscle. Am J Physiol 215:716–722
Poole AR, Howell JI, Lucy JA (1970) Lysolecithin and cell fusion. Nature 227:810–813
Porayko O, Smith RS (1968) Morphology of muscle spindles in the rat. Experientia 24:588–589
Porter KR (1954) The myo-tendon junction in larval forms of *Amblystoma punctatum* (Abstr). Anat Rec 118:342
Porter KR (1956) The sarcoplasmic reticulum in muscle cells of amblystoma larvae. J Biophys Biochem Cytol [Suppl] 2:163–170
Porter KR (1961) The sarcoplasmic reticulum. Its recent history and present status. J Biophys Biochem Cytol [Suppl] 10:219–226
Porter KR, Palade GE (1957) Studies on the endoplasmic reticulum. III. Its form and distribution in striated muscle cells. J Biophys Biochem Cytol 3:269–300
Poste G, Papahadjopoulos D (1976) Lipid vesicles as carriers for introducing materials into cultured cells: Influence of vesicle lipid composition on mechanism(s) of vesicle incorporation into cells. Proc Natl Acad Sci USA 73:1603–1607
Potter LF (1970) Synthesis, storage and release of [^{14}C] acetylcholine in isolated rat diaphragm muscles. J Physiol (Lond) 206:145–166

Poulsen FR, Lowy J (1983) Small-angle X-ray scattering from myosin heads in relaxed and rigor frog skeletal muscles. Nature 303:146–152

Prendergast FJ, McGeachie JK, Edis RH, Allbrook D (1977) Whole-muscle reimplantation with microneurovascular anastomoses. Ann R Coll Surg Engl 59:393–400

Prince FP, Hikida RS, Hagerman FC, Staron RS, Allen WH (1981) A morphometric analysis of human muscle fibers with relation to fiber types and adaptations to exercise. J Neurol Sci 49:165–179

Pringle JWS (1974) Locomotion: Flight. In: Rockstein M (ed) The physiology of insecta, vol III, 2nd edn. Academic Press, New York London, pp 433–476

Prives J, Shinitzky M (1977) Increased membrane fluidity precedes fusion of muscle cells. Nature 268:761–763

Prives J, Silman I, Amsterdam A (1976) Appearance and disappearance of acetylcholine receptor during differentiation of chick skeletal muscle in vitro. Cell 7:543–550

Prives JM, Paterson BM (1974) Differentiation of cell membranes in cultures of embryonic chick breast muscle. Proc Natl Acad Sci USA 71:3208–3211

Prochniewicz-Nakayama E, Yanagida T, Oosawa F (1983) Studies on conformation of F-actin in muscle fibers in the relaxed state, rigor, and during contraction using fluorescent phalloidin. J Cell Biol 97:1663–1667

Proske U (1969) The innervation of muscle spindles in the lizard *Tiliqua nigrolutea*. J Anat 105:217–230

Proske U, Vaughan P (1968) Histological and electrophysiological investigation of lizard skeletal muscle. J Physiol (Lond) 199:495–509

Przybyla A, Strohman RC (1974) Myosin heavy chain messenger RNA from myogenic cell cultures. Proc Natl Acad Sci USA 71:662–666

Pumplin DW, Fambrough DM (1982) Turnover of acetylcholine receptors in skeletal muscle. Annu Rev Physiol 44:319–335

Pumplin DW, Fambrough DM (1983) ($Na^+ + K^+$)-ATPase correlated with a major group of intramembrane particles in freeze-fracture replicas of cultured chick myotubes. J Cell Biol 97:1214–1225

Rambourg A, Leblond CP (1967) Electron microscope observations on the carbohydrate-rich cell coat present at the surface of cells in the rat. J Cell Biol 32:27–53

Rambourg A, Segretain D (1980) Three-dimensional electron microscopy of mitochondria and endoplasmic reticulum in the red muscle fiber of the rat diaphragm. Anat Rec 197:33–48

Rambourg A, Neutra M, Leblond CP (1966) Presence of a "cell coat" rich in carbohydrate at the surface of cells in the rat. Anat Rec 154:41–72

Ramirez BU, Pette D (1974) Effects of long-term electrical stimulation on sarcoplasmic reticulum of fast rabbit muscle. FEBS Lett 49:188–190

Ramsey RW, Street SF (1940) The isometric length-tension diagram of isolated skeletal muscle fibers of the frog. J Cell Comp Physiol 15:11–34

Ranatunga KW, Wylie SR (1983) Temperature-dependent transitions in isometric contractions of rat muscle. J Physiol (Lond) 339:87–95

Ranvier L (1873) Propriétés et structures différentes des muscles rouges et des muscles blancs, chez les lapins et chez les raies. C R Acad Sci [D] (Paris) 77:1030–1034

Ranvier L (1874) Note sur les vaisseaux sanguins et la circulation dans les muscles rouges. C R Soc Biol (Paris) 26:28–31

Rapoport SI (1973) The anisotropic elastic properties of the sarcolemma of the frog semitendinosus muscle fiber. Biophys J 13:14–36

Rapoport SI, Peachey LD, Goldstein DA (1969) Swelling of the transverse tubular system in frog sartorius. J Gen Physiol 54:166–177

Rash JE, Ellisman MH (1974) Studies of excitable membranes. I. Macromolecular specializations of the neuromuscular junction and the nonjunctional sarcolemma. J Cell Biol 63:567–586

Rash JE, Fambrough D (1973) Ultrastructural and electrophysiological correlates of cell coupling and cytoplasmic fusion during myogenesis *in vitro*. Dev Biol 30:166–186

Rash JE, Staehelin LA (1974) Freeze-cleave demonstration of gap junctions between skeletal myogenic cells *in vivo*. Dev Biol 36:455–461

Rash JE, Shay JW, Biesele JJ (1968) Urea extraction of Z bands, intercalated disks, and desmosomes. J Ultrastruct Res 24:181–189

Rash JE, Ellisman MH, Staehelin LA, Porter KR (1974) Molecular specializations of excitable membranes in normal, chronically denervated, and dystrophic muscle fibers. In: Milhorat AT (ed) Exploratory concepts in muscular dystrophy II. Excerpta Medica, Amsterdam, pp 271–291

Rash JE, Graham WF, Hudson CS (1979) Sources and rates of contamination in a conventional Balzers freeze-etch device. In: Rash JE, Hudson CS (eds) Freeze fracture: Methods, artifacts and interpretations. Raven Press, New York, pp 111–122

Rash JE, Hudson CS, Graham WF, Mayer RF, Warnick JE, Albuquerque EX (1981) Freeze fracture studies of human neuromuscular junctions. Membrane alterations observed in myasthenia gravis. Lab Invest 44:519–530

Ravdin P, Axelrod D (1977) Fluorescent tetramethyl rhodamine derivatives of α-bungarotoxin: preparation, separation, and characterization. Anal Biochem 80:585–592

Rayns DG (1972) Myofilaments and cross bridges as demonstrated by freeze-fracturing and etching. J Ultrastruct Res 40:103–121

Rayns DG, Simpson FO, Bertaud WS (1968) Surface features of striated muscle. II. Guinea-pig skeletal muscle. J Cell Sci 3:475–482

Redfern PA (1970) Neuromuscular transmission in new-born rats. J Physiol (Lond) 209:701–709

Reed R, Rudall KM (1948) Electron microscope studies of muscle structure. Biochim Biophys Acta 2:19–26

Reedy MK (1964a) Z band structure in vertebrate skeletal muscle (Abstr). Third European Regional Conference on Electron Microscopy. Publishing House of the Czechoslovak Academy of Sciences, Prague, p 587

Reedy MK (1964b) The structure of actin filaments and the origin of the axial periodicity in the I-substance of vertebrate striated muscle. Proc R Soc Lond [Biol] 160:458–460

Reedy MK (1967) Cross-bridges and periods in insect flight muscle. Am Zoologist 7:465–481

Reedy MK (1968) Ultrastructure of insect flight muscle. I. Screw sense and structural grouping in the rigor cross-bridge lattice. J Mol Biol 31:155–176

Reedy MK, Holmes KC, Tregear RT (1965) Induced changes in orientation of the cross-bridges of glycerinated insect flight muscle. Nature 207:1276–1280

Reedy MK, Goody RS, Hofmann W, Rosenbaum G (1983) Co-ordinated electron microscopy and X-ray studies of glycerinated insect flight muscle. I. X-ray diffraction monitoring during preparation for electron microscopy of muscle fibres fixed in rigor, in ATP and in AMPPNP. J Muscle Res Cell Motil 4:25–53

Regenstein JM, Szent-Györgyi AG (1975) Regulatory proteins of lobster striated muscle. Biochemistry 14:917–925

Reger JF (1954) Electron microscopy of the motor end-plate in intercostal muscle of the rat (Abstr). Anat Rec 118:344

Reger JF (1958) The fine structure of neuromuscular synapses of gastrocnemii from mouse and frog. Anat Rec 130:7–24

Reger JF (1959) Studies in the fine structure of normal and denervated neuromuscular junctions from mouse gastrocnemius. J Ultrastruct Res 2:269–282

Reger JF (1967a) A comparative study on striated muscle fibers of the first antenna and the claw muscle of the crab pinnixia sp. J Ultrastruct Res 20:72–82

Reger JF (1967b) A comparative study on sub-filament organization in primary myofilaments of basalar direct flight and tibial extensor muscles of the lepidopteran, *Achalarus lyciades*. Z Zellforsch Mikrosk Anat 81:361–365

Reger JF, Cooper DP (1967) A comparative study on the fine structure of the basalar muscle of the wing and the tibial extensor muscle of the leg of the lepidopteran *Achalarus lyciades*. J Cell Biol 33:531–542

Reichel H (1960) Muskelphysiologie. Springer, Berlin Heidelberg New York

Reichmann H, Pette D (1982) A comparative microphotometric study of succinate dehydrogenase activity levels in type I, IIA and IIB fibres of mammalian and human muscles. Histochemistry 74:27–41

Reiness CG, Hall ZW (1981) The developmental change in immunological properties of the acetylcholine receptor in rat muscle. Dev Biol 81:324–331

Reiness CG, Weinberg CB (1981) Metabolic stabilization of acetylcholine receptors at newly formed neuromuscular junctions in rat. Dev Biol 84:247–254

Reinking RM, Stephens JA, Stuart DG (1975) The motor units of cat medial gastrocnemius: Problem of their categorisation on the basis of mechanical properties. Exp Brain Res 23:301–313

Reis DJ, Moorhead D, Wooten GF, Hollenberg M (1969) Some dynamic aspects of blood flow in red and white skeletal muscle in cat. In: Locke S (ed) Modern neurology. Papers in tribute to D. Denny-Brown. Little, Brown and Company, Boston, pp 187–201

Reissig D, Schippel K (1966) Elektronenmikroskopische Befunde am Muskel-Sehnenübergang im Zwerchfell von Ratte und Kaninchen. Z Mikrosk Anat Forsch 76:1–11

Repasky EA, Granger BL, Lazarides E (1982) Widespread occurrence of avian spectrin in nonerythroid cells. Cell 29:821–833

Revel JP (1962) The sarcoplasmic reticulum of the bat cricothyroid muscle. J Cell Biol 12:571–588

Revel JP (1964) Electron microscopy of glycogen. J Histochem Cytochem 12:104–114

Rexed B, Therman PO (1948) Calibre spectra of motor and sensory nerve fibres to flexor and extensor muscles. J Neurophysiol 11:133–139

Reznik M (1969a) Thymidine-^{3}H uptake by satellite cells of regenerating skeletal muscle. J Cell Biol 40:568–571

Reznik M (1969b) Origin of myoblasts during skeletal muscle regeneration. Electron microscopic observations. Lab Invest 20:353–363

Reznik M, Engel WK (1970) Ultrastructural and histochemical correlations of experimental muscle regeneration. J Neurol Sci 11:167–185

Rheuben MB, Kammer AE (1982) Structural and electrotonic connections between developing moth muscle fibers. J Neurobiol 13:559–565

Rice MJ (1973) Supercontracting striated muscle in a vertebrate. Nature 243:238–240

Rice RV (1961a) Conformation of individual macromolecular particles from myosin solutions. Biochim Biophys Acta 52:602–604

Rice RV (1961b) An electron microscopical examination of configurations of ribonucleic acid and other polyelectrolytes. Biochim Biophys Acta 53:29–43

Richler C, Yaffe D (1970) The *in vitro* cultivation and differentiation capacities of myogenic cell lines. Dev Biol 23:1–22

Richmond FJR, Abrahams VC (1975) Morphology and enzyme histochemistry of dorsal muscles of the cat neck. J Neurophysiol 38:1312–1321

Richmond FJR, Stuart DG (1985) Distribution of sensory receptors in the flexor carpi radialis muscle of the cat. J Morphol 183:1–13

Rifenberick DH, Gamble JG, Max SR (1973) Response of mitochondrial enzymes to decreased muscular activity. Am J Physiol 225:1295–1299

Riley DA (1973) Histochemical changes in ATPase activity during regeneration of adult skeletal muscle fibers. Exp Neurol 41:690–704

Riley DA, Allin EF (1973) The effects of inactivity, programmed stimulation, and denervation on the histochemistry of skeletal muscle fiber types. Exp Neurol 40:391–413

Ringel SP, Bender AN, Festoff BW, Engel WK, Vogel Z, Daniels MP (1975) Ultrastructural demonstration and analytical application of extrajunctional receptors of denervated human and rat skeletal muscle fibres. Nature 255:730–731

Ringel SP, Bender AN, Engel WK (1976) Extrajunctional acetylcholine receptors. Alterations in human and experimental neuromuscular diseases. Arch Neurol 33:751–758

Ringqvist M (1973) Histochemical enzyme profiles of fibres in human masseter muscles with special regard to fibres with intermediate myofibrillar ATPase reaction. J Neurol Sci 18:133–141

Ringqvist M, Ringqvist I, Eriksson PO, Thornell L-E (1982) Histochemical fibre-type profile in the human masseter muscle. J Neurol Sci 53:273–282

Robbins N, Karpati G, Engel WK (1969) Histochemical and contractile properties in the cross-innervated guinea pig soleus muscle. Arch Neurol 20:318–329

Robertson JD (1954) Electron microscope observations on a reptilian myoneural junction (Abstr). Anat Rec 118:346

Robertson JD (1956) Some features of the ultrastructure of reptilian skeletal muscle. J Biophys Biochem Cytol 2:369–380

Rogers AW, Barnard EA (1969) Quantitative studies on enzymes in structures in striated muscles by labeled inhibitor methods. II. Confirmation of radioautographic measurement by liquid-scintillation counting. J Cell Biol 41:686–695

Rogers AW, Darzynkiewicz Z, Salpeter MM, Ostrowski K, Barnard EA (1969) Quantitative studies on enzymes in structures in striated muscles by labeled inhibitor methods. 1. The number of acetylcholinesterase molecules and of other DFP-reactive sites at motor endplates, measured by radioautography. J Cell Biol 41:665–685

Romanul FCA (1965) Capillary supply and metabolism of muscle fibers. Arch Neurol 12:497–509

Romanul FCA, Pollock M (1969) The parallelism of changes in oxidative metabolism and capillary supply of skeletal muscle fibers. In: Locke S (ed) Modern neurology. Papers in tribute to D. Denny-Brown. Little, Brown and Company, Boston, pp 203–213

Romanul FCA, Van Der Meulen JP (1967) Slow and fast muscles after cross innervation. Arch Neurol 17:387–402

Rome E (1967) Light and X-ray diffraction studies of the filament lattice of glycerol-extracted rabbit psoas muscle. J Mol Biol 27:591–602

Rome E (1968) X-ray diffraction studies of the filament lattice of striated muscle in various bathing media. J Mol Biol 37:331–344

Rome E, Offer G, Pepe FA (1973a) X-ray diffraction of muscle labelled with antibody to C-protein. Nature New Biology 244:152–154

Rome EM, Hirabayashi T, Perry SV (1973b) X-ray diffraction of muscle labelled with antibody to troponin-C. Nature New Biology 244:154–155

Roos DS, Robinson JM, Davidson RL (1983) Cell fusion and intramembrane particle distribution in polyethylene glycol-resistant cells. J Cell Biol 97:909–917

Rootman DS, Tatton WB, Hay M (1981) Postnatal histogenetic death of rat forelimb motoneurons. J Comp Neurol 199:17–27

Rosati G (1967) Enzyme treatment of glycogen particles in rat liver and muscle. J Ultrastruct Res 18:444–455

Rosenbluth J (1969) Sarcoplasmic reticulum of an unusually fast-acting crustacean muscle. J Cell Biol 42:534–547

Rosenfalck P, Buchthal F (1970) On the concept of the motor subunit. Int J Neurosci 1:27–37

Rosenthal JL, Taraskevich PS (1977) Reduction of multiaxonal innervation at the neuromuscular junction of the rat during development. J Physiol (Lond) 270:299–310

Roth S, McGuire EJ, Roseman S (1971) Evidence for cell-surface glycosyltransferases. Their potential role in cellular recognition. J Cell Biol 51:536–547

Rotundo RL, Fambrough DM (1980) Synthesis, transport and fate of acetylcholinesterase in cultured chick embryo muscle cells. Cell 22:583–594

Rowe RW (1971) Ultrastructure of the Z line of skeletal muscle fibers. J Cell Biol 51:674–685

Rowe RWD (1973) The ultrastructure of Z disks from white, intermediate, and red fibers of mammalian striated muscles. J Cell Biol 57:261–277

Rowe RWD (1981) Morphology of perimysial and endomysial connective tissue in skeletal muscle. Tissue Cell 13:681–690

Rowe RWD, Goldspink G (1968) Surgically induced hypertrophy in skeletal muscles of the laboratory mouse. Anat Rec 161:69–76

Rowlerson A (1980) Differentiation of muscle fibre types in fetal and young rats studied with a labelled antibody to slow myosin (Abstr). J Physiol (Lond) 301:19P

Rowlerson A, Pope B, Murray J, Whalen RB, Weeds AG (1981) A novel myosin present in cat jaw-closing muscles. J Muscle Res Cell Motil 2:415–438

Rubin LL, Schuetze SM, Weill CL, Fischbach GD (1980) Regulation of acetylcholinesterase appearance at neuromuscular junction *in vitro*. Nature 283:264–267

Rubinstein NA, Holtzer H (1979) Fast and slow muscles in tissue culture synthesise only fast myosin. Nature 280:323–325

Rubinstein NA, Kelly AM (1978) Myogenic and neurogenic contributions to the development of fast and slow twitch muscles in the rat. Dev Biol 62:473–485

Rubinstein NA, Kelly AM (1981) Development of muscle fiber specialization in the rat hindlimb. J Cell Biol 90:128–144

Rubinstein NA, Pepe FA, Holtzer H (1977) Myosin types during the development of embryonic chicken fast and slow muscles. Proc Natl Acad Sci USA 74:4524–4527

Rubio R, Sperelakis N (1972) Penetration of horseradish peroxidase into the terminal cisternae of frog skeletal muscle fibers and blockade of caffeine contracture by Ca^{++} depletion. Z Zellforsch Mikrosk Anat 124:57–71

Rumpelt H-J, Schmalbruch H (1969) Zur Morphologie der Bauelemente von Muskelspindeln bei Mensch und Ratte. Z Zellforsch Mikrosk Anat 102:601–630
Ruska C, Ruska H (1971) Beobachtungen über Strukturen in zellulären Neutralfetten bei der Darstellung durch die Gefrierätztechnik. In: Auwärter M (ed) Ergebnisse der Hochvakuumtechnik und der Physik dünner Schichten, Bd 2. Wissenschaftl Verlagsges, Stuttgart, pp 59–80
Ruska H (1954) Elektronenmikroskopischer Beitrag zur Histologie des Skelettmuskels kleiner Säugetiere. Z Naturforsch [B] 9b:358–371
Ruska H (1958) The morphology of muscle fibers and muscle cells with different properties of conduction of excitation. Exp Cell Res [Suppl] 5:560–567
Ruska H, Edwards GA (1957) A new cytoplasmic pattern in striated muscle fibers and its possible relation to growth. Growth 21:73–88
Ruska H, Edwards GA, Caesar R (1958) A concept of intracellular transmission of excitation by means of the endoplasmic reticulum. Experientia 14:117–120
Ruskell GL (1984) Spiral nerve endings in human extraocular muscles terminate in motor end plates. J Anat 139:39–43
Russel RG, Oteruelo FT (1981) An ultrastructural study of the differentiation of skeletal muscle in the bovine fetus. Anat Embryol 162:403–417
Rutishauser U, Grumet M, Edelman GM (1983) Neural cell adhesion molecule mediates initial interactions between spinal cord neurons and muscle cells in culture. J Cell Biol 97:145–152
Rutz R, Haney C, Hauschka S (1982) Spatial analysis of limb bud myogenesis: a proximodistal gradient of muscle colony-forming cells in chick embryo leg buds. Dev Biol 90:399–411
Ryan DM, Shafiq SA (1980) A freeze-fracture study of the anterior and posterior latissimus dorsi muscles of the chicken. Anat Rec 198:147–161
Rybicka K (1981) Binding of glycosomes to endoplasmic reticulum and to intermediate filaments in cardiac conduction fibers. J Histochem Cytochem 29:553–560
Sacktor B, Shimada Y (1972) Degenerative changes in the mitochondria of flight muscle from aging blowflies. J Cell Biol 52:465–477
Sakmann B, Brenner HR (1978) Change in synaptic channel gating during neuromuscular development. Nature 276:401–402
Salafsky B (1971) Functional studies of regenerated muscles from normal and dystrophic mice. Nature 229:270–272
Salmons S, Sreter FA (1976) Significance of impulse activity in the transformation of skeletal muscle type. Nature 263:30–34
Salmons S, Vrbova G (1969) The influence of activity on some contractile characteristics of mammalian fast and slow muscles. J Physiol (Lond) 201:535–549
Salpeter MM (1969) Electron microscope radioautography as a quantitative tool in enzyme cytochemistry. II. The distribution of DFP-reactive sites at motor endplates of a vertebrate twitch muscle. J Cell Biol 42:122–134
Salpeter MM, Eldefrawi ME (1973) Sizes of end plate compartments, densities of acetylcholine receptor and other quantitative aspects of neuromuscular transmission. J Histochem Cytochem 21:769–778
Salpeter MM, Plattner H, Rogers AW (1972) Quantitative assay of esterases in end plates of mouse diaphragm by electron microscope autoradiography. J Histochem Cytochem 20:1059–1068
Saltin B (1977) The interplay between peripheral and central factors in the adaptive response to exercise and training. Ann N Y Acad Sci 301:224–231
Saltin B, Nazar K, Costill DL, Stein E, Jansson E, Essén B, Gollnick PD (1976) The nature of the training response; peripheral and central adaptations to one-legged exercise. Acta Physiol Scand 96:289–305
Saltin B, Henriksson J, Nygaard E, Andersen P, Jansson E (1977) Fiber types and metabolic potentials of skeletal muscles in sedentary man and endurance runners. Ann N Y Acad Sci 301:3–39
Salviati G, Betto R, Danieli Betto D (1982) Polymorphism of myofibrillar proteins of rabbit skeletal-muscle fibres. Biochem J 207:261–272
Samaha FJ, Yunis EJ (1973) Quantitative and histochemical demonstration of a calcium activated mitochondrial ATPase in skeletal muscle. Exp Neurol 41:431–439
Sandborn EB, Côté MG, Roberge J, Bois P (1967) Microtubules et filaments cytoplasmiques dans le muscle de mammifères. J Microscop (Paris) 6:169–178
Sandow A (1952) Excitation-contraction coupling in muscular response. Yale J Biol Med 25:176–201

Sandritter W, Scomazzoni G (1964) Deoxyribonucleic acid content (Feulgen photometry) and dry weight (interference microscopy) of normal and hypertrophic heart muscle fibres. Nature 202:100–101

Sanes JR (1982) Laminin, fibronectin, and collagen in synaptic and extrasynaptic portions of muscle fiber basement membrane. J Cell Biol 93:442–451

Sanes JR, Cheney JM (1982) Lectin binding reveals a synapse-specific carbohydrate in skeletal muscle. Nature 300:646–647

Sanes JR, Hall ZW (1979) Antibodies that bind specifically to synaptic sites on muscle fiber basal lamina. J Cell Biol 83:357–370

Sanes JR, Marshall LM, McMahan UJ (1978) Reinnervation of muscle fiber basal lamina after removal of myofibers. Differentiation of regenerating axons at original synaptic sites. J Cell Biol 78:176–198

Sanger JW (1974) The use of cytochalasin B to distinguish myoblasts from fibroblasts in cultures of developing chick striated muscle. Proc Natl Acad Sci USA 71:3621–3625

Santa T, Engel AG (1973) Histometric analysis of neuromuscular junction ultrastructure in rat red, white and intermediate muscle fibers. In: Desmedt JE (ed) New developments in EMG and clinical neurophysiology, vol 1. Elsevier, Amsterdam, pp 41–54

Santa T, Engel AG, Lambert EH (1972) Histometric study of neuromuscular junction ultrastructure. I. Myasthenia gravis. Neurology (NY) 22:71–82

Sarkar S, Sreter FA, Gergely J (1971) Light chains of myosins from white, red, and cardiac muscles. Proc Natl Acad Sci USA 68:946–950

Sartore S, Tarone G, Cantini M, Schiaffino S, Comoglio PM (1979) Cell surface changes during muscle differentiation in vitro: A study with the probe 2,4,6-trinitrobenzene sulphonate. Cell Differ 8:1–9

Sarzala MG, Pilarska M, Zubrzycka E, Michalak M (1975) Changes in the structure, composition and function of sarcoplasmic-reticulum membrane during development. Eur J Biochem 57:25–34

Savay G, Csillik B (1959) Lead reactive substances in peripheral synapses. Experientia 15:396–397

Sawka MN, Petrofsky JS, Phillips CA (1981) Energy cost of submaximal isometric contractions in cat fast and slow twitch muscles. Pflugers Arch 390:164–168

Scales D, Inesi G (1976) Assembly of ATPase protein in sarcoplasmic reticulum membranes. Biophys J 16:735–751

Scales DJ, Yasumura T (1983) Transmembrane displacement of the Ca^{2+}-ATPase in the sarcoplasmic reticulum membrane (Abstr). J Cell Biol 97:313a

Scalzi HA, Price HM (1972) Electron-microscopic observations of the sensory region of the mammalian muscle spindle. In: Banker BQ, Przybylski RJ, Van Der Meulen JP, Victor M (eds) Research in muscle development and the muscle spindle. Excerpta Medica, Amsterdam, pp 254–266

Scarlato G, Cerri C (eds) (1983) Mitochondrial pathology in muscle diseases. Piccin, Padua

Schantz P (1982) Capillary supply in hypertrophied human skeletal muscle. Acta Physiol Scand 114:635–637

Schantz P, Fox ER, Norgren P, Tydén A (1981) The relationship between the mean muscle fibre area and the muscle cross-sectional area of the thigh in subjects with large differences in thigh girth. Acta Physiol Scand 113:537–539

Schantz P, Randall-Fox E, Hutchison W, Tydén A, Åstrand P-O (1983) Muscle fibre type distribution, muscle cross-sectional area and maximal voluntary strength in humans. Acta Physiol Scand 117:219–226

Schiaffino S (1974) Histochemical enzyme profile of the masseter muscle in different mammalian species. Anat Rec 180:53–62

Schiaffino S, Hanzlikova V (1972a) Autophagic degradation of glycogen in skeletal muscles of the newborn rat. J Cell Biol 52:41–51

Schiaffino S, Hanzlikova V (1972b) Studies on the effect of denervation in developing muscle. II. The lysosomal system. J Ultrastruct Res 39:1–14

Schiaffino S, Margreth A (1969) Coordinated development of the sarcoplasmic reticulum and T system during postnatal differentiation of rat skeletal muscle. J Cell Biol 41:855–875

Schiaffino S, Pierobon Bormioli S (1973) Adaptive changes in developing rat skeletal muscle in response to functional overload. Exp Neurol 40:126–137

Schiaffino S, Hanzlikova V, Pierobon S (1970) Relations between structure and function in rat skeletal muscle fibers. J Cell Biol 47:107–119

Schiaffino S, Pierobon Bormioli S, Aloisi M (1972) Cell proliferation in rat skeletal muscle during early stages of compensatory hypertrophy. Virchows Arch [Cell Pathol] 11:268–273
Schiaffino S, Sjöström M, Thornell LE, Nyström B, Håkelius L (1975) The process of survival of denervated and freely autotransplantated skeletal muscle. Experientia 31:1328–1330
Schiaffino S, Cantini M, Sartore S (1977) T-system formation in cultured rat skeletal tissue. Tissue Cell 9:437–446
Schiaffino S, Pierobon Bormioli S, Aloisi M (1979) Fiber branching and formation of new fibers during compensatory muscle hypertrophy. In: Mauro A (ed) Muscle regeneration. Raven Press, New York, pp 177–188
Schmalbruch H (1967a) Kristalloide in menschlichen Skelettmuskelfasern. Naturwissenschaften 54:519
Schmalbruch H (1967b) Fasertypen in der Unterschenkelmuskulatur der Maus. Z Zellforsch Mikrosk Anat 79:64–75
Schmalbruch H (1968) Noniusperioden und Längenwachstum der quergestreiften Muskelfaser. Z Mikrosk Anat Forsch 79:493–507
Schmalbruch H (1970) Die quergestreiften Muskelfasern des Menschen. Springer, Berlin
Schmalbruch H (1971) "Rote" Muskelfasern. Z Zellforsch Mikrosk Anat 119:120–146
Schmalbruch H (1973) Contracture knots in normal and diseased muscle fibres. Brain 96:637–640
Schmalbruch H (1974) The sarcolemma of skeletal muscle fibres as demonstrated by a replica technique. Cell Tissue Res 150:377–387
Schmalbruch H (1975) Segmental fibre breakdown and defects of the plasmalemma in diseased human muscles. Acta Neuropathol (Berlin) 33:129–141
Schmalbruch H (1976a) Muscle fibre splitting and regeneration in diseased human muscle. Neuropathol Appl Neurobiol 2:3–19
Schmalbruch H (1976b) The morphology of regeneration of skeletal muscles in the rat. Tissue Cell 8:673–692
Schmalbruch H (1977) Regeneration of soleus muscles of rat autografted in toto as studied by electron microscopy. Cell Tissue Res 177:159–180
Schmalbruch H (1978a) Satellite cells of rat muscles as studied by freeze-fracturing. Anat Rec 191:371–376
Schmalbruch H (1978b) Early changes in chlorphentermine myopathy of rat studied by freeze fracturing. Muscle Nerve 1:421–422
Schmalbruch H (1979a) The membrane systems in different fibre types of the triceps surae muscle of cat. Cell Tissue Res 204:187–200
Schmalbruch H (1979b) Manifestations of regeneration in myopathic muscles. In: Mauro A (ed) Muscle regeneration. Raven Press, New York, pp 201–213
Schmalbruch H (1979c) 'Square arrays' in the sarcolemma of human skeletal muscle fibres. Nature 281:145–146
Schmalbruch H (1980a) The early changes in experimental myopathy induced by chloroquine and chlorphentermine. J Neuropathol Exp Neurol 39:65–81
Schmalbruch H (1980b) Delayed fixation alters the pattern of intramembrane particles in mammalian muscle fibers. J Ultrastruct Res 70:15–20
Schmalbruch H (1982a) Freeze-fracture studies of experimentally damaged skeletal muscle fibers. In: Schotland DL (ed) Disorders of the motor unit. Wiley, New York, pp 503–515
Schmalbruch H (1982b) The muscular dystrophies. In: Mastaglia FL, Walton Sir John (eds) Skeletal muscle pathology. Churchill Livingstone, Edinburgh, pp 235–265
Schmalbruch H (1982c) Skeletal muscle fibers of newborn rats are coupled by gap junctions. Dev Biol 91:485–490
Schmalbruch H (1983) The fine structure of mitochondrial abnormalities in muscle diseases. In: Scarlato G, Cerri C (eds) Mitochondrial pathology in muscle diseases. Piccin, Padua, pp 39–56
Schmalbruch H (1984) Motoneuron death after sciatic nerve section in newborn rats. J Comp Neurol 224:252–258
Schmalbruch H, Hellhammer U (1976) The number of satellite cells in normal human muscle. Anat Rec 185:279–288
Schmalbruch H, Hellhammer U (1977) The number of nuclei in adult rat muscles with special reference to satellite cells. Anat Rec 189:169–176

Schmalbruch H, Kamieniecka Z (1974) Fiber types in the human brachial biceps muscle. Exp Neurol 44:313–328
Schmalbruch H, Kamieniecka Z (1975) Histochemical fiber typing and staining intensity in cat and rat muscles. J Histochem Cytochem 23:395–401
Schmid H, Taylor DCM, Pierau F-K (1983) Tracing of sensory neurones and spinal motoneurones of the pigeon by injection of fluorescent dyes into peripheral nerves. Cell Tissue Res 232:9–19
Schmidt WJ (1937) Die Doppelbrechung von Karyoplasma, Zytoplasma und Metaplasma. Borntraeger, Berlin
Schmitt FO, Bear RS, Hall CE, Jakus MA (1947) Electron microscope and X-ray diffraction studies of muscle structure. Ann NY Acad Sci 47:799–812
Schön FA (1978) Licht- und elektronenmikroskopische Befunde am Musculus vastus lateralis und ihr Bezug zu physiologischen Meßgrößen bei Normalpersonen, Sportstudenten und Ausdauertrainierten. (Thesis). Deutsche Sporthochschule, Köln
Schotland DL (ed) (1982) Disorders of the motor unit. Wiley, New York
Schotland DL, Bonilla E, VanMeter M (1977) Duchenne dystrophy: Alteration in muscle plasma membrane structure. Science 196:1005–1007
Schotland DL, Bonilla E, Wakayama Y (1981) Freeze fracture studies of muscle plasma membrane in human muscular dystrophy. Acta Neuropathol (Berl) 54:189–197
Schultz E (1974) A quantitative study of the satellite cell population in postnatal mouse lumbrical muscle. Anat Rec 180:589–596
Schultz E (1976) Fine structure of satellite cells in growing skeletal muscle. Am J Anat 147:49–70
Schultz E (1978) Changes in the satellite cells of growing muscle following denervation. Anat Rec 190:299–312
Schultz E (1979) Quantification of satellite cells in growing muscle using electron microscopy and fiber whole mounts. In: Mauro A (ed) Muscle regeneration. Raven Press, New York, pp 131–135
Schulze ML (1955) Die absolute und relative Zahl der Muskelspindeln in den kurzen Daumenmuskeln des Menschen. Anat Anz 102:290–291
Schulze W, Wollenberger A (1967) Zytochemische Lokalisation und Charakterisierung von phosphatabspaltenden Fermenten im sarkotubulären System quergestreifter Muskeln. Histochemie 10:140–153
Schumann HJ, Berger W (1974) Überlastungsnekrosen der Herz- und Skelettmuskulatur bei kleinen Zugvögeln. Beitr Vogelkd (Leipzig) 20:301–309
Schwann T (1839) Mikroskopische Untersuchungen über die Uebereinstimmung in der Struktur und dem Wachsthum der Thiere und Pflanzen. Reimer, Berlin
Schwarzacher HG (1959) Über die Länge und Anordnung der Muskelfasern in menschlichen Skeletmuskeln. Acta Anat (Basel) 37:217–231
Schwarzacher HG (1960a) Untersuchungen über den Cholinesterasegehalt der Skeletmuskel-Sehnenverbindung. Arch Int Pharmacodyn Ther 128:330–342
Schwarzacher HG (1960b) Untersuchungen über die Skeletmuskel-Sehnenverbindung. I. Elektronenmikroskopische und lichtmikroskopische Untersuchungen über den Feinbau der Muskelfaser-Sehnenverbindung. Acta Anat (Basel) 40:59–86
Schwarzacher HG (1960c) Untersuchungen über die Skeletmuskel-Sehnenverbindung. II. Histochemische Lokalisation der Acetylcholinesterase und Untersuchungen über ihre mögliche Funktion an der Muskelfaser-Sehnenverbindung. Acta Anat (Basel) 42:318–332
Schwarzacher HG (1960d) Untersuchungen über die Skeletmuskel-Sehnenverbindung. III. Die Form der Muskelfaserenden. Acta Anat (Basel) 43:144–157
Schwarzacher HG (1961) Acetylcholinesterase in mammalian myo-tendinous junctions. Bibl Anat 2:220–227
Seibel HR, Dolwick MF, Bush FM, Craig SS (1978) Electron-microscopic study of the rat masseter muscle following injection of lidocaine. Acta Anat (Basel) 100:354–364
Seiden D (1971) Slow muscle fibers in the tensor tympani muscle of the guinea pig. Am J Anat 132:267–273
Seinsch W (1974) Die Entwicklung der Nackenmuskulatur bei Mäuseembryonen im Elektronenmikroskop. Anat Anz 135:151–163
Seinsch W, Schweichel J-U (1974) Physiologic cell necroses during the early development of muscles of the back in embryonic mice. Z Anat Entwicklungsgesch 145:101–112
Selby CC, Bear RS (1956) The structure of actin-rich filaments of muscles according to X-ray diffraction. J Biophys Biochem Cytol 2:71–85

Selinger Z, Klein M, Amsterdam A (1969) Properties of particles prepared from sarcoplasmic reticulum by deoxycholate. Biochim Biophys Acta 183:19–26

Sen A, Williams WP, Brain APR, Dickens MJ, Quinn PJ (1981) Formation of inverted micelles in dispersions of mixed galactolipids. Nature 293:488–490

Shafiq SA, Gorycki M, Goldstone L, Milhorat AT (1966) Fine structure of fiber types in normal human muscle. Anat Rec 156:283–302

Shafiq SA, Gorycki MA, Mauro A (1968) Mitosis during postnatal growth in skeletal and cardiac muscle of the rat. J Anat 103:135–141

Shafiq SA, Gorycki MA, Milhorat AT (1969) An electron microscope study of fibre types in normal and dystrophic muscles of the mouse. J Anat 104:281–293

Shafiq SA, Asiedu S, Ryan D, Milhorat AT (1974) Effect of denervation and hereditary muscular dystrophy on the differentiation of chicken fiber types with some observations on early changes in muscular dystrophy. In: Milhorat AT (ed) Exploratory concepts in muscular dystrophy II. Excerpta Medica, Amsterdam, pp 230–245

Shafiq SA, Leung B, Schutta HS (1979) A freeze-fracture study of fibre types in normal human muscle. J Neurol Sci 42:129–138

Shainberg A, Yagil G, Yaffe D (1969) Control of myogenesis in vitro by Ca^{2+} concentration in nutritional medium. Exp Cell Res 58:163–167

Shainberg A, Yagil G, Yaffe D (1971) Alterations of enzymatic activities during muscle differentiation *in vitro*. Dev Biol 25:1–29

Shainberg A, Cohen SA, Nelson PG (1976) Induction of acetylcholine receptors in muscle cultures. Pflugers Arch 361:255–261

Shamarina NM (1962) Electric response of 'tonic' muscle fibres of the frog skeletal musculature. Nature 193:783–784

Shamarina NM (1966) Synaptic transmission in different fibers of tonic frog skeletal muscle. Fed Proc 25:T589–T594

Shaw MD, Baker R (1983) The locations of stapedius and tensor tympani motoneurons in the cat. J Comp Neurol 216:10–19

Shear CR, Goldspink G (1971) Structural and physiological changes associated with the growth of avian fast and slow muscle. J Morphol 135:351–372

Sheetz MP, Spudich JA (1983) Movement of myosin-coated fluorescent beads on actin cables in vitro. Nature 303:31–35

Shelton GD, Bandman E (1983) Development of a unique myosin isozyme pattern in the chicken gastrocnemius (Abstr). J Cell Biol 97:50a

Sherrington C (1930) Some functional problems attaching to convergence. Proc R Soc Lond [Biol] 105:332–362

Shimada Y (1972) Early stages in the reorganization of dissociated embryonic chick skeletal muscle cells. Z Anat Entwicklungsgesch 138:255–264

Shimada Y, Fischman DA (1973) Morphological and physiological evidence for the development of functional neuromuscular junctions *in vitro*. Dev Biol 31:200–225

Shimada Y, Kano M (1971) Formation of neuromuscular junctions in embryonic cultures. Arch Histol Jpn 33:95–114

Shimada Y, Obinata T (1977) Polarity of actin filaments at the initial stage of myofibril assembly in myogenic cells in vitro. J Cell Biol 72:777–785

Shimada Y, Fischman DA, Moscona AA (1967) The fine structure of embryonic chick skeletal muscle cells differentiated in vitro. J Cell Biol 35:445–453

Shimada Y, Fischman DA, Moscona AA (1969) The development of nerve-muscle junctions in monolayer cultures of embryonic spinal cord and skeletal muscle cells. J Cell Biol 43:382–387

Shotton DM (1982) Quantitative freeze-fracture electron microscopy of dystrophic muscle membranes. J Neurol Sci 57:161–190

Sica REP, McComas AJ (1971) Fast and slow twitch units in a human muscle. J Neurol Neurosurg Psychiatry 34:113–120

Sickles DW, Pinkstaff CA (1981a) Comparative histochemical study of prosimian primate hindlimb muscles. I. Muscle fiber types. Am J Anat 160:175–186

Sickles DW, Pinkstaff CA (1981b) Comparative histochemical study of prosimian primate hindlimb muscles. II. Populations of fiber types. Am J Anat 160:187–194

Siebeck R, Krüger P (1955) Die histologische Struktur der äußeren Augenmuskeln als Ausdruck ihrer Funktion. Graefes Arch Ophthalmol 156:637–652

Sigel P, Pette D (1969) Intracellular localization of glycogenolytic and glycolytic enzymes in white and red rabbit skeletal muscle. A gel film method for coupled enzyme reactions in histochemistry. J Histochem Cytochem 17:225–237

Silbermann M, Finkelbrand S, Weiss A, Gershon D, Reznick A (1983) Morphometric analysis of aging skeletal muscle following endurance training. Muscle Nerve 6:136–142

Silverman H, Atwood HL (1980) Surface density of t tubules in normal and dystrophic mouse muscle. Exp Neurol 70:40–46

Simpson Jr SB, Cox PG (1967) Vertebrate regeneration system: culture in vitro. Science 157:1330–1332

Simpson Jr SB, Cox PG (1972) Studies on lizard myogenesis. In: Banker BQ, Przybylski RJ, Van Der Meulen JP, Victor M (eds) Research in muscle development and the muscle spindle. Excerpta Medica, Amsterdam, pp 72–87

Singer II (1979) The fibronexus: a transmembrane association of fibronectin-containing fibers and bundles of 5 nm microfilaments in hamster and human fibroblasts. Cell 16:675–685

Sisto Daneo L, Filogamo G (1974) Ultrastructure of developing myo-neural junctions. Evidence for two patterns of synaptic area differentiation. J Submicrosc Cytol 6:219–228

Sisto Daneo L, Filogamo G (1975) Differentiation of synaptic area in "slow" and "fast" muscle fibres. J Submicrosc Cytol 7:121–131

Sitaramayya C (1951) The finer structure of sarcolemma (Abstr). J Physiol (Lond) 114:27P–28P

Sjøgaard G, Houston ME, Nygaard E, Saltin B (1978a) Subgrouping of fast twitch fibres in skeletal muscles of man: a critical appraisal. Histochemistry 58:79–87

Sjøgaard G, Houston M, Nygaard-Jensen E, Saltin B (1978b) Human skeletal muscle: subgrouping of fast twitch fibres (Abstr). Acta Physiol Scand 102:40A–41A

Sjöstrand FS (1962) The connections between A- and I-band filaments in striated frog muscle. J Ultrastruct Res 7:225–246

Sjöstrand FS (1964) Ultrastructural changes in skeletal muscle myofibrils in connexion with contraction. Nature 201:45–47

Sjöstrand FS, Andersson-Cedergren E (1957) The ultrastructure of the skeletal muscle myofilaments at various states of shortening. J Ultrastruct Res 1:74–108

Sjöstrand FS, Jagendorf-Elfvin M (1967) Ultrastructure studies of the contraction-relaxation cycle of glycerinated rabbit psoas muscle. I. The ultrastructure of glycerinated fibers contracted by treatment with ATP. J Ultrastruct Res 17:348–378

Sjöström M, Squire JM (1977a) Fine structure of the A-band in cryo-sections. The structure of the A-band of human skeletal muscle fibres from ultra-thin cryo-sections negatively stained. J Mol Biol 109:49–68

Sjöström M, Squire JM (1977b) Cryo-ultramicrotomy and myofibrillar fine structure: A review. J Microsc 111:239–278

Sjöström M, Kidman S, Henriksson-Larsén K, Ängqvist K-A (1982a) Z- and M-band appearance in different histochemically defined types of human skeletal muscle fibers. J Histochem Cytochem 30:1–11

Sjöström M, Ängqvist K-A, Bylund A-C, Fridén J, Gustavsson L, Scherstén T (1982b) Morphometric analysis of human muscle fiber types. Muscle Nerve 5:538–553

Sjöström M, Fridén J, Ekblom B (1982c) Fine structural details of human muscle fibres after fibre type specific glycogen depletion. Histochemistry 76:425–438

Skoglund CR (1961) Functional analysis of swim-bladder muscles engaged in sound production of the toadfish. J Biophys Biochem Cytol [Suppl] 10:187–200

Slack JR, Pockett S, MacClement BAE (1983) Regulation of postnatal growth of motor end plates in rat soleus muscle. Exp Neurol 80:321–328

Slayter HS, Lowey S (1967) Substructure of the myosin molecule as visualized by electron microscopy. Proc Natl Acad Sci 58:1611–1618

Sloper JC, Bateson RB, Hindle D, Warren J (1970) Muscle regeneration in man and the mouse; evidence derived from tissue culture and from the evolution of experimental and surgical injuries in the irradiated and non-irradiated subject. In: Mauro A, Shafiq SA, Milhorat AT (eds) Regeneration of striated muscle, and myogenesis. Excerpta Medica, Amsterdam, pp 157–164

Smith DS (1961) Reticular organizations within the striated muscle cell. An historical survey of light microscopic studies. J Biophys Biochem Cytol [Suppl] 10:61–87

Smith DS (1966) The organization and function of the sarcoplasmic reticulum and T-system of muscle cells. Prog Biophys Mol Biol 16:107–142

Smith DS (1983) 100 Hz remains upper limit of synchronous muscle contraction – an anomaly resolved. Nature 303:539–540
Smith DS, Gupta BL, Smith CU (1966) The organization and myofilament array of insect visceral muscles. J Cell Sci 1:49–57
Smith DS, Baerwald RJ, Hart MA (1975) The distribution of orthogonal assemblies and other intercalated particles in frog sartorius and rabbit sacrospinalis muscle. Tissue Cell 7:369–382
Smith JL, Betts B, Edgerton VR, Zernicke RF (1980) Rapid ankle extension during paw shakes: Selective recruitment of fast ankle extensors. J Neurophysiol 43:612–620
Smith RD, Marcarian HQ (1966) The muscle spindles of the dorsal and palmar interosseus muscles. Anat Anz 119:409–414
Smith RG, Appel SH (1983) Extracts of skeletal muscle increase neurite outgrowth and cholinergic activity of fetal rat spinal motor neurons. Science 219:1079–1081
Smith RS, Ovalle Jr WK (1973) Varieties of fast and slow extrafusal muscle fibres in amphibian hind limb muscles. J Anat 116:1–24
Snow DH, Billeter R, Jenny E (1981) Myosin types in equine skeletal muscle fibres. Res Vet Sci 30:381–382
Snow MH (1977a) Myogenic cell formation in regenerating rat skeletal muscle injured by mincing. II. An autoradiographic study. Anat Rec 188:201–218
Snow MH (1977b) The effects of aging on satellite cells in skeletal muscles of mice and rats. Cell Tissue Res 185:399–408
Snow MH (1979) Origin of regenerating myoblasts in mammalian skeletal muscle. In: Mauro A (ed) Muscle regeneration. Raven Press, New York, pp 91–100
Snow MH (1983) A quantitative ultrastructural analysis of satellite cells in denervated fast and slow muscles of the mouse. Anat Rec 207:593–604
Sohal GS, Boydston WR (1982) Influence of neostigmine treatment on embryonic development of acetylcholine receptors and neuromuscular junctions. J Cell Biol 94:540–548
Sola OM, Martin AW (1953) Denervation hypertrophy and atrophy of the hemidiaphragm of the rat. Am J Physiol 172:324–332
Sola OM, Christensen DL, Martin AW (1973) Hypertrophy and hyperplasia of adult chicken anterior latissimus dorsi muscles following stretch with and without denervation. Exp Neurol 41:76–100
Somlyo AV (1979) Bridging structures spanning the junctional gap at the triad of skeletal muscle. J Cell Biol 80:743–750
Somlyo AV, Gonzalez-Serratos H, Shuman H, McClellan G, Somlyo AP (1981) Calcium release and ionic changes in the sarcoplasmic reticulum of tetanized muscle: An electron-probe study. J Cell Biol 90:577–594
Sommerkamp H (1928) Das Substrat der Dauerverkürzung am Froschmuskel. (Physiologische und pharmakologische Sonderstellung bestimmter Muskelfasern). Arch Exp Pathol Pharmacol 128:99–115
Somogyi E, Sotonyi P (1969) Data to the electron microscopic demonstration of ATP-ase. Acta Histochem 34:70–85
Spalteholz W (1888) Die Vertheilung der Blutgefässe im Muskel. Des XIV. Bandes der Abhandlungen der mathematisch-physischen Classe der Königl. Sächs. Gesellschaft der Wissenschaften No XI. Hirzel, Leipzig, pp 506–528
Speidel CC (1939) Studies of living muscles; Histological changes in single fibers of striated muscle during contraction and clotting. Am J Anat 65:471–529
Spencer M, Worthington CR (1960) A hypothesis of contraction in striated muscle. Nature 187:388–391
Squire JM (1973) General model of myosin filament structure. III. Molecular packing arrangements in myosin filaments. J Mol Biol 77:291–323
Squire J (1981) The structural basis of muscular contraction. Plenum Press, New York
Squire JM, Sjöström M, Luther P (1976) Fine structure of the A-band in cryo-sections. II. Evidence for a length-determining protein in the thick filaments of vertebrate skeletal muscle. In: Ben Shaul Y (ed) Proc 6th Eur Congr Elec Microsc, Jerusalem. Tal International, Israel, pp 91–95
Squire J, Edman A-C, Freundlich A, Harford J, Sjöström M (1982a) Muscle structure, cryo-methods and image analysis. J Microsc 125:215–225
Squire JM, Harford JJ, Edman AC, Sjöström M (1982b) Fine structure of the A-band in cryo-sections. III. Crossbridge distribution and the axial structure of the human C-zone. J Mol Biol 155:467–494

Sreter FA, Gergely J, Salmons S, Romanul F (1973) Synthesis by fast muscle of myosin light chains characteristic of slow muscle in response to long-term stimulation. Nature [New Biol] 241:17–19
Sreter FA, Pinter K, Jolesz F, Mabuchi K (1982) Fast to slow transformation of fast muscles in response to long-term phasic stimulation. Exp Neurol 75:95–102
Staehelin LA (1972) Three types of gap junctions interconnecting intestinal epithelial cells visualized by freeze-etching. Proc Natl Acad Sci USA 69:1318–1321
Stålberg E, Gath L (1977) Measurements of the innervation ratio in the human brachial biceps (Abstr). Electroencephalogr Clin Neurophysiol 43:597
Stålberg E, Thiele B (1975) Motor unit fibre density in the extensor digitorum communis muscle. J Neurol Neurosurg Psychiatry 38:874–880
Stålberg E, Schwartz MS, Thiele B, Schiller HH (1976) The normal motor unit in man. A single fibre EMG multielectrode investigation. J Neurol Sci 27:291–301
Staron RS, Hikida RS, Hagerman FC (1983) Reevaluation of human muscle fast-twitch subtypes: Evidence for a continuum. Histochemistry 78:33–39
Stauber WT, Trout JJ, Schottelius BA (1981a) Exocytosis of intact lysosomes from skeletal muscle after chloroquine treatment. Exp Mol Pathol 34:87–93
Stauber WT, Hedge AM, Trout JJ, Schottelius BA (1981b) Inhibition of lysosomal function in red and white skeletal muscles by chloroquine. Exp Neurol 71:295–306
Stefani E, Chiarandini DJ (1982) Ionic channels in skeletal muscle. Annu Rev Physiol 44:357–372
Stefani E, Steinbach A (1968) Persistence of excitation contraction coupling in "slow" muscle fibres after a treatment that destroys transverse tubules in "twitch" fibres. Nature 218:681–682
Stein JM, Padykula HA (1962) Histochemical classification of individual skeletal muscle fibers of the rat. Am J Anat 110:103–123
Steinbach JH (1974) Role of muscle activity in nerve-muscle interaction *in vitro*. Nature 248:70–71
Steinbach JH (1981) Neuromuscular junctions and α-bungarotoxin-binding sites in denervated and contralateral cat skeletal muscles. J Physiol (Lond) 313:513–528
Stensen N (Nicolaus Steno) (1667) Elementarum myologiae specimen, seu musculi descripto geometrica. Florence; cited from: Needham DM (1971) Machina Carnis. University Press, Cambridge, p 18
Stephens JA, Stuart DG (1974) The classification of motor units in cat medial gastrocnemius muscle (Abstr). J Physiol (Lond) 240:43P–44P
Stephens JA, Garnett R, Buller NP (1978) Reversal of recruitment order of single motor units produced by cutaneous stimulation during voluntary muscle contraction in man. Nature 272:362–364
Stephens RE (1965) Analysis of muscle contraction by ultraviolet microbeam disruption of sarcomere structure. J Cell Biol 25:129–139
Sterling RJ (1974) Muscle fibre types and collateral innervation in the frog spindle (Abstr). J Anat 119:203–204
Stewart DM, Sola OM, Martin AW (1972) Hypertrophy as a response to denervation in skeletal muscle. Z Vergl Physiol 76:146–167
Stewart M, Ashton FT, Lieberson R, Pepe FA (1981) The myosin filament. IX. Determination of subfilament positions by computer processing of electron micrographs. J Mol Biol 153:381–392
Stickland NC (1981) Muscle development in the human fetus as exemplified by m. sartorius: a quantitative study. J Anat 132:557–579
Stickland NC (1983) Growth and development of muscle fibres in the rainbow trout (*Salmo gairdneri*). J Anat 137:323–333
Stockdale F, Okazaki K, Nameroff M, Holtzer H (1964) 5-Bromodeoxyuridine: effect on myogenesis in vitro. Science 146:533–535
Stockdale FE (1970) Changing levels of DNA polymerase activity during the development of skeletal muscle tissue in vivo. Dev Biol 21:462–474
Stockdale FE, Holtzer H (1961) DNA synthesis and myogenesis. Exp Cell Res 24:508–520
Stockdale FE, Raman N, Baden H (1981) Myosin light chains and the developmental origin of fast muscle. Proc Natl Acad Sci USA 78:931–935
Stolov WC, Weilepp TG (1966) Passive length-tension relationship of intact muscle, epimysium, and tendon in normal and denervated gastrocnemius of the rat. Arch Phys Med Rehabil 47:612–620
Stonnington HH, Engel AG (1973) Normal and denervated muscle. A morphometric study of fine structure. Neurology (NY) 23:714–724

Storti RV, Rich A (1976) Chick cytoplasmic actin and muscle actin have different structural genes. Proc Natl Acad Sci USA 73:2346–2350
Street SF, Ramsey RW (1965) Sarcolemma: Transmitter of active tension in frog skeletal muscle. Science 149:1379–1380
Strehler BL, Konigsberg IR, Kelley JET (1963) Ploidy of myotube nuclei developing in vitro as determined with a recording double beam micro-spectrophotometer. Exp Cell Res 32:232–241
Strickholm A (1962) Excitation currents and impedence of a small electrically isolated area of the muscle cell surface. J Cell Comp Physiol 60:149–167
Strickholm A (1966) Local sarcomere contraction in fast muscle fibres. Nature 212:835–836
Strohman RC, Micou-Eastwood J, Glass CA, Matsuda R (1983) Human fetal muscle and cultured myotubes derived from it contain a fetal-specific myosin light chain. Science 221:955–957
Stromer MH, Hartshorne DJ, Rice RV (1967) Removal and reconstitution of Z-line material in a striated muscle. J Cell Biol 35:C23–C28
Stromer MH, Hartshorne DJ, Mueller H, Rice RV (1969) The effect of various protein fractions on Z- and M-line reconstitution. J Cell Biol 40:167–178
Stromer MH, Tabatabai LB, Robson RM, Goll DE, Zeece MG (1976) Nemaline myopathy, an integrated study: selective extraction. Exp Neurol 50:402–421
Studitsky AN (1963) Dynamics of the development of myogenic tissue under conditions of explantation and transplantation. In: Rose GG (ed) Cinemicrography in cell biology. Academic Press, New York, pp 171–200
Studitsky AN (1964) Free auto- and homografts of muscle tissue in experiments on animals. Ann N Y Acad Sci 120:789–801
Studitsky AN (1974) The neural factor in the development of transplanted muscles. In: Milhorat AT (ed) Exploratory concepts in muscular dystrophy II. Excerpta Medica, Amsterdam, pp 351–367
Stygall K, Mirsky R, Mowbray J (1979) The effect of local anaesthetics and barbiturates on myogenesis and myotube integrity in rat skeletal muscle cultures. J Cell Sci 37:231–241
Sugi H (1974) Inward spread of activation in frog muscle fibres investigated by means of high-speed microcinematography. J Physiol (Lond) 242:219–235
Sugi H, Ochi R (1967) The mode of transverse spread of contraction initiated by local activation in single frog muscle fibers. J Gen Physiol 50:2167–2176
Suzuki S, Sugi H (1983) Extensibility of the myofilaments in vertebrate skeletal muscle as revealed by stretching rigor muscle fibers. J Gen Physiol 81:531–546
Suzuki A, Goll DE, Singh I, Allen RE, Robson RM, Stromer MH (1976) Some properties of purified skeletal muscle α-actinin. J Biol Chem 251:6860–6870
Swash M, Fox KP (1972) Muscle spindle innervation in man. J Anat 112:61–80
Swatland HJ (1978) Comparison of red and white muscles by cytophotometry of their muscle fibre populations. Histochem J 10:349–360
Swatland HJ, Cassens RG (1972) Muscle growth: the problem of muscle fibers with an intrafascicular termination. J Anim Sci 35:336–344
Sytkowski AJ, Vogel Z, Nirenberg MW (1973) Development of acetylcholine receptor clusters on cultured muscle cells. Proc Natl Acad Sci USA 70:270–274
Szarski H (1976) Cell size and nuclear DNA content in vertebrates. Int Rev Cytol 44:93–111
Szent-Györgyi A (1942) The contraction of myosin threads. Stud Inst Med Chem Univ Szeged 1:17–26
Szent-Györgyi AG (1953) Meromyosins, the subunits of myosin. Arch Biochem Biophys 42:305–320
Szent-Györgyi AG (1968) The role of actin-myosin interaction in contraction. Symposium of the Society for Experimental Biology 22:17–42
Szent-Györgyi AG, Cohen C, Philpott DE (1960) Light meromyosin fraction I: A helical molecule from myosin. J Mol Biol 2:133–142
Szent-Györgyi AG, Holtzer H (1960) Fixation of muscle proteins with antibodies. Biochim Biophys Acta 41:14–19
Tabary JC, Tardieu C (1983) Effect of denervation on sarcomere numbers and tendon length adaptation in young rabbit soleus muscles immobilized in the lengthened position (Abstr). J Physiol (Lond) 345:54P
Tabary JC, Tabary C, Tardieu C, Tardieu G, Goldspink G (1972) Physiological and structural

changes in the cat's soleus muscle due to immobilization at different lengths by plaster casts. J Physiol (Lond) 224:231–244
Tachikawa T, Clementi F (1979) Early effects of denervation on the morphology of junctional and extrajunctional sarcolemma. Neuroscience 4:437–451
Tada M, Yamamoto T, Tonomura Y (1978) Molecular mechanism of active calcium transport by sarcoplasmic reticulum. Physiol Rev 58:1–79
Taniguchi M, Ishikawa H (1982) *In situ* reconstitution of myosin filaments within the myosin-extracted myofibril in cultured skeletal muscle cells. J Cell Biol 92:324–332
Tanji J, Kato M (1973) Firing rate of individual motor units in voluntary contraction of abductor digiti minimi muscle in man. Exp Neurol 40:771–783
Tasaki I, Mizutani K (1944) Comparative studies on the activities of the muscle evoked by two kinds of motor nerve fibers. Part I. Myographic studies. Jpn J Med Sci III Biophys 10:237–244
Taxt T, Ding R, Jansen JKS (1983) A note on the elimination of polyneuronal innervation of skeletal muscles in neonatal rats. Acta Physiol Scand 117:557–560
Taylor A, Cody FWJ, Bosley MA (1973) Histochemical and mechanical properties of the jaw muscles of the cat. Exp Neurol 38:99–109
Taylor AW, Essén B, Saltin B (1974) Myosin ATPase in skeletal muscle of healthy men. Acta Physiol Scand 91:568–570
Taylor SR, Rüdel R, Blinks JR (1975) Calcium transients in amphibian muscle. Fed Proc 34:1379–1381
Teig E (1972a) Force and contraction velocity of the middle ear muscles in the cat and the rabbit. Acta Physiol Scand 84:1–10
Teig E (1972b) Tension and contraction time of motor units of the middle ear muscles in the cat. Acta Physiol Scand 84:11–21
Teig E, Dahl HA (1972) Actomyosin ATPase activity of middle ear muscles in the cat. Histochemie 29:1–7
Telerman-Toppet N, Coërs C (1973) L'histochimie des fibres musculaires normales. Ann Anat Pathol (Paris) 18:35–48
Teräväinen H (1967) Electron microscopic localization of cholinesterases in the rat myoneural junction. Histochemie 10:266–271
Teräväinen H (1968) Electron microscopic and histochemical observations on different types of nerve endings in the extraocular muscles of the rat. Z Zellforsch Mikrosk Anat 90:372–388
Teräväinen H (1969a) Localization of acetylcholinesterase activity in myotendinous and myomyous junctions of the striated skeletal muscles of the rat. Experientia 25:524–525
Teräväinen H (1969b) Axonal protrusions in the small multiple endings in the extraocular muscles of the rat. Z Zellforsch Mikrosk Anat 96:206–211
Teräväinen H (1970) Satellite cells of striated muscle after compression injury so slight as not to cause degeneration of the muscle fibres. Z Zellforsch Mikrosk Anat 103:320–327
Tesch PA, Larsson L (1982) Muscle hypertrophy in bodybuilders. Eur J Appl Physiol 49:301–306
Thompson GC, Igarashi M, Stach BA (1985) Identification of stapedius muscle motoneurons in squirrel monkey and bush baby. J Comp Neurol 231:270–279
Thornell LE, Edström L, Eriksson A, Henriksson KG, Ängqvist KA (1980) The distribution of intermediate filament protein (skeletin) in normal and diseased human skeletal muscle. An immunohistochemical and electronmicroscopic study. J Neurol Sci 47:153–170
Thornell L-E, Eriksson A, Edström L (1983) Intermediate filaments in human myopathies. In: Dowben RM, Shay JW (eds) Cell and muscle motility. Plenum, New York, pp 85–136
Thornell L-E, Billeter R, Eriksson P-O, Ringqvist M (1984) Heterogeneous distribution of myosin in human masticatory muscle fibres as shown by immunocytochemistry. Arch Oral Biol 29:1–5
Tice LW, Barrnett RJ (1962) Fine structural localization of adenosinetriphosphatase activity in heart muscle myofibrils. J Cell Biol 15:401–416
Tillack TW, Boland R, Martonosi A (1974) The ultrastructure of the developing sarcoplasmic reticulum. J Biol Chem 249:624–633
Timson BF (1982) The effect of varying postnatal growth rate on skeletal muscle fiber number in the mouse. Growth 46:36–45
Tonomura Y, Yagi K, Kubo S, Kitagawa S (1961) A molecular mechanism of muscle contraction. J Res Inst Catalysis Hokkaido Univ 9:256–286
Traeger L, Goldstein MA (1983) Thin filaments are not of uniform length in rat skeletal muscle. J Cell Biol 96:100–103

Tregear RT, Marston SB (1979) The crossbridge theory. Annu Rev Physiol 41:723–736
Tregear RT, Squire JM (1973) Myosin content and filament structure in smooth and striated muscle. J Mol Biol 77:279–290
Tribe MA, Ashhurst DE (1972) Biochemical and structural variations in the flight muscle mitochondria of ageing blowflies *Calliphora erythrocephala*. J Cell Sci 10:443–469
Trombitas K, Tigyi-Sebes A (1975) The Z line of the flight muscle of honey-bee. Acta Biochim Biophys Acad Sci Hung 10:83–93
Trotter JA, Corbett K, Avner BP (1981) Structure and function of the murine muscle-tendon junction. Anat Rec 201:293–302
Trotter JA, Eberhard S, Samora A (1983) Structural domains of the muscle-tendon junction. 1. The internal lamina and the connecting domain. Anat Rec 207:573–591
Trout JJ, Stauber WT, Schottelius BA (1979) Cytochemical observations of two distinct acid phosphatase-reactive structures in anterior latissimus dorsi muscle of the chicken. Histochem J 11:223–230
Trout JJ, Stauber WT, Schottelius BA (1981) Increased autophagy in chloroquine-treated tonic and phasic muscles: An alternative view. Tissue Cell 13:393–401
Trupin GL (1976) The satellite cells of normal anuran skeletal muscle. Dev Biol 50:517–524
Tschiriew S (1879) Sur les terminaisons nerveuses dans les muscles striés. Arch Physiol Norm Pathol 6:89–116
Tunik BD (1971) Cytochemical localization of myofibrillar adenosine triphosphatase activity in sarcomeres of glycerinated muscle by the calcium precipitation method. A novel control is needed to demonstrate artifacts. J Histochem Cytochem 19:75–84
Turner DC (1978) Differentiation in cultures derived from embryonic chicken muscle. The postmitotic, fusion-capable myoblast as a distinct cell type. Differentiation 10:81–93
Turner RS, Burger MM (1973) Involvement of a carbohydrate group in the active site for surface guided reassociation of animal cells. Nature 244:509–510
Tyrrell DA, Heath TD, Colley CM, Ryman BE (1976) New aspects of liposomes. Biochim Biophys Acta 457:259–302
Uhrik B, Schmidt H (1973) Distribution of slow muscle fibres in the frog rectus abdominis muscle. An electron microscopical investigation. Pflugers Arch 340:361–366
Ullrick WC (1967) A theory of contraction for striated muscle. J Theor Biol 15:53–69
Ullrick WC, Toselli PA, Chase D, Dasse K (1977a) Are there extensions of thick filaments to the Z line in vertebrate and invertebrate striated muscle? J Ultrastruct Res 60:263–271
Ullrick WC, Toselli PA, Saide JD, Phear WPC (1977b) Fine structure of the vertebrate Z-disc. J Mol Biol 115:61–74
Van Harreveld A (1947) On the force and size of motor units in the rabbit's sartorius muscle. Am J Physiol 151:96–106
Van de Graaff KM, Frederick ED, Williamson RG, Goslow GE (1977) Motor units and fiber types of primary ankle extensors of the skunk (*Mephitis mephitis*). J Neurophysiol 40:1424–1431
Van der Bosch J, Schudt C, Pette D (1973) Influence of temperature, cholesterol, dipalmitoyllecithin and Ca^{2+} on the rate of muscle cell fusion. Exp Cell Res 82:433–438
Van Linge B (1962) The response of muscle to strenuous exercise. An experimental study in the rat. J Bone Joint Surg [Br] 44B:711–721
Venable JH (1966a) Constant cell populations in normal, testosterone-deprived and testosterone-stimulated levator ani muscles. Am J Anat 119:263–270
Venable JH (1966b) Morphology of the cells of normal, testosterone-deprived and testosterone-stimulated levator ani muscles. Am J Anat 119:271–302
Veneroni G, Murray MR (1969) Formation *de novo* and development of neuromuscular junctions *in vitro*. J Embryol Exp Morphol 21:369–382
Veratti E (1902) Investigations on the fine structure of striated muscle fiber (Italian). Translated and reprinted in: J Biophys Biochem Cytol [Suppl] 10:1–59 (1961)
Verkleij AJ, Mombers C, Leunissen-Bijvelt J, Ververgaert PHJT (1979) Lipidic intramembranous particles. Nature 279:162–163
Vetterlein F, Schmidt G (1983) Histological demonstration of capillaries, interstitial space and muscle fibers in heart and skeletal muscle with fluorescent dyes. Acta Anat (Basel) 116:285–289
Vihko V, Rantamaki J, Salminen A (1978) Exhaustive physical exercise and acid hydrolase activity in mouse skeletal muscle. A histochemical study. Histochemistry 57:237–249
Vita G, Muglia U, Germana G, Pennica F, Carfi F (1983) Histochemical characteristics of rabbit stapedius muscle. Exp Neurol 81:511–516

Vogel Z, Daniels MP (1976) Ultrastructure of acetylcholine receptor clusters on cultured muscle fibers. J Cell Biol 69:501–507
Vogel Z, Sytkowski AJ, Nirenberg MW (1972) Acetylcholine receptors of muscle grown in vitro. Proc Natl Acad Sci USA 69:3180–3184
Volkmann R (1893) Über die Regeneration des quergestreiften Muskelgewebes beim Menschen und Säugethier. Beitr PathAnat 12:233–332
Voss H (1971) Tabelle der absoluten und relativen Muskelspindelzahlen der menschlichen Skelettmuskulatur. Anat Anz 129:562–572
Vracko R, Benditt EP (1972) Basal lamina: The scaffold for orderly cell replacement. Observations on regeneration of injured skeletal muscle fibers and capillaries. J Cell Biol 55:406–419
Vye MV, Fischman DA, Hansen JL (1969) Ultrastructural localization of adenosinetriphosphatase activity in skeletal muscle by calcium precipitation at high pH. Virchows Arch [Cell Pathol] 3:307–323
Wachholder K, Von Ledebur J (1930) Untersuchungen über "tonische" und "nichttonische" Wirbeltiermuskeln. I. Mitteilung. Die Umklammerungshaltung des Frosches und die Schutzhaltung der Schildkröte; ihre spezifische Nachahmung durch muskuläre Acetylcholinwirkung. Pflugers Arch 225:627–642
Wachsberger P, Lampson L, Pepe FA (1983) Non-uniform staining of myofibril A bands by a monoclonal antibody to skeletal muscle myosin S1 heavy chain. Tissue Cell 15:341–349
Wachstein M, Meisel E (1955) The distribution of histochemically demonstrable succinic dehydrogenase and of mitochondria in tongue and skeletal muscles. J Biophys Biochem Cytol 1:483–488
Wakshull E, Bayne EK, Chiquet M, Fambrough DM (1983) Characterization of a plasma membrane glycoprotein common to myoblasts, skeletal muscle satellite cells, and glia. Dev Biol 100:464–477
Walcott B, Ridgway EB (1967) The ultrastructure of myosin-extracted striated muscle fibers. Am Zoologist 7:499–504
Walker SM, Edge MB (1971) The sarcoplasmic reticulum and development of Z lines in skeletal muscle fibers of fetal and postnatal rats. Anat Rec 169:661–678
Walker SM, Schrodt GR (1966) Evidence for connections between mitochondria and the sarcoplasmic reticulum and evidence for glycogen granules within the sarcoplasmic reticulum. Am J Phys Med 45:25–44
Walker SM, Schrodt GR (1967) Triads in foetal skeletal muscle. Nature 216:985–988
Walker SM, Schrodt GR (1968) Triads in skeletal muscle fibers of 19-day fetal rats. J Cell Biol 37:564–569
Walker SM, Schrodt GR, Bingham M (1968) Electron microscope study of the sarcoplasmic reticulum at the Z line level in skeletal muscle fibers of fetal and newborn rats. J Cell Biol 39:469–475
Walker SM, Schrodt GR, Currier GJ, Yuen JW (1975) Development of the triadic junction in skeletal muscle fibers of fetal and postnatal rats. Ann J Phys Med 54:61–79
Wallimann T, Szent-Györgyi AG (1981 a) An immunological approach to myosin light-chain function in thick filament linked regulation. I. Characterization, specificity, and cross-reactivity of anti-scallop myosin heavy- and light-chain antibodies by competitive, solid-phase radioimmunoassay. Biochemistry 20:1176–1187
Wallimann T, Szent-Györgyi AG (1981 b) An immunological approach to myosin light-chain function in thick filament linked regulation. 2. Effects of anti-scallop myosin light-chain antibodies. Possible regulatory role for the essential light chain. Biochemistry 20:1188–1197
Wallimann T, Pelloni G, Turner DC, Eppenberger HM (1979) Removal of the M-line by treatment with FAB' fragments of antibodies against MM-creatine kinase. In: Pepe FA, Sanger JW, Nachmias VT (eds) Motility in cell function. Academic Press, New York, pp 415–417
Walls EW (1960) The microanatomy of muscle. In: Bourne GH (ed) The structure and function of muscle, vol 1. Academic Press, New York, pp 21–61
Walro JM, Kucera J (1984) Nonselective motor innervation of intrafusal fibers in muscle spindles of the rat. Exp Brain Res 56:187–192
Walro JM, Hikida RS, Frangowlakis TM (1982) Patterns of avian muscle fiber type regeneration. Evidence for a myotrophic influence. J Neurol Sci 55:327–338
Walsh JV, Burke RE, Rymer WZ, Tsairis P (1978) Effect of compensatory hypertrophy studied in individual motor units in medial gastrocnemius muscle of the cat. J Neurophysiol 41:496–508
Walsh FS, Ritter MA (1981) Surface antigen differentiation during human myogenesis in culture. Nature 289:60–64

Walter WG (1947) The tensile strength of striated muscle, investigated on the gastrocnemius muscle of the frog. Arch Neerl Physiol 28:655–669

Wang K, McClure J, Tu A (1979) Titin: Major myofibrillar components of striated muscle. Proc Natl Acad Sci USA 76:3698–3702

Wang K, Ramirez-Mitchell R (1983) A network of transverse and longitudinal intermediate filaments is associated with sarcomeres of adult vertebrate skeletal muscle. J Cell Biol 96:562–570

Wang K, Williamson CL (1980) Identification of an N_2line protein of striated muscle. Proc Natl Acad Sci USA 77:3254–3258

Wanson J-C, Drochmans P (1968) Rabbit skeletal muscle glycogen. A morphological and biochemical study of glycogen β-particles isolated by the precipitation-centrifugation method. J Cell Biol 38:130–150

Warner DT (1970) A proposal for the mechanism of muscle contraction at the molecular level. J Theor Biol 26:289–313

Warren RH (1968) The effect of colchicine on myogenesis in vivo in *Rana pipiens* and *Rhodnius prolixus* (Hemiptera). J Cell Biol 39:544–555

Warren RH (1973) Interaction of the sarcoplasmic reticulum with Z-lines during myogenesis in amphibian skeletal muscle. Anat Rec 177:225–242

Waser PG, Reller J (1965) Bestimmung der Zahl aktiver Zentren der Acetylcholinesterase in motorischen Endplatten. Experientia 21:402–403

Watanabe M, Muramatsu T, Shirane H, Ugai K (1981) Discrete distribution of binding sites for dolichos biflorus agglutinin (DBA) and for peanut agglutinin (PNA) in mouse organ tissues. J Histochem Cytochem 29:779–790

Watt DJ, Lambert K, Morgan JE, Partridge TA, Sloper JC (1982) Incorporation of donor muscle precursor cells into an area of muscle regeneration in the host mouse. J Neurol Sci 57:319–331

Webb JN (1972) The development of human skeletal muscle with particular reference to muscle cell death. J Pathol 106:221–228

Weber A (1959) On the role of calcium in the activity of adenosine 5′-triphosphate hydrolysis by actomyosin. J Biol Chem 234:2764–2769

Weber A, Herz R (1968) The relationship between caffeine contracture of intact muscle and the effect of caffeine on reticulum. J Gen Physiol 52:750–759

Weber A, Murray JM (1973) Molecular control mechanisms in muscle contraction. Physiol Rev 53:612–673

Weber A, Winicur S (1961) The role of calcium in superprecipitation of actomyosin. J Biol Chem 236:3198–3202

Weber A, Herz R, Reiss I (1969) The role of magnesium in the relaxation of myofibrils. Biochemistry 8:2266–2270

Weber HH (1934a) Der Feinbau und die mechanischen Eigenschaften des Myosinfadens. Pflugers Arch 235:205–233

Weber HH (1934b) Die Muskeleiweißkörper und der Feinbau des Skelettmuskels. Ergeb Physiol 36:109–150

Weber HH (1951) Die Aktomyosinmodelle und der Kontraktionszyklus des Muskels. Z Elektrochem Angew Phys Chem 55:511–518

Weeds AG (1976) Light chains from slow-twitch muscle myosin. Eur J Biochem 66:157–173

Weeds AG, Pope B (1971) Chemical studies on light chains from cardiac and skeletal muscle myosins. Nature 234:85–88

Weeds AG, Trentham DR, Kean CJC, Buller AJ (1974) Myosin from cross-reinnervated cat muscles. Nature 247:135–139

Weinberg CB, Hall ZW (1979) Junctional form of acetylcholinesterase restored at nerve-free endplates. Dev Biol 68:631–635

Weinberg CB, Sanes JR, Hall ZW (1981) Formation of neuromuscular junctions in adult rats: Accumulation of acetylcholine receptors, acetylcholinesterase, and components of synaptic basal lamina. Dev Biol 84:255–266

Weismann A (1861) Über das Wachsen der quergestreiften Muskeln nach Beobachtungen am Frosch. Zeitschrift für rationelle Medizin 10:263–284

Whalen RG, Butler-Browne GS, Gros F (1976) Protein synthesis and actin heterogeneity in calf muscle cells in culture. Proc Natl Acad Sci USA 73:2018–2022

Whalen RG, Butler-Browne GS, Gros F (1978) Identification of a novel form of myosin light chain present in embryonic muscle tissue and cultured muscle cells. J Mol Biol 126:415–431
Whalen RG, Schwartz K, Bouveret P, Sell SM, Gros F (1979) Contractile protein isozymes in muscle development: identification of an embryonic form of myosin heavy chain. Proc Natl Acad Sci USA 76:5197–5201
Whalen RG, Sell SM, Butler-Browne GS, Schwartz K, Bouveret P, Pinset-Harström J (1981) Three myosin heavy-chain isozymes appear sequentially in rat muscle development. Nature 292:805–809
Whittaker VP (1977) The electromotor system of *Torpedo*. A model cholinergic system. Naturwissenschaften 64:606–611
Wibo M, Poole B (1974) Protein degradation in cultured cells. II. The uptake of chloroquine by rat fibroblasts and the inhibition of cellular protein degradation and cathepsin B_1. J Cell Biol 63:430–440
Wier ML, Lennon VA (1981) Differentiation of skeletal muscle from dissociated optic nerve cells. Immunocytochemical observations. J Neuroimmunology 1:61–68
Wilkins JA, Lin S (1982) High-affinity interaction of vinculin with actin filaments in vitro. Cell 28:83–90
Winand R, Luzzati D (1975) Cell surface changes during myoblast differentiation: preparation and carbohydrate composition of plasma membranes. Biochimie 57:765–771
Winegrad S (1965) Role of intracellular calcium movements in excitation-contraction coupling in skeletal muscle. Fed Proc 24:1146–1152
Winegrad S (1968) Intracellular calcium movements of frog skeletal muscle during recovery from tetanus. J Gen Physiol 51:65–83
Wirtz P, Loermans H, Rutten E (1982) Effects of irradiation on regeneration in dystrophic mouse leg muscles. Br J Exp Pathol 63:671–679
Witkowski JA, Dubowitz V (1976) Letters to the editor. J Neurol Sci 29:117–121
Wittig M (1968) Die Morphogenese der lumbalen Hypaxialmuskulatur des Hauskaninchens (*Oryctolagus cuniculus* L.). II. Licht-, polarisations- und elektronenmikroskopische Untersuchungen zur Histo- und Cytogenese. Z Mikrosk Anat Forsch 79:27–81
Wöhlisch E, Du Mesnil de Rochemont R, Gerschler H (1926) Untersuchungen über die elastischen Eigenschaften tierischer Gewebe; Elastizitätsmodul, Zerreißfestigkeit, Arbeitsvermögen und elastische Vollkommenheit. Z Biol 85:325–341
Wood DS, Zollman J, Reuben JP, Brandt PW (1975) Human skeletal muscle: Properties of the "chemically skinned" fiber. Science 187:1075–1076
Wood DS, Willner JH, Salviati G (1982) Malignant hyperthermia: The pathogenesis of abnormal caffeine contracture. In: Schotland DL (ed) Disorders of the motor unit. Wiley, New York, pp 597–610
Worthington CR (1959) Large axial spacings in striated muscle. J Mol Biol 1:398–401
Worthington CR, Liu SC (1973) Structure of sarcoplasmic reticulum membranes at low resolution (17 Å). Arch Biochem Biophys 157:573–579
Wray JS (1979) X-ray diffraction studies of myosin filament structures in crustacean muscles. In: Pepe FA, Sanger JW, Nachmias VT (eds) Motility in cell function. Academic Press, New York, pp 347–350
Wray JS, Holmes KC (1981) X-ray diffraction studies of muscle. Annu Rev Physiol 43:553–565
Wray SH (1969) Innervation ratios for large and small limb muscles in the baboon. J Comp Neurol 137:227–250
Wright WE (1984a) Induction of muscle genes in neural cells. J Cell Biol 98:427–435
Wright WE (1984b) Control of differentiation in heterokaryons and hybrids involving differentiation - defective myoblast variants. J Cell Biol 98:436–443
Wuerker RB, McPhedran AM, Henneman E (1965) Properties of motor units in a heterogeneous pale muscle (m. gastrocnemius) of the cat. J Neurophysiol 28:85–99
Yablonka Z, Yaffe D (1977) Synthesis of myosin light chains and accumulation of translatable mRNA coding for light chain-like polypeptides in differentiating muscle cultures. Differentiation 8:133–143
Yaffe D (1968) Retention of differentiated potentialities during prolonged cultivation of myogenic cells. Proc Natl Acad Sci USA 61:477–483
Yaffe D (1971) Developmental changes preceding cell fusion during muscle differentiation in vitro. Exp Cell Res 66:33–48

Yaffe D, Dym H (1972) Gene expression during differentiation of contractile muscle fibers. Cold Spring Harbor Symp Quant Biol 37:543–547
Yaffe D, Feldman M (1965) The formation of hybrid multinucleated muscle fibers from myoblasts of different genetic origin. Dev Biol 11:300–317
Yaffe D, Gershon D (1967) Multinucleated muscle fibers: induction of DNA synthesis and mitosis by polyoma virus infection. Nature 215:421–424
Yaffe D, Shainberg A, Dym H (1972) Studies on the prefusion stage during formation of multinucleated muscle fibers in vitro. In: Banker BQ, Przybylski RJ, Van Der Meulen JP, Victor M (eds) Research in muscle development and the muscle spindle. Excerpta Medica, Amsterdam, pp 110–133
Yagi N, Matsubara I (1980) Myosin heads do not move on activation in highly stretched vertebrate striated muscle. Science 207:307–308
Yamaguchi M, Robson RM, Stromer MH, Dahl DS, Oda T (1978) Actin filaments form the backbone of nemaline myopathy rods. Nature 271:265–267
Yamamoto T, Takisawa H, Tonomura Y (1979) Reaction mechanisms for ATP hydrolysis and synthesis in the sarcoplasmic reticulum. Curr Top Bioenerg 9:179–236
Yanagida T, Nakase M, Nishiyama K, Oosawa F (1984) Direct observation of motion of single F-actin filaments in the presence of myosin. Nature 307:58–60
Yasumura T, Scales DJ (1982) II. The three dimensional rearrangement of mammalian sarcoplasmic reticulum after contraction. J Ultrastruct Res 81:27–36
Yee AG, Fischbach GD, Karnovsky MJ (1978) Clusters of intramembranous particles on cultured myotubes at sites that are highly sensitive to acetylcholine. Proc Natl Acad Sci USA 75:3004–3008
Yellin H (1967) Neural regulation of enzymes in muscle fibers of red and white muscle. Exp Neurol 19:92–103
Yellin H (1969a) Unique intrafusal and extraocular muscle fibers exhibiting dual actomyosin ATPase activity. Exp Neurol 25:153–163
Yellin H (1969b) A histochemical study of muscle spindles and their relationship to extrafusal fiber types in the rat. Am J Anat 125:31–46
Yellin H (1975) Limitations to the neuroregulation of enzymes in mammalian skeletal muscle. Anat Rec 182:479–497
Yellin H, Guth L (1970) The histochemical classification of muscle fibers. Exp Neurol 26:424–432
Yeoh GCT, Holtzer H (1977) The effect of cell density, conditioned medium and cytosine arabinoside on myogenesis in primary and secondary cultures. Exp Cell Res 104:63–78
Yoshioka M, Okuda R (1977) Human skeletal muscle fibers in normal and pathological states; Freeze-etch replica observations. J Electron Microsc (Tokyo) 26:103–110
Young A, Hughes I, Round JM, Edwards RHT (1982) The effect of knee injury on the number of muscle fibres in the human quadriceps femoris. Clin Sci 62:227–234
Young D, Josephson RK (1984) 100 Hz is not the upper limit of synchronous muscle contraction. Nature 309:286–287
Yu LC, Dowben RM, Kornacker K (1970) The molecular mechanism of force generation in striated muscle. Proc Natl Acad Sci USA 66:1199–1205
Yu LC, Lymn RW, Podolsky RJ (1977) Characterization of a non-indexible equatorial X-ray reflection from frog sartorius muscle. J Mol Biol 115:455–464
Zacks SI, Blumberg JM (1961) Observations on the fine structure and cytochemistry of mouse and human intercostal neuromuscular junction. J Biophys Biochem Cytol 10:517–528
Zacks SI, Sheff MF, Saito A (1973) Structure and staining characteristics of myofiber external lamina. J Histochem Cytochem 21:703–714
Zalin RJ (1977) Prostaglandins and myoblast fusion. Dev Biol 59:241–248
Zebe E (1965) Zur Lokalisation ATP-spaltender Reaktionen im "Sarcoplasmatischen Reticulum" quergestreifter Muskeln. Histochemie 5:32–43
Zenker FA (1864) Über die Veränderungen der willkührlichen Muskeln im Typhus abdominalis. Vogel, Leipzig
Zenker W (1966) Zur Frage der Endigung von Muskelfasern am Bindegewebsapparat des menschlichen Stimmbandes. Acta Anat (Basel) 64:198–213
Zenker W, Anzenbacher H (1964) On the different forms of myo-neural junction in two types of muscle fiber from the external ocular muscles of the rhesus monkey. J Cell Comp Physiol 63:273–285

Zenker W, Gruber H (1967) Über Form, Anordnung, Zahl und Größe der myoneuralen Synapsen multipel innervierter Skelettmuskelfasern. Z Mikrosk Anat Forsch 76:361–377
Zenker W, Hohberg E (1973) α-Motorische Nervenfasern, Axonquerschnittsfläche von Stammfaser und Endästen. Z Anat Entwickl-Gesch 139:163–172
Zevin-Sonkin D, Yaffe D (1980) Accumulation of muscle-specific RNA sequences during myogenesis. Dev Biol 74:326–334
Ziskind-Conhaim L, Bennett JI (1982) The effects of electrical inactivity and denervation on the distribution of acetylcholine receptors in developing rat muscle. Dev Biol 90:185–197
Ziskind-Conhaim L, Dennis MJ (1981) Development of rat neuromuscular junctions in organ culture. Dev Biol 85:243–251
Zobel CR, Baskin RJ, Wolfe SL (1967) Electron microscope observations on thick filaments isolated from striated muscle. J Ultrastruct Res 18:637–650
Zobel CR, Carlson FD (1963) An electron microscopic investigation of myosin and some of its aggregates. J Mol Biol 7:78–89

Author Index

Page numbers in *italics* refer to the bibliography

Subject Index

MIX
Papier aus verantwortungsvollen Quellen
Paper from responsible sources
FSC® C105338

If you have any concerns about our products,
you can contact us on
ProductSafety@springernature.com

In case Publisher is established outside the EU,
the EU authorized representative is:
Springer Nature Customer Service Center GmbH
Europaplatz 3, 69115 Heidelberg, Germany

Printed by Libri Plureos GmbH
in Hamburg, Germany